T0172098

Frontiers in Mathematics

Jorge Picado
Aleš Pultr

Frames and Locales

Topology without points

 Birkhäuser

Jorge Picado
CMUC, Department of Mathematics
University of Coimbra
3001-454 Coimbra
Portugal
picado@mat.uc.pt

Aleš Pultr
Department of Applied Mathematics
MFF Karlovy University
Malostranské nám. 25
11800 Praha 1
Czech Republic
pultr@kam.mff.cuni.cz

2010 Mathematics Subject Classification: 06D22, 54-XX, 18B30

ISSN 1660-8046   e-ISSN 1660-8054
ISBN 978-3-0348-0153-9   e-ISBN 978-3-0348-0154-6
DOI 10.1007/978-3-0348-0154-6

Library of Congress Control Number: 2011941088

Printed on acid-free paper

Springer Basel AG is part of Springer Science+Business Media

www.birkhauser-science.com

To our wives Jitka and Isabel,

and to our teacher and friend Bernhard Banaschewski

Contents

Preface xiii

Introduction xv

I. **Spaces and Lattices of Open Sets**
  1. Sober spaces 2
  2. The axiom T_D: another case of spaces easy to reconstruct 5
  3. Summing up 6
  4. Aside: several technical properties of T_D-spaces 7

II. **Frames and Locales. Spectra**
  1. Frames 9
  2. Locales and localic maps 11
  3. Points 13
  4. Spectra 15
  5. The unit σ and spatiality 18
  6. The unit λ and sobriety 20

III. **Sublocales**
  1. Extremal monomorphisms in **Loc** 23
  2. Sublocales 25
  3. The co-frame of sublocales 27
  4. Images and preimages 29
  5. Alternative representations of sublocales 30
  6. Open and closed sublocales 33
  7. Open and closed localic maps 36
  8. Closure 39
  9. Preimage as a homomorphism 41
  10. Other special sublocales: one-point sublocales,
       and Boolean ones 42

11. Sublocales as quotients. Factorizing frames is
    surprisingly easy 44

IV. **Structure of Localic Morphisms. The Categories Loc and Frm**
    1. Special morphisms. Factorizing in **Loc** and **Frm** 49
    2. The down-set functor and free constructions 51
    3. Limits and a colimit in **Frm** 54
    4. Coproducts of frames 56
    5. More on the structure of coproduct 59
    6. Epimorphisms in **Frm** 64

V. **Separation Axioms**
    1. Instead of T_1: subfit and fit 73
    2. Mimicking the Hausdorff axiom 79
    3. I-Hausdorff frames and regular monomorphisms 83
    4. Aside: Raney identity 86
    5. Quite like the classical case: Regular, completely
       regular and normal 88
    6. The categories **RegLoc**, **CRegLoc**, **HausLoc** and **FitLoc** 92

VI. **More on Sublocales**
    1. Subspaces and sublocales of spaces 99
    2. Spatial and induced sublocales 101
    3. Complemented sublocales of spaces are spatial 103
    4. The zero-dimensionality of $\mathcal{S}\ell(L)^{\mathrm{op}}$ and a few consequences 105
    5. Difference and pseudodifference, residua 108
    6. Isbell's Development Theorem 112
    7. Locales with no non-spatial sublocales 116
    8. Spaces with no non-induced sublocales 120

VII. **Compactness and Local Compactness**
    1. Basics, and a technical lemma 125
    2. Compactness and separation 127
    3. Kuratowski-Mrówka characterization 128
    4. Compactification 131
    5. Well below and rather below. Continuous completely
       regular frames 134
    6. Continuous is the same as locally compact.
       Hofmann-Lawson duality 137
    7. One more spatiality theorem 140

8.   Supercompactness. Algebraic, superalgebraic
      and supercontinuous frames 141

VIII. (Symmetric) Uniformity and Nearness
1.   Background 145
2.   Uniformity and nearness in the point-free context 148
3.   Uniform homomorphisms. Modelling embeddings. Products ... 152
4.   Aside: admitting nearness in a weaker sense 157
5.   Compact uniform and nearness frames. Finite covers 159
6.   Completeness and completion 159
7.   Functoriality. **CUniFrm** is coreflective in **UniFrm** 164
8.   An easy completeness criterion 167

IX. Paracompactness
1.   Full normality 169
2.   Paracompactness, and its various guises 173
3.   An elegant, specifically point-free, characterization
      of paracompactness 176
4.   A pleasant surprise: paracompact (co)reflection 180

X. More about Completion
1.   A variant of the completion of uniform frames 183
2.   Two applications 188
3.   Cauchy points and the resulting space 189
4.   Cauchy spectrum 193
5.   Cauchy completion. The case of countably
      generated uniformities 199
6.   Generalized Cauchy points 201

XI. Metric Frames
1.   Diameters and metric diameters 203
2.   Metric spectrum 207
3.   Uniform Metrization Theorem 211
4.   Metrization theorems for plain frames 214
5.   Categories of metric frames 218

XII. Entourages. Asymmetric Uniformity
1.   Entourages 228
2.   Uniformities via entourages 231
3.   Entourages versus covers 233
4.   Asymmetric uniformity: the classical case 236

5.   Biframes   239
6.   Quasi-uniformity in the point-free context via paircovers   241
7.   The adjunction **QUnif** ⇄ **QUniFrm**   248
8.   Quasi-uniformity in the point-free context
     via entourages   250

XIII.   Connectedness
1.   A few observations about sublocales   253
2.   Connected and disconnected locales   254
3.   Locally connected locales   258
4.   A weird example   260
5.   A few notes   267

XIV.   The Frame of Reals and Real Functions
1.   The frame $\mathfrak{L}(\mathbb{R})$ of reals   269
2.   Properties of $\mathfrak{L}(\mathbb{R})$ 272
3. $\mathfrak{L}(\mathbb{R})$ versus the usual space of reals   274
4.   The metric uniformity of $\mathfrak{L}(\mathbb{R})$ 276
5.   Continuous real functions   278
6.   Cozero elements   284
7.   More general real functions   287
8.   Notes   294

XV.   Localic Groups
1.   Basics   297
2.   The category of localic groups   302
3.   Closed Subgroup Theorem   305
4.   The multiplication μ is open. The semigroup of open parts . .   306
5.   Uniformities   310
6.   Notes   312

Appendix I. Posets
1.   Basics   315
2.   Zorn's Lemma   318
3.   Suprema and infima   319
4.   Semilattices, lattices and complete lattices. Completion   320
5.   Galois connections (adjunctions)   323
6.   (Semi)lattices as algebras. Distributive lattices   325
7.   Pseudocomplements and complements.
     Heyting and Boolean algebras   330

Appendix II. Categories

  1.   Categories   337
  2.   Functors and natural transformations   340
  3.   Some basic constructions   343
  4.   More special morphisms. Factorization   346
  5.   Limits and colimits   350
  6.   Adjunction   352
  7.   Adjointness and (co)limits   356
  8.   Reflective and coreflective subcategories   359
  9.   Monads   361
 10.   Algebras in a category   365

Bibliography   367

List of Symbols   381

List of Categories   385

Index   387

Preface

Point-free topology is based, roughly speaking, on the fact that the abstract lattice of open sets can contain a lot of information about a topological space, and that an algebraic treatment can provide new insights into the nature of spaces. The intuition behind this approach is quite natural: one thinks of a space as consisting of "realistic places" of non-trivial extent endowed with the information of how they meet and how a bigger place is composed from smaller ones.

During the time this theory has been cultivated (the concepts as we use them today – frame, frame homomorphism, etc. – were coined in the late 1950s) the abstract lattice approach proved to be able to replace the classical one, and in many respects to advantage: it turned out that some results are in a sense better. For instance, one has fully constructive counterparts of facts that are, in classical topology by necessity, heavily choice dependent, such as the compactness of products of compact spaces. Sometimes one obtains satisfactory facts instead of classically unsatisfactory ones, for instance, the behaviour of paracompact spaces. And one gains new and useful insights into many space phenomena.

Twenty years after the theory started to develop there appeared, in 1983, Johnstone's excellent monograph "Stone Spaces" which is still, after a quarter of a century, the standard reference book. Since then a lot has happened in the area; however, it is all in (numerous) journal articles, and with the exception of two chapters in handbooks and a slim booklet on the covariant approach, there is no new comprehensive literature available. The aim of the present book is to fill this gap.

We would like to acknowledge the support we have had while working on the text: The excellent conditions and the active and friendly atmosphere in the institutions we are happy to work in – the Department of Mathematics and the Centre for Mathematics of the University of Coimbra, the Department of Applied Mathematics and ITI of the Charles University in Prague – and the financial assistance provided by CMUC/FCT and grants 1M0545 and MSM 0021620838 of the Ministry of Education of the Czech Republic that enabled our cooperation. Further, our thanks go to the numerous members of the international community of categorical topology and related topics for their kind encouragement. And we wish to express most warmly our thanks to our families for support, understanding and patience.

— J. Picado and A. Pultr

Coimbra and Prague, February 2011

Introduction

Suppose you compute $\sqrt{2}$. You compute it with increasing precision. This can be viewed as producing a sequence of crisp numbers

$$1., \quad 1.4, \quad 1.41, \quad 1.414, \quad \ldots$$

stopping with some (say)

$$1.414\ldots 3$$

satisfactory for your purposes. Or you can view the progress as narrowing down spots of non-trivial extent

$$1.(\text{further blurred}), \quad 1.4(\text{further blurred}),$$

$$1.41(\text{further blurred}), \quad 1.414(\text{further blurred}), \quad \ldots$$

stopping with a satisfactorily small grain. In the latter case you in fact adopt the point-free attitude: the space (in our case the line) is a conglomerate of realistic places, spots of non-trivial extent, somehow related with each other; the points are just abstractions of "centers" of diminishing systems of spots. They are not very realistic (measurements do not give you point values), but we badly need them for theoretical purposes. Or do we? Yes, sometimes we indeed cannot do without. But there are cases where the points are not needed at all. Moreover, in most everyday spaces the points can be reconstructed upon request from the "realistic" data (and surprisingly enough, the spaces in which points do not make enough sense turn out to be beneficent for the theory, too; but about this later). This is what point-free topology is concerned with, and what the present book is devoted to.

Authors who are not specialists in history walk on thin ice when trying to outline the evolvement of a field. Nevertheless we have to try, only hoping that we will be pardoned for omitting important milestones by ignorance. For a more accurate treatment of the historical development of point-free topology we can advise the reader to see [152] (see also the excellent surveys in Johnstone [146] or [151]).

Geometry, the study of spatial relations, is as old as mathematics, but the idea of a space as we see it today is relatively recent. The development of modern analysis in the 19th century needed a precise understanding of continuity, convergence, and various related concepts. A very useful metric view of spaces developed. Later it turned out, however, that the metric (distance) is often somewhat superfluous, and that in fact one has sometimes structures with clearly space-like features that cannot be treated using distance at all. A fundamental break came in 1914 in Hausdorff's "*Mengenlehre*" ([125]) that indisputably originated modern topology (it is worth noting that this article was preceded by one year by Carathéodory's [59] viewing points as entities localized by special systems of diminishing sets). The structure of a space was based on the brilliant idea of a neighbourhood (surrounding as opposed to just containing). In the 1920s and 1930s the concept of an open set (first, as a derived notion and then gradually becoming a primary one) became more and more favoured (see, e.g., Alexandroff [2], or Sierpiński [243]). There are at least two reasons that makes it so important, in particular for our purposes:

- the notion of open set is very near to the intuition of a "realistic spot", and

- open sets constitute a very nicely behaving lattice (the properties of which have later become the basis of the techniques of the theory).

For various reasons, already in the 1930s the question repeatedly arose whether one can treat spaces as the lattices of their open sets (the famous Stone duality representing – albeit very special – spaces as Boolean algebras [249, 250] dates as far back as 1934 resp. 1936), or simply somehow without points (Wallman 1938 [266], Menger 1940 [183], McKinsey and Tarski 1940 [182], Pereira Gomes 1946 [198]). It took, however, more than a decade until the concepts and perspectives started to stabilize to the contemporary one (in particular one should mention the work of the Ehresmann seminar [82, 53, 191] in the late 1950s; here the term *frame* for point-free space was coined – and became accepted usage, similarly like the later term *locale*); it should be noted that (independent) proofs that the lattice of open sets determines the space in a really broad class appeared only as late as 1962 (Bruns [57], Thron [253]). After that, many authors became interested (C.H. Dowker, D. Papert (Strauss), J. Isbell, B. Banaschewski, etc.) and the field started to develop rapidly. The pioneering paper [136] by J. Isbell merits particular mention for opening several important topics (in this paper we first encounter the term *locale*, today used typically when emphasizing a covariant approach). In 1983 appeared the monograph by Johnstone ("Stone Spaces" [145]) which is still a primary source of reference.

Since then more than a quarter of a century has elapsed and a lot has happened in the area. Shortly after [145], in 1985, there appeared a very specialized book [265]. This was followed, much later, by two chapters in handbooks ([220, 209]), and a slim booklet [206] promoting a covariant approach. But there is no comprehensive text, and this is a gap we would like to fill, at least partly, by this monograph.

We hope that the book will be of interest also for people who are already in the area, or close to it. But of course we would like to address readers for whom the field is more or less new. And particularly to those we owe a few words on the usefulness of the extended area.

Extending or generalizing an already established concept calls for justification. There are haunting questions that should be answered: Is the broader range of spaces desirable at all? That is, is the theory in some sense more satisfactory? Is the algebraic technique appropriate and does it not obscure the geometric contents? And, when abandoning points, do we not lose too much information?

Let us start with the last one. Of course there are spaces that are not adequately representable: for instance those that are not T_0. But the more interesting ones, starting with the separation axiom T_D (much weaker than T_1) can be fully reconstructed. And the same holds for sober spaces (sobriety is independent of T_1 and implied by the Hausdorff property).

Does one obtain a better theory? In some areas one does. To give an example: paracompactness, a very important property, is classically very badly behaved (even a product of a paracompact space with a metric one is not necessarily paracompact). In the point-free setting one has no trouble with the standard constructions: the category of paracompact frames is coreflective in that of completely regular ones.

To give another example: somewhat surprisingly, it has turned out that some facts that are heavily non-constructive in the classical context are now fully constructive (no choice principle and no excluded middle needed). And this does not concern esoteric hair-splitting questions. For instance, one does not need choice for proving that products of compact "generalized spaces" are compact (the reader will see a very transparent choice-free construction of Stone-Čech compactification in Chapter VII); similarly, a completion of a uniform locale is choice-free.

A very important class of spaces (the locally compact ones) is represented equivalently, that is, there are no unwanted new spaces. But again, there are facts that are constructive and in the classical context not so.

The technique may seem to be at the beginning somewhat less intuitive, but often it turns out to be actually simpler. Also it opens new perspectives hidden in the classical approach (structure of subspaces, compactifications and completions, etc.).

———————————

It is often the case that the structure relating the individual parts of a monograph is by necessity intricate. The architecture of the text then cannot be linear and it is profitable to advise the reader what is to be read before this or that chapter and what one can skip at the first reading. Here it so happens that a linear sequence of events is quite natural and we think that the reader can read the text in the order presented (of course, sometimes skipping proofs). Also, we tried hard to choose the headings of the individual chapters and paragraphs to be as informative as possible. Thus, presenting the subject matter of the text as a succession of items might easily degenerate into a mere stylistic adaptation of the Table of Contents. It will be perhaps better to make a few loose comments instead.

The first two chapters (besides, of course, introducing the basic concepts) delimit the relations with classical topology and provide a formal bond between the two basic categories in the form of a spectrum adjunction. Note that the two weak conditions, T_D and sobriety, under which the spaces and "generalized spaces" relate particularly well, were not tailored for the purposes of point-free topology, and are of interest in themselves.

Sublocales: these generalized subspaces constitute a far more intricate and interesting subject than subspaces of classical spaces. We have split the matter into two parts, the general theory, and several special items that are better introduced after the reader is already informed on some categorical facts and about special classes of frames (locales). In the first part we pay particular attention to the inconspicuous quotient theorem; it will play an important role repeatedly.

In among the categorical facts it is advisable to stop for a moment at the product (coproduct of frames). The difference between this and the classical product of spaces (which is in fact not quite as big as it looks at first sight), rather than creating troubles, has in fact often rather pleasing consequences.

Amid the separation axioms we do not meet a natural counterpart of T_1. On the other hand we encounter two conditions, fitness and subfitness, that are not much investigated in the pointy context (although they make good sense even there). The subfitness condition deserves a particular interest (the terminology developed historically so that "fit" sounds like the basic property and "subfit" is somehow derived; in fact "subfit" is more basic, "fit" is "hereditarily subfit" – important enough, though).

Compactness does not need much comment. Most of the basic facts are quite parallel to the classical ones. But note the very specific construction of compactification and the absence of choice assumption, and on the other hand the necessity of choice in the duality between locally compact spaces and locally compact frames.

From Chapter VIII on, the leitmotiv is enrichment by uniformity and related structures (the weaker nearness, and the (dia)metric). After the first uniformity and completeness chapter, however, a chapter on paracompactness is inserted. Why: in view of some of its characteristics, paracompactness is a non-enriched matter and therefore it should be treated as soon as possible. But on the other hand it is closely connected with the possible uniformizations. This is particularly so in the point-free context where we have the beautiful – and extremely useful – fact that a locale is paracompact iff it admits a complete uniformity. Thus, it could not be introduced sooner.

What meets the eye first at the uniformities defined by entourages (Chapter XII) is the possibility of a non-symmetric generalization. This is indeed very important, but it should be noted that even the symmetric variant (equivalent with the cover approach from the previous pages) can be of a great technical advantage; this will be seen, for instance, in the chapter on localic groups.

Chapter XIII, concerning connectedness, may look out of place. Indeed, it is a plain space or frame (locale) matter; we do not need any extra structure, and very little special properties. It can be easily read already after Chapter VII. But we have decided to put it into the concluding triad of chapters concerned with special facts (because of a construction of a very special example, and also because this is a topic that still deserves a lot of further investigation, also in view of extra structures).

In the last two chapters we present some basic information on the point-free theory of reals (Chapter XIV), and, in Chapter XV, localic groups, the analogon of topological ones, including (among a number of others) the somewhat surprising fact that subgroups are in the point-free context always closed.

For the reader's convenience we add two short (but, we hope, informative enough) appendices containing the necessary facts about partially ordered sets and categories.

After thus presenting the contents we perhaps owe to the reader a comment on some points that are not included. First, there are a few topics in "Stone Spaces" to which there is, even after a quarter of a century, nothing to be added (e.g., the subject of coherent frames); about such we can only advise the reader to consult this classical monograph. Further, our information about the frame of reals is rather concise because there exists an excellent specialized scriptum [16] that may soon grow into a book. Similarly we have not tried to introduce the topic of σ-frames where one can expect better qualified authors to produce a specialized text.

Partly because of the size we have avoided (with one exception) lengthy technical constructions of examples and counterexamples that can be better appreciated in their article form.

Convention

A cross-reference inside a chapter is indicated by the number of section or subsection ("see 6", "by 6.3"). If it refers to an item in another chapter, the (roman) number of the chapter is prefixed ("see IX.6", "by IX.6.3"). For the appendices we use the prefixes AI resp. AII.

Chapter I

Spaces and Lattices of Open Sets

It is our intention to forget about points in spaces as often as possible. Rather, we would like to work with "places of non-trivial extent".

To start with, think of a standard topological space given by a system of open sets. Now recall the classical Euclidean geometry where one views a line p as a basic entity, not as the collection of points incident with p, and try to see our space similarly: to view the open sets U as basic entities, and the points sitting in U as other type of entities in some way incident with U.

> **Note.** In fact, it is quite realistic to think, rather, *of a point as of a collection of open sets*, namely the collection of all its neighbourhoods, viewed as approximations of the ideal point by smaller and smaller spots.

In this short chapter we will discuss the question of how much of the information on a space can be recovered from the abstract structure of its open sets. It will turn out that in fact for spaces from large classes (for instance, all T_1 ones), no information is lost.

Let us agree, once for ever, that

> *all the spaces discussed will be T_0,*

not only in this chapter, but throughout the whole of the text: of course the distinction of points that cannot be separated in any way cannot be recovered by reasoning on open sets. This convention will be now and then recalled.

Notation. Throughout the book, the (complete) lattice of open sets of a space X will be denoted by

$$\Omega(X).$$

(Of course this convention will not prevent us from sometimes using expressions like "a topological space (X, τ)", etc.)

All that we want to tell the reader at this stage is contained in Sections 1–3. Section 4 contains a few technical facts about a special topological property. We have included it because otherwise these facts would have to appear scatteredly throughout the later text although they are best treated together. For the time being, the reader should skip it. It will be duly recalled when needed.

1. Sober spaces

1.1. In the lattice $\Omega(X)$, the elements of the form

$$X \smallsetminus \overline{\{x\}}$$

have the following property:

if $U \cap V \subseteq X \smallsetminus \overline{\{x\}}$ *then either* $U \subseteq X \smallsetminus \overline{\{x\}}$ *or* $V \subseteq X \smallsetminus \overline{\{x\}}$.

(Indeed, we have to have $x \notin U$ or $x \notin V$, say the first; then $x \in X \smallsetminus U$, hence $\overline{\{x\}} \subseteq X \smallsetminus U$ and $U \subseteq X \smallsetminus \overline{\{x\}}$.)

This is usually expressed by saying that they are *meet-irreducible open sets* (and the reader certainly sees that if we view the intersection as a sort of multiplication, these open sets behave like prime numbers).

A space X is said to be *sober* if there are no other meet-irreducible $U \neq X$ in $\Omega(X)$.

1.2. Notes. (1) The original definition by Grothendieck was formulated in the language of closed sets: a (T_0-) space is sober if the closures of points $\overline{\{x\}}$ are precisely the *join-irreducible closed sets*. This formulation is perhaps simpler, but we are interested (for the motivations already mentioned) in the lattice of open sets. First of all, however, we will use another, very expedient (and transparent), equivalent formulation anyway.

(2) Sobriety is not a very rare property. For instance,

each Hausdorff space is sober.

Indeed, suppose $U \neq X$ to be meet-irreducible. Then we cannot have $x_1 \neq x_2$, $x_1, x_2 \notin U$ because for disjoint open $U_i \ni x_i$ we had $U = (U \cup U_1) \cap (U \cup U_2)$ with $U \neq U \cup U_i$.

For more see Section 3 below.

1.3. Recall (AI.6.4.1) that a (proper) filter F in a lattice L is *prime* if $a_1 \vee a_2 \in F$ implies that $a_i \in F$ for some i. In a *completely prime filter* we require that this holds for any join, not only for the finite ones. That is,

$$\text{for any } J \text{ and } a_i \in L, \ i \in J, \quad \bigvee_{i \in J} a_i \in F \ \Rightarrow \ \exists i, \ a_i \in F.$$

A typical completely prime filter in $\Omega(X)$ is the

$$\mathcal{U}(x) = \{U \mid x \in U\},$$

the system of all the open neighbourhoods of a point x.

1.3.1. Proposition. *A space X is sober if and only if the filters $\mathcal{U}(x)$ are precisely the completely prime ones.*

Proof. $\Rightarrow$: Let X be sober and let $\mathcal{F}$ be a completely prime filter in $\Omega(X)$. Then

$$W = \bigcup \{U \in \Omega(X) \mid U \notin \mathcal{F}\}$$

is meet-irreducible: by complete primeness, $W \notin \mathcal{F}$ (and hence in particular $W \neq X$), and if $W \supseteq V_1 \cap V_2$ one cannot have both $V_i \in \mathcal{F}$; thus, say, $V_1 \notin \mathcal{F}$ and $V_1 \subseteq W$. Hence, $W = X \smallsetminus \overline{\{x\}}$ for some $x \in X$ and we obtain

$$U \notin \mathcal{F} \quad \text{iff} \quad U \subseteq W = X \smallsetminus \overline{\{x\}} \quad \text{iff} \quad x \notin U,$$

and $\mathcal{F} = \mathcal{U}(x)$.

$\Leftarrow$: If X is not sober then there is a meet-irreducible open W that is not $X \smallsetminus \overline{\{x\}}$ for any x. Set

$$\mathcal{F} = \{U \in \Omega(X) \mid U \nsubseteq W\}.$$

$\mathcal{F}$ is a completely prime filter: if $U \cap V \subseteq W$ we have either $U \subseteq W$ or $V \subseteq W$; obviously $V \supseteq U \in \mathcal{F}$ implies $V \in \mathcal{F}$, and finally if $\bigcup_{i \in J} U_i \nsubseteq W$ we have $U_i \nsubseteq W$ for some i. But $\mathcal{F}$ is none of the $\mathcal{U}(x)$, else we had

$$U \nsubseteq V \quad \text{iff} \quad x \in U \quad \text{iff} \quad U \nsubseteq X \smallsetminus \overline{\{x\}}$$

and $W = X \smallsetminus \overline{\{x\}}$. $\qquad\square$

1.4. Corollary. *Let L be a complete lattice isomorphic to the $\Omega(X)$ of a sober space X. Then X is homeomorphic to*

$$\left(\{F \mid F \text{ completely prime filter in } L\}, \tau_L = \{\tilde{a} \mid a \in L\} \right)$$

where $\tilde{a} = \{F \mid a \in F\}$.

Thus, the space X can be reconstructed from the lattice L by purely lattice theoretical means. $\qquad\square$

(Of course: $\mathcal{U}(x) \in \tilde{U}$ iff $U \in \mathcal{U}(x)$ iff $x \in U$.)

1.5. Notes. (1) The property from 1.3.1 is often more handy than the original definition of sobriety. We will often use it instead.

(2) In some texts, sobriety is listed in among the "separation axioms" (such as T_1, regularity, etc.). As it was rightfully pointed out by Marcel Erné, this is misleading.

Rather, it is a property of completeness type, similar, e.g., to the completeness of a metric space. Namely, some filters of open sets that "look like systems of neighbourhoods" (Cauchy filters in metric spaces, the completely prime filters in our case) are required to be really the neighbourhood filters of points.

1.6. Reconstructing continuous mappings. We can expect that maps sending in some way completely prime filters to completely prime filters will be

- – in some way connected with continuous maps, and
- – easily described in the language of (complete) lattices.

Indeed we have

1.6.1. Lemma. *Let L, M be complete lattices and let $h\colon L \to M$ be a mapping such that*

$$h(0) = 0, h(1) = 1, h(a \wedge b) = h(a) \wedge h(b), \text{ and } h(\bigvee_{i \in J} a_i) = \bigvee_{i \in J} h(a_i). \qquad (1.6.1)$$

Then for each completely prime filter F in M the set $h^{-1}[F]$ is a completely prime filter in L.

Proof. It is straightforward. Note that $h^{-1}[F]$ is proper because of $h(0) = 0$ and non-empty because of $h(1) = 1$. □

1.6.2. Proposition. *Let Y be sober and let X be arbitrary. Then the continuous maps $f\colon X \to Y$ are in a one-one correspondence, given by*

$$f \mapsto h = \Omega(f), \quad \Omega(f)(U) = f^{-1}[U],$$

with the mappings $h\colon \Omega(Y) \to \Omega(X)$ satisfying the conditions (1.6.1).

Proof. The $\Omega(f)$ obviously has the properties listed under (1.6.1). Now let them be satisfied by an $h\colon \Omega(Y) \to \Omega(X)$. For an $x \in X$ consider the completely prime filter $\mathcal{U}(x)$. Then $h^{-1}[\mathcal{U}(x)]$ is by 1.6.1 a completely prime filter and hence, by 1.3, equal to some $\mathcal{U}(y)$ in $\Omega(Y)$. The point y is uniquely determined (our spaces are T_0); denote it by $f(x)$. Now we have

$$x \in h(V) \quad \text{iff} \quad V \in h^{-1}[\mathcal{U}(x)] \quad \text{iff} \quad f(x) \in V \quad \text{iff} \quad x \in f^{-1}[V]$$

and we see that $h = \Omega(f)$ (and that the f is continuous). □

1.6.3. Convention. We have already introduced the symbol $\Omega(X)$ for the lattice of open sets of X. From now on, we will also use the Ω for continuous mappings, as in the

$$\Omega(f)$$

above.

2. The axiom T_D: another case of spaces easy to reconstruct

2.1. Unlike sobriety, the axiom T_D to be discussed is really of a separation type. It appeared in the article by Aull and Thron in 1963 ([3]) as one of the separation axioms between T_0 and T_1. In [253] published one year earlier by one of the authors, and independently in the paper [57] by Bruns, it was shown that under this very weak condition, again, the lattice $L = \Omega(X)$ determines X up to homeomorphism.

What does T_D (sometimes, $T_{\frac{1}{2}}$, to indicate its position among the separation axioms) require? It is as follows.

(T_D) *For each $x \in X$ there is an open $U \ni x$ such that $U \smallsetminus \{x\}$ is open as well.*

Obviously $T_1 \Rightarrow T_D$, and T_D is weaker since for instance each quasidiscrete (Alexandroff) space (obtained from a poset $(X, \leq)$ by declaring all the up-sets open) is T_D: we have both $\uparrow x$ and $\uparrow x \smallsetminus \{x\}$ open.

Further, $T_D \Rightarrow T_0$: if $y \neq x$ take the U from the definition; then either $y \notin U \ni x$, or $y \in U \smallsetminus \{x\}$. And T_0 does not imply T_D: see the space Y in 3.1 below.

2.2. An element b immediately precedes a in a poset (in our case, a lattice) if $b < a$ and if $b < c \leq a$ implies that $c = a$.

A *slicing filter* is a prime filter F in a lattice L such that

there exist a, b with $b \notin F \ni a$, and b immediately preceding a in L.

2.3. Proposition. *Each slicing filter in $\Omega(X)$ is of the form*

$$\mathcal{U}(x) = \{U \in \Omega(X) \mid x \in U\}.$$

If X is a T_D-space, the filters $\mathcal{U}(x)$ are precisely the slicing ones.

Proof. Let $\mathcal{F}$ be a slicing filter witnessed by open sets $B \notin \mathcal{F} \ni A$, B immediately preceding A. Choose an $x \in A \smallsetminus B$.

Let $U \ni x$. Then $A \supseteq A \cap (B \cup U) \supset B$, the $\supset$ because $A \cap (B \cup U)$ contains x. Thus, $A = A \cap (B \cup U) = B \cup (A \cap U)$, and since $\mathcal{F}$ is prime and $B \notin \mathcal{F}$, we have $A \cap U \in \mathcal{F}$ and finally $U \in \mathcal{F}$. Thus, $\mathcal{U}(x) \subseteq \mathcal{F}$.

On the other hand, let $U \in \mathcal{F}$. We have $A \supseteq A \cap (B \cup U) = B \cup (A \cap U) \supsetneq B$, the $\supsetneq$ because $A \cap U \in \mathcal{F}$. Thus, $A = A \cap (B \cup U)$, and hence $x \in U$.

Now if X is a T_D-space then each $\mathcal{U}(x)$ is slicing: choose a $U \ni x$ with $V = U \smallsetminus \{x\}$ open; then V immediately precedes U, and $V \notin \mathcal{U}(x) \ni U$. $\square$

2.3.1. Note. For T_0-spaces we have that

if V immediately precedes U then $V = U \smallsetminus \{x\}$ for some $x \in U$.

(Indeed, suppose there are $x \neq y$, $x, y \in U \smallsetminus V$. By T_0 we have an open W such that, say, $y \notin W \ni x$. Then $V \subsetneq V \cup (U \cap W) \subsetneq U$.)

Thus, for T_0-spaces, if each $\mathcal{U}(x)$ is a slicing filter then X is T_D.

2.4. Proposition. [Thron] *Let X be a T_D-space, let Y be a T_0-space and let*

$$\phi \colon \Omega(Y) \cong \Omega(X)$$

be a lattice isomorphism. Then there is precisely one continuous mapping $f \colon X \to Y$ such that $\phi = \Omega(f)$.

Consequently, if both X and Y are T_D then the f is a homeomorphism.

Proof. Take an $x \in X$. By 2.3, $\mathcal{U}(x)$ is slicing and hence so is also its isomorphic image under ϕ^{-1},

$$\{\phi^{-1}(U) \mid x \in U\} = \{V \mid x \in \phi(V)\},$$

and hence by 2.3, $\{V \mid x \in \phi(V)\} = \{V \mid y \in V\}$ for some $y \in X$. Now since Y is T_0 this y is uniquely determined; set $y = f(x)$. Thus we have

$$x \in \phi(V) \quad \text{iff} \quad f(x) \in V \quad \text{iff} \quad x \in f^{-1}[V]$$

so that $\phi(V) = f^{-1}[V] = \Omega(f)(V)$ (the equality also makes f continuous, of course). $\qquad\Box$

3. Summing up

3.1. Let $\mathbb{N}$ be the set of natural numbers, $\omega \notin \mathbb{N}$. Consider the spaces

- $X = (\mathbb{N}, \tau)$ with τ consisting of $\emptyset$ and all the complements of finite subsets, and
- $Y = (\mathbb{N} \cup \{\omega\}, \theta)$ with θ consisting of $\emptyset$ and all the U containing ω and such that $\mathbb{N} \smallsetminus U$ is finite.

Then obviously

$$\Omega(X) \cong \Omega(Y) \quad \text{while } X \text{ and } Y \text{ are not homeomorphic.}$$

3.2. In the example above, both the spaces are T_0. Hence we see that we cannot have a reconstruction theorem similar to 1.4 or 2.4 for all T_0-spaces. But we see more: we cannot have such a theorem holding for all the sober and T_D-spaces simultaneously. Namely,

$$X \text{ is } T_D \text{ (in fact, } T_1),$$

and

$$Y \text{ is sober (the suspect meet-irreducible } \emptyset \text{ is } Y \smallsetminus \overline{\{\omega\}}).$$

3.3. The example also shows that

> T_1 *(not to speak of T_D) does not imply sober*

(if X were sober we could apply 1.4), and

> *sober does not imply T_D (not to speak of T_1)*

(if Y were T_D we could apply 2.4).

3.4. Anyway, however, each of the two classes is large enough to justify studying spaces in terms of their lattices of open sets.

The reader may hesitate about the class of sober spaces: so far one only sees that it is just somewhat larger than the class of all Hausdorff ones; the fact that also all the finite T_0-spaces are sober may be not quite convincing (and of the infinite quasidiscrete (Alexandroff) spaces – recall 2.1 – only the Noetherian ones are sober, see II.6.4.4 below). But posets with the Scott topologies (spaces much more important in theoretical computer science than the quasidiscrete ones) are typically sober; finding a non-sober Scott example was rather hard ([143]).

In fact, the point-free topology to be presented in the sequel will mimic the sober case rather than the T_D one. But T_D will be important in interpreting some of the facts, too.

4. Aside: several technical properties of T_D-spaces

The reader is, again, encouraged to skip this section. We will recall it later, whenever some of the facts will be needed.

4.1. Let $X = (X, \tau)$ be a topological space. Recall that an element x of a subset $S \subseteq X$ is said to be *isolated* in S if there is an open set U in X such that $U \cap S = \{x\}$. It is *weakly isolated* in S if there is an open set U in X such that $x \in U \cap S \subseteq \overline{\{x\}}$.

4.1.1. For any subset $A \subseteq X$ define an equivalence E_A on $\Omega(X)$ by setting

$$(U, V) \in E_A \quad \text{iff} \quad U \cap A = V \cap A.$$

Obviously

$$A \subseteq B \quad \Rightarrow \quad E_B \subseteq E_A. \tag{4.1.1}$$

4.1.2. Recall that the *Skula topology* (sometimes called the *front topology*) of (X, τ) (notation $\mathrm{Sk}(\tau)$) is obtained from the basis

$$\{U \cap (X \smallsetminus V) \mid U, V \in \tau\}.$$

4.2. Here are a few statements equivalent with the T_D property.

Proposition. *Let* $X = (X, \tau)$ *be a topological space. Then the following statements are equivalent:*

(1) X *is* T_D.

(2) *For every* $x \in X$, *the set* $W(x) = (X \smallsetminus \overline{\{x\}}) \cup \{x\}$ *is open.*

(3) *For any* $S \subseteq X$, *every weakly isolated point in* S *is isolated in* S.

(4) *For any closed* $S \subseteq X$, *every weakly isolated point in* S *is isolated in* S.

(5) *For all* $A, B \subseteq X$, $E_B \subseteq E_A \;\Rightarrow\; A \subseteq B$.

(6) *For all* $A, B \subseteq X$, $E_B = E_A \;\Rightarrow\; A = B$.

(7) *The Skula topology* $\mathrm{Sk}(\tau)$ *is discrete.*

Proof. (1)$\Rightarrow$(2): Let U be open, $x \in U$, and let also $U \smallsetminus \{x\}$ be open. Then $U \smallsetminus \{x\} = U \smallsetminus \overline{\{x\}}$. Now $(X \smallsetminus \overline{\{x\}}) \cup \{x\}$ is a neighbourhood of all the $y \in X \smallsetminus \overline{\{x\}}$, and it is a neighbourhood of x as well: we have $x \in U = (U \smallsetminus \overline{\{x\}}) \cup \{x\} \subseteq (X \smallsetminus \overline{\{x\}}) \cup \{x\}$.

(2)$\Rightarrow$(3): If $x \in U \cap S \subseteq \overline{\{x\}}$ then $S \cap U \cap W(x) = \{x\}$.

(3)$\Rightarrow$(4) is trivial.

(4)$\Rightarrow$(1): x is trivially weakly isolated in $\overline{\{x\}}$. Therefore it is isolated in $\overline{\{x\}}$ and there is an open U such that $U \cap \overline{\{x\}} = \{x\}$; hence $U \smallsetminus \{x\} = U \smallsetminus \overline{\{x\}}$ is open.

(1)$\Rightarrow$(5): Let $A \not\subseteq B$. Then there is an $x \in A \smallsetminus B$. By T_D we have an open $U \ni x$ such that $V = U \smallsetminus \{x\}$ is open. Then $(U, V) \notin E_A$ while $(U, V) \in E_B$.

(5)$\Rightarrow$(6) is trivial.

(6)$\Rightarrow$(1): Let $x \in X$. Since E_X is trivial and $E_{X \smallsetminus \{x\}} \neq E_X$ there have to exist open $U \neq V$ such that $U \cap (X \smallsetminus \{x\}) = V \cap (X \smallsetminus \{x\})$. That is possible only if one of them, say U, contains x and $V = U \smallsetminus \{x\}$.

(2)$\Rightarrow$(7): $\{x\} = W(x) \cap \overline{\{x\}}$ is open in $\mathrm{Sk}(\tau)$.

(7)$\Rightarrow$(1): For $x \in X$ there are open U, V such that $\{x\} = U \cap (X \smallsetminus V)$. Then $\overline{\{x\}} \subseteq X \smallsetminus V$ and $\{x\} = U \cap \overline{\{x\}}$ and finally $U \smallsetminus \{x\} = U \smallsetminus \overline{\{x\}}$. $\square$

Chapter II

Frames and Locales. Spectra

Is it worthwhile to try to base topology on the algebra of open sets (and to neglect points)? We hope that this question has been answered in the previous chapter satisfactorily by showing that there are two large classes of spaces in which no information is lost, the class of sober spaces, and the class of T_D-ones.

The point-free topology as we will present it will lean slightly to the sober side. There are various reasons. One of them is that for reconstructing continuous maps we have a very simple condition. Not only simple, also a very natural one: preserving all joins and all finite meets by the maps on the algebraic side – and there is no doubt that in topology one likes to think in terms of general unions and finite intersections. And here is another reason: although this class does not go as far down in separation as the other one (not even all the T_1-spaces are covered), the Scott spaces (appearing in many modern applications) are typically (not always, but typically) sober, and seldom T_D. The T_D axiom will, nevertheless, also play a role later.

In this chapter we will introduce the basic algebraic background we will work with: frames and locales representing spaces, and frame homomorphisms resp. localic morphisms representing continuous maps. Further, we will discuss in some detail the reconstruction of points (spectra of frames) and some aspects of the spectrum adjunction.

1. Frames

1.1. The category of topological spaces and continuous maps will be denoted by

$$\mathbf{Top}$$

and we will write

$$\mathbf{Sob}$$

for its full subcategory generated by sober spaces.

1.2. The complete lattice of open sets satisfies the distributive law

$$(\bigvee_{i \in J} U_i) \wedge V = \bigvee_{i \in J} (U_i \wedge V)$$

since an arbitrary join coincides with the union, and the meet $U \wedge V$ is the intersection $U \cap V$.

(Note that the general meet in $\Omega(X)$ typically *does not* coincide with the intersection. Hence there is no surprise that we generally do not have $(\bigwedge_{i \in J} U_i) \vee V = \bigwedge_{i \in J} (U_i \vee V)$.)

Thus, aiming for "generalized spaces" represented as special complete lattices (mimicking the lattices of open sets) we are led to the following definition.

A *frame* is a complete lattice L satisfying the distributivity law

$$(\bigvee A) \wedge b = \bigvee \{a \wedge b \mid a \in A\} \tag{frm}$$

for any subset $A \subseteq L$ and any $b \in L$.

Recalling I.1.6 we define *frame homomorphisms* $h\colon L \to M$ between frames L, M as maps $L \to M$ preserving all joins (including the bottom 0) and all finite meets (including the top 1). The resulting category will be denoted by

Frm.

1.2.1. The lattice of open sets of the void space is the one-element lattice $\{0 = 1\}$. It will be denoted by

O.

The lattice of open sets of the one-point space is the two-element Boolean algebra $\{0 < 1\}$. It will be denoted by

2   or sometimes by   P.

1.2.2. Later on (for instance for the structure of generalized subspaces of a frame) we will also need the dual notion of a *co-frame*, that is of a complete lattice satisfying

$$(\bigwedge A) \vee b = \bigwedge \{a \vee b \mid a \in A\}. \tag{co-frm}$$

1.3. Recall the symbol Ω introduced in the preamble to Chapter I and extended for continuous maps in I.1.6.3. We have frames $\Omega(X)$ and frame homomorphisms $\Omega(f)\colon \Omega(Y) \to \Omega(X)$ for continuous maps $f\colon X \to Y$ resulting in a contravariant functor

$$\Omega\colon \mathbf{Top} \to \mathbf{Frm}.$$

1.3.1. A frame L is said to be *spatial* if it is isomorphic to an $\Omega(X)$.

There exist non-spatial frames (we will present some soon) and hence we have "more spaces than before". This may not seem a priori desirable; it has turned out, however, that it makes the theory, in some respects, more satisfactory.

Note. Nevertheless, the question naturally arises whether one can further narrow down the class of lattices by algebraic means so as to obtain just spatial frames, or at least much fewer non-spatial ones. There is a stricter distributivity rule (still holding in all the $\Omega(X)$) under which it is really hard to find a non-spatial example. See V.4.

1.4. The rule (frm) says that the mappings $-\wedge b = (x \mapsto x \wedge b)$ preserve all suprema. Hence, they are left Galois adjoints, and the corresponding right adjoints $b \to -$ produce a Heyting operation in L (see Appendix I). Thus,

a frame is automatically a complete Heyting algebra.

The Heyting operation will be frequently used. The reader has to have in mind, though, that this operation is not generally preserved by frame homomorphisms.

If a frame happens to be a Boolean algebra we speak of a *Boolean frame*. Note that frame homomorphisms between Boolean frames are complete Boolean homomorphisms (because lattice homomorphisms preserve complements – see AI.7.4.5), and that a Boolean frame is a co-frame as well.

2. Locales and localic maps

2.1. The contravariant functor Ω from 1.3, restricted to the category **Sob** of sober spaces is, by I.1.6.2, a full embedding (for any two sober spaces X, Y we have a one-one correspondence between the continuous maps $f\colon X \to Y$ and the frame homomorphisms $h\colon \Omega(Y) \to \Omega(X)$ provided by sending f to $\Omega(f)$). That is, it would be a full embedding if only it were covariant which it is not. This can be easily helped by introducing the *category of locales*

Loc

as the dual category **Frm**^{op}, and this category can be viewed as an extension of the category of sober spaces (and the frames – now referred to as *locales* – as generalized spaces).

2.2. It is not very intuitive to have morphisms $A \to B$ in a category represented as mappings $A \leftarrow B$. But this can be mended:

frame homomorphisms $h\colon M \to L$, since they preserve all suprema, have uniquely defined right Galois adjoints $h_*\colon L \to M$.

Thus, we can represent the morphisms in **Loc** as

the infima-preserving $f\colon L \to M$ such that the corresponding left adjoints $f^*\colon M \to L$ preserve finite meets.

Such f will be referred to as *localic maps* and we will think of the category **Loc** as having these for morphisms.

2.2.1. Note. We have not introduced localic maps this way only for the sake of having actual maps where we had, in the formal categorical construction, in essence only symbols. It will turn out later that it often really helps understanding further notions and constructions.

2.2.2. The locale O from 1.2.1 behaves indeed as the void (generalized) space should. There is precisely one localic map $O \to L$ for any locale L, and none $L \to O$ unless L is void (that is, O itself).

Also, the locale P behaves like a point. So far we immediately see that for any locale L there is precisely one localic map $L \to P$ (since there is precisely one frame homomorphism $P \to L$, namely $(0 \mapsto 0, 1 \mapsto 1)$). As for the localic maps $P \to L$ we will see shortly (as soon as we will introduce points in the next section) that they can be viewed as embeddings of points.

2.3. The following characteristics of localic maps will be sometimes useful.

Proposition. *Let* $f: L \to M$ *have a left adjoint* f^*. *Then*

(a) $f^*(1) = 1$ *iff* $f[L \smallsetminus \{1\}] \subseteq M \smallsetminus \{1\}$, *and*

(b) f^* *preserves binary meets iff*

$$f(f^*(a) \to b) = a \to f(b).$$

Proof. (a): If $f^*(1) = 1$ and $1 \leq f(x)$ then $1 = f^*(1) \leq x$; on the other hand, if the inclusion holds, use the inequality $f(f^*(1)) \geq 1$.

(b) $\Rightarrow$: $x \leq f(f^*(a) \to b)$ iff $f^*(x \wedge a) = f^*(x) \wedge f^*(a) \leq b$ iff $x \wedge a \leq f(b)$ iff $x \leq a \to f(b)$.

$\Leftarrow$: $f^*(a \wedge b) \leq x$ iff $a \leq b \to f(x) = f(f^*(b) \to x)$ iff $f^*(a) \leq f^*(b) \to x$ iff $f^*(a) \wedge f^*(b) \leq x$. $\square$

2.4. Now, when the category **Loc** has this concrete representation, it is worthwhile to modify the functor Ω accordingly. We will make it to a functor

$$\mathsf{Lc} \colon \mathbf{Top} \to \mathbf{Loc}$$

by setting

$$\mathsf{Lc}(X) = \Omega(X), \quad \mathsf{Lc}(f) = \Omega(f)_*.$$

From the standard adjunction

$$f^{-1}[B] \subseteq A \quad \text{iff} \quad B \subseteq Y \smallsetminus f[X \smallsetminus A]$$

of preimage we immediately learn that

$$\mathsf{Lc}(f)(U) = Y \smallsetminus \overline{f[X \smallsetminus U]}.$$

This does not seem to be a very transparent formula, but it becomes more accept-able after the following observation.

Represent a point $x \in X$ by the element

$$\widetilde{x} = X \smallsetminus \overline{\{x\}}$$

of $\mathsf{Lc}(X)$. Then we have

$$\mathsf{Lc}(f)(\widetilde{x}) = \widetilde{f(x)}$$

(indeed, $\mathsf{Lc}(f)(\widetilde{x}) = Y \smallsetminus \overline{f[X \smallsetminus (X \smallsetminus \overline{\{x\}})]} = Y \smallsetminus \overline{f[\overline{\{x\}}]} = Y \smallsetminus \overline{\{f(x)\}} = \widetilde{f(x)}$).

3. Points

3.1. The intuitive concept of a diminishing system of spots. We believe that this is what most people do when visualizing a point: one thinks of a very small spot; then admits it may be still too large, thinks of a smaller one inside, etc. It should be noted that this view of a point was around decades before there was anything like a point-free topology (for instance we can go as far back as to Carathéodory 1913 [59]).

Thus we can think of a point in a locale as a specific filter. We are still in trouble in telling what "diminishing size" of spots should mean (later on, in Chapter X we will be able to do it in various ways, with various resulting definitions of points, but now we do not want to employ any extra structure). A general filter, for instance $\{x \in L \mid x \geq a\}$ will hardly do. But take the hint from sober spaces: in I.1.3 we have seen that in this fairly broad class of spaces the points can be detected (represented by their sets of neighbourhoods $\mathcal{U}(x)$) as the completely prime filters. This leads to the following definition:

(P1) a *point* in a frame (locale) L is a completely prime filter $F \subseteq L$.

3.2. Another, but practically the same representation. The points x in a space X are in a natural one-one correspondence with the (continuous) mappings $f = (0 \mapsto x) \colon \{0\} \to X$ and hence, in the sober case, with the frame homomorphisms $h \colon \Omega(X) \to \mathbf{2}$. Thus,

(P2) a *point* in L can be viewed as a frame homomorphism $h \colon L \to \mathbf{2}$

(or as a *localic map* $\mathsf{P} \to L$, with the "one-point locale" $\mathsf{P} = \mathbf{2}$).

The translation between (P1) and (P2) is obvious: given a completely prime filter $F \subseteq L$ define $h \colon L \to \mathbf{2}$ by setting $h(x) = 1$ iff $x \in F$, and given an $h \colon L \to \mathbf{2}$ we have the completely prime filter $F = \{x \mid h(x) = 1\}$.

3.3. Yet another, this time not so obvious representation of points. Recall that an element $p \neq 1$ in a lattice L is *meet-irreducible* (also, *prime*) if for any $a, b \in L$,

$a \wedge b \leq p$ implies that either $a \leq p$ or $b \leq p$. (This means that the principal ideal $\downarrow p$ is a prime ideal – see AI.6.4.1).

Given a completely prime filter $F \subseteq L$ define

$$p_F = \bigvee \{x \mid x \notin F\}.$$

Then

p_F is a meet-irreducible element.

(Indeed, by the complete primeness, $p_F \notin F$ and hence $p_F \neq 1$, and if $a \wedge b \leq p_F$ then we cannot have both $a, b \in F$ and hence, say, $a \notin F$ and $a \leq p_F$.)

On the other hand, if $p \in L$ is meet-irreducible, set

$$F_p = \{x \mid x \not\leq p\}.$$

Then

F_p is a completely prime filter.

(In fact, if $x, y \in F_p$ then $x \wedge y \in F_p$ by meet-irreducibility; obviously $y \geq x \in F_p$ implies $y \in F_p$, and if $\bigvee_{i \in J} x_i \not\leq p$ then $x_i \not\leq p$ for some i; finally, $1 \not\leq p$ and $0 \leq p$ so that F_p is non-void and proper.)

Finally, $p_{F_p} = \bigvee \{x \mid x \leq p\} = p$ and $x \in F_{p_F}$ iff $x \not\leq \bigvee \{y \mid y \notin F\}$ iff $x \in F$.

(If $x \in F$ we cannot have $x \leq \bigvee \{y \mid y \notin F\}$ by complete primeness, and if $x \notin F$ then $x \leq \bigvee \{y \mid y \notin F\}$ trivially.)

Thus,

(P3)   a *point* in L can also be viewed as a meet-irreducible element $p \in L$.

3.4. Localic maps preserve points in the sense of (P3). We have

Lemma. *Localic maps send meet-irreducible elements to meet-irreducible ones again.*

Proof. First, localic maps reflect tops, that is, if $f(a) = 1$ then $a = 1$: indeed, if $1 \leq f(a)$ then $1 = f^*(1) \leq a$.

Now let $f \colon L \to M$ be a localic map and let a be meet-irreducible in L. Let $x \wedge y \leq f(a)$. Then $f^*(x) \wedge f^*(y) = f^*(x \wedge y) \leq a$ so that, say, $f^*(x) \leq a$ and $x \leq f(a)$ (which cannot be the top, by the preceding paragraph).   □

3.5. Now we can better explain what we said in 2.2.2 about the behaviour of the locale P as a one-point space. Namely, the localic maps $f \colon \mathsf{P} \to L$ are in a natural one-one correspondence with the meet-irreducibles in L:

 - $\mathsf{P} = \mathbf{2}$ has precisely one meet-irreducible element, namely 0; associate with f the meet-irreducible $f(0)$;
 - if $p \in L$ is meet-irreducible we have the localic map $(0 \mapsto a, 1 \mapsto 1)$, adjoint to the frame homomorphism $h(x) = 0$ iff $x \leq p$.

4. Spectra

Adopting the idea of a point as in 3.1 we implicitly think of the filter F also as of the system of neighbourhoods of the "ideal point in the center". This makes the system of points of L to a space, the *spectrum* of L. It will be discussed in the remaining sections of this chapter.

We will abbreviate "completely prime filter" to "c.p. filter", and, as before, we will write

$$f^* \text{ for the left Galois adjoint of } f.$$

4.1. For an element a of a locale L set

$$\Sigma_a = \{F \text{ c.p. filter in } L \mid a \in F\}.$$

We easily check the

Observation. $\Sigma_0 = \emptyset$ *and* $\Sigma_1 = \{$*all c.p. filters*$\}$ *(note that our filters are non-void and proper)*, $\Sigma_{a \wedge b} = \Sigma_a \cap \Sigma_b$, *and* $\Sigma_{\bigvee_{i \in J} a_i} = \bigcup_{i \in J} \Sigma_{a_i}$.

4.2. Consequently we have a topological space

$$\mathsf{Sp}(L) = \Big(\{\text{all c.p. filters in } L\}, \{\Sigma_a \mid a \in L\} \Big).$$

This space is called the *spectrum* of L.

4.3. The functor Sp. If $f \colon L \to M$ is a localic map we define

$$\mathsf{Sp}(f) \colon \mathsf{Sp}(L) \to \mathsf{Sp}(M)$$

by setting

$$\mathsf{Sp}(f)(F) = (f^*)^{-1}[F]$$

(since f^* is a frame homomorphism the preimage is a c.p. filter).

4.3.1. Lemma. $(\mathsf{Sp}(f))^{-1}[\Sigma_a] = \Sigma_{f^*(a)}$. Consequently, $\mathsf{Sp}(f)$ is a continuous map.

Proof. $(\mathsf{Sp}(f))^{-1}[\Sigma_a] = \{F \mid \mathsf{Sp}(f)(F) = (f^*)^{-1}[F] \in \Sigma_a\} = \{F \mid a \in (f^*)^{-1}[F]\}$
$= \{F \mid f^*(a) \in F\} = \Sigma_{f^*(a)}$. $\qquad\qquad\qquad\qquad\qquad\qquad\qquad\qquad\qquad\qquad\square$

4.3.2. Obviously $\mathsf{Sp}(\mathrm{id}_L) = \mathrm{id}_{\mathsf{Sp}(L)}$ and $\mathsf{Sp}(g \cdot f) = \mathsf{Sp}(g) \cdot \mathsf{Sp}(f)$. Hence we have

Fact. *The formulas for* $\mathsf{Sp}(L)$ *and* $\mathsf{Sp}(f)$ *define a functor*

$$\mathsf{Sp} \colon \mathbf{Loc} \to \mathbf{Top}.$$

One speaks of the *spectrum functor*.

4.4. Note. Usually one considers as the spectrum functor the contravariant

$$\Sigma\colon \mathbf{Frm} \to \mathbf{Top}, \quad \Sigma L = \mathsf{Sp}(L), \quad \Sigma h(F) = h^{-1}[F],$$

as a counterpart to the $\Omega\colon \mathbf{Top} \to \mathbf{Frm}$. We have modified the notation to emphasize the covariance. Shortly we will show that Sp is a right adjoint to Lc. Of course, the adjunction can also be described in terms of an adjunction of Ω and Σ; but which of the functors is to the left and which is to the right is much more transparent for covariant functors.

4.5. The spectrum in terms of meet-irreducibles. If we represent the points by meet-irreducible elements as in 3.3 we obtain the spectrum functor in a particularly simple form. We set, of course,

$$\mathsf{Sp}'(L) = \Big(\{\text{all meet-irreducible elements in } L\}, \{\Sigma_a' \mid a \in L\}\Big)$$

where $\Sigma_a' = \{p \mid a \not\le p\}$ (in the translation from 3.2: $F_p \in \Sigma_a$ iff $a \in F_p$ iff $a \not\le p$ iff $p \in \Sigma_a'$) and using 3.4 we can define, for a localic map $f\colon L \to M$,

$$\mathsf{Sp}'(f)\colon \mathsf{Sp}'(L) \to \mathsf{Sp}'(M)$$

simply as the restriction, that is, by the formula $\mathsf{Sp}'(f)(q) = f(q)$.

> (This is indeed what the $\mathsf{Sp}(f)$ above translates to. Let $q \in \mathsf{Sp}'(L)$. By the formulas from 3.3 we obtain $(f^*)^{-1}[F_q] = \{a \mid f^*(a) \not\le q\} = \{a \mid a \not\le f(q)\}$ and hence $p_{\mathsf{Sp}(f)(F_q)} = p_{\{a \mid a \not\le f(q)\}} = \bigvee\{a \mid a \le f(q)\} = f(q).$)

The reader may rightfully ask why we have not preferred this description of spectrum to the one above. The reason is that we want to keep the intuition of a point as a "limit of a diminishing system of spots", and this is, of course, fairly explicitly visible when using the filters, and not at all when using the meet-irreducibles. But we will use the construction Sp' as well.

4.6. The spectrum adjunction. (Recall AII.6) The functors Lc and Sp are not inverse to each other (and certainly have not been expected to be). But we have the next best: they are adjoint to each other, and as we will see in the following sections, the adjunction units have very nice properties.

4.6.1. By 4.1 we have frame homomorphisms

$$\phi_L\colon L \to \mathsf{Lc}(\mathsf{Sp}(L))$$

given by $\phi_L(a) = \Sigma_a$. Consider their right Galois adjoints, the localic maps

$$\sigma_L = (\phi_L)_*\colon \mathsf{Lc}(\mathsf{Sp}(L)) \to L.$$

Lemma. (1) *The system $(\sigma_L)_L$ constitutes a natural transformation* $\mathsf{LcSp} \overset{\cdot}{\to} \mathrm{Id}$.
(2) *Each σ_L is one-one.*

Proof. (1) Let $f\colon L \to M$ be a localic map. We have

$$(\sigma_M \cdot \mathsf{LcSp}(f))^*(a) = \mathsf{LcSp}(f)^*(\phi_M(a)) = \Omega(\mathsf{Sp}(f))(\Sigma_a)$$
$$= \{F \mid \mathsf{Sp}(f)(F) \ni a\} = \{F \mid (f^*)^{-1}[F] \ni a\} = \{F \mid f^*(a) \in F\}$$
$$= \Sigma_{f^*(a)} = \phi_L(f^*(a)) = (\sigma_L)^*(f^*(a) = (f \cdot \sigma_L)^*(a).$$

(2) The mapping ϕ_L is onto. Hence, from the standard Galois equation (see AI.5.3.1) $\phi_L \sigma_L \phi_L = \phi_L$ we obtain $\phi_L \sigma_L = \mathrm{id}$ which makes σ_L one-one. $\qquad \square$

4.6.2. For a space X consider the mapping

$$\lambda_X \colon X \to \mathsf{Sp}(\mathsf{Lc}(X))$$

defined by

$$\lambda_X(x) = \mathcal{U}(x) = \{U \mid x \in U\}.$$

Lemma. (1) λ_X *is a continuous map.*

(2) *The system* $(\lambda_X)_X$ *constitutes a natural transformation* $\mathrm{Id} \xrightarrow{.} \mathsf{SpLc}$.

Proof. (1) Each open set in $\mathsf{SpLc}(X)$ is of the form Σ_U with U open in X. We have $\lambda_X^{-1}[\Sigma_U] = \{x \mid \mathcal{U}(x) \in \Sigma_U\} = \{x \mid U \in \mathcal{U}(x)\} = \{x \mid x \in U\} = U$.

(2) Let $f\colon X \to Y$ be continuous. Since $\Omega(f)$ is the left adjoint to $\mathsf{Lc}(f)$ we have

$$(\mathsf{SpLc}(f) \cdot \lambda_X)(x) = \mathsf{SpLc}(f)(\mathcal{U}(x)) = (\Omega(f))^{-1}(\mathcal{U}(x))$$
$$= \{U \mid f^{-1}[U] \in \mathcal{U}(x)\} = \{U \mid x \in f^{-1}[U]\}$$
$$= \{U \mid f(x) \in U\} = \lambda_Y(f(x)). \qquad \square$$

4.6.3. Theorem. *The functors* Lc *and* Sp *are adjoint,* Lc *to the left and* Sp *to the right, with units* $\lambda\colon \mathrm{Id} \xrightarrow{.} \mathsf{SpLc}$ *and* $\sigma\colon \mathsf{LcSp} \xrightarrow{.} \mathrm{Id}$.

Proof. Consider

$$\mathsf{Sp}(\sigma_L) \cdot \lambda_{\mathsf{Sp}L} \colon \mathsf{Sp}L \to \mathsf{Sp}(\mathsf{Lc}(\mathsf{Sp}L)) \to \mathsf{Sp}L.$$

Since $(\sigma_L)^* = \phi_L$ we have

$$(\mathsf{Sp}(\sigma_L) \cdot \lambda_{\mathsf{Sp}L})(F) = \mathsf{Sp}(\sigma_L))\mathcal{U}(F)) = \phi_L^{-1}[\mathcal{U}(F)]$$
$$= \{U \mid \Sigma_U \in \mathcal{U}(F)\} = \{U \mid F \in \Sigma_U\} = \{U \mid U \in F\} = F.$$

Consider

$$\sigma_{\mathsf{Lc}(X)} \cdot \mathsf{Lc}(\lambda_X) \colon \mathsf{Lc}X \to \mathsf{Lc}(\mathsf{Sp}(\mathsf{Lc}X)) \to \mathsf{Lc}X.$$

We have

$$(\sigma_{\mathsf{Lc}X} \cdot \mathsf{Lc}(\lambda_X))^*(U) = (\mathsf{Lc}(\lambda_X)^* \cdot \phi_{\mathsf{Lc}(X)})(U) = (\mathsf{Lc}(\lambda_X))^*(\Sigma_U)$$
$$= \Omega(\lambda_X)(\Sigma_U) = \lambda_X^{-1}[\Sigma_U] = \{x \mid \mathcal{U}(x) \ni U\} = \{x \mid x \in U\} = U. \qquad \square$$

4.6.4. Observation. We already know that each σ_L is one-one. Now we have also learned that

$$\sigma_{\mathsf{Lc}(X)} \cdot \mathsf{Lc}(\lambda_X) = \mathrm{id}_{\mathsf{Lc}(X)}$$

which makes $\sigma_{\mathsf{Lc}(X)}$ an onto map and we conclude that

for each X, the localic map $\sigma_{\mathsf{Lc}(X)} \colon \mathsf{Lc}(\mathsf{Sp}(\mathsf{Lc}X)) \to \mathsf{Lc}X$ *is an isomorphism.*

4.7. The units in meet-irreducibles. In the spectrum description Sp' the formulas for the units appear as

$$\lambda_X(x) = X \smallsetminus \overline{\{x\}} \quad \text{and} \quad (\sigma_X)^*(a) = \Sigma'_a = \{p \mid a \nleq p\}.$$

5. The unit σ and spatiality

5.1. Proposition. *The following statements on a locale L are equivalent.*

(1) *L is spatial,*

(2) *$\sigma_L \colon \mathsf{Lc}(\mathsf{Sp}(L)) \to L$ is a complete lattice isomorphism,*

(3) *$\sigma_L^* \colon L \to \mathsf{Lc}(\mathsf{Sp}(L))$ is a complete lattice isomorphism,*

(4) *σ_L is onto,*

(5) *σ_L^* is one-one.*

Proof. The implications $(2)\Rightarrow(1)$, $(2)\Rightarrow(4)$ and $(3)\Rightarrow(5)$ are obvious. By 4.6.1 any of (4), (5) implies any of (2), (3), (4), (5) (σ_L and σ_L^* are monotone maps and any of (4), (5) makes them inverse to each other, and hence complete lattice isomorphisms).

It remains to prove that $(1)\Rightarrow(2)$. Recall 4.6.4. If $h \colon L \to \mathsf{Lc}(X)$ is an isomorphism we have the isomorphism $\sigma_L = h^{-1} \cdot \sigma_{\mathsf{Lc}(X)} \cdot \mathsf{LcSp}(h)$. $\qquad\square$

5.2. Thus the question whether L is spatial or not reduces to checking whether it is (isomorphic to) the topology of its own spectrum. If it is not, it is not isomorphic to the topology of any space.

5.3. Consider the reformulation of the spectrum in terms of meet-irreducibles. Then the equivalence $(1)\Leftrightarrow(5)$ in 5.1 concerns the implication

$$\Sigma'_a = \{p \mid a \nleq p\} \neq \Sigma'_b = \{p \mid b \nleq p\} \quad \Rightarrow \quad a \neq b. \tag{5.3.1}$$

We obtain the following criterion.

Proposition. *A locale L is spatial if and only if each element $a \in L$ is a meet of meet-irreducible ones.*

Proof. 5.1(5) can be reformulated as

$$b \nleq a \;\Rightarrow\; \Sigma'_b \nleq \Sigma'_a.$$

Thus, if $b \not\leq a$ there is a point $p \in \Sigma'_b \smallsetminus \Sigma'_a$, that is, $a \leq p$ and $b \not\leq p$. Consequently, $a = \bigwedge\{p \text{ points} \mid a \leq p\}$.

Conversely, if $a = \bigwedge\{p \text{ points} \mid a \leq p\}$ then for $b \not\leq a$ there is a point p such that $a \leq p$ and $b \not\leq p$, hence $p \in \Sigma'_b \smallsetminus \Sigma'_a$. $\qquad\square$

5.4. Aside: spatial Boolean algebras. Recall that an *atom* (resp. a *co-atom*) in L is an element $a > 0$ (resp. $a < 1$) such that for each x, $a \geq x > 0$ implies that $x = a$ (resp. $a \leq x < 1$ implies that $x = a$).

The following is a trivial

5.4.1. Observation. *In a Boolean algebra, a is an atom iff its complement a^c is a co-atom.*

5.4.2. A Boolean algebra is said to be *atomic* if each of its elements is a join of atoms. The following is a well-known fact.

Proposition. *The following statements on a Boolean algebra B are equivalent.*

(1) *B is atomic,*

(2) *each element of B is a meet of co-atoms,*

(3) *B is isomorphic to the Boolean algebra $\mathfrak{P}(X)$ of all subsets of a set X.*

Proof. (1) and (2) are equivalent by De Morgan formulas. Now if B is atomic take X the set of all atoms and represent $b \in B$ by the set $A(b) = \{x \in X \mid x \leq b\}$. This is obviously an isomorphism (the only fact to realize is that $A(a) = A(b)$ implies that $a = b$ and this is obtained as follows: if $x \in X$ and $x \leq a = \bigvee A(a)$ then $0 \neq x = \bigvee\{x \wedge y \mid y \in A(a)\}$ and hence, by atomicity, $x = y$ for some $y \in A(b)$). $\qquad\square$

5.4.3. Proposition. *Every meet-irreducible element in a Boolean algebra is a co-atom.*

Proof. Let p be meet-irreducible in a Boolean algebra B. Let $p < x$. If y is the complement of x we have $0 = x \wedge y \leq p$ and by meet-irreducibility $y \leq p$ but then $y \leq x$ and $1 = x \vee y = x$. $\qquad\square$

5.4.4. Now by 5.3, every element of a spatial Boolean locale is a meet of co-atoms. Consequently, by 5.4.2 a Boolean locale is spatial only if it is atomic; also, only if it is (isomorphic to) the discrete topology of a space. Hence,

the spatial Boolean locales correspond to the discrete spaces.

The non-atomic complete Boolean algebras are, hence, examples of non-spatial locales. To present a concrete one, take the real line and consider the Boolean algebra B of all the open subsets U such that $U = \operatorname{int}\overline{U}$. Here we have no atoms at all (each non-void element contains a strictly smaller non-void one) so that $\mathsf{Sp}(B) = \emptyset$ although B is fairly large.

Frames of this type play a prominent role as we will see later (III.8.3, VI.6).

6. The unit λ and sobriety

6.1. Proposition. *The space* $\mathsf{Sp}(L)$ *is always sober.*

Proof. It is T_0 because if c.p. filters F, G are distinct there is, say, an $a \in F \smallsetminus G$ and then $G \notin \Sigma_a \ni F$.

Now let $\mathcal{F}$ be a completely prime filter in $\mathsf{Sp}(L)$. Set $F = \{a \mid \Sigma_a \in \mathcal{F}\}$. By Observation 4.1, F is a c.p. filter in L. If $\Sigma_a \in \mathcal{F}$ then $a \in F$ and $F \in \Sigma_a$; if $F \in \Sigma_a$ then $a \in F$ and $\Sigma_a \in \mathcal{F}$. Thus, $\mathcal{F} = \{\Sigma_a \mid F \in \Sigma_a\} = \mathcal{U}(F)$. $\qquad\square$

6.2. Proposition. *The following statements about a space* X *are equivalent.*

(1) X *is sober,*

(2) $\lambda_X \colon X \to \mathsf{SpLc}(X)$ *is one-one onto,*

(3) $\lambda_X \colon X \to \mathsf{SpLc}(X)$ *is a homeomorphism.*

Proof. (1)$\Rightarrow$(2) is in I.1.3.1.

(2)$\Rightarrow$(3): We have to prove that $\lambda_X[U]$ is open whenever U is. But $\lambda_X[U]$ is the set $\{\mathcal{U}(x) \mid x \in U\} = \{\mathcal{U}(x) \mid U \in \mathcal{U}(X)\} = \{F \text{ c.p. filter} \mid U \in F\} = \Sigma_U$.

(3)$\Rightarrow$(1) is in 6.1. $\qquad\square$

6.3. Hence, $\mathsf{SpLc}(X)$ is sober, and for a sober X we have $\mathsf{SpLc}(X) \cong X$. Thus, $\mathsf{SpLc}[\mathbf{Top}] = \mathbf{Sob}$ and we can decompose

$$\mathbf{Top} \xrightarrow{\ \mathsf{Lc}\ } \mathbf{Loc} \xrightarrow{\ \mathsf{Sp}\ } \mathbf{Top} \quad \text{as} \quad \mathbf{Top} \xrightarrow{\ R\ } \mathbf{Sob} \xrightarrow{\ J=\subseteq\ } \mathbf{Top}$$

with R identical on $\mathbf{Sob}$.

This makes $\mathbf{Sob}$ a reflective subcategory in $\mathbf{Top}$ (see AII.8) with the reflection

$$\lambda_X \colon X \to JR(X).$$

6.3.1. Remarks. (1) Note that the reflection in 6.3 (the *sobrification*, as it is often called) does the same as the completion of metric spaces: it fills in the special filters that have not been, so far, neighbourhood systems as new points; they become neighbourhood systems "of themselves".

(2) From 4.6.4 we see that for a spatial frame L, the X such that $L \cong \Omega(X)$ is not uniquely determined: the lattice of open sets of a space and that of its sobrification are isomorphic. But nothing worse can happen: if there is an isomorphism $\alpha \colon \mathsf{Lc}(X) \to \mathsf{Lc}(Y)$ we have an isomorphism $\mathsf{Sp}(\alpha) \colon \mathsf{SpLc}(X) \to \mathsf{SpLc}(Y)$. Thus, the sobrifications are isomorphic. This is, of course, no surprise. We know since I.1.4 that a sober space can be reconstructed from its lattice of open sets.

(3) The fact that the open set lattice of a space is isomorphic with that of its sobrification shows that the sobriety cannot be expressed as an algebraic property of a frame. Other "completeness properties" can, however, as we will see later.

6.4. Aside: sober posets. We have already said in I.3.4 that the Alexandroff space

$$(X, \mathfrak{Up}(X, \leq))$$

(here $\mathfrak{Up}(X, \leq)$ is the set of all up-sets in $(X, \leq)$) is not very often sober. We will now present a characteristics of those that are. Furthermore, we will see that the sobrification of an Alexandroff space is its set of ideals endowed with the *Scott topology* (recall that a subset S of a poset is *Scott-open* if it is an upper set and all directed sets D that have a supremum in S have non-empty intersection with S).

6.4.1. We have already had three representations of points. For $\mathfrak{Up}(X, \leq)$ there is yet another, and a very handy one.

For a completely prime filter $\mathcal{F}$ in $\mathfrak{Up}(X, \leq)$ define

$$J_{\mathcal{F}} = \{x \in X \mid {\uparrow}x \in \mathcal{F}\}.$$

Fact. *$J_{\mathcal{F}}$ is an ideal in $(X, \leq)$.*

(Indeed, if ${\uparrow}x \in \mathcal{F}$ and $y \leq x$ then ${\uparrow}x \subseteq {\uparrow}y$. If ${\uparrow}x, {\uparrow}y \in \mathcal{F}$ then $\emptyset \neq {\uparrow}x \cap {\uparrow}y \in \mathcal{F}$ and we have a $z \geq x, y$. Further, $J_{\mathcal{F}}$ is non-empty by the complete primeness: take any $U = \bigcup\{{\uparrow}x \mid x \in U\} \in \mathcal{F}$.)

On the other hand, for an ideal J in $(X, \leq)$ define

$$\mathcal{F}_J = \{U \in \mathfrak{Up}(X, \leq) \mid U \cap J \neq \emptyset\}.$$

We have

Fact. *$\mathcal{F}_J$ is a completely prime filter in $\mathfrak{Up}(X, \leq)$.*

(Obviously $\emptyset \notin \mathcal{F}_J$, $\bigcup U_i \in \mathcal{F}_J \Rightarrow \exists i, U_i \in \mathcal{F}_J$, and $V \supseteq U \in \mathcal{F}_J \Rightarrow V \in \mathcal{F}_J$. If $x \in U \cap J$ and $y \in V \cap J$ choose a $z \in J$ such that $z \geq x, y$; then $z \in (U \cap V) \cap J$.)

6.4.2. Lemma. *$\mathcal{F} \mapsto J_{\mathcal{F}}$ and $J \mapsto \mathcal{F}_J$ are mutually inverse correspondences.*

Proof. If $U \in \mathcal{F}_{J_{\mathcal{F}}}$ then $U \cap J_{\mathcal{F}} \neq \emptyset$ and we have an x with ${\uparrow}x \in \mathcal{F}$ and $x \in U$. Then ${\uparrow}x \in U$ and $U \in \mathcal{F}$. Conversely, if $U \in \mathcal{F}$ then $U = \bigcup\{{\uparrow}x \mid x \in U\} \in \mathcal{F}$ and since $\mathcal{F}$ is completely prime there is an $x \in U$ with ${\uparrow}x \in \mathcal{F}$; hence $x \in J_{\mathcal{F}}$, $U \cap J_{\mathcal{F}} \neq \emptyset$, and $U \in \mathcal{F}_{J_F}$.

If $x \in J_{\mathcal{F}_J}$ then ${\uparrow}x \in \mathcal{F}_J$ and we have a $y \in J$ such that $y \in {\uparrow}x$, that is, $x \leq y$, and $x \in J$. On the other hand, if $x \in J$ then ${\uparrow}x \in \mathcal{F}_J$ and $x \in J_{\mathcal{F}_J}$. □

6.4.3. Lemma. *A c.p. filter $\mathcal{F}$ is a $\mathcal{U}(x) = \{U \mid x \in U\}$ if and only if x is the maximum of $J_{\mathcal{F}}$.*

Proof. If $\mathcal{F} = \mathcal{U}(x)$ then for all $U \in \mathcal{F}$, $U \supseteq {\uparrow}x$, and in particular for $y \in J_{\mathcal{F}}$, $x \geq y$. If x is the maximum of $J_{\mathcal{F}}$ then $U \in \mathcal{F}$ iff $\exists y \in J_{\mathcal{F}}$, $y \in U$ iff $x \in U$. □

6.4.4. Recall that a poset $(X, \leq)$ is *Noetherian* if there is no strictly increasing sequence

$$a_1 < a_2 < \cdots < a_n < \cdots . \tag{$*$}$$

Proposition. $\mathfrak{Up}(X, \leq)$ *is sober if and only if* $(X, \leq)$ *is Noetherian.*

Proof. It immediately follows from 6.4.3. Each ideal has a maximum iff there is no sequence of the form $(*)$: if there were we had the ideal $\{x \mid \exists i, \ x \leq a_i\}$. $\quad\square$

6.4.5. Thus, the sober reflection of a poset $(X, \leq)$ (more precisely, of its Alexandroff space) can be represented as

$$(\mathsf{Idl}(X, \leq), \widetilde{\tau})$$

where $\mathsf{Idl}(X, \leq)$ is the set of ideals of $(X, \leq)$ and $\widetilde{\tau}$ consists of the $\widetilde{U} = \{J \mid U \cap J \neq \emptyset\}$ with $U \in \mathfrak{Up}(X, \leq)$ (use 6.4.2: Σ_U translates to $\{J \mid \mathcal{F}_J \ni U\} = \{J \mid U \cap J \neq \emptyset\}$). Note that $x \in U$ iff $\downarrow x \in \widetilde{U}$.

6.4.6. Proposition. $\widetilde{\tau}$ *is the Scott topology of* $(\mathsf{Idl}(X, \leq), \subseteq)$.

Proof. Obviously each $\widetilde{U}$ is an up-set in $\subseteq$. Let $J = \bigvee_i^{\uparrow} J_i \in \widetilde{U}$. The directed supremum $\bigvee_i^{\uparrow} J_i$ of ideals is obviously the union $\bigcup_i J_i$, hence $\bigcup_i J_i \cap U \neq \emptyset$ and there is an i such that $J_i \cap U \neq \emptyset$. Thus, $\widetilde{U}$ is Scott open.

Now let $\mathcal{U}$ be Scott open in $(\mathsf{Idl}(X, \leq), \subseteq)$. Set

$$U = \{x \mid \downarrow x \in \mathcal{U}\}.$$

Let $J \in \mathcal{U}$. We have $J = \bigvee^{\uparrow} \{\downarrow x \mid x \in J\}$, hence there is an $x \in J$ such that $\downarrow x \in \mathcal{U}$; this x is in $J \cap U$, and $\downarrow x \in U$. If J is in $\widetilde{U}$ then $J \cap U \neq \emptyset$ so there is an $x \in J$ with $\downarrow x \in \mathcal{U}$; but $J \supseteq \downarrow x$, and hence $J \in \mathcal{U}$. $\quad\square$

Chapter III

Sublocales

The morphisms in the category of locales are, admittedly, not quite standard. Even replacing the "inversely viewed" frame homomorphisms by the localic maps as we did in II.2.2 may seem to be a bit ad hoc. Therefore it is worthwhile to be thorough in explaining why the definition of a sublocale to be presented in this section is what it is. We will see that it is in fact most natural, and that it hardly can be chosen otherwise.

1. Extremal monomorphisms in Loc

1.1. In a general category we have the notion of a *monomorphism* that (roughly speaking, but not really very roughly) models the idea of a one-one mapping (that is, one-one mapping well behaved with respect to the structure in question). It is a morphism m such that $mf = mg$ implies $f = g$ (see AII.1.3; it is an easy exercise to check that this requirement characterizes the one-one mappings in the category of sets, but also the one-one continuous mappings in **Top**, one-one homomorphisms in everyday categories of algebras, one-one graph homomorphisms, etc.). A one-one morphism is not necessarily an embedding of a subobject: see for instance the one-one graph homomorphism $m: A \to B$ sending i to i in Figure 1.

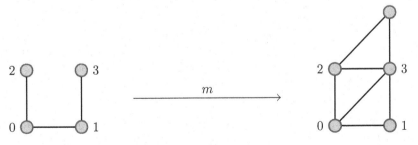

Figure 1: One-one graph homomorphism m.

Why is it *not* an embedding? The structure of the source A is weaker than that of its image in B – in other words, on the vertices of A there is a strictly stronger structure under which the formula still constitutes a graph homomorphism. That is, we can insert a graph C, a *one-one onto* homomorphism $e\colon A \to C$ that *is not* an isomorphism, and an $m'\colon C \to B$ such that $m = m'e$, as in Figure 2.

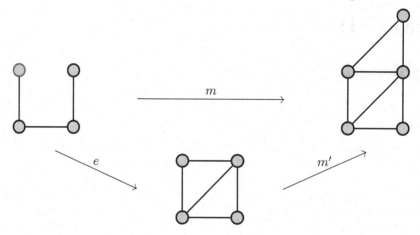

Figure 2: m is not an embedding.

An embedding, hence, is a monomorphism for which such an insertion is impossible. This leads to the definition of *extremal monomorphism*: a monomorphism m such that whenever $m = m'e$ with e an epimorphism then e is an isomorphism (this e is necessarily a monomorphism, so that we speak of an e that is both an epi- and a monomorphism) – see AII.4.2.

  Dually, one defines an *extremal epimorphism* as an epimorphism e such that if $e = me'$ with m a monomorphism then m is an isomorphism. Obviously f is an extremal monomorphism in a category $\mathcal{C}$ if and only if it is an extremal epimorphism in the dual category $\mathcal{C}^{\mathrm{op}}$.

1.1.1. Lemma. *Monomorphisms in* **Frm** *are precisely the one-one frame homomorphisms.*

Proof. A one-one homomorphism h is obviously a monomorphism: if $h(f(x)) = h(g(x)$ then $f(x) = g(x)$. Now let $h\colon L \to M$ not be one-one. Then there are $a, b \in L$, $a \neq b$ such that $h(a) = h(b)$. Consider the frame $\mathbf{3} = \{0 < c < 1\}$ and the maps $f, g\colon \mathbf{3} \to L$ sending 0 to 0, 1 to 1, and $f(c) = a$, $g(c) = b$. Obviously f and g are frame homomorphisms, and $hf = hg$ while $f \neq g$.   □

1.1.2. Lemma. *Let $f\colon L \to M$ be a frame homomorphism. Then $f[L]$ is a subframe of M (that is, a subset closed under all joins and finite meets; in particular, it contains 0_M and 1_M) and $f = j \cdot g$ where $j\colon f[L] \to M$ is the embedding mapping, and g is a frame homomorphism defined by $g(x) = f(x)$.*

Proof. It is trivial (if $y_i = f(x_i)$ then

$$\bigvee_{i \in J} y_i = \bigvee_{i \in J} f(x_i) = f(\bigvee_{i \in J} x_i)$$

etc.). $\square$

1.1.3. Proposition. *Extremal epimorphisms in* **Frm** *are precisely the onto frame homomorphisms.*

Proof. Let $f: L \to M$ be an onto frame homomorphism. Then, first, it is obviously an epimorphism. Now let $f = mg$ with $m: N \to M$ a monomorphism. Then m is a one-one onto frame homomorphism, and hence an isomorphism (as any invertible homomorphism of algebras), and hence f is extremal.

Now let $f: L \to M$ not be onto. Then we have the decomposition $f = jg$ from 1.1.2, with j a non-isomorphic monomorphism. $\square$

1.2. Corollary. *Extremal monomorphisms in* **Loc** *are precisely the one-one localic maps.* $\square$

(Indeed, for Galois adjoints f, g we have $fgf = f$ and $gfg = g$ – see AI.5. Thus, if f is onto, $fg = $ id and g has to be one-one; if g is one-one we obtain that $fg = $ id which makes f onto.)

1.3. Remarks. (1) Note that we have not learned what the epimorphisms in **Frm** resp. monomorphisms in **Loc** are. They constitute, in fact, a rather wild class, see IV.6 below.

(2) The reader may be surprised that the embeddings of subobjects will be represented by just onto frame homomorphisms (and one-one localic maps), not some special ones. But realize what happens with embedding of spaces $j: Y \subseteq (X, \tau)$: the subspace topology is defined as $\{U \cap Y \mid U \in \tau\}$ making for the onto frame homomorphism $(U \mapsto U \cap Y): \Omega(X) \to \Omega(Y)$. If we had Y endowed with a finer topology, $\Omega(j)$ would not be onto (and the corresponding localic maps would not be one-one).

2. Sublocales

A one-one localic map that is onto is an isomorphism (because the corresponding frame homomorphism is one), and a poset isomorphic to a locale is a locale as well. Thus, discussing embeddings of locales into other ones, that is (by 1.2) one-one localic maps, can be reduced to the discussion of actual sub-locales, that is, embeddings $j: S \subseteq L$ of subsets of L such that

- S is a locale in the order induced by that of L (which does not say that it should inherit the lattice structure), and
- j is a localic map.

2.1. A subset $S \subseteq L$ is a *sublocale* if

(S1) it is closed under all meets, and

(S2) for every $s \in S$ and every $x \in L$, $x \to s \in S$.

Note that a sublocale is always non-empty: $1 = \bigwedge \emptyset$ is in every S satisfying (S1). The least sublocale $\{1\}$ will be denoted by

$$\mathsf{O}.$$

2.2. Proposition. *Let L be a locale. A subset $S \subseteq L$ is a sublocale if and only if it is a locale in the induced order and the embedding map $j \colon S \subseteq L$ is a localic map.*

Proof. $\Rightarrow$: Let S be a sublocale. By (S1) it is a complete lattice and by (S2), in particular, it is closed under the Heyting operation, hence each of the $(x \mapsto x \wedge a)$ has the right adjoint $(x \mapsto (a \to x)) \colon S \to S$, and the frame distributivity holds. Thus, S is a locale. Now because of (S1), j preserves meets, and hence it has a right adjoint $h \colon L \to S$. We have

$$\forall x \in L, \forall s \in S, \quad h(x) \le s \quad \text{iff} \quad x \le s.$$

We want to prove that h preserves finite meets. For any $a, b \in L$, $h(a) \wedge h(b) \le s \in S$ iff $h(a) \le h(b) \to s$ iff $a \le h(b) \to s$ iff $h(b) \le a \to s$ iff $b \le a \to s$ iff $a \wedge b \le s$ iff $h(a \wedge b) \le s$ and hence $h(a) \wedge h(b) = h(a \wedge b)$. Since $x \le h(x)$ for any x (as $h(x) \le h(x)$), we have $1 \le h(1)$.

$\Leftarrow$: Let S be a locale and let $j \colon S \subseteq L$ be a localic map. Then, first, S is a complete lattice and j preserves all meets so that (S1) holds. Further, S is a Heyting algebra; denote by $\xrightarrow{S}$ its Heyting operation. For the frame homomorphism h adjoint to j, and for any $x, a \in L$ and $s \in S$, we have $x \le h(a) \xrightarrow{S} s$ iff $h(x) \le h(a) \xrightarrow{S} s$ iff $h(x \wedge a) = h(x) \wedge h(a) \le s$ iff $x \wedge a \le s$ iff $x \le a \to s$ and hence $a \to s = h(a) \xrightarrow{S} s \in S$. $\square$

2.2.1. Thus, a sublocale $S \subseteq L$ is itself a locale. Since it is closed under meets in L, it follows from (S2) that

the Heyting operation in S coincides with that in L.

2.3. Corollary. *Let S be a sublocale of L and let T be a sublocale of S. Then T is a sublocale of L.* $\square$

2.4. Unlike the meets and the Heyting operation, the joins in a sublocale $S \subseteq L$ in general differ from those in L. For a simple formula see 5.4 below.

3. The co-frame of sublocales

3.1. We will point out a few very simple formulas concerning Heyting operation. Some of them will be used right away, and some of them will find an application later (in particular when we will discuss special sublocales).

First, the Galois adjunction in the basic formula

$$a \wedge b \leq c \quad \text{iff} \quad a \leq b \rightarrow c \tag{H}$$

immediately yields

$$a \rightarrow \bigwedge_{i \in J} b_i = \bigwedge_{i \in J} (a \rightarrow b_i) \tag{$\wedge$-distr}$$

since the right adjoint $(a \rightarrow -)$ preserves meets (preserving joins by the left adjoint resulting in the frame distributivity has been already mentioned in II.1.4).

Further, note that the Heyting operation is antitone in the first variable, that is,

$$x \leq y \quad \Rightarrow \quad y \rightarrow a \leq x \rightarrow a \tag{anti}$$

(if $z \leq y \rightarrow a$, that is, $z \wedge y \leq a$ and $x \leq y$ then $z \wedge x \leq a$ and hence $z \leq x \rightarrow a$).

Finally, using the commutativity of meet we obtain from (H) the contravariant adjunction

$$a \leq b \rightarrow c \quad \text{iff} \quad b \leq a \rightarrow c \tag{H*}$$

resulting in the formula

$$\left(\bigvee_{i \in J} a_i \right) \rightarrow b = \bigwedge_{i \in J} (a_i \rightarrow b). \tag{$\bigvee\wedge$-dist}$$

3.1.1. Proposition. *In any Heyting lattice L*

(H1) *there is a largest element* 1, *and* $1 \rightarrow a = a$ *for all* a,

(H2) $a \leq b$ *iff* $a \rightarrow b = 1$,

(H3) $a \leq b \rightarrow a$,

(H4) $a \rightarrow b = a \rightarrow (a \wedge b)$,

(H5) $a \wedge (a \rightarrow b) = a \wedge b$,

(H6) $a \wedge b = a \wedge c$ *iff* $a \rightarrow b = a \rightarrow c$,

(H7) $(a \wedge b) \rightarrow c = a \rightarrow (b \rightarrow c)$ *and* $a \rightarrow (b \rightarrow c) = b \rightarrow (a \rightarrow c)$,

(H8) $a = (a \vee b) \wedge (b \rightarrow a)$,

(H9) $a \leq (a \rightarrow b) \rightarrow b$,

(H10) $((a \rightarrow b) \rightarrow b) \rightarrow b = a \rightarrow b$.

Proof. We will use (H) repeatedly without mentioning.

(H1): Choose an arbitrary $x \in L$ and set $1 = x \to x$. For any $a \in L$ we have $a \wedge x \le x$ and hence $a \le x \to x$. Now for any y we have $y \le 1 \to a$ iff $y = y \wedge 1 \le a$, and hence $1 \to a = a$.

(H2): $1 \le a \to b$ iff $a = 1 \wedge a \to b$.

(H3): $a \wedge b \le a$, hence $a \le b \to a$.

(H4): By ($\bigwedge$-distr), $a \to (a \wedge b) = (a \to a) \wedge (a \to b) = 1 \wedge (a \to b)$.

(H5): $a \to b \le a \to b$ and hence $(a \to b) \wedge a \le b$ and consequently $a \wedge (a \to b) \le a \wedge b$. On the other hand, $a \wedge b \le a \wedge (a \to b)$ by (H3).

(H6) follows from (H4) and (H5).

(H7): We have $x \le (a \wedge b) \to c$ iff $x \wedge a \wedge b \le c$ iff $x \wedge a \le b \wedge c$ iff $x \le a \to (b \to c)$. The second formula now follows from $(a \wedge b) \to c = (b \wedge a) \to c$.

(H8): $a \le (a \vee b) \wedge (b \to a)$ by (H3); on the other hand, by distributivity, $(a \vee b) \wedge (b \to a) = (a \wedge (b \to a)) \vee (b \wedge (b \to a)) \le a \vee a = a$ by (H3) and (H5).

(H9) follows from (H5).

(H10): The inequality $\ge$ follows from (H9) while $\le$ follows from (H9) and (anti). $\qquad\square$

3.2. Obviously, an arbitrary intersection

$$\bigcap_{i \in J} S_i$$

of sublocales S_i is a sublocale. Hence (recall AI.4.3) the system

$$\mathcal{S}\ell(L)$$

of all sublocales of a locale L is a complete lattice.

The join of sublocales is (of course) not the union, but we have a very simple formula

$$\bigvee_{i \in J} S_i = \{\bigwedge A \mid A \subseteq \bigcup_{i \in J} S_i\}.$$

(Indeed, each sublocale containing all the S_i has to contain all the meets $\bigwedge A$ with $A \subseteq \bigcup_{i \in J} S_i$. On the other hand,

$$T = \{\bigwedge A \mid A \subseteq \bigcup_{i \in J} S_i\}$$

is obviously closed under meets, and for an arbitrary $x \in L$, by ($\bigwedge$-distr), $x \to \bigwedge A = \bigwedge\{x \to a \mid a \in A\}$; each such $x \to a$ is in $\bigcup_{i \in J} S_i$ and hence the meet is in T.)

3.2.1. Theorem. *$\mathcal{S}\ell(L)$ is a co-frame (that is, the dual poset $\mathcal{S}\ell(L)^{\mathrm{op}}$ is a frame).*

Proof. The inclusion $(\bigcap_{i \in J} A_i) \vee B \subseteq \bigcap_{i \in J}(A_i \vee B)$ is trivial.

Now let $x \in \bigcap_{i \in J}(A_i \vee B)$. Thus, for each i there are $a_i \in A_i$ and $b_i \in B$ such that $x = a_i \wedge b_i$. Set $b = \bigwedge_{i \in J} b_i$. Then

$$x = \bigwedge_{j \in J} a_j \wedge \bigwedge_{j \in J} b_j = \bigwedge_{j \in J} a_j \wedge b \leq a_i \wedge b \leq a_i \wedge b_i \leq x$$

and hence $x = a_i \wedge b$ for all i. By (H6), all the $b \to a_i$ coincide; if we denote by a the common value we have, by (H5), $x = b \wedge a_i = b \wedge (b \to a_i) = b \wedge a \in B \vee \bigcap_{i \in J} A_i$. □

(The first short proof of such distributivity in a similar vein is due to Dana Scott, as communicated to us by P.T. Johnstone.)

4. Images and preimages

4.1. The image. Let $f \colon L \to M$ be a localic map. Then we have

Proposition. *For each sublocale S of L, the set-theoretic image $f[S] = \{f(s) \mid s \in S\}$ is a sublocale of M.*

Proof. (S1) is immediate since f, as a right adjoint, preserves meets.

(S2): Here we have to recall II.2.3. Let $s \in S$ and let $y \in M$ be arbitrary. Then $y \to f(s) = f(f^*(y) \to s) \in f[S]$. □

The sublocale $f[S]$ will be referred to as

the image of S

under (the localic map) f.

4.2. The preimage. Unlike $f[S]$, the set-theoretic preimage $f^{-1}[T]$ of a sublocale $T \subseteq M$ is not necessarily a sublocale of L. To obtain a concept of a preimage suitable for our purposes we will, first, make the following

Observation. *Let $A \subseteq L$ be a subset closed under meets. Then $\{1\} \subseteq A$ and if $S_i \subseteq A$ for $i \in J$ then $\bigvee_{i \in J} S_i \subseteq A$.*

Consequently there exists the largest sublocale contained in an A closed under meets. It will be denoted by

$$A_{\mathsf{sloc}}.$$

4.2.1. The set-theoretic preimage $f^{-1}[T]$ of a sublocale T is closed under meets (indeed, $f(1) = 1$, and if $x_i \in f^{-1}[T]$ then $f(x_i) \in T$, and hence $f(\bigwedge_{i \in J} x_i) = \bigwedge_{i \in J} f(x_i) \in T$, and $\bigwedge_{i \in J} x_i \in f^{-1}[T]$) and we have the sublocale

$$f_{-1}[T] = (f^{-1}[T])_{\mathsf{sloc}}.$$

It will be referred to as

the preimage of T

under (the localic map) f.

4.3. Proposition. *For every localic map $f\colon L \to M$, the preimage function $f_{-1}[-]$ is a right Galois adjoint of the image function $f[-]\colon \mathcal{S\!l}(L) \to \mathcal{S\!l}(M)$.*

Proof. If $f[S] \subseteq T$ then $S \subseteq f^{-1}[T]$. Since S is a sublocale, $S \subseteq f_{-1}[T]$, the largest sublocale contained in $f^{-1}[T]$. On the other hand, if $S \subseteq f_{-1}[T]$ then $S \subseteq f^{-1}[T]$ and hence $f[S] \subseteq T$. □

5.   Alternative representations of sublocales

5.1. Sublocale homomorphisms. Often, following the fact that extremal monomorphisms in **Loc** are the adjoints to the extremal epimorphisms in **Frm**, sublocales of L are represented as onto frame homomorphisms $g\colon L \to M$ (recall 1.1.3); one speaks of *sublocale homomorphisms*. This is a straightforward use of the extremal monomorphism idea, but has a lot of disadvantages. First of all, a sublocale homomorphism is only one of many representatives of the "generalized subspace" in question; if there is an isomorphism ϕ such that $\phi g_1 = g_2$ then the "subspace" represented by g_1 and g_2 is the same. The order (more precisely, preorder) is obvious: $g_1\colon L \to M_1$ is smaller or equal $g_2\colon L \to M_2$ if $g_1 = hg_2$ for some frame homomorphism $h\colon M_2 \to M_1$ (which is then, automatically, itself onto). This is transparent, but the lattice operations not quite so: the meet of $(g_i\colon L \to M_i)_{i \in J}$ is the canonical $g\colon L \to M$ into the colimit of the diagram; thus for instance, the meet of $g_1\colon L \to M_1$ and $g_2\colon L \to M_2$ is the diagonal of the pushout

(something theoretically lucid, but not very easy to compute). The join is even more entangled. Compare this with the lattice operations in $\mathcal{S\!l}(L)$!

The translation of sublocale homomorphisms to sublocales as above, and vice versa, is as follows:

$h \mapsto h_*[M]$ for an onto $h\colon L \to M$ and h_* its right adjoint, and

$S \mapsto j_S^*\colon L \to S$ for $j_S\colon S \subseteq L$.

5.2. Frame congruences. Using *frame congruences* (that is, equivalences respecting all joins and finite meets) is more handy. A sublocale homomorphism $g\colon L \to M$

induces a frame congruence

$$E_g = \{(x, y) \mid g(x) = g(y)\}$$

and a frame congruence gives rise to the sublocale homomorphism

$$(x \mapsto Ex) \colon L \to L/E$$

(where L/E denotes the quotient frame defined by the congruence E, just as quotients are always defined for algebraic systems, and Ex denotes the E-class $\{y \mid (y, x) \in E\}$ of $x \in L$).

This correspondence is of course not one-one, but $E_h = E_g$ precisely if there is an isomorphism ϕ such that $\phi g = h$; thus, a frame congruence represents precisely the equivalence class of sublocale homomorphisms representing the same sublocale, and hence the frame congruences represent well the sublocales themselves.

Furthermore, intersections of frame congruences are frame congruences which makes the structure of the lattice fairly transparent.

Attention. The natural subset order in the lattice of frame congruences is inverse to the natural order of sublocales (the bigger the congruence the smaller the subspace). Thus, by 3.2.1 the ensuing lattice is a frame. One speaks of the *congruence frame* of L and writes

$$\mathfrak{C}(L).$$

It will play an interesting role later.

5.3. Nuclei. Another important alternative representation of sublocales is provided by the following notion.

A *nucleus* in a locale L is a mapping $\nu \colon L \to L$ such that

(N1) $a \leq \nu(a)$,

(N2) $a \leq b \;\Rightarrow\; \nu(a) \leq \nu(b)$,

(N3) $\nu\nu(a) = \nu(a)$, and

(N4) $\nu(a \wedge b) = \nu(a) \wedge \nu(b)$.

The translation between nuclei and frame congruences resp. sublocale homomorphisms is straightforward:

$$\nu \mapsto E_\nu = \{(x, y) \mid \nu(x) = \nu(y)\},$$

$$E \mapsto \nu_E = (x \mapsto \bigvee Ex) \colon L \to L;$$

$$\nu \mapsto h_\nu = \nu \text{ restricted to } L \to \nu[L],$$

$$h \mapsto \nu_h = (x \mapsto h_* h(x)) \colon L \to L.$$

Since the nucleus representation is very important it is worthwhile to briefly discuss the relation of sublocales and nuclei directly.

For a sublocale $S \subseteq L$ set

$$\nu_S(a) = j_S^*(a) = \bigwedge\{s \in S \mid a \leq s\} \quad (j_S \colon S \subseteq X),$$

and for a nucleus $\nu \colon L \to L$ set

$$S_\nu = \nu[L].$$

5.3.1. Lemma. (a) *For every $a, b \in L$ we have $\nu(a \to \nu(b)) = a \to \nu(b)$.*

(b) *For every $a, b \in L$, $b \to \nu(a) = \nu(b) \to \nu(a)$.*

Proof. (a): Using, in this order, (N1), (N4), (H5) from 3.1.1 and (N2), (N3), we obtain
$$\nu(a \to \nu(b)) \wedge a \leq \nu(a \to \nu(b)) \wedge \nu(a)$$
$$= \nu((a \to \nu(b)) \wedge a) \leq \nu(\nu(b)) = \nu(b).$$

Thus, by the basic Heyting formula, $\nu(a \to \nu(b)) \leq a \to \nu(b)$. Finally, use (N1) again.

(b): Since $b \to \nu(a) \leq b \to \nu(a)$ we have by (H) in 3.1, $b \to \nu(a) \wedge b \leq \nu(a)$ and, by(N4) and (N3), $\nu(b \to \nu(a)) \wedge \nu(b) \leq \nu(a)$. By (H) and (N1), $b \to \nu(a) \leq \nu(b) \to \nu(a)$. Since $u \to v$ is antitone in the first variable, we also have $\nu(b) \to \nu(a) \leq b \to \nu(a)$. $\qquad\square$

5.3.2. Proposition. *The formulas $S \mapsto \nu_S$ and $\nu \mapsto S_\nu$ above constitute a one-one correspondence between sublocales of L and nuclei in L.*

Proof. Let ν be a nucleus and let $a_i = \nu(b_i) \in \nu[L]$. Then $\nu(\bigwedge a_i) \leq \bigwedge \nu(a_i) = \bigwedge \nu\nu(b_i) = \bigwedge \nu(b_i) = \bigwedge a_i$. If a is general and $s = \nu(b) \in \nu[L]$ then $a \to s = \nu(a \to s) \in S$ by 5.3.1(a). Thus, $\nu[L]$ is a sublocale.

Let S be a sublocale. Then ν_S obviously satisfies (N1), (N2), (N3) and the inequality $\nu_S(a \wedge b) \leq \nu_S(a) \wedge \nu_S(b)$. Now we have $a \wedge b \leq \nu_S(a \wedge b)$, hence $a \leq b \to \nu_S(a \wedge b)$ and, by (S2), $\nu_S(a) \leq b \to \nu_S(a \wedge b)$; further $b \leq \nu_S(a) \to \nu_S(a \wedge b)$, and by (S2) again, $\nu_S(b) \leq \nu_S(a) \to \nu_S(a \wedge b)$ and finally $\nu_S(a) \wedge \nu_S(b) \leq \nu_S(a \wedge b)$.

$\nu_S[L] = S$ since $\nu_S(a) \in S$ by (S1), and for $s \in S$ trivially $\nu_S(s) = s$.

Finally, $\nu_{S_\nu}(a) = \bigwedge\{\nu(b) \mid a \leq \nu(b), \ b \in L\} = \nu(a)$ since $a \leq \nu(a)$ and if $a \leq \nu(b)$ then $\nu(a) \leq \nu\nu(b) = \nu(b)$. $\qquad\square$

5.3.3. Corollary. The congruence E_S associated with a sublocale S is given by the formula

$$a E_S b \quad \text{iff} \quad \forall s \in S \ (a \leq s \text{ iff } b \leq s).$$

5.4. Recall 2.4. The joins $\bigsqcup$ in a sublocale $S \subseteq L$ are given by

$$\bigsqcup_{i \in J} a_i = \nu_S(\bigvee_{i \in J} a_i).$$

Indeed: $\nu(\bigvee a_i)$ is in S and $\geq a_i$; on the other hand, if $b \in S$ and $b \geq a_i$ for all i then $b \geq \bigvee a_i$ and hence $b = \nu(b) \geq \nu(\bigvee a_i)$.

6.   Open and closed sublocales

6.1. In this and the following sections we will often use the formulas from 3.1 like (H), ($\bigwedge$-dist), (Hj). Repeated recalling the address 3.1 will be omitted.

6.1.1. Open sublocales. The frame $\Omega(X)$ associated with a space is the lattice of all open subsets of X. Thus, an element $a \in L$ of a locale (frame) can be expected to have also something to do with a "generalized open subspace". If $U \subseteq X$ is an open subset, the frame representing it as a "generalized space" is

$$\Omega(U) = \{V \text{ open in } U \ \equiv \ \text{open in } X \mid V \subseteq U\},$$

that is, $\Omega(U) = {\downarrow}U$ in $\Omega(X)$. Thus, one might expect a representation of open sublocales of L as the subsets ${\downarrow}a \subseteq L$, $a \in L$.

Such a subset ${\downarrow}a$, however, *is not a sublocale*. What went wrong? There is a natural *sublocale homomorphism*

$$\widehat{a} = (x \mapsto x \wedge x) \colon L \to {\downarrow}a$$

but we have forgotten that we have to do with the associated localic map, not with the frame homomorphism $\widehat{a}$. This f is determined by the adjunction formula

$$a \wedge x \le y \quad \text{iff} \quad x \le f(y),$$

that is, $x \le a \to y$ iff $x \le f(y)$, yielding $f(y) = a \to y$. Thus, the *open sublocale* associated with an $a \in L$, the (isomorphic) image of ${\downarrow}a$ under f, is

$$\mathfrak{o}(a) = \{a \to x \mid x \in L\}$$

(by (H4), $a \to a \wedge x = a \to x$, hence we could replace in the formula the $x \le a$ by the more expedient $x \in L$). As an image of a locale under a localic map, $\mathfrak{o}(a)$ is indeed a sublocale, but checking the properties directly is also very easy.

By (H7), $a \to (a \to x) = a \to x$ and hence we have alternatively the formula

$$\mathfrak{o}(a) = \{x \in L \mid a \to x = x\},$$

often useful in computations.

6.1.2. Closed sublocales. These are technically simpler to describe. Here the natural representation of the "complement of a" (we will see shortly that it is really the complement of $\mathfrak{o}(a)$ in $\mathcal{S\ell}(L)$) is

$$\mathfrak{c}(a) = {\uparrow}a,$$

already a sublocale of L as it is.

6.1.3. Proposition. $\mathfrak{o}(a)$ *and* $\mathfrak{c}(a)$ *are complements of each other in* $\mathcal{S\ell}(L)$.

Proof. If y is in $\mathfrak{c}(a) \cap \mathfrak{o}(a)$ we have $a \le a \to x = y$ for some $x \in L$; hence by (H) $a \le x$ and by (H2), $y = a \to x = 1$. On the other hand, for any $x \in L$, $x = (x \vee a) \wedge (a \to x)$ by (H8) so that $x \in \mathfrak{c}(a) \vee \mathfrak{o}(a)$.   □

6.1.4. Corollary. $a \le b$ *iff* $\mathfrak{c}(a) \supseteq \mathfrak{c}(b)$ *iff* $\mathfrak{o}(a) \subseteq \mathfrak{o}(b)$.   □

(The first equivalence is obvious and the second one follows from the complementarity.)

6.1.5. Proposition. *We have*

$$\mathfrak{o}(a) \cap \mathfrak{o}(b) = \mathfrak{o}(a \wedge b), \qquad \bigvee_{i \in J} \mathfrak{o}(a_i) = \mathfrak{o}\left(\bigvee_{i \in J} a_i \right),$$

$$\bigcap_{i \in J} \mathfrak{c}(a_i) = \mathfrak{c}\left(\bigvee_{i \in J} a_i \right), \qquad \mathfrak{c}(a) \vee \mathfrak{c}(b) = \mathfrak{c}(a \wedge b).$$

Proof. We have $x \in \bigcap \mathfrak{c}(a_i)$ iff for every i, $x \ge a_i$ iff $x \ge \bigvee a_i$, and $\mathfrak{c}(a) \vee \mathfrak{c}(b) = \{x \wedge y \mid x \ge a, \; y \ge b\} = \mathfrak{c}(a \wedge b)$. The first formula follows from the last one by complementation; the second formula follows from the third, also by complementation (note that in any co-frame the second De Morgan law $(\bigwedge x_i)^* = \bigvee x_i^*$ holds – see AI.7.3.3).   □

6.2. Open and closed sublocales of sublocales. They behave similarly like in spaces, as could be expected. We have

6.2.1. Proposition. *Let S be a sublocale of L and let ν_S be the associated nucleus. Then the closed sublocales $\mathfrak{c}_S(a)$ in S are precisely the intersections with S of the sublocales closed in L and we have $\mathfrak{c}(a) \cap S = \mathfrak{c}_S(\nu_S(a))$.*

Similarly the open sublocales $\mathfrak{o}(a)$ in S are precisely the intersections with S of the sublocales open in L and we have $\mathfrak{o}(a) \cap S = \mathfrak{o}_S(\nu_S(a))$.

Proof. Let $\mathfrak{c}_S(a) = \{x \in S \mid x \ge a\}$ be a closed sublocale in S. Then $\mathfrak{c}_S(a) = \uparrow a \cap S = \mathfrak{c}(a) \cap S$. On the other hand, let $\mathfrak{c}(a)$ be an arbitrary closed sublocale in L. Consider $a' = \bigwedge(\mathfrak{c}(a) \cap S)$. Then $a' \in S$ and

$$x \in S \text{ and } x \ge a' \quad \text{iff} \quad x \in S \text{ and } x \ge a,$$

that is, $x \in \mathfrak{c}_S(a')$ iff $x \in \mathfrak{c}(a) \cap S$. Finally, we have

$$a' = \bigwedge(\uparrow a \cap S) = \bigwedge\{s \in S \mid a \le s\} = \nu(a).$$

The statement on open sublocales follows from the fact that $\mathfrak{o}(a)$ resp. $\mathfrak{o}_S(a)$ is a complement of $\mathfrak{c}(a)$ resp. $\mathfrak{c}_S(a)$ and from the unicity of complements.   □

6.2.2. Remark. The reader may wonder about these formulas; if Y is a subspace of X and (say) U is open in Y then it is typically not open in X while we have here an open sublocale of Y represented by $\nu(a)$ indicating an open set in X. But we should not forget that the subspace Y, as a sublocale of $\Omega(X)$ is not represented as $\Omega(Y)$ (that is, as the frame of the $U \cap Y$, $U \in \Omega(X)$), but as the frame of the largest U in $\Omega(X)$ such that $U \cap Y = V \cap Y$ for some $V \in \Omega(X)$. In particular, the closed sublocales A resp. $A \cap Y$ are represented by the up-sets of open sets, and we have

$$U \cap Y \supseteq (X \smallsetminus A) \cap Y \quad \text{iff} \quad U \supseteq Y \smallsetminus (A \cap Y).$$

6.3. Preimages of open and closed sublocales. Here, again we have a situation similar with that of spaces. Note that in the closed case, the preimage even coincides with the set-theoretical one.

Proposition. *The preimage of a closed (resp. open) sublocale under a localic map is closed (resp. open). Specifically, we have*

$$f_{-1}[\mathfrak{c}(a)] = f^{-1}[\mathfrak{c}(a)] = \mathfrak{c}(f^*(a)) \quad \text{and} \quad f_{-1}[\mathfrak{o}(a)] = \mathfrak{o}(f^*(a)).$$

Proof. The adjunction formula

$$f^*(a) \leq x \quad \text{iff} \quad a \leq f(x)$$

can be rewritten as

$$x \in \mathfrak{c}(f^*(a)) \quad \text{iff} \quad x \in f^{-1}[{\uparrow}a].$$

Thus, $f^{-1}[\mathfrak{c}(a)] = \mathfrak{c}(f^*(a))$, a sublocale, and hence further equal to $f_{-1}[\mathfrak{c}(a)]$.

Now since $f(f^*(a) \to y) = a \to f(y)$ by II.2.3 we have $\mathfrak{o}(f^*(a)) \subseteq f^{-1}[\mathfrak{o}(a)]$. Let S be a sublocale, $S \subseteq f^{-1}[\mathfrak{o}(a)]$. We will prove that $S \subseteq \mathfrak{o}(f^*(a))$. If $s \in S$ than any $x \to s$ is in S, and $f(x \to s)$ is in $\mathfrak{o}(a)$. Hence by II.2.3 and (H7)

$$f(x \to s) = a \to f(x \to s) = f(f^*(a) \to (x \to s)) = f((f^*(a) \wedge x) \to s)$$

and in particular for $x = f^*(a) \to s$ we obtain, by (H7) and (H2),

$$f((f^*(a) \to s) \to s) = f((f^*(a) \wedge s) \to s) = f(1) = 1$$

and by II.2.3 and (H3), $f^*(a) \to s = s$, that is, $s \in \mathfrak{o}(f^*(a))$. □

6.4. Notes. (1) The nuclei associated with $\mathfrak{c}(a)$ resp. $\mathfrak{o}(a)$ are

$$\nu_{\mathfrak{c}(a)}(x) = a \vee x \quad \text{resp.} \quad \nu_{\mathfrak{o}(a)}(x) = a \to x.$$

(The first is straightforward. Now $\nu_{\mathfrak{o}(a)}(x) = \bigwedge\{a \to y \mid x \leq a \to y, \; y \in L\}$; if $x \leq a \to y$ then $a \to x \leq a \to (a \to y) = a \to y$, hence $\nu_{\mathfrak{o}(a)}(x) \geq a \to x$, and on the other hand, $a \to x \leq a \to x$ and by (H7) $a \to (a \to x) = a \to x$ proving the other inequality.)

Thus (recall 5.4), the non-empty joins in a closed sublocale coincide with the original ones. This, of course, is an exceptional case.

(2) For the congruences associated with $\mathfrak{c}(a)$ resp. $\mathfrak{o}(a)$ one uses the standard symbols

$$\nabla_a \quad \text{resp.} \quad \Delta_a.$$

From the first note above we infer that

$$x\nabla_a y \quad \text{iff} \quad x \vee a = y \vee a, \quad \text{and}$$
$$x\Delta_a y \quad \text{iff} \quad x \wedge a = y \wedge a.$$

6.4.1. Recall from 5.2 the congruence frame $\mathfrak{C}(L)$, isomorphic to $\mathcal{S}\ell(L)^{\mathrm{op}}$. Thus (in the natural inclusion order of the congruences) we have the one-one frame homomorphism

$$\nabla_L = (a \mapsto \nabla_a)\colon L \to \mathfrak{C}(L).$$

This homomorphism will play an important role later.

6.5. Each sublocale can be obtained from closed and open ones using intersections (meets) and finite joins. We have

Proposition. *Let S be a sublocale. Then*

$$S = \bigcap\{\mathfrak{c}(x) \vee \mathfrak{o}(y) \mid \nu_S(x) = \nu_S(y)\}.$$

Proof. I. Let $a \in S$ and let $\nu_S(x) = \nu_S(y)$. We have $\nu_S(a) = a$ and hence by 5.3.1(b), $x \to a = \nu(x) \to a = \nu(y) \to a = y \to a$ and by (H8), $a = (a \vee x) \wedge (x \to a) = (a \vee x) \wedge (y \to a) \in \mathfrak{c}(x) \vee \mathfrak{o}(y)$.

II. Let a be in $\bigcap\{\mathfrak{c}(x) \vee \mathfrak{o}(y) \mid \nu(x) = \nu(y)\}$. Then in particular $a \in \mathfrak{c}(\nu(a)) \vee \mathfrak{o}(a)$. Hence, $a = y \wedge (a \to z)$ for some $y \geq \nu(a) (\geq a)$ and $z \in L$, and by ($\wedge$-distr) and further by (H7) and (H2), $1 = a \to a = (a \to y) \wedge (a \to (a \to z)) = 1 \wedge (a \to z) = a \to z$ and hence $a = y \wedge (a \to z) = y \geq \nu(a)$. Thus, $a = \nu(a) \in S$. $\qquad\square$

This can be translated into the language of frame congruences as follows

6.5.1. Corollary. *Each frame congruence E can be expressed as*

$$E = \bigvee\{\nabla_a \cap \Delta_b \mid aEb\}. \qquad\qquad\square$$

7. Open and closed localic maps

We will analyze the counterparts of open *continuous* maps and closed *continuous* maps in classical spaces.

7.1. Lemma. *Let $f\colon L \to M$ be a localic map and let S be a sublocale. Then we have for the congruences associated with S and $f[S]$ the formula*

$$aE_{f[S]}b \quad \text{iff} \quad f^*(a)E_S f^*(b).$$

Proof. Recall 5.3.3. We have

$$aE_{f[S]}b \quad \text{iff} \quad \forall s \in S, \ (a \leq f(s) \text{ iff } b \leq f(s))$$
$$\text{iff} \quad \forall s \in S, \ (f^*(a) \leq s \text{ iff } f^*(b) \leq s) \quad \text{iff} \quad f^*(a)E_S f^*(b). \qquad \square$$

7.2. A localic map $f: L \to M$ is said to be *open* if the image $f[S]$ of each open sublocale $S \subseteq L$ is open.

Proposition. [Joyal & Tierney] *A localic map $f: L \to M$ is open if and only if f^* is a complete Heyting homomorphism.*

Proof. By 6.4 the congruence Δ_a associated with an open sublocale $\mathfrak{o}(a)$ is given by

$$x\Delta_a y \quad \text{iff} \quad x \wedge a = y \wedge a. \tag{$*$}$$

Now $f: L \to M$ is open iff for each $a \in L$ there is a $b \in M$ such that $f[\mathfrak{o}(a)] = \mathfrak{o}(b)$. Denote such a b by $\phi(a)$. Thus, by $(*)$ and 7.1 we have for $x, y \in M$

$$x \wedge \phi(a) = y \wedge \phi(a) \quad \text{iff} \quad f^*(x) \wedge a = f^*(y) \wedge a.$$

This formula can obviously be rewritten as

$$x \wedge \phi(a) \leq y \wedge \phi(a) \quad \text{iff} \quad f^*(x) \wedge a \leq f^*(y) \wedge a$$

and since $x \wedge \phi(a) \leq \phi(a)$ resp. $f^*(x) \wedge a \leq a$ anyway this in turn is equivalent to

$$x \wedge \phi(a) \leq y \quad \text{iff} \quad f^*(x) \wedge a \leq f^*(y). \tag{$**$}$$

Setting $x = 1$ we obtain

$$\phi(a) \leq y \quad \text{iff} \quad a \leq f^*(y).$$

Thus, f^* also has a *left* adjoint ϕ and hence it preserves all meets (besides all the joins), that is, it is a complete homomorphism. Now, using $(**)$ with general x we obtain, for any a,

$$a \leq f^*(x) \to f^*(y) \quad \text{iff} \quad f^*(x) \wedge a \leq f^*(y)$$
$$\text{iff} \quad x \wedge \phi(a) \leq y \quad \text{iff} \quad \phi(a) \leq x \to y \quad \text{iff} \quad a \leq f^*(x \to y).$$

Thus, f^* preserves the Heyting operation.

On the other hand, let f^* be a complete Heyting homomorphism. Then f^* has a left Galois adjoint ϕ, because of the completeness, and

$$f^*(x) \wedge a \leq f^*(y) \quad \text{iff} \quad a \leq f^*(x) \to f^*(y) = f^*(x \to y)$$
$$\text{iff} \quad \phi(a) \leq x \to y \quad \text{iff} \quad x \wedge \phi(a) \leq y,$$

and $(**)$ holds. $\qquad \square$

7.2.1. Remarks. (1) A simple translation using adjoints provides two alternative characterizations for localic open maps $f\colon L \to M$ (see [206] for details):

– f^* admits a left adjoint $f_!$ that satisfies the (Frobenius) identity

$$f_!(a \wedge f^*(b)) = f_!(a) \wedge b \text{ for all } a \in L \text{ and } b \in M.$$

– f^* admits a left adjoint $f_!$ that satisfies the identity

$$f(a \to f^*(b)) = f_!(a) \to b \text{ for all } a \in L \text{ and } b \in M.$$

(2) The open localic maps model the open continuous maps, but not without a proviso. Namely, there are some anomalies in representing subspaces as sublocales: if a space is not T_D then distinct subspaces can be represented by the same sublocale (and an open one can be represented equally with one that is not open). But if we consider T_D-spaces (which is not a very serious restriction) then indeed a continuous $f\colon X \to Y$ is an open continuous map iff the corresponding localic map $\mathsf{Lc}(f)\colon \mathsf{Lc}(X) \to \mathsf{Lc}(Y)$ is open. See [227].

(3) Comparing frame homomorphisms and open frame homomorphisms ($\equiv$ complete Heyting homomorphisms) the question naturally arises about an interpretation of complete lattice homomorphisms (disregarding the Heyting operation). It may come as a surprise that in a very broad class of locales (including the $\mathsf{Lc}(X)$ of all T_1-spaces, and more), such a homomorphism is automatically Heyting and hence open. See V.1.8 below.

7.3. A localic map $f\colon L \to M$ is said to be *closed* if the image of each closed sublocale is closed.

Proposition. *For a localic map $f\colon L \to M$ the following statements are equivalent.*

(1) *f is closed.*

(2) *For each $a \in L$, $f[\uparrow a] = \uparrow f(a)$.*

(3) *For every $a \in L$ and $b \in M$, $f(a \vee f^*(b)) = f(a) \vee b$.*

(4) *For every $a \in L$ and $b, c \in M$, $c \leq f(a) \vee b$ iff $f^*(c) \leq a \vee f^*(b)$.*

(5) *For every $a \in L$ and $b, c \in M$, $f(a) \vee b = f(a) \vee c$ iff $a \vee f^*(b) = a \vee f^*(c)$.*

Proof. (1)$\Leftrightarrow$(2): $f[\uparrow a] = \uparrow b$ for some b and since $f(a)$ is obviously smallest in $f[\uparrow a]$, $b = f(a)$.

(2)$\Leftrightarrow$(3): We always have $f(a \vee f^*(b)) \geq f(a) \vee ff^*(b) \geq f(a) \vee b$. If f is closed, $f(a) \vee b = f(x)$ for some $x \geq a$. As $f(x) \geq b$ we have $x \geq f^*(b) \vee a$, and $f(a \vee f^*(b)) \leq f(x) = f(a) \vee b$.

Conversely, if $x \geq f(a)$ then $x = f(a) \vee x = f(a \vee f^*(x))$.

(3)$\Leftrightarrow$(4): $(f(a) \vee b = f(a \vee f^*(b))) \Leftrightarrow (c \leq f(a) \vee b)$ iff $c \leq (f(a \vee f^*(b))) \Leftrightarrow (c \leq f(a) \vee b$ iff $f^*(c) \leq a \vee f^*(b))$.

(4)$\Rightarrow$(5) is obvious.

(5)$\Rightarrow$(4): If $c \leq f(a) \vee b$ then $f^*(c) \leq f^*(f(a) \vee b) = f^*f(a) \vee f^*(b) \leq a \vee f^*(b)$.
On the other hand, if $f^*(c) \leq a \vee f^*(b)$ then $f^*(c) \vee f^*(b) \vee a = a \vee f^*(b)$, that is, $f^*(c \vee b) \vee a = a \vee f^*(b)$. Thus, by hypothesis, $f(a) \vee b = f(a) \vee b \vee c$ and finally $f(a) \vee b \geq c$. $\qquad\qquad\qquad\qquad\qquad\qquad\qquad\qquad\qquad\qquad\qquad\qquad\qquad\square$

7.3.1. Notes. (1) Confronting closed continuous maps and closed localic maps, the same proviso holds as in 7.2.1(2). See [227].

(2) The reader may check as an exercise that a sublocale S of L is open iff the embedding $j_S \colon S \to L$ is an open localic map (so open maps generalize open sublocales). In order to have a similar result for closed sublocales one has to restrict the class of closed maps to proper maps (the point-free version of the classical notion of perfect map): a localic map $f \colon L \to M$ is *proper* if it is closed and the right adjoint of f^* preserves directed joins (see [263] and [264] for alternative descriptions). Then a sublocale S of L is closed iff the embedding $j_S \colon S \to L$ is a proper map. For more information on open, closed and proper maps consult [157], [255] and [263].

8. Closure

8.1. Closure of a sublocale. Let S be a sublocale. Each sublocale $T \supseteq S$ has to contain $\bigwedge S$ and on the other hand, obviously $S \subseteq \uparrow(\bigwedge S)$. Thus, we have the *closure* of S, the least closed sublocale containing S, given by the formula

$$\overline{S} = \uparrow(\bigwedge S). \qquad (8.1.1)$$

Proposition. (1) *We have* $\overline{\mathsf{O}} = \mathsf{O}$, $\overline{\overline{S}} = \overline{S}$ *and* $\overline{S \vee T} = \overline{S} \vee \overline{T}$.

(2) *The closure of an open sublocale is* $\overline{\mathfrak{o}(a)} = \mathfrak{c}(a^*)$ $(= \uparrow(a^*))$.

Proof. (1) The first two formulas are trivial, and the third one is very easy: Set $a = \bigwedge S$, $b = \bigwedge T$. Then

$$\overline{S} \vee \overline{T} = \uparrow a \vee \uparrow b = \{x \wedge y \mid x \geq a, y \geq b\} = \uparrow(a \wedge b) = \uparrow\bigwedge(S \vee T) = \overline{S \vee T}.$$

(2) By ($\bigwedge$-distr), $\bigwedge_{x \in L}(a \to x) = a \to 0 = a^*$ and thus $\overline{\mathfrak{o}(a)} = \uparrow(a^*)$. $\qquad\square$

8.2. Density. As in the classical topology, a sublocale S is *dense* in L if $\overline{S} = L$. By the formula (8.1.1),

$$S \text{ is dense in } L \text{ iff } 0 \in S.$$

We will also speak of a *dense localic map* $f \colon L \to M$ if $0 \in f[S]$ (which, since f is monotone, amounts to stating that $f(0) = 0$), and of the associated frame homomorphism as of *dense frame homomorphism*. The latter is characterized by the implication

$$h(x) = 0 \quad \Rightarrow \quad x = 0$$

(since $h(x) \leq 0$ iff $x \leq f(0) = 0$).

8.3. Isbell's Density Theorem. Set

$$B_L = \{x \to 0 \mid x \in L\}.$$

It is a sublocale since $\bigwedge(x_i \to 0) = (\bigvee x_i \to 0)$ by ($\bigvee \bigwedge$-distr), and $y \to (x \to 0) = (y \wedge x) \to 0$ by (H7). Consequently we have

Proposition. *Each locale contains the least dense sublocale, namely B_L.*

Proof. By (S2) each dense sublocale of L contains B_L. On the other hand, B_L itself is dense by (H1).   □

Note. This may be a surprising fact and certainly has no counterpart in classical topology. Note that we see, in particular, that a spatial locale can contain a sublocale that is not induced by a subspace of the corresponding space. For instance, in the frame of open sets of the real line, the B_L, certainly non-empty, is contained in the sublocales corresponding to the subspaces of the rationals and the irrationals. This example also shows that the correspondence associating the sublocales with subspaces does not respect meets. The relation of classical subspaces and sublocales will be discussed in more detail later (in Chapter VI).

8.4. Closure under image. This is quite analogous like in spaces. We have a trivial

Observation. *Let $S \subseteq L$ be a sublocale and let $f : L \to M$ be a localic map. Then*

$$f[\overline{S}] \subseteq \overline{f[S]}.$$

(Indeed, since f preserves meets, $f(\bigwedge S) = \bigwedge f[S]$ and hence $f[\uparrow \bigwedge S] \subseteq \uparrow \bigwedge f[S]$.)

8.5. Closure in a sublocale. Let $T \subseteq S$ be sublocales in L. Trivially, $\{x \in S \mid x \geq \bigwedge T\} = \{x \in L \mid x \geq \bigwedge T\} \cap S$, that is, for the closure of T in S we have

$$\overline{T}^S = \overline{T} \cap S,$$

similarly as in spaces.

9.   Preimage as a homomorphism

9.1. Lemma. *Let $f\colon L \to M$ be a localic map. Then for every $a, b \in L$,*

$$f_{-1}[\mathfrak{c}(a) \vee \mathfrak{o}(b)] = f_{-1}[\mathfrak{c}(a)] \vee f_{-1}[\mathfrak{o}(b)].$$

Proof. The preimage is a right adjoint and hence it preserves meets. Thus, we have $f_{-1}[\mathfrak{c}(a) \vee \mathfrak{o}(b)] \cap f_{-1}[\mathfrak{o}(a) \cap \mathfrak{c}(b)] = f_{-1}[\mathsf{O}] = \mathsf{O}$.

On the other hand

$$\begin{aligned}
f_{-1}[\mathfrak{c}(a) \vee \mathfrak{o}(b)] \vee f_{-1}[\mathfrak{o}(a) \cap \mathfrak{c}(b)] &= f_{-1}[\mathfrak{c}(a) \vee \mathfrak{o}(b)] \vee (f_{-1}[\mathfrak{o}(a)] \cap f_{-1}[\mathfrak{c}(b)]) \\
&\geq (f_{-1}[\mathfrak{c}(a)] \vee (f_{-1}[\mathfrak{o}(a)])) \cap (f_{-1}[\mathfrak{o}(b)] \vee f_{-1}[\mathfrak{c}(b)]))
\end{aligned}$$

by distributivity and monotony. We proceed, by 6.3,

$$\cdots = (\mathfrak{c}(f^*(a) \vee \mathfrak{o}(f^*(a)) \cap (\mathfrak{o}(f^*(b) \vee \mathfrak{c}(f^*(b)) = L \cap L = L.$$

Thus, $f_{-1}[\mathfrak{c}(a) \vee \mathfrak{o}(b)]$ is the complement of $(f_{-1}[\mathfrak{o}(a) \cap \mathfrak{c}(b)]$, that is,

$$(f_{-1}[\mathfrak{o}(a)] \cap f_{-1}[\mathfrak{c}(b)])^* = f_{-1}[\mathfrak{c}(a)] \vee f_{-1}[\mathfrak{o}(b)],$$

using 6.3 again. $\square$

9.2. Proposition. *Let $f\colon L \to M$ be a localic map. Then the preimage map*

$$f_{-1}[-]\colon \mathcal{S\!\ell}(M) \to \mathcal{S\!\ell}(L)$$

is a co-frame homomorphism (that is, preserves all meets and all finite joins).

Proof. Preserving meets is given by the adjunction, and $f_{-1}[\mathsf{O}] = \mathsf{O}$ since $\mathsf{O} = \mathfrak{c}(1)$. Thus, we have to prove that f_{-1} preserves binary joins. Let S, T be in $\mathcal{S\!\ell}(M)$. By 6.5, S, T can be written as $S = \bigcap_{i \in I}(\mathfrak{c}(x_i) \vee \mathfrak{o}(y_i))$ and $T = \bigcap_{j \in J}(\mathfrak{c}(u_j) \vee \mathfrak{o}(v_j))$ with suitable x_i, y_i, u_j and v_j. Thus, by 9.1, 6.3, 6.1.5, and 6.3 and 9.1 again

$$\begin{aligned}
f_{-1}[S \vee T] &= f_{-1}\Big[\bigcap_{i,j}(\mathfrak{c}(x_i \wedge u_j) \vee \mathfrak{o}(y_i \vee v_j))\Big] \\
&= \bigcap_{i,j}(f_{-1}[\mathfrak{c}(x_i \wedge u_j)] \vee f_{-1}[\mathfrak{o}(y_i \vee v_j)]) \\
&= \bigcap_{i,j}(\mathfrak{c}(f^*(x_i \wedge u_j)) \vee \mathfrak{o}(f^*(y_i \vee v_j))) \\
&= \bigcap_{i,j}(\mathfrak{c}(f^*(x_i) \wedge f^*(u_j)) \vee \mathfrak{o}(f^*(y_i) \vee f^*(v_j))) \\
&= \bigcap_{i,j}(\mathfrak{c}(f^*(x_i)) \vee \mathfrak{c}(f^*(u_j)) \vee \mathfrak{o}(f^*(y_i)) \vee \mathfrak{o}(f^*(v_j))) \\
&= \bigcap_{i,j}(f_{-1}[\mathfrak{c}(x_i)] \vee f_{-1}[\mathfrak{c}(u_j)] \vee f_{-1}[\mathfrak{o}(y_i)] \vee f_{-1}[\mathfrak{o}(v_j)]) \\
&= \bigcap_{i} f_{-1}[\mathfrak{c}(x_i) \vee \mathfrak{o}(y_i)] \vee \bigcap_{j} f_{-1}[\mathfrak{c}(u_j) \vee \mathfrak{o}(v_j)] = f_{-1}[S] \vee f_{-1}[T].\quad \square
\end{aligned}$$

9.3. Corollary. *The preimage function preserves complements.*   □

9.4. Note. Recall the congruence frame $\mathfrak{C}(L)$ from 5.2, isomorphic to $\mathcal{S}\!\ell(L)^{\mathrm{op}}$. The suitably modified preimage function $f_{-1}[-]'$ is then a frame homomorphism $\mathfrak{C}(L) \to \mathfrak{C}(M)$, and by 4.3 we see that the respective modification of the image function is a localic map

$$f[-]': \mathfrak{C}(L) \to \mathfrak{C}(M).$$

10. Other special sublocales: one-point sublocales, and Boolean ones

10.1. One-point sublocales. Each sublocale contains the top, and the least sublocale $\mathsf{O} = \{1\}$ (recall 2.1) plays the role of the void subspace. Therefore, the smallest non-trivial sublocales are those that contain precisely one element $a \neq 1$.

10.1.1. Lemma. *Let $a \in L$ be meet-irreducible. Then for any $x \in L$ either $x \to a = 1$ or $x \to a = a$.*

Proof. If $x \to a \neq 1$, that is, $x \not\leq a$, we have to have $x \to a \leq a$ since $x \wedge (x \to a) \leq a$ by (H5) and hence $x \to a = a$ by (H3).   □

10.1.2. Proposition. *Let $a \in L$, $a \neq 1$. Then $\{a, 1\}$ is a sublocale if and only if a is meet-irreducible, that is, a point in the sense of the representation (P3) in II.3.3.*

Proof. If a is meet-irreducible then $\{a, 1\}$ is a sublocale by the lemma.

If $\{a, 1\}$ is a sublocale and $a \neq 1$, and if $x \wedge y \leq a$ then $x \leq y \to a$, and if $y \not\leq a$, that is, $y \to a \neq 1$, $x \leq a = y \to a$.   □

The sublocales $\{p, 1\}$, $p \neq 1$ will be called *one-point sublocales*. Note that the relation between points and one-point sublocales,

$$p \quad \text{and} \quad \{p, 1\},$$

is the same as that between x and $\{x\}$ in classical spaces.

10.2. The sublocale $B_L \subseteq L$ from 8.3 is, by (S2), the smallest sublocale of L containing 0. More generally, for each $a \in L$ we have

$$\mathfrak{b}(a) = \{x \to a \mid x \in L\},$$

the smallest sublocale containing a (it is a sublocale for the same reason B_L is: ($\bigvee \bigwedge$-dist) yields (S1) since $\bigwedge(x_i \to a) = (\bigvee x_i) \to a$ and (H7) yields (S2) since $y \to (x \to a) = (x \wedge y) \to a$; $a \in \mathfrak{b}(a)$ as $a = 1 \to a$).

10.2.1. Recall from AI.7.3 that in a Heyting algebra H the pseudocomplement can be expressed as

$$x^* = x \to 0. \tag{psc}$$

We have

Lemma. (a) $\bigwedge \mathfrak{b}(a) = a$,

(b) *the pseudocomplement in $\mathfrak{b}(a)$ is given by $x^* = x \to a$, and*

(c) $\mathfrak{b}(a) = \{x \mid x = (x \to a) \to a\}$.

Proof. (a): a is in $\mathfrak{b}(a)$ (since $a = 1 \to a$) and by (H3) $a \leq x \to a$.

(b) follows immediately from (a) and (psc).

(c): Trivially $\{x \mid x = (x \to a) \to a\} \subseteq \mathfrak{b}(a)$. Now let $x \in L$. We have, for any y,

$$y \leq (y \to a) \to a \tag{$*$}$$

since $y \wedge (y \to a) \leq a$ by (H5). Substituting $y = x \to a$ we obtain $x \to a \leq ((x \to a) \to a) \to a$ and using (anti) from 3.1 for $(*)$ and $(- \to a)$ we get $x \to a \geq ((x \to a) \to a) \to a$ so that

$$x \to a = ((x \to a) \to a) \to a \tag{$**$}$$

and hence $x \to a \in \{y \mid y = (y \to a) \to a\}$. $\square$

10.3. Generalizing the notion of density from 8.2 we say that a sublocale S is dense in a sublocale T if

$$S \subseteq T \subseteq \overline{S}.$$

Proposition. $\mathfrak{b}(\bigwedge S)$ *is the smallest sublocale dense in S.*

Proof. Set $a = \bigwedge S$. Then $a = \bigwedge \mathfrak{b}(a)$ and $S \subseteq \overline{\mathfrak{b}(a)} = \overline{S}$. Since $a \in S$, $\mathfrak{b}(a) \subseteq S$ being the smallest sublocale containing a. $\square$

10.4. Recall from AI.7.4 that

a Heyting algebra is a Boolean algebra iff $x^{**} = x$ for all $x \in L$

(to refresh the fact: using the De Morgan formula $(x \vee y)^* = x^* \wedge y^*$ we obtain $x \vee x^* = ((x \vee x^*)^*)^* = (x^* \wedge x^{**})^* = 0^* = 1$). Consequently, by 10.2.1(c),

each $\mathfrak{b}(a)$ is a Boolean algebra.

We have more, namely

Proposition. *A sublocale $S \subseteq L$ is a Boolean algebra iff $S = \mathfrak{b}(a)$ for some $a \in L$.*

Proof. Let $S \subseteq L$ be a Boolean sublocale. Set $a = \bigwedge S$. Let x be in S. The pseudocomplement in S is a complement and hence, since $x^* = x \to a$, we have $x = (x \to a) \to a \in \mathfrak{b}(a)$, by 10.2.1(c). On the other hand, if $x \in \mathfrak{b}(a)$ then $x \in S$ since $a = \bigwedge S \in S$. Thus, $S = \mathfrak{b}(a)$. $\square$

10.5. From the fact that $\mathfrak{b}(a)$ is the smallest sublocale of L containing a follows immediately that

Corollary. *Each sublocale S of L is a join in $\mathcal{Sl}(L)$ (indeed a union)*

$$S = \bigvee\{\mathfrak{b}(a) \mid a \in S\} \ (= \bigcup\{\mathfrak{b}(a) \mid a \in S\})$$

of Boolean sublocales.   □

11. Sublocales as quotients. Factorizing frames is surprisingly easy

11.1. Although we mostly prefer the description of sublocales as in Section 2, that is, really as sub-locales, subsets the embeddings of which are localic maps, the alternative representations (as in Section 5) are often very useful.

It often happens in constructions that we have given, instead of a congruence, a set $R \subseteq L \times L$ of couples of elements to be identified, and we would like to determine the sublocale associated with the congruence generated by R. In fact, it will turn out that we can describe this sublocale more or less directly, without constructing the congruence first. This will be discussed in this section.

> It should be noted that in frames an explicit extension of a relation to a congruence is not quite such a hard task as in other type of algebras. The fact that each congruence class has a maximal element helps to make it fairly transparent.

11.2. Recall that a sublocale $S \subseteq L$ is associated with the congruence

$$x E_S y \ \text{iff} \ (\forall s \in S, x \leq s \ \text{iff} \ x \leq s)$$

and with the nucleus
$$\nu_S(x) = \bigwedge\{s \in S \mid x \leq s\} \tag{11.2.1}$$
(so that $x E_S y$ iff $\nu_S(x) = \nu_S(y)$), and the sublocale S can be reconstructed from the congruence by setting

$$S = \{\max_{Ex} \mid x \in L\} \ \text{where} \ \max_{Ex} = \bigvee Ex = \bigvee\{y \mid y Ex\}. \tag{11.2.2}$$

Now let R be a binary relation on a sublocale L. An element $s \in S$ is said to be

$$R\text{-saturated (briefly, } saturated\text{)}$$

if
$$\forall a, b, c \quad aRb \ \Rightarrow \ (a \wedge c \leq s \ \text{iff} \ b \wedge c \leq s).$$

The set of all saturated elements will be denoted by

$$L/R.$$

Define a mapping

$$\mu_R = (x \mapsto \bigwedge\{s \text{ saturated } \mid x \leq s\} \colon L \to L/R.$$

We have

11.2.1. Proposition. $S = L/R$ *is a sublocale of L and $\mu_R = j^*$ where $j \colon S \subseteq L$ is the embedding map. For the associated nucleus we have $\nu_S(a) = \mu_R(a)$ and the implication*

$$aRb \Rightarrow \nu_S(a) = \nu_S(b).$$

Proof. Obviously L/R is closed under meets. Now let $s \in L/R$. Then for any $x \in L$

$$a \wedge c \leq x \to c \quad \text{iff} \quad a \wedge c \wedge x \leq s \quad \text{iff} \quad b \wedge c \wedge x \leq s \quad \text{iff} \quad b \wedge c \leq x \to s.$$

Finally, since $a \leq \nu(a) \in L/R$ we have $b \leq \nu(a)$ and hence $\nu(b) \leq \nu(a)$, and similarly $\nu(a) \leq \nu(b)$.

We have $\mu_R(x) \leq s$ iff $x \leq s = j(s)$ for any $x \in L$ and $s \in S$ and the formula for $\mu_R(x)$ coincides with that for $\nu_S(x)$ in (11.2.1). $\square$

11.2.2. Proposition. *Let S be a sublocale and let E be the associated congruence. Then*

$$S = L/E.$$

Proof. We have to prove that the E-saturated elements are precisely the elements of the form $\max_{Ex}$ from (11.2.2).

Since E respects meets (and consequently, of course, the order) we have

$$aEb \quad \Rightarrow \quad (a \wedge c)E(b \wedge c) \quad \Rightarrow \quad (a \wedge c \leq \max_{Ex} \text{ iff } b \wedge c \leq \max_{Ex}).$$

Thus, each $\max_{Ex}$ is saturated. On the other hand, if s is E-saturated then we have in particular, since $\max_{Es}Es$ and $s \leq s$, $\max_{Es} \leq s$. $\square$

Note. The symbol L/E was used in 5.2 only slightly differently: in the transition to what we have now, the class Ex is represented by its maximum.

11.3. Theorem. *Let R be a binary relation on a locale L. Let a localic map $f \colon M \to L$ be such that*

$$aRb \quad \Rightarrow \quad f^*(a) = f^*(b).$$

Then $f[M] \subseteq L/R$ and hence there is a localic map $\overline{f} \colon M \to L/R$ such that

commutes (where $j \colon L/R \to L$ is the embedding).

Proof. It suffices to show that each $f(x)$ is saturated. For aRb we have

$$a \wedge c \leq f(x) \quad \text{iff} \quad f^*(a) \wedge f^*(c) = f^*(a \wedge c) \leq x$$
$$\text{iff} \quad f^*(b) \wedge f^*(c) = f^*(b \wedge c) \leq x \quad \text{iff} \quad b \wedge c \leq f(x). \qquad \square$$

11.3.1. Note that for the left adjoint $\overline{f}^*$ of $\overline{f}$ we have $\overline{f}^*(a) = f^*(a)$ for all $a \in L/R$. Indeed, $\overline{f}^*(a) = \overline{f}^*(\nu(a)) = \overline{f}^* j^*(a) = f^*(a)$.

In the language of frame homomorphisms we can summarize the facts above as follows:

Theorem. *Let R be a binary relation on a locale L. Then $\mu_R \colon L \to L/R$ is a frame homomorphism such that*

$$aRb \Rightarrow \mu_R(a) = \mu_R(b),$$

and for every frame homomorphism $h \colon L \to M$ such that

$$aRb \Rightarrow h(a) = h(b)$$

there is a frame homomorphism $\overline{h} \colon L/R \to M$ such that

$$
\begin{array}{ccc}
L & \xrightarrow{\mu_R} & L/R \\
 & \searrow{\scriptstyle h} & \downarrow{\scriptstyle \overline{h}} \\
 & & M
\end{array}
$$

commutes; moreover, $\overline{h} = h|_{(L/R)}$, that is, for every $a \in L/R$, $\overline{h}(a) = h(a)$. $\qquad \square$

11.4. The relations occurring in constructions of sublocales, although typically by far not congruences, are sometimes of a special nature which yields a more transparent formula for saturatedness.

Proposition. *Let C be a join-basis of L and let $R \subseteq L \times L$ be such that*

$$\forall a, b \in L \; \forall c \in C \quad aRb \; \Rightarrow \; (a \wedge c)R(b \wedge c).$$

Then $s \in L$ is R-saturated iff

$$aRb \quad \Rightarrow \quad (a \leq s \quad \text{iff} \quad b \leq s).$$

If moreover $aRb \; \Rightarrow \; a \leq b$ and if $R' \subseteq R$ is such that for each $(a, b) \in R$ there is an $(a', b) \in R'$ such that $a' \leq a$ this reduces to

$$aR'b \quad \Rightarrow \quad (a \leq s \quad \Rightarrow \quad b \leq s),$$

or, trivially rewritten, to

$$aR'b \; \& \; (a \leq s) \quad \Rightarrow \quad b \leq s.$$

Proof. Let $c \in L$ be general. Take a join $c = \bigvee_{i \in J} c_i$ with $c_i \in C$. Then $a \wedge c \le s$ iff

$$\bigvee_{i \in J} (a \wedge c_i) \le s \text{ iff } \forall i, a \wedge c_i \le s \text{ iff } \forall i, b \wedge c_i \le s$$

$$\text{iff } \bigvee_{i \in J} (b \wedge c_i) \le s \text{ iff } b \wedge c \le s. \qquad \square$$

11.4.1. The join of congruences is a very complicated relation. However, in the factorization procedure creating the L/R the situation is very easy. We have

Proposition. *Let R_i, $i \in J$, be binary relations on a frame L. Then*

$$\bigwedge_{i \in J} L/R_i = \bigcap_{i \in J} L/R_i = L/(\bigcup_{i \in J} R_i). \qquad \square$$

(Indeed, the implication $\forall a, b, c \; a(\bigcup R_i)b \;\Rightarrow\; (a \wedge c \le s \text{ iff } b \wedge c \le s)$ is obviously equivalent to $\forall i, \forall a, b, c \; aR_i b \;\Rightarrow\; (a \wedge c \le s \text{ iff } b \wedge c \le s)$.)

11.5. Prenuclei. Sometimes the relation R in question determines very transparently the desired extensions of the elements to approach the saturation. Typically, however, one does not achieve the saturation (the $\mu_R(a)$) in one step and one has to repeat the procedure again and again, till it stops after sufficiently many steps (we will see an instructive example in the construction of coproduct in IV.4.3). This leads to the definition of *prenucleus*, a mapping $\nu\colon L \to L$ such that

(N1) $a \le \nu(a)$,
(N2) $a \le b \;\Rightarrow\; \nu(a) \le \nu(b)$, and
(PN) $\nu(a) \wedge b \le \nu(a \wedge b)$.

We will see that such a mapping naturally creates a nucleus.

11.5.1. First of all, the reader marks the formally weaker (PN) instead of (N4) (the absence of (N3) is no surprise); this is just making the checking simpler: once the idempotency is achieved, (N4) is implied.

Observation. *A prenucleus satisfying $\nu\nu(a) = \nu(a)$ is a nucleus.*

(Indeed, we have $\nu(a) \wedge \nu(b) \le \nu(a \wedge \nu(b)) \le \nu(\nu(a \wedge b)) = \nu(a \wedge b)$, and the other inequality is trivial.)

11.5.2. Given a prenucleus $\nu^1\colon L \to L$ we proceed to obtain a nucleus (which will be referred to as the *nucleus generated by ν^1*) as follows. For ordinals α set

$$\nu^0(a) = a, \quad \nu^{\alpha+1}(a) = \nu^1(\nu^\alpha(a)), \quad \text{and}$$

$$\nu^\lambda(a) = \bigvee_{\alpha < \lambda} \nu^\alpha(a) \quad \text{for limit } \lambda.$$

Then for a sufficiently large α we have $\nu^1(\nu^\alpha) = \nu^\alpha$ – and hence $\nu^\alpha\nu^\alpha = \nu^\alpha$ – and by the observation above it suffices to prove (PN). This follows by transfinite induction:

$$\nu^{\alpha+1}(a) \wedge b = \nu^1(\nu^\alpha(a)) \wedge b$$
$$\leq \nu^1(\nu^\alpha(a) \wedge b) \leq \nu^1(\nu^\alpha(a \wedge b)) = \nu^{\alpha+1}(a \wedge b),$$

$$\nu^\lambda(a) \wedge b = (\bigvee_{\alpha<\lambda} \nu^\alpha(a)) \wedge b = \bigvee_{\alpha<\lambda} (\nu^\alpha(a) \wedge b) \leq \bigvee_{\alpha<\lambda} \nu^\alpha(a \wedge b).$$

11.5.3. Note. Thus the nucleus obtained is the least one that majorizes the repeated closing under the relation in question. This is fairly intuitive and the procedure is useful. It should be noted, however, that it heavily depends on ordinal numbers. A more constructive (that is, choice-free) approach, if available, is preferred – but sometimes it is not available, or rather complicated.

Chapter IV

Structure of Localic Morphisms.
The Categories Loc and Frm

1. Special morphisms. Factorizing in Loc and Frm

1.1. A few simple expedient facts. (1) For maps f, g that are Galois adjoint one has $fgf = f$ and $gfg = g$. Consequently,

$$f \text{ is one-one iff } g \text{ is onto}$$

(if f is onto than $fg = \mathrm{id}$, if it is one-one then $gf = \mathrm{id}$).

(2) Recall the following standard fact on homomorphisms of algebras. If A, B, C are algebras – in particular frames – and $f\colon A \to B$, $g\colon B \to C$ mappings such that gf is a homomorphism then

- if g is a one-one homomorphism then f is a homomorphism, and
- if f is a homomorphism onto then g is a homomorphism.

Now localic maps are not (algebraic) homomorphisms, but from (1) and from the obvious fact that if f_i are (say, right) adjoints to g_i ($i = 1, 2$) then $f_2 f_1$ is an adjoint to $g_1 g_2$ we immediately infer that

- *if L, M, N are locales and $f\colon L \to M$, $g\colon M \to N$ are mappings such that gf is a localic map then*
  - *if g is a one-one localic map then f is a localic map, and*
  - *if f is a localic map onto then g is a localic map.*

  Consequently, if $f\colon L \to M$ is a localic map, $S \subseteq M$ a sublocale, and if $f[L] \subseteq S$ then

  $$f' = (x \mapsto f(x))\colon L \to S$$

  is a localic map.

(3) Recall the translation of sublocales to sublocale homomorphisms and nuclei in III.5.1 and 5.3. In particular we saw that the left adjoint $j^*\colon L \to S$ to the embedding $j\colon S \subseteq L$ is given by

$$j^*(x) = \nu_S(x)$$

and hence in particular $j^*(j^*(x)) = j^*(x)$.

1.2. Monomorphisms in Frm. (Recall AII.1.3) Obviously, one-one frame homomorphisms are monomorphisms. Moreover we have made in III.1.1.1 an easy

Observation. *The monomorphisms in* **Frm** *are precisely the one-one homomorphisms. Consequently, the epimorphisms in* **Loc** *are precisely the onto localic maps.*

1.3. More about onto homomorphisms. We already know that the onto homomorphisms are precisely the extremal epimorphisms in **Frm**. In fact they have a stronger property.

Proposition. (Recall AII.4) *In* **Frm** *the extremal, strong and regular epimorphisms coincide and they are precisely the onto homomorphisms.*

Consequently, in **Loc** *the one-one localic maps are precisely the extremal resp. strong resp. regular monomorphisms.*

Proof. It suffices to show that each homomorphism $h\colon L \to M$ that is onto is a coequalizer. Define

$$E = \{(x,y) \mid x,y \in L,\ h(x) = h(y)\} \quad \text{and} \quad g_i = ((x_1,x_2) \mapsto x_i)\colon E \to L;$$

obviously E with $\bigvee$ and $\wedge$ taken coordinatewise is a frame, g_i are homomorphisms, and $hg_1 = hg_2$. Now if $\phi\colon L \to N$ is a homomorphism such that $\phi g_1 = \phi g_2$ we can define $\overline{\phi}\colon M \to N$ by setting $\overline{\phi}(h(x)) = \phi(x)$: this is correct since if $h(x) = h(y)$ we have $(x,y) \in E$ and $\phi(x) = \phi g_1(x,y) = \phi g_2(x,y) = \phi(y)$. Now $\overline{\phi}h = \phi$ and by 1.1(2) $\overline{\phi}$ is a homomorphism, unique since h is onto. $\qquad\square$

1.3.1. Note. Thus, the monomorphisms and extremal (strong, regular) epimorphisms in **Frm** are fairly transparent. Not so the plain epimorphisms (about which we will learn more in Section 6 below) and special monomorphisms (here, the extremal and regular ones do not even coincide; for some information on the regular ones see V.3).

1.4. Natural factorizations in Frm and Loc. A frame homomorphism $h\colon L \to M$ can be naturally decomposed as

$$L \xrightarrow{\ \overline{h}=(x \mapsto h(x))\ } h[L] \xrightarrow{\ k=\subseteq\ } M \qquad\qquad (1.4.1)$$

with $\overline{h}$ onto and k one-one. By 1.2 and 1.3 it is a decomposition with k a monomorphism and $\overline{h}$ a strong epimorphism. Since strong epimorphisms are orthogonal to monomorphisms we see that the decompositions (1.4.1) yield a factorization system (recall AII.4.3) $(\mathcal{E}, \mathcal{M})$ in **Frm** with $\mathcal{E}$ the class of all strong (extremal, regular) epimorphisms and $\mathcal{M}$ the class of all monomorphisms.

Similarly we have the factorization system $(\mathcal{E}, \mathcal{M})$ in **Loc** with $\mathcal{E}$ the class of all epimorphisms and $\mathcal{M}$ the class of all strong monomorphisms, given by the decompositions (of localic maps $f\colon M \to L$)

$$M \xrightarrow{\;g=(x\mapsto f(x))\;} f[M] \xrightarrow{\;j=\subseteq\;} L \;. \tag{1.4.2}$$

1.5. Let us take a localic map $f\colon M \to L$ decomposed as in (1.4.2) and the left adjoint $h = f^*$ written as

$$L \xrightarrow{\;j^*\;} f[M] \xrightarrow{\;g^*\;} M \;,$$

again with g^* one-one and j^* onto, similarly like in (1.4.1). Now $f[M]$ does not coincide with $h[L]$, but the unicity from AII.4.3 is (of course) not violated. We have the connection isomorphism $\alpha\colon f[M] \to h[L]$ given by $\alpha(x) = h(x)$. Indeed:

$$\alpha(j^*(x)) = h(j^*(x)) = g^* j^* j^*(x) = g^* j^*(x) = f^*(x) = h(x) = \overline{h}(x)$$

(by 1.1(2)), and $k\alpha j^* = k\overline{h} = h = g^* j^*$ and since j^* is onto, $k\alpha = g^*$. The inverse isomorphism $\beta\colon h[L] \to f[M]$ is given by the formula $\beta(y) = f(y)$: we have $\beta\alpha(x) = fh(x) = fh(f(y)) = f(y) = x$ and $\alpha\beta(y) = hf(y) = hf(h(x)) = h(x) = y$.

2. The down-set functor and free constructions

2.1. In the sequel, an important role will be played by *meet-semilattices with top* 1(thus, we assume infima of *all* finite subsets, including $\emptyset$) and their $(\wedge, 1)$-homomorphisms (that is, mappings preserving all finite infima). To simplify the language we will speak simply of *semilattices* and *homomorphisms*. The ensuing category will be denoted by

$$\mathbf{SLat}_1.$$

2.2. We will consider the *down-set functor* $\mathfrak{D}\colon \mathbf{SLat}_1 \to \mathbf{Frm}$ defined by

$$\mathfrak{D}S = \{X \subseteq S \mid \,\downarrow X = X\}, \text{ ordered by inclusion, and}$$

$$\mathfrak{D}h(X) = \,\downarrow h[X] \quad \text{for} \quad h\colon S \to T.$$

Note that $\mathfrak{D}S$ is really a frame, with $X \wedge Y = X \cap Y$, and $\bigvee_{i \in J} X_i = \bigcup_{i \in J} X_i$ (the intersection distributes over unions).

Next, $\mathfrak{D}h(\emptyset) = \emptyset$, $\mathfrak{D}h(S) = {\downarrow}h(S) \supseteq {\downarrow}1 = T$,

$$\mathfrak{D}h(\bigcup X_i) = {\downarrow}h[\bigcup X_i] = {\downarrow}\bigcup h[X_i] = \bigcup{\downarrow}h[X_i] = \bigcup\mathfrak{D}h(X_i),$$

and

$$\begin{aligned}
\mathfrak{D}h(X) \cap \mathfrak{D}h(Y) &= {\downarrow}h(X) \cap {\downarrow}h(Y) \\
&= \{z \mid \exists x \in X, y \in Y,\ z \le h(x) \wedge h(y) = h(x \wedge y)\} \\
&= \{z \mid \exists u \in h[X \cap Y],\ z \le u\} = \mathfrak{D}h(X \cap Y),
\end{aligned}$$

the penultimate equality because X and Y are down-sets and hence $x \wedge y$ is in both X and Y.

Further, we will consider the mappings

$$\alpha_S \colon S \to \mathfrak{D}S, \quad \alpha_S(x) = {\downarrow}x.$$

Note that

$$\text{\textit{each } } \alpha_S \text{ \textit{ is a semilattice homomorphism}}$$

$({\downarrow}x \cap {\downarrow}y = {\downarrow}(x \wedge y)$ from the definition of infimum and ${\downarrow}1 = S = 1_{\mathfrak{D}(S)})$.

2.3. Proposition. *Let S be a semilattice and L a frame. Let $f \colon S \to L$ be a semilattice homomorphism (L viewed, for a moment, as the semilattice $(L, \wedge, 1)$). Then there exists precisely one frame homomorphism $h \colon \mathfrak{D}M \to L$ such that $h \cdot \alpha_S = f$.*

Proof. There is at most one such h: since it has to preserve joins we have to have

$$h(X) = h(\bigcup\{{\downarrow}x \mid x \in X\}) = \bigvee_{x \in X} h({\downarrow}x) = \bigvee_{x \in X} f(x).$$

On the other hand, define $h \colon \mathfrak{D}S \to L$ by the formula $h(X) = \bigvee_{x \in X} f(x)$. Then $h(\emptyset) = 0$ (void join), $h(S) = \bigvee_{x \in X} f(x) \ge 1$ (and hence $=1$), and obviously also $h(\bigcup X_i) = \bigvee h(X_i)$ for non-empty joins. Finally,

$$\begin{aligned}
h(X) \wedge h(Y) &= \bigvee_{x \in X} f(x) \wedge \bigvee_{y \in Y} f(y) = \bigvee\{f(x) \wedge f(y) \mid x \in X,\ y \in Y\} \\
&= \bigvee\{f(x \wedge y) \mid x \in X,\ y \in Y\} \le \bigvee\{f(z) \mid z \in X \cap Y\} \\
&= h(X \cap Y) \le h(X) \wedge h(Y),
\end{aligned}$$

the first $\le$ since X and Y are down-sets, and the second one since h is obviously monotone. $\qquad\square$

2.3.1. Remark. Often we will have the down-set in question defined as ${\downarrow}X$ for X that is not a down-set itself. Note that for the homomorphism h we have $h({\downarrow}X) = \bigvee_{x \in X} f(x)$.

2.3.2. Thus, $\mathfrak{D}\colon \mathbf{SLat_1} \to \mathbf{Frm}$ is a left adjoint to the $U\colon \mathbf{Frm} \to \mathbf{SLat_1}$ forgetting the join structure, with the adjunction units α and β,

$$\beta_L\colon \mathfrak{D}U(L) \to L, \quad \beta_L(X) = \bigvee X,$$

that is, the homomorphism from 2.3 obtained by lifting the identity map and hence $\beta_L \cdot \alpha_{U(L)}$, more exactly $U(\beta_L) \cdot \alpha_{U(L)}$ is the identity, and $\beta_{\mathfrak{D}S}(\mathfrak{D}\alpha_S(X)) = \bigvee {\downarrow}\alpha_S[X] = \bigvee \alpha_S[X] = \bigvee\{{\downarrow}x \mid x \in X\} = X$. Recall AII.6.3.

2.4. Thus, $\mathfrak{D}$ (together with the α_S's) provides a *free construction* of frames from semilattices (see AII.6.2).

The reader may well ask for a free frame constructed from a *set* (in other words, for a left adjoint to the forgetful functor $\mathbf{Frm} \to \mathbf{Set}$). This is easy: now it suffices to construct free semilattices and this can be done as follows.

For a set M consider

$$\mathfrak{S}M = \{X \subseteq M \mid X \text{ finite}\}$$

ordered *inversely* by inclusion, and setting $0 \leq X$ for all the finite $X \subseteq M$. Thus, $X \wedge Y = X \cap Y$; the intuition is sometimes being helped by writing a set $\{x_1, \ldots, x_n\}$ as a formal expression $\bigwedge_{i=1}^{n} x_i$, and

$$\mathfrak{S}f(X) = f[X].$$

One then considers the maps

$$\varepsilon_M\colon M \to \mathfrak{S}M, \quad \varepsilon_M(x) = \{x\}$$

and easily verifies that for any mapping $f\colon M \to S$ of a set M into a semilattice S one has precisely one semilattice homomorphism $h\colon \mathfrak{S}M \to S$ such that $h \cdot \varepsilon_M = f$, namely one given by $h(X) = \bigwedge\{f(x) \mid x \in X\}$.

Thus, we obtain a free functor $\mathfrak{F} = \mathfrak{D} \cdot \mathfrak{S}\colon \mathbf{Set} \to \mathbf{Frm}$ (with the unit $\phi_M = \alpha_{\mathfrak{S}M} \cdot \varepsilon_M\colon M \to \mathfrak{F}M$).

2.4.1. Similarly like in other classes of algebras, the free construction allows us to describe frames "by generators and defining relations". Namely, we start with a meet-semilattice M and a system of couples (τ_i, θ_i), $i \in J$, of terms in the elements of M and formal join symbols. Then we obtain the desired frame as

$$\mathfrak{D}(M)/R \quad \text{with} \quad R = \{(\overline{\tau}_i, \overline{\theta}_i) \mid i \in J\}$$

where $\overline{\tau}_i$ resp. $\overline{\theta}_i$ are obtained from τ_i resp. θ_i by replacing each $a \in M$ by ${\downarrow}a$.

Similarly, of course, we can define frames from *sets of generators* instead of semilattices of generators by using the combined free construction $\mathfrak{F} = \mathfrak{D}\mathfrak{S}$ as above.

3.   Limits and a colimit in Frm

3.1. Products in Frm. This is a straightforward affair. Namely, if L_i, $i \in J$, are frames, we endow the cartesian product

$$\prod_{i \in J} L_i$$

with the structure of frame coordinatewise (which is the same as defining the order by

$$(x_i)_{i \in J} \le (y_i)_{i \in J} \quad \text{iff} \quad \forall i, \ x_i \le y_i \ ).$$

The projections

$$p_j = ((x_i)_{i \in J} \to x_j) \colon \prod_{i \in J} L_i \to L_j$$

are then frame homomorphisms and we immediately see that for each system $(h_j \colon M \to L_j)_{j \in J}$ of frame homomorphisms there is precisely one frame homomorphism $h \colon M \to \prod_{i \in J} L_i$ such that $p_j h = h_j$ for all $j \in J$, namely the one given by $h(x) = (h_j(x))_{j \in J}$.

3.1.1. Those were the products of non-empty systems. The empty product is the one-element frame

$$\mathbf{1} = \{0_1 = 1_1\} \quad \text{(this is, of course, the O from III.2.1).}$$

The constant mappings $L \to \mathbf{1}$ are obviously frame homomorphisms.

3.1.2. Since **Frm** has products, **Loc** has coproducts. Let us present the coproduct injections explicitly. For a fixed $j \in J$ define

$$\alpha_j \colon L_j \to \prod_{i \in J} L_i \quad \text{by setting} \quad (\alpha_j(x))_i = \begin{cases} x & \text{if } \ i = j, \\ 1 & \text{otherwise.} \end{cases}$$

We immediately see that

$$p_j((x_i)_{i \in J}) \le x \quad \text{iff} \quad (x_i)_{i \in J} \le \alpha_j(x).$$

Thus

$$\left(\alpha_j \colon L_j \to \prod_{i \in J} L_i \right)_{j \in J}$$

constitutes the coproduct in **Loc**. Needless to say, the one-element locale **1** is the initial object in **Loc**, with the unique localic maps $(1_1 \mapsto 1_L) \colon \mathbf{1} \to L$.

3.2. The equalizers in Frm. Again there is no problem. If $h_1, h_2 \colon L \to M$ are frame homomorphisms consider

$$E = E(h_1, h_2) = \{x \in L \mid h_1(x) = h_2(x)\}.$$

Obviously E is a subframe of L and the embedding $j\colon E \subseteq L$ has the property that whenever $h_1 g = h_2 g$ for a frame homomorphism $g\colon N \to L$ then we have (precisely one) frame homomorphism $\bar{g}\colon N \to E$ with $j\bar{g} = g$ (namely, the one given by $\bar{g}(x) = g(x)$).

Consequently, the category **Loc** has coequalizers (if $f_1, f_2\colon M \to L$ are localic maps, the coequalizer can be obtained as

$$(x \mapsto \nu_{E(f_1^*, f_2^*)}(x))\colon L \to E(f_1^*, f_2^*),$$

where ν is the corresponding nucleus – see III.4.3.).

Thus we can conclude that (see AII.5.3.2)

the category **Frm** *is complete, and (hence) the category* **Loc** *is cocomplete.*

3.3. Coequalizers in Frm (and equalizers in Loc). For frame homomorphisms $h_1, h_2\colon L \to M$ consider the binary relation

$$R = \{(h_1(x), h_2(x)) \mid x \in L\} \subseteq M \times M.$$

Then for a frame homomorphism $g\colon M \to N$,

$$g h_1 = g h_2 \quad \text{iff} \quad \forall (a, b) \in R, \ g(a) = g(b).$$

Thus (recall (III.11.3.1), whenever $g h_1 = g h_2$, we have (precisely one) $\bar{g}\colon M/R \to N$ such that $\bar{g}\mu_R = g$; hence, $\mu_R\colon M \to M/R$ is a coequalizer of h_1, h_2 in **Frm**.

3.3.1. Consequently we have equalizers in **Loc**. They can be also described in a way more explicit then the embedding of the M/R from the construction above.

Let $f_1, f_2\colon M \to L$ be localic maps. The set

$$E = \{x \in M \mid f_1(x) = f_2(x)\}$$

is generally not a sublocale. But since the f_i preserve meets it is closed under meets and hence, by III.4.2, we have the largest sublocale of M contained in E, the E_{sloc}. Consider the embedding map

$$j\colon E_{\mathsf{sloc}} \subseteq M.$$

If $g\colon N \to M$ is a localic map such that $f_1 g = f_2 g$ then

- $g[N] \subseteq E$, and
- by III.4.1, $g[N]$ is a sublocale.

Thus, $g[N] \subseteq E_{\mathsf{sloc}}$ and by 1.1(2) we have the localic map $\bar{g} = (x \mapsto g(x))\colon N \to E_{\mathsf{sloc}}$ such that $j\bar{g} = g$.

4.   Coproducts of frames

4.1. Some notation. To prove that the category of frames is cocomplete it remains to be shown that it has coproducts. Constructing these will be more involved than the simple observations in Section 3.

The following notation will be used for both cases. Let L_i, $i \in J$, be frames or semilattices. Set

$$\prod_{i \in J}' L_i = \{(x_i)_{i \in J} \in \prod_{i \in J} L_i \mid \text{up to finitely many } i, \ x_i = 1\}.$$

Let $j \in J$ be fixed. For $x \in L_j$ and $u \in \prod_{i \in J}' L_i$ set

$$x *_j u = v, \quad \text{where} \ v_i = \begin{cases} x \text{ for } i = j \\ u_i \text{ for } i \neq j. \end{cases}$$

Finally define $\bar{1} \in \prod_{i \in J}' L_i$ by

$$\bar{1}_i = 1 \ \text{ for all } i$$

(note that the $\alpha_j(x)$ in 3.1.2 is $x *_j \bar{1}$).

4.2. Coproducts in SLat₁. Let L_i, $i \in J$, be a system of semilattices. Consider the mappings

$$\kappa_j = (x \mapsto x *_j \bar{1}) \colon L_j \to \prod_{i \in J}' L_i.$$

Obviously κ_j are homomorphisms. We have

Proposition. *The system* $\left(\kappa_j \colon L_j \to \prod_{i \in J}' L_i\right)_{j \in J}$ *constitutes a coproduct in* **SLat₁**.

Proof. We have to prove that for any system $h_j \colon L_j \to M$ of homomorphisms there is precisely one homomorphism $h \colon \prod_{i \in J}' L_i \to M$ such that $h\kappa_j = h_j$ for all j.

First, we see that there is at most one such h. For $(x_i)_i \in \prod_{i \in J}' L_i$ let $x_{j_1}, \dots, x_{j_n}$ be all the coordinates that are not 1. Then necessarily

$$h((x_i)_i) = h\left(\bigwedge_{k=1}^{n} (x_{j_k} *_{j_k} \bar{1})\right) = \bigwedge_{k=1}^{n} h_{j_k}(x_{j_k}).$$

Now define

$$h((x_i)_i) = \bigwedge_{i \in J} h_i(x_i) \tag{4.2.1}$$

(this is, in essence, a finite meet since all but finitely many of the $h_i(x_i)$ are 1). Then $h(\kappa_j(x)) = h_j(x)$, $h(1) = 1$, and if $x = (x_i)_i, y = (y_i)_i$ then

$$h(x \wedge y) = h((x_i \wedge y_i)_i) = \bigwedge_{i \in J} h_i(x_i \wedge y_i)$$

$$= \bigwedge_{i \in J} h_i(x_i) \wedge \bigwedge_{i \in J} h_i(y_i) = h(x) \wedge h(y). \qquad \square$$

4.3. Now let L_i, $i \in J$, be frames. On the down-set frame $\mathfrak{D}(\prod'_{i \in J} L_i)$ consider the binary relation R consisting of all the couples

$$\left(\bigcup_{k \in K} \downarrow(x^k *_j u), \downarrow((\bigvee_{k \in K} x^k) *_j u) \right) \tag{R}$$

with $u \in \prod'_{i \in J} L_i$ arbitrary, $j \in J$ and $\{x^k \mid k \in K\} \subseteq L_j$; the index set K can be void, and hence we have, in particular

$$R \ni (\emptyset, \downarrow(0 *_j u)) \quad \text{for all } j \text{ and } u. \tag{4.3.1}$$

This relation R satisfies the conditions of III.11.4 and consequently the saturated elements are the down-sets $U \subseteq \prod'_{i \in J} L_i$ such that

$$\forall j, \ \forall \{x^k \mid k \in K\} \subseteq L_j \text{ and } \forall u \in \prod'_{i \in J} L_i$$
$$\{x^k *_j u \mid k \in K\} \subseteq U \ \Rightarrow \ \left(\bigvee_{k \in K} x^k\right) *_j u \in U.$$

Now set

$$\bigoplus_{i \in J} L_i = \mathfrak{D}\left(\prod'_{i \in J} L_i\right) \Big/ R$$

(thus, $\bigoplus_{i \in J} L_i$ is the frame of all the down-sets $U \subseteq \prod'_{i \in J} L_i$ satisfying the saturation condition above).

Similarly like in III.11.2 we will use the symbol

$$\mu \colon \mathfrak{D}\left(\prod'_{i \in J} L_i\right) \to \bigoplus_{i \in J} L_i$$

for the frame homomorphism given by the nucleus, and

$$\alpha \colon \prod'_{i \in J} L_i \to \mathfrak{D}\left(\prod'_{i \in J} L_i\right)$$

will be the semilattice homomorphism from 2.2. Finally define

$$\iota_j \colon L_j \to \bigoplus_{i \in J} L_i \quad \text{by} \quad \iota_j = \mu \alpha \kappa_j.$$

4.3.1. Lemma. *Each ι_j is a frame homomorphism.*

Proof. μ, α and κ_j preserve finite meets and hence also ι_j does. For the joins we have (see III.11.3.1)

$$\iota_j\left(\bigvee_{k \in K} x^k\right) = \mu(\downarrow \bigvee_{k \in K}(x^k *_j \overline{1})) = \mu(\bigcup_{k \in K} \downarrow(x_k *_j \overline{1})) = \bigvee_{k \in K} \mu((\downarrow(x_k *_j \overline{1})) = \bigvee_{k \in K} \iota_j(x^k)$$

(note that this includes also the case of $K = \emptyset$). $\qquad \square$

4.3.2. Proposition. *The system*

$$\left(\iota_j \colon L_j \to \bigoplus_{i \in J} L_i \right)_{j \in J}$$

is a coproduct of L_i, $i \in J$, *in the category* **Frm**.

Proof. Let $h_j \colon L_j \to M$ be frame homomorphisms. First, view for a moment the L_j as semilattices (forgetting the join structure). By 4.2 we have a semilattice homomorphism

$$f \colon {\prod}'_{i \in J} L_i \to M,$$

defined by $f((x_i)_{i \in J}) = \bigwedge_{i \in J} f_i(x_i)$, such that

$$\forall i, \quad f \cdot \kappa_i = h_i.$$

This in turn gives rise to a frame homomorphism

$$g \colon \mathfrak{D}\left({\prod}'_{i \in J} L_i \right) \to M,$$

defined by setting $g(X) = \bigvee \{ f(x) \mid x \in X \}$, such that $g \cdot \alpha = f$. Now take a $j \in J$, $u \in \prod'_{i \in J} L_i$ and $\{ x^k \mid k \in K \} \subseteq L_j$. We have (recall III.11.3.1 and note that $x *_j u = (x *_j \overline{1}) \wedge (1 *_j u)$):

$$
\begin{aligned}
g({\textstyle\bigcup}{\downarrow}(x^k *_j u)) &= \bigvee_k f(x^k *_j u) = \bigvee_k f((x^k *_j \overline{1}) \wedge (1 *_j u)) \\
&= \bigvee_k f(x^k *_j \overline{1}) \wedge f(1 *_j u) = \bigvee_k (f_j(x^k) \wedge f(1 *_j u)) \\
&= ({\textstyle\bigvee_k} f_j(x^k)) \wedge f(1 *_j u) = f_j({\textstyle\bigvee_k} x^k) \wedge f(1 *_j u) \\
&= f(({\textstyle\bigvee_k} x^k) *_j \overline{1}) \wedge f(1 *_j u) = f(({\textstyle\bigvee_k} x^k) *_j u) = g({\downarrow}({\textstyle\bigvee_k} x_k) *_j u).
\end{aligned}
$$

Thus, g respects the relation R and by III.11.3.1 there is a frame homomorphism $h \colon \bigoplus_{i \in J} L_i \to M$ such that $h\mu = g$. Now we have

$$h \cdot \iota_j = h\mu\alpha\kappa_j = g\alpha\kappa_j = f\kappa_j = h_j.$$

Finally, a down-set $U \subseteq \prod'_{i \in J} L_i$ is a join of elements ${\downarrow}(x_i)_i$ each of which is a finite meet of the ${\downarrow}\kappa_j(x_j)$ with $x_j \neq 1$. Thus, $\mathfrak{D}(\prod'_{i \in J} L_i)$ is generated (by finite meets and joins) by the elements $\alpha\kappa_j(x)$, $j \in J$, $x \in L_j$, and hence $\bigoplus_i L_i$ is generated by the $\iota_i(x)$. Hence, a frame homomorphism h such that $h\iota_i = h_i$ for all i is uniquely determined. $\qquad \square$

4.3.3. Notation. For finite systems of frames we often use the symbols

$$L \oplus M, \; L_1 \oplus L_2 \oplus L_3, \; L_1 \oplus \cdots \oplus L_n, \; \text{etc.}.$$

4.3.4. Remark. Look at a coproduct $L_1 \oplus L_2$ of two frames. Note the striking similarity with the construction of tensor products of Abelian groups. First, one takes the cartesian product

$$A_1 \times A_2$$

that happens to be both product and coproduct (similarly like the $L_1 \times L_2$ in **SLat$_1$**). Then one takes the free Abelian group $\mathcal{F}(A_1 \times A_2)$ (similarly, we have taken the $\mathfrak{D}(L_1 \times L_2)$), and finally factorizes by the relations

$$(a_1 + a_2, b) \sim (a_1, b) + (a_2, b), \quad (a, b_1 + b_2) \sim (a, b_1) + (a, b_2)$$

(similarly like we did,

$$\downarrow(a_1 \vee a_2, b) \sim \downarrow(a_1, b) \cup \downarrow(a_2, b), \quad \downarrow(a, b_1 \vee b_2) \sim \downarrow(a, b_1) \cup \downarrow(a, b_2)).$$

This similarity is not accidental. The reader interested in categories of algebras will probably know that this is how the coproduct in the category of commutative rings with unit is constructed: the tensor product of the underlying Abelian groups is only endowed with a suitable multiplication. Here we have the join $\bigvee$ as the "additive part of the structure", and the meet is the multiplication (distributing over it). For the general pattern see [38].

4.3.5. Combining 4.3.2 with the results of Section 3 we obtain (consult also AII.5.3.2)

Corollary. *Categories* **Frm** *and* **Loc** *are complete and cocomplete.* $\qquad\square$

5. More on the structure of coproduct

5.1. Because of the possible void K in the definition of the relation R,

every u such that there is an i with $u_i = 0$ is in every saturated U.

Set

$$\mathbf{n} = \left\{ u \in {\textstyle\prod}'_{i \in J} L_i \mid \exists i, \ u_i = 0 \right\}.$$

Since obviously $\mathbf{n}$ is itself saturated we have

5.1.1. Observation. $\mathbf{n}$ *is the smallest saturated set, hence, the bottom of the coproduct.*

5.1.2. Proposition. *For every $a \in \prod'_{i \in J} L_i$ the set*

$$\downarrow a \cup \mathbf{n}$$

is the $\mu(\downarrow A)$, the least saturated U containing $\downarrow a$.

Proof. By 5.1.1, $\mu(\downarrow a) \supseteq \downarrow a \cup \mathbf{n}$. Thus, we have to prove that $\downarrow a \cup \mathbf{n}$ is itself saturated.

Let $x^k *_j u \in \downarrow a \cup \mathbf{n}$ for all $k \in K$. Then either $u_i = 0$ for some $i \neq j$, and $\bigvee_k x^k *_j u \in \mathbf{n}$. Or, $u_i \neq 0$ for all $i \neq j$. Then either $\bigvee_k x^k = 0$ and again $\bigvee_k x^k *_j u \in \mathbf{n}$, or $x^t \neq 0$ for some t, and hence $x^t *_j u \in \downarrow a$, consequently $u_i \leq a_i$ for all $i \neq j$, and $x^k \leq a_j$ for all k (some of the x^k can be 0, of course). Then $\bigvee_k x^k *_j u \leq a$. $\square$

The element $\downarrow(a_i)_i \cup \mathbf{n}$ will be denoted by

$$\oplus_i a_i,$$

in case of small systems we write

$$a_1 \oplus a_2, \quad a \oplus b, \quad a_1 \oplus a_2 \oplus a_3, \quad \text{etc.}$$

5.2. The following easy facts are very useful.

Proposition.

(1) *For each* $U \in \bigoplus_i L_i$,

$$U = \bigcup\{\oplus_i a_i \mid \oplus_i a_i \leq U\} = \bigvee\{\oplus_i a_i \mid \oplus_i a_i \leq U\}.$$

(2) $\oplus_i a_i = \bigwedge_{i \in J} \iota_i(a_i)$.

(3) *In the case of coproducts of two frames we have* $\bigvee_{i \in J}(a_i \oplus b) = (\bigvee_{i \in J} a_i) \oplus b$ *and* $\bigvee_{i \in J}(a \oplus b_i) = a \oplus (\bigvee_{i \in J} b_i)$; *consequently,*

$$\left(\bigvee_{i \in J} a_i\right) \oplus \left(\bigvee_{i \in J} b_i\right) = \bigvee_{i,j \in J} (a_i \oplus b_j).$$

(4) *If* $\oplus_i a_i \neq \mathbf{n}$ *(which is the same as saying that* $a = (a_i)_i \notin \mathbf{n}$*) then*

$$\oplus_i a_i \leq \oplus_i b_i \quad \Rightarrow \quad \forall i, \ a_i \leq b_i.$$

(5) *The codiagonal homomorphism* $\nabla: {}^J L \to L$ *(that is,* $\nabla: {}^J L = \bigoplus_{i \in J} L_i \to L$ *with* $L_i = L$ *and* $\nabla \cdot \iota_i = \mathrm{id}_L$ *for all* i*) is given by the formula*

$$\nabla\left(\bigoplus_{i \in J} a_i\right) = \bigwedge_{i \in J} a_i.$$

Proof. (1): U is a down-set and hence $U = \bigcup\{\downarrow a \mid a \in U\} = \bigcup\{\downarrow a \mid \downarrow a \subseteq U\}$. Since it is saturated, $\downarrow a \subseteq U$ implies $\mu(\downarrow a) = \oplus_i a_i \leq U$. Thus,

$$U \leq \bigcup\{\oplus_i a_i \mid \oplus_i a_i \leq U\} \leq U$$

and since the union is, hence, saturated, it is the join.

(2): For a better transparency, let $J(a)$ be the finite set of indices $i \in J$ for which $a_i \neq 1$. We have $a = \bigwedge_{i \in J(a)} \kappa_i(a_i)$ and hence (as α and μ preserve finite meets)

$$\bigoplus_{i \in J(a)} a_i = \mu(\downarrow a) = \mu\alpha\left(\bigwedge_{i \in J(a)} \kappa_i(a_i)\right) = \bigwedge_{i \in J(a)} \mu\alpha\kappa_i(a_i) = \bigwedge_{i \in J(a)} \iota_i(a_i).$$

(3) follows immediately from (2), using the fact that the mappings ι_i are frame homomorphisms.

(4): If $\oplus_i a_i \leq \oplus_i b_i$ then in particular $a = (a_i)_i \in {\downarrow}b \cup \mathbf{n}$. Since $a \notin \mathbf{n}$, we have $a \leq b$.

(5) follows immediately from (2). $\qquad\qquad\qquad\qquad\qquad\qquad\qquad\qquad\qquad\quad$ $\square$

5.3. Also, we will find expedient to have an explicit formula for the diagonal $\Delta\colon L \to L^J$ in **Loc** (of course, this L^J is the $^J L$ from **Frm**). Here it is:

$$\Delta(a) = \{(u_i)_i \mid \bigwedge u_i \leq a\}. \tag{Δ}$$

Thus, in case of the diagonal $\Delta\colon L \to L \oplus L$ we have

$$\Delta(a) = \{(x,y) \mid x \wedge y \leq a\}. \tag{$\Delta 2$}$$

It is an easy exercise to prove that each such $\Delta(a)$ is saturated, and we have

$$\nabla(U) \leq a \quad \text{iff} \quad \forall u \in U,\ \nabla(u) = \bigwedge u_i \leq a$$
$$\text{iff} \quad \forall u \in U,\ u \in \Delta(a) \quad \text{iff} \quad U \leq \Delta(a).$$

5.4. The product of locales compared with the product of spaces. Recall the spectrum adjunction (Chapter III). Since Sp is a right adjoint it preserves limits, and hence in particular products.

The reader is, however, probably more curious about the relation of the just constructed (co)product with the product topology in spaces, that is, about the relation between $\Omega(\prod_{i \in J} X_i)$ and $\bigoplus_{i \in J} \Omega(X_i)$. There is the canonical frame homomorphism

$$\pi\colon \bigoplus_{i \in J} \Omega(X_i) \to \Omega\left(\prod_{i \in J} X_i\right) \quad \text{determined by} \quad \pi\iota_i = \Omega(p_i)$$

where $p_j\colon \prod_{i \in J} X_i \to X_j$ are the projections. It is not necessarily an isomorphism, but it is always onto and dense, and it is an isomorphism in some important special cases – see 5.4.1 and 5.4.2 below.

First, however, let us look at what happens in a product $X \times Y$ of two spaces. The open sets in the product are unions of the oblongs $U \times V$ with U open in X and V open in Y. Such a $U \times V$ is represented by the element $U \oplus V \in \Omega(X) \oplus \Omega(Y)$ and a system of oblongs $U_i \times V_i$, $i \in J$, is equivalent to $U'_i \times V'_i$, $i \in J'$, if

$$\bigcup_{i \in J} (U_i \times V_i) = \bigcup_{i \in J'} (U'_i \times V'_i) \tag{5.4.1}$$

which one would wish to be represented by the equality

$$\bigvee\{U_i \oplus V_i \mid i \in J\} = \bigvee\{U'_i \oplus V'_i \mid i \in J'\}$$

in $\Omega(X) \oplus \Omega(Y)$. The formula (5.4.1) is based on points and one can hardly expect it to be represented adequately. The relation (identification) R in the construction of $\Omega(X) \oplus \Omega(Y)$ corresponds to the very special cases of (5.4.1) concerning just the system of oblongs $U_i \times V$, $i \in J$, equivalent to $U_i' \times V'$, $i \in J'$, and $U \times V_i$, $i \in J$ equivalent to $U' \times V_i'$, $i \in J'$. It is rather surprising that in fact in some important cases like for instance in the finite products of locally compact spaces, or even the general products of regular compact spaces such a restricted identification after all really produces the desired (see VII.6.5).

5.4.1. Proposition. *The homomorphism π is always onto and dense.*

Proof. We have, for an $(x_i)_i \in \prod_{i \in J}' L_i$ – that is a system of $x_i \in L_i$ such that $x_i = 1$ for $i \notin K$, $K \subseteq J$, finite,

$$\pi(\oplus_i x_i) = \bigcap_{i \in K} p_i^{-1}(x_i).$$

Since the sets $\bigcap_{i \in K} p_i^{-1}(x_i)$ generate $\Omega(\prod_{i \in J} X_i)$ by unions (joins), π is onto.

Now if $u \neq 0$ is in $\bigoplus_{i \in J} \Omega(X_i)$ we have an $(x_i)_i \in \prod_{i \in J}' L_i$ as above, moreover with all the x_i with $i \in K$ non-empty such that $\oplus_i x_i \leq u$. Thus,

$$\emptyset \neq \bigcap_{i \in K} p_i^{-1}(x_i) \subseteq \pi(u). \qquad \square$$

5.4.2. Proposition. *If X_i are sober spaces and if $\bigoplus_i \Omega(X_i)$ is spatial then π is an isomorphism. Thus, in such a case the product $\prod_i X_i$ is being sent to the coproduct of $\Omega(X_i)$ (product of $\mathsf{Lc}(X_i)$ in **Loc**).*

Proof. We have an isomorphism

$$\phi \colon \bigoplus_i \Omega(X_i) \to \Omega(Y);$$

choose Y sober. By I.1.6.2, $\phi \cdot \iota_i = \Omega(q_i)$ for (uniquely determined) continuous $q_i \colon Y \to X_i$. If $f_i \colon \mathbf{2} \to X_i$ are continuous maps consider the homomorphisms $\Omega(f_i) \colon \Omega(X_i) \to \Omega(\mathbf{2})$, and the resulting

$$h \colon \bigoplus_i \Omega(X_i) \to \Omega(\mathbf{2})$$

such that $h \cdot \iota_i = \Omega(f_i)$. Now we have $h \cdot \phi^{-1} \colon \Omega(Y) \to \Omega(\mathbf{2})$, and a uniquely defined $f \colon \mathbf{2} \to Y$ such that $\Omega(f) = h \cdot \phi^{-1}$. Thus,

$$\Omega(q_i f) = f_i = h\phi^{-1}\phi\iota_i = h\iota_i = \Omega(f_i)$$

and hence $q_i f = f_i$. The unicity of such a map f is easily checked and hence we have that $(q_i \colon Y \to X_i)_{i \in J}$ constitutes a product in **Top**. Hence there is an isomorphism $g \colon X \to Y$ such that $q_i g = p_i$. Then

$$\Omega(g) \cdot \phi \cdot \iota_i = \Omega(q_i \cdot g) = \Omega(p_i) = \pi \cdot \iota_i$$

and hence $\pi = \Omega(g) \cdot \phi$ is an isomorphism. $\qquad \square$

5.5. One may wish for a more explicit description of the product projections

$$p_j \colon \bigoplus_{i \in J} L_i \to L_j$$

in **Loc**. We cannot do much. Of course we have, using the adjunction, the formula $p_j(U) = \bigvee\{x \mid \iota_j(x) \subseteq U\}$ and since the set U is saturated we have $\mu\alpha\kappa_j(x) \subseteq U$ iff $\downarrow\kappa_j(x) \subseteq U$ iff $x *_j \overline{1} \in U$ so that we obtain

$$p_j(U) = \bigvee\{x \mid x *_j \overline{1} \in U\}.$$

Here is a formula that is sometimes of help.

5.5.1. Proposition. *Let $f_i \colon M \to L_i$, $i = 1, 2$, be localic maps. Then the associated $f \colon M \to L_1 \oplus L_2$ is given by*

$$f(a) = \{(x, y) \mid y \le f_2(f_1^*(x) \to a)\} = \{(x, y) \mid x \le f_1(f_2^*(y) \to a)\}.$$

Proof. By the formula (for g and hence for f^*) in the proof of 4.3.2 we have

$$f^*(U) = \bigvee\{f_1^*(x) \wedge f_2^*(y) \mid (x, y) \in U\}.$$

Let f be as above. If $f^*(U) \le a$ and $(x, y) \in U$ then $f_1^*(x) \wedge f_2^*(y) \le a$, hence $f_1^*(x) \le f_2^*(y) \to a$ and hence $x \le f_1(f_2^*(y) \to a)$ and $(x, y) \in f(a)$. On the other hand, let $U \subseteq f(a)$ and $(x, y) \in U$. Then $x \le f_1(f_2^*(y) \to a)$, hence $f_1^*(x) \le f_2^*(y) \to a$ and $f_1^*(x) \wedge f_2^*(y) \le a$. Thus, $f^*(U) \subseteq a$.   □

5.5.2. Corollary. *In particular, the diagonal morphism $\Delta \colon L \to L \oplus L$ is given by the formula*

$$\Delta(a) = \{(x, y) \mid x \wedge y \le a\}.$$ □

5.5.3. Corollary. *For any localic maps $f_i \colon M_i \to L_i$ $(i = 1, 2)$, the unique $f_1 \oplus f_2$ that makes the diagram*

$$
\begin{array}{ccccc}
M_1 & \xleftarrow{\;p_{M_1}\;} & M_1 \oplus M_2 & \xrightarrow{\;p_{M_2}\;} & M_2 \\
{\scriptstyle f_1}\downarrow & & {\scriptstyle f_1 \oplus f_2}\downarrow & & \downarrow{\scriptstyle f_2} \\
L_1 & \xleftarrow[\;p_{L_1}\;]{} & L_1 \oplus L_2 & \xrightarrow[\;p_{L_2}\;]{} & L_2
\end{array}
$$

commutative is given by the formula $(f_1 \oplus f_2)(U) = \bigvee\{f_1(a) \oplus f_2(b) \mid a \oplus b \le U\}$.

Proof. By 5.5.1

$$
\begin{aligned}
(f_1 \oplus f_2)(U) &= \{(a, b) \mid b \le f_2 \, p_{M_2}((f_1 \, p_{M_1})^*(a) \to U)\} \\
&= \{(a, b) \mid (f_1 \, p_{M_1})^*(a) \cap (f_2 \, p_{M_2})^*(b) \le U\} \\
&= \{(a, b) \mid (f_1^*(a) \oplus 1) \cap (1 \oplus f_2^*(b)) \le U\} \\
&= \{(a, b) \mid (f_1^*(a) \oplus f_2^*(b)) \le U\}.
\end{aligned}
$$

The latter set, being saturated, coincides with $\bigvee\{a \oplus b \mid f_1^*(a) \oplus f_2^*(b) \le U\}$, which is easily seen to be equal to $\bigvee\{f_1(a) \oplus f_2(b) \mid a \oplus b \le U\}$ by the adjunction $f^* \dashv f$. $\qquad\square$

5.6. Coproduct of two frames obtained using a prenucleus. Recall the frame quotient

$$\mu \colon \mathfrak{D}(L_1 \times L_2) \to L_1 \oplus L_2$$

from 4.3 above. By abuse of notation we will also use the symbol μ for the corresponding nucleus $\mathfrak{D}(L_1 \times L_2) \to \mathfrak{D}(L_1 \times L_2)$. Now (recall III.11.5) we immediately see that our μ is generated by the prenucleus $\nu = \pi_2\pi_1$ where

$$\pi_1(U) = \{(\bigvee A, b) \mid A \times \{b\} \subseteq U\},$$

$$\pi_2(U) = \{(a, \bigvee B) \mid \{a\} \times B \subseteq U\}.$$

The following lemma will come handy

Lemma. *Let* $(a_1, b), \dots, (a_n, b) \in \nu(U) = \pi_2\pi_1(U)$. *Then* $(\bigvee_{i=1}^{n} a_i, b) \in \nu(U)$.

Proof. We have $B_i \subseteq L_2$ such that $\{a_i\} \times B_i \subseteq \pi_1(U)$ and $b = \bigvee_{i=1}^{n} B_i$. Since $\pi_1(U)$ is a down-set, $B = B_1 \wedge \dots \wedge B_n$ has, now independently of i, the property that $\{a_i\} \times B \subseteq \pi_1(U)$ and $b = \bigvee B$. For each $c \in B$ we have $A_i(c)$ such that $\bigvee A_i(c) = a_i$, and $A_i(c) \times \{c\} \subseteq U$.

Choose an arbitrary $y \in B$ and $i \in \{1, \dots, n\}$. Then for an $x \in A_i(y)$, $x \le a_i$, and since $(a_i, c) \in \pi_1(U)$, which is a down-set, $(x, c) \in \pi_1(U)$ and hence

$$\bigcup_y A_i(y) \times \{c\} \subseteq \pi_1(U)$$

and since $\bigvee_{i=1}^{n} (\bigcup_y A_i(y)) = a$, $(a, c) \in \pi_1\pi_1(U) = \pi_1(U)$. Thus, $\{a\} \times B \subseteq \pi_1(U)$ and finally $(a, b) \in \pi_2\pi_1(U)$. $\qquad\square$

6. Epimorphisms in Frm

6.1. We have seen in Chapter III that monomorphisms in **Frm** are easy to describe (one-one frame homomorphisms). On the contrary, epimorphisms constitute a rather wild class, as we will now see. We start with an example that shows that not every epimorphism in **Frm** is surjective.

6.1.1. Example. Let X be a T_1-space. The insertion

$$e \colon \Omega(X) \hookrightarrow \mathfrak{P}(X)$$

is certainly monic, and is surjective only when X is discrete. However e is an epimorphism as we now prove. Since X is T_1, $F_x = \{x\}$ is closed for every $x \in X$.

Let $U_x = X \smallsetminus F_x \in \Omega(X)$. Note that for any morphism $f \colon \mathfrak{P}(X) \to L$ into an arbitrary frame L, the value $f(F_x)$ is uniquely determined by $f(U_x)$ (since $f(U_x)$ and $f(F_x)$ are complements in L). Consider now two morphisms $f, g \colon \mathfrak{P}(X) \to L$ such that $fe = ge$. Of course $f(U_x) = g(U_x)$ for every $x \in X$, and hence $f(F_x) = g(F_x)$. Consider any $S \in \mathfrak{P}(X)$. We have $S = \bigcup \{F_x \mid x \in S\}$. Thus $f(S) = \bigvee \{f(F_x) \mid x \in S\} = \bigvee \{g(F_x) \mid x \in S\} = g(S)$, which shows that $f = g$.

6.2. Recall the one-one frame homomorphism

$$\nabla_L = (a \mapsto \nabla_a) \colon L \to \mathfrak{C}(L)$$

from III.6.4.1. From Corollary III.6.5.1 it follows that

6.2.1. Proposition. *The embedding $\nabla_L \colon L \to \mathfrak{C}(L)$ is an epimorphism in* **Frm**.

Proof. Indeed if $f, g \colon \mathfrak{C}(L) \to M$ coincide in the ∇_a's, they coincide, by complementation, also in the Δ_a's, and hence by Corollary III.6.5.1 in every $E \in \mathfrak{C}(L)$. $\qquad\square$

Therefore ∇_L is a *bimorphism* (i.e., both mono and epic). However

6.2.2. Proposition. $\nabla_L \colon L \to \mathfrak{C}(L)$ *is an isomorphism if and only if L is a Boolean algebra.*

Proof. Suppose first that ∇_L is an isomorphism. Then every $E \in \mathfrak{C}(L)$ is closed. In particular, for every $a \in L$, $\Delta_a = \nabla_b$ for some $b \in L$. But then $\nabla_a \vee \nabla_b = 1$ and $\nabla_a \wedge \nabla_b = 0$, that is, $a \vee b = 1$ and $a \wedge b = 0$, which shows that b is the complement of a.

Conversely, assume that L is Boolean. Then each $E \in \mathfrak{C}(L)$ is equal to ∇_a for $a = \bigvee \{x \in L \mid (x, 0) \in E\}$. Indeed: for every $x, y \in L$, $(x, x \vee a)$ and $(y \vee a, y)$ are in E and, consequently, if $(x, y) \in \nabla_a$ then $(x, y) \in E$; conversely, if $(x, y) \in E$ then $(x \vee a, y) \in E$ and $((x \vee a) \wedge y^*, 0) \in E$; so $(x \vee a) \wedge y^* \leq a$, that is, $x \vee a \leq y \vee a$.

This proves that ∇_L is surjective. Being always an embedding, it is then an isomorphism. $\qquad\square$

This shows that the category **Frm** is not balanced. It has bimorphisms that are not isomorphisms. In fact, as we will see in the sequel, the situation is even more complicated.

6.3. The functor $\mathfrak{C}$. The next result is crucial in establishing functorial properties of $\mathfrak{C}(L)$.

6.3.1. Proposition. *Let $f \colon L \to M$ be a frame homomorphism such that each member of the image of L under f is complemented in M. There exists a unique map $\overline{f} \colon \mathfrak{C}(L) \to M$ such that $\overline{f} \cdot \nabla_L = f$.*

Proof. Uniqueness follows from the fact that ∇_L is an epimorphism, so it suffices to prove the existence of such $\overline{f}$. For each $E \in \mathfrak{C}(L)$ set

$$\overline{f}(E) = \bigvee\{f(a)^* \wedge f(b) \mid (a,b) \in E, a \leq b\}$$

where $f(a)^*$ is the given complement of $f(a)$ in M. We must check that $\overline{f}$ is indeed a frame homomorphism. From the definition of $\overline{f}$ and the frame distributive law it immediately follows that $\overline{f}$ is order-preserving and preserves finite meets. Turning to arbitrary joins, it suffices to show that $\overline{f}$ has a right (Galois) adjoint $f_\#$. For $m \in M$, define

$$f_\#(m) = \{(x,y) \mid f(x) \vee m = f(y) \vee m\}.$$

This is easily seen to be a congruence on L and $m_1 \leq m_2$ implies $f_\#(m_1) \subseteq f_\#(m_2)$. Moreover, $\overline{f}(C) \leq m$ iff $C \subseteq f_\#(m)$. Indeed, suppose $a \leq b$ and $(a,b) \in C$. Then

$$f(a)^* \wedge f(b) \leq m \text{ iff } f(b) \leq m \vee f(a) \text{ iff } m \vee f(b) = m \vee f(a)$$

(since $f(a) \leq f(b)$), as required.

Finally, we check that $\overline{f} \cdot \nabla_L = f$. Let $c \in L$. Then

$$\overline{f}(\nabla_L(c)) = \overline{f}(\nabla_c) = \bigvee\{f(a)^* \wedge f(b) \mid a \vee c = b \vee c, a \leq b\}.$$

Since $0 \vee c = c \vee c$, $\overline{f}(\nabla_c) \geq f(0)^* \wedge f(c) = f(c)$. On the other hand, suppose $a \vee c = b \vee c$ and $a \leq b$. Then

$$(f(a)^* \wedge f(b)) \vee f(c) = (f(a)^* \vee f(c)) \wedge f(b \vee c)$$
$$= (f(a)^* \vee f(c)) \wedge f(a \vee c) = (f(a)^* \wedge f(a)) \vee f(c) = f(c),$$

thus $\overline{f}(\nabla_c) \leq f(0)^* \wedge f(c) = f(c)$. □

6.3.2. Corollary. *The assignment $L \mapsto \mathfrak{C}(L)$ is the object part of a functor*

$$\mathfrak{C} \colon \mathbf{Frm} \to \mathbf{Frm}.$$

Moreover, $\nabla_L \colon L \to \mathfrak{C}(L)$ defines a natural transformation $\nabla \colon \mathrm{Id} \overset{\cdot}{\to} \mathfrak{C}$.

Proof. Let $f \colon L \to M$ be a frame homomorphism. Then $\nabla_M \cdot f \colon L \to \mathfrak{C}(M)$ is also a frame homomorphism in the conditions of the proposition. It suffices then to define $\mathfrak{C}(f) \colon \mathfrak{C}(L) \to \mathfrak{C}(M)$ as the unique map, given by the proposition, for which the following diagram commutes.

$$
\begin{array}{ccc}
L & \overset{\nabla_L}{\longrightarrow} & \mathfrak{C}L \\
{\scriptstyle f}\downarrow & & \downarrow{\scriptstyle \nabla_M \cdot f} \\
M & \underset{\nabla_M}{\longrightarrow} & \mathfrak{C}M
\end{array}
$$

□

6.4. The congruence tower. Given a frame L, define an ordinal sequence of frames $\mathfrak{C}^\alpha(L)$ and a doubly-indexed ordinal sequence of frame homomorphisms $\nabla_L^{\beta,\alpha} \colon \mathfrak{C}^\beta(L) \to \mathfrak{C}^\alpha(L)$ $(\beta \leq \alpha)$ as follows:

$$\mathfrak{C}^0(L) = L, \quad \nabla_L^{0,0} = \mathrm{id}_L,$$

$$\mathfrak{C}^{\alpha+1}(L) = \mathfrak{C}(\mathfrak{C}^\alpha(L)), \quad \nabla_L^{\beta,\alpha+1} = \nabla_L \cdot \nabla_L^{\beta,\alpha},$$

and for a limit ordinal λ,

$$\mathfrak{C}^\lambda(L) = \mathrm{colim}_{\alpha<\lambda}\mathfrak{C}^\alpha(L), \quad \nabla_L^{\beta,\lambda} = q_\beta^\lambda \colon \mathfrak{C}^\beta(L) \to \mathrm{colim}_{\alpha<\lambda}\mathfrak{C}^\alpha(L)$$

where $(q_\alpha^\lambda \colon \mathfrak{C}^\alpha(L) \to \mathrm{colim}_{\alpha<\lambda}\mathfrak{C}^\alpha(L))_{\alpha<\lambda}$ is the colimit of the diagram

$$D_\lambda \colon \{\alpha \mid \alpha < \lambda\} \to \mathbf{Frm}$$

defined by $D_\lambda(\alpha) = \mathfrak{C}^\alpha(L)$ for $\alpha < \lambda$ and $D_\lambda(\beta \to \alpha) = \nabla_L^{\beta,\alpha}$ for $\beta \leq \alpha < \lambda$ (see AII.5.1). Note that, by induction, $\nabla_L^{\gamma,\alpha} = \nabla_L^{\beta,\alpha} \cdot \nabla_L^{\gamma,\beta}$ for all $\gamma \leq \beta \leq \alpha$.

Now for a frame homomorphism $f \colon L \to M$ define

$$\mathfrak{C}^\alpha(f) \colon \mathfrak{C}^\alpha(L) \to \mathfrak{C}^\alpha(M)$$

by recursion, as follows: $\mathfrak{C}^0(f) = f$, $\mathfrak{C}^{\alpha+1}(f) = \mathfrak{C}(\mathfrak{C}^\alpha(f))$; for a limit ordinal λ, since $\{\nabla_M^{\alpha,\lambda} \cdot \mathfrak{C}^\alpha(f) \mid \alpha < \lambda\}$ is an upper bound of D_λ, there is a (unique) q_λ such that $q_\lambda \nabla_L^{\alpha,\lambda} = \nabla_M^{\alpha,\lambda} \cdot \mathfrak{C}^\alpha(f)$ for all $\alpha < \lambda$, which is taken as the definition for $\mathfrak{C}^\lambda(f)$. It is a straightforward exercise to see (by induction, using the universal property of the colimit at limit stages) that each $\mathfrak{C}^\alpha$ becomes a functor $\mathbf{Frm} \to \mathbf{Frm}$. One can again show by induction that if $\beta \leq \alpha$ then

$$\mathfrak{C}^\alpha(f) \cdot \nabla_L^{\beta,\alpha} = \nabla_M^{\beta,\alpha} \cdot \mathfrak{C}^\beta(f),$$

and so $(\nabla_L^{\beta,\alpha})_{L\in\mathbf{Frm}}$ is a natural transformation $\mathfrak{C}^\beta \dot{\to} \mathfrak{C}^\alpha$.

6.4.1. Note. $\nabla_{\mathfrak{C}^\gamma L}^{\beta,\alpha} = \nabla_L^{\gamma+\beta,\gamma+\alpha}$ and $\mathfrak{C}^\gamma(\nabla_L^{\beta,\alpha}) = \nabla_L^{\gamma+\beta,\gamma+\alpha}$.

6.5. Directed colimits of frames. Let $D \colon K \to \mathbf{Frm}$ be a diagram of frames indexed by a (up-)directed poset K (viewed as a category, recall AII.1.1.1). The following lemma, that we shall need in 6.6.1, is the particular case of a more general result about filtered colimits of frames ([139], [270]). It asserts that the colimit of such a diagram D is computed by taking the directed limit of the underlying sets and right adjoint maps.

6.5.1. Lemma. *Let $D(\alpha)$ be a monomorphism for every morphism α of K. The colimit of D is the frame L of all $\sigma \in \prod_{a\in\mathrm{obj}K} D(a)$ such that for every morphism $\alpha \colon a \to b$ of K, $D(\alpha)_*(\sigma(b)) = \sigma(a)$, ordered pointwise. For each $a \in \mathrm{obj}K$ the canonical injection $D(a) \to L$ is the left adjoint to the projection $g_a \colon L \to D(a)$ and it is a monomorphism as well.*

Proof. For each $\alpha\colon a \to b$ in K let $f_{ab} = D(\alpha)$ and let $g_{ab}\colon D(b) \to D(a)$ denote its right adjoint. Since each g_{ab} preserves arbitrary meets, it follows that L is a complete lattice with meets defined pointwise. In order to prove that it is a frame it suffices to show that it has a Heyting operation. For each $\sigma, \gamma \in L$ and $b \in \mathrm{obj}K$ define

$$\delta_b = \bigwedge_{a \geq b} g_{ba}(g_a(\sigma) \to g_a(\gamma)). \tag{6.5.1}$$

Since for each $a \geq c \geq b$,

$$g_{ba}(g_a(\sigma) \to g_a(\gamma)) = g_{bc}g_{ca}(g_a(\sigma) \to g_a(\gamma))$$
$$\leq g_{bc}(g_{ca}g_a(\sigma) \to g_{ca}g_a(\gamma)) \leq g_{bc}(g_c(\sigma) \to g_c(\gamma)),$$

the meet in (6.5.1) may be taken over any cofinal subset of K all of whose elements are greater than b. Therefore, for $d \geq b$ we may write

$$g_{bd}(\delta_d) = g_{bd}(\bigwedge_{a \geq d} g_{da}(g_a(\sigma) \to g_a(\gamma))) = \bigwedge_{a \geq d} g_{bd}g_{da}(g_a(\sigma) \to g_a(\gamma))$$
$$= \bigwedge_{a \geq d} g_{ba}(g_a(\sigma) \to g_a(\gamma)) = \bigwedge_{a \geq b} g_{bd}g_{da}(g_a(\sigma) \to g_a(\gamma)) = \delta_b$$

which shows that $\delta = (\delta_b)_{b \in \mathrm{obj}K}$ is an element of L. Moreover

$$g_b(\delta \wedge \sigma) = \delta_b \wedge g_b(\sigma) = g_b(\sigma) \wedge \bigwedge_{a \geq b} g_{ba}(g_a(\sigma) \to g_a(\gamma))$$
$$\leq g_b(\sigma) \wedge (g_b(\sigma) \to g_b(\gamma)) \leq g_b(\sigma)$$

and thus $\sigma \wedge \delta \leq \gamma$. We now show that $\delta = \sigma \to \gamma$. Let $\lambda \wedge \sigma \leq \gamma$ in L. Then $g_a(\lambda) \wedge g_a(\sigma) \leq g_a(\gamma)$, that is, $g_a(\lambda) \leq g_a(\sigma) \to g_a(\gamma)$, and if $a \geq b$, $g_b(\lambda) = g_{ba}g_a(\lambda) \leq g_{ba}(g_a(\sigma) \to g_a(\gamma))$. Hence $g_b(\lambda) \leq \bigwedge_{a \geq b} g_{ba}(g_a(\sigma) \to g_a(\gamma)) = g_b(\delta)$ and so $\lambda \leq \delta$ and $\delta = \sigma \to \gamma$ in L. Thus L is a frame.

Since the frame homomorphisms f_{ba} satisfy $g_{ab}f_{ab} = \mathrm{id}_{D(a)}$, we have

$$g_{cb}((f_a(x))_b) = g_{cb}f_{ab}(x) = g_{cb}f_{cb}f_{ac}(x) = f_{ac}(x) = (f_a(x))_c$$

for any $b \geq c \geq a$ and we may define $f_a\colon D(a) \to L$ by $(f_a(x))_b = f_{ab}(x)$. Clearly f_a is the left adjoint to g_a (since $g_af_a(x) = x$ and if $b \geq a$, $(g_af_a(y))_b = f_{ab}g_a(y) = f_{ab}g_{ab}g_b(y) \leq g_b(y)$ and so $f_ag_a(y) \leq y$) and it is one-one. Moreover f_a preserves finite meets (and so it is a frame map and g_a is a localic map):

$$(f_a(x \wedge z))_b = f_{ab}(x \wedge z) = f_{ab}(x) \wedge f_{ab}(z) = (f_a(x))_b \wedge (f_a(z))_b.$$

Finally, if $r_a\colon D(a) \to M$ are frame homomorphisms such that $r_bf_{ab} = r_a$ let $r\colon L \to M$ be given by $r(\sigma) = \bigvee_{b \in \mathrm{obj}K} r_b(\sigma(b))$. It clearly preserves arbitrary joins and satisfies $rf_a = r_a$ for every a. So,

$$r(\sigma \wedge \gamma) = \bigvee_{a \in \mathrm{obj}K} r_a(\sigma(a) \wedge \gamma(a)) = \bigvee_{a \in \mathrm{obj}K} r_a(\sigma(a)) \wedge r_a(\gamma(a))$$

and, on the other hand,

$$r(\sigma) \wedge r(\gamma) = \left(\bigvee_{a \in \mathrm{obj}K} r_a(\sigma(a)) \right) \wedge \left(\bigvee_{b \in \mathrm{obj}K} r_b(\gamma(b)) \right) = \bigvee_{a,b \in \mathrm{obj}K} r_a(\sigma(a)) \wedge r_b(\gamma(b)).$$

Since K is up-directed and for $a \leq c$, $r_a(\sigma(a)) = r_c f_{ac}(\sigma(a)) \leq r_c(\sigma(c))$ (because $\sigma(a) \leq \sigma(a) = g_{ac}(\sigma(c))$ and thus $f_{ab}(\sigma(a)) \leq \sigma(c)$), then $r(\sigma \wedge \gamma) = r(\sigma) \wedge r(\gamma)$ and r is a frame homomorphism. $\square$

6.6. Weird epimorphisms

6.6.1. Lemma. (a) *For all ordinals α, β with $\beta \leq \alpha$, the morphism $\nabla_L^{\beta,\alpha}$ is both a monomorphism and an epimorphism in* **Frm**.

(b) *For every frame homomorphism $f \colon L \to M$ with M Boolean, and every ordinal α, there exists a unique frame homomorphism $\overline{f} \colon \mathfrak{C}^\alpha L \to M$ such that $f = \overline{f} \cdot \nabla_L^{0,\alpha}$.*

Proof. (a) By the first equation of 6.4.1 we can assume that $\beta = 0$. For each L, ∇_L is mono and epi, and compositions of monos or epis are likewise mono or epi. Thus it suffices to show (by induction) that $\nabla_L^{0,\lambda} = q_0^\lambda \colon \mathfrak{C}^0(L) = L \to \mathrm{colim}_{\alpha < \lambda} \mathfrak{C}^\alpha(L)$ is also mono and epi, for any limit ordinal λ.

That q_0^λ is epi follows from the universal property of the colimit (AII.5.1): if $f, g \colon \mathrm{colim}_{\alpha < \lambda} \mathfrak{C}^\alpha(L) \to M$ are frame homomorphisms such that $f \cdot q_0^\lambda = g \cdot q_0^\lambda$, then for every $\alpha < \lambda$, we have

$$f \cdot q_\alpha^\lambda \cdot D_\lambda(0 \to \alpha) = f \cdot q_0^\lambda = g \cdot q_0^\lambda = g \cdot q_\alpha^\lambda \cdot D_\lambda(0 \to \alpha),$$

and therefore $f \cdot q_\alpha^\lambda = g \cdot q_\alpha^\lambda$ (since $D_\lambda(0 \to \alpha) = \nabla_L^{0,\alpha}$ is epi); consequently $(f \cdot q_\alpha^\lambda = g \cdot q_\alpha^\lambda)_{\alpha < \lambda}$ is an upper bound of D_λ and thus $f = g$ by the universal property of the colimit.

For monos this is an immediate consequence of 6.5.1.

(b) Uniqueness of such $\overline{f}$ follows from the fact in (a) that each $\nabla_L^{0,\alpha}$ is epi. For every ordinal α, morphisms $\overline{f_\alpha} \colon \mathfrak{C}^\alpha(L) \to M$ can be constructed by recursion on α by setting $\overline{f_0} = f$, using the universal property 6.3.1 of $\mathfrak{C}$ for all successor ordinals (since every element of M is complemented), and using the (unique) morphism from the colimit, for all limit ordinals. $\square$

6.6.2. Proposition. *If there exists an ordinal α such that $\nabla_L^{\alpha,\alpha+1}$ is an isomorphism, then $\mathfrak{C}^\alpha(L)$ is the free complete Boolean algebra generated by L.*

Proof. 6.2.2 shows that $\mathfrak{C}^\alpha(L)$ must be Boolean. Then 6.6.1(b) ensures that every frame homomorphism from L to a complete Boolean algebra factors uniquely through $\nabla_L^{0,\alpha}$. $\square$

6.6.3. Corollary. *There exists a frame L for which none of the maps*

$$\nabla_L^{\alpha,\alpha+1} \colon \mathfrak{C}^\alpha(L) \to \mathfrak{C}^{\alpha+1}(L)$$

is an isomorphism.

Proof. Let L be the free frame on $\aleph_0$ generators. If $\nabla_L^{\alpha,\alpha+1}$ is an isomorphism for some ordinal α then, by 6.6.2, $\mathfrak{C}^\alpha(L)$ would be the free complete Boolean algebra on $\aleph_0$ generators, which is impossible by the result (proved independently by Gaifman [110] and Hales [124]) that the free complete Boolean algebra on a countably infinite set is a proper class. $\square$

This shows how weird the structure of epimorphisms in **Frm** is:

One has epimorphisms $\nabla_L^{0,\alpha} \colon L \to \mathfrak{C}^\alpha(L)$ for all ordinals α (that is, the $\mathfrak{C}^\alpha(L)$ are all subobjects of L in **Loc**; though not, of course, sublocales). For the frame L of the corollary the $\mathfrak{C}^\alpha(L)$ never stop growing. Thus, for such a frame L one has epimorphisms $L \to M$ with arbitrarily large M (in the usual categorical terminology, the category **Frm** is not *co-wellpowered*).

In all the known cases either the $\mathfrak{C}^\alpha(L)$ never stop growing, or the growth stops before the fourth step (Plewe [213]); whether it can be otherwise is still an open problem.

6.7. How one recognizes epimorphisms? There is a useful characterization of frame epimorphisms due to Madden and Molitor [178] that we present next. Roughly, $f \colon L \to M$ is an epimorphism if and only if every element of M is accessible from L via iterated complementation in an appropriate extension of L.

6.7.1. Background: Simplification of certain colimits in algebraic categories $\mathcal{A} = \mathrm{Mod}(\Omega, E)$. If $\mathcal{A}$ is the category of models of an equational theory (Ω, E) and every $\omega \in \Omega$ has $|\iota(\omega)| < \kappa$ for some fixed regular cardinal κ, then certain colimit constructions on $\mathcal{A}$ become simplified (one says that $\mathcal{A}$ is *locally κ-presentable* [1]). For example, one says that K is a *κ-filtered category* if any diagram $D \colon K' \to K$ has a lower bound where K' has less than κ morphisms. Note that a poset X is κ-filtered as a category just when it is *κ-directed* as a poset (i.e., given any index set J with $|J| < \kappa$ and $(x_i)_{i \in J}$ in X, there exists $z \in X$ such that $x_i \leq z$ for all i). Then, if K is a κ-filtered category, the forgetful functor $\mathcal{A} \to \mathbf{Set}$ *creates colimits of diagrams $K \to \mathbf{Set}$* (i.e., the colimit of any diagram $D \colon K \to \mathcal{A}$ is calculated as in **Set**). In particular, if K is a chain (as a category) such that every $S \subseteq K$ with $|S| < \kappa$ (a *κ-subset*) has an upper bound, and $D \colon K \to \mathcal{A}$ is a diagram such that for every $j \leq k$, $D(j \to k)$ is an inclusion, then the colimit of D is just the union of the algebras $D(j)$, $j \in K$.

6.7.2. Background: κ-frames and κ-complete Boolean algebras. Let κ be a regular cardinal. A *κ-frame* is a poset with finite meets and κ-joins (i.e., every subset with cardinality strictly less than κ has a join), and binary meet distributes over these joins; κ-frame homomorphisms preserve finite meets and κ-joins. We denote the category of κ-frames and κ-morphisms by

$$\kappa\text{-}\mathbf{Frm}.$$

The universal property of $\mathfrak{C}$ on **Frm** can be easily generalized to κ-**Frm**. Thus, for a κ-frame L we let $d_L \colon L \to \mathfrak{B}L$ be the result in κ-**Frm** of freely complementing the elements of L. This can be constructed as a quotient of a free extension of L by new elements $\{a^* \mid a \in L\}$, where we divide by the congruence determined by equations saying that a^* is the complement of a. Just as with the congruence tower for frames, we can iterate the functor $\mathfrak{B}$ to produce an ordinal sequence of κ-frames $\mathfrak{B}^\alpha(L)$ and a doubly-indexed ordinal sequence of κ-morphisms

$$d_L^{\beta,\alpha} \colon \mathfrak{B}^\beta(L) \to \mathfrak{B}^\alpha(L),$$

with analogous properties. One can also define functors

$$\mathfrak{B}^\alpha \colon \kappa\text{-}\mathbf{Frm} \to \kappa\text{-}\mathbf{Frm}$$

in such a way that $(d_L^{\beta,\alpha})_{L \in \kappa\text{-}\mathbf{Frm}}$ establish natural transformations

$$\mathfrak{B}^\beta \overset{\cdot}{\to} \mathfrak{B}^\alpha \quad (\beta \le \alpha).$$

However, when it comes to the reflection problem, there is an important difference: by 6.7.1, since the ordinal κ (by regularity) is κ-filtered as a category and κ-**Frm** is locally κ-presentable, the colimit used to construct $\mathfrak{B}^\kappa(L)$ is just the union of the $\mathfrak{B}^\alpha(L)$ for $\alpha < \kappa$ (assuming that we identify each κ-frame in the tower with its image under d). Thus the result is a Boolean κ-frame, and it follows that the full subcategory of κ-complete Boolean algebras is reflective, with reflection functor $\mathfrak{B}^\kappa$. The advantage of this stems from the following result of LaGrange ([171]):

In the category of κ-complete Boolean algebras, every epimorphism is surjective.

In the following, to simplify notation, we shall use the obvious forgetful functor **Frm** $\to \kappa$-**Frm** without explicit mention. We shall also denote $\nabla_L^{0,\alpha}$ and $d_L^{0,\alpha}$ just by ∇_L^α and d_L^α respectively.

6.7.3. Lemma. (a) *Let L be a frame. For each ordinal α there is a κ-frame homomorphism $e_L^\alpha \colon \mathfrak{B}^\alpha(L) \to \mathfrak{C}^\alpha(L)$ such that $\nabla_L^\alpha = e_L^\alpha \cdot d_L^\alpha$.*

(b) *Let $f \colon L \to M$ be a frame epimorphism. Then for sufficiently large κ, f is an epimorphism in the category of κ-frames.*

Proof. (a) This is easy to verify by induction using the universal mapping property of d_L.

(b) Suppose f is not an epimorphism in any κ-**Frm**. Fix $\kappa > 2^{|M|}$. Let $g, h \colon M \to K$ be κ-frame homomorphisms such that $g \ne h$ and $gf = hf$. We may assume that K is generated as a κ-frame by the union of the images of g and h, and therefore that $|K| < \kappa$. Then, evidently, K is a frame. Similarly, g and h are actually frame homomorphisms. Then f is not an epimorphism in **Frm**, a contradiction. $\qquad\square$

6.7.4. α-epimorphisms. For an ordinal α, we say that a frame homomorphism $f: L \to M$ is α-*epi* if the image of $\mathfrak{C}^{\alpha}(f)$ contains the image of $\nabla_M^{\alpha}: M \to \mathfrak{C}^{\alpha}(M)$, that is,

$$\nabla_M^{\alpha}(M) \subseteq \mathfrak{C}^{\alpha}(f)(\mathfrak{C}^{\alpha}(L)).$$

Clearly, 0-epi means onto, and if f is α-epi then it is β-epi for all $\beta \geq \alpha$.

Theorem. [Madden & Molitor] *A morphism in* **Frm** *is an epimorphism if and only if it is α-epi for some ordinal α.*

Proof. Let $f: L \to M$ be a frame homomorphism. Using 6.7.3(b), take κ such that f is an epimorphism in κ-**Frm**. Then

$$\mathfrak{B}^{\kappa}(f): \mathfrak{B}^{\kappa}(L) \to \mathfrak{B}^{\kappa}(M)$$

is an epimorphism of κ-complete Boolean algebras because $\mathfrak{B}^{\kappa}$, being a reflection, preserves epimorphisms (recall AII.8.3). Thus, by the result of LaGrange quoted above, it is surjective. Now, by (a),

$$e_M^{\kappa} \mathfrak{B}^{\kappa}(f) d_L^{\kappa} = \mathfrak{C}^{\kappa}(f) e_L^{\kappa} d_L \kappa$$

and so, since d_L^{κ} is epic, $e_M^{\kappa} \mathfrak{B}^{\kappa}(f) = \mathfrak{C}^{\kappa}(f) e_L^{\kappa}$. Thus, by the surjectivity of $\mathfrak{B}^{\kappa}(f)$, $e_M^{\kappa}(M) \subseteq \mathfrak{C}^{\kappa}(f)(\mathfrak{C}^{\alpha}(L))$. Since $\nabla_M^{\kappa}(M) \subseteq e_M^{\kappa}(M)$, then finally

$$\nabla_M^{\kappa}(M) \subseteq \mathfrak{C}^{\kappa}(f)(\mathfrak{C}^{\alpha}(L)).$$

Conversely, suppose that $\nabla_M^{\alpha}(M) \subseteq \mathfrak{C}^{\alpha}(f)(\mathfrak{C}^{\alpha}(L))$, and that $gf = hf$ for frame homomorphisms $g, h: M \to K$. Since $\mathfrak{C}^{\alpha}(g)$ and $\mathfrak{C}^{\alpha}(h)$ agree on the image of $\mathfrak{C}^{\alpha}(f)$ we have

$$\nabla_K^{\alpha} \cdot g = \mathfrak{C}^{\alpha}(g) \cdot \nabla_M^{\alpha} = \mathfrak{C}^{\alpha}(h) \cdot \nabla_M^{\alpha} = \nabla_K^{\alpha} \cdot h.$$

But ∇_K^{α} is one-one, thus $g = h$. $\qquad\qquad\square$

Chapter V

Separation Axioms

In this chapter we will discuss the counterparts (or, almost counterparts) of the separation axioms T_i. Although the standardly used definitions seem to be heavily dependent on points, it is only partly so: for instance regularity and complete regularity can be perfectly reformulated in the point-free language. Instead of T_1 we can consider a weaker version that has, however a very good interpretation in classical topology. What is hard to replace is the Hausdorff axiom; however, there are formally very similar conditions (not quite corresponding to the classical concept factually) that turn out to be very useful.

Of course, T_0 will be dismissed right away, being absolutely irrelevant in the point-free context.

1. Instead of T_1: subfit and fit

1.1. Subfit locales. It is not quite difficult to formulate a property of a locale L that would in the lattice $\Omega(X)$ (with a T_0-space X) correspond precisely to T_1. The problem is in the tendency of such formulas to make the locale spatial. Therefore one starts, rather, with a weaker condition that has a very good meaning both in topology and in algebra (and in logic, too).

A locale (frame) L is said to be *subfit* if

$$a \not\leq b \quad \Rightarrow \quad \exists c, \ a \vee c = 1 \neq b \vee c. \tag{Sfit}$$

We will say that a *space* X is *subfit* if the frame $\Omega(X)$ is.

1.1.1. Facts. (1) *Every T_1-space is subfit.*

(2) *For a space X, $\quad T_1 \equiv T_D$ & subfit.*

(3) *X is subfit iff for each open U and each $x \in U$ there is a $y \in \overline{\{x\}}$ such that $\overline{\{y\}} \subseteq U$.*

Proof. (1): If $U \not\subseteq V$ take an $x \in U \smallsetminus V$. Then $U \cup (X \smallsetminus \{x\}) = X \neq V \cup (X \smallsetminus \{x\})$.

(2): T_1 trivially implies T_D (recall I.2.1), and by (1) it also implies subfitness. Now let X be T_D and subfit. For $x \in X$ choose an open U containing x such that $U \smallsetminus \{x\}$ is open. Since $U \not\subseteq U \smallsetminus \{x\}$ there is an open V such that $U \cup V = X \neq (U \smallsetminus \{x\}) \cup V$. By the latter, $x \notin V$ and hence $X \smallsetminus \{x\} = (U \cup V) \smallsetminus \{x\} = (U \smallsetminus \{x\}) \cup V$ is open.

(3): Let the statement hold and let $U \not\subseteq V$. For an $x \in U \smallsetminus V$ choose $y \in \overline{\{x\}}$ such that $\overline{\{y\}} \subseteq U$. Then $U \cup (X \smallsetminus \overline{\{y\}}) = X$. On the other hand, $x \notin V$, hence $\overline{\{x\}} \cap V = \emptyset$ and $y \notin V$, so that $y \in V \cup (X \smallsetminus \overline{\{y\}})$.

Conversely let X be subfit and $x \in U \in \Omega(X)$. Then $U \not\subseteq X \smallsetminus \overline{\{x\}}$ and there is a V such that $V \cup U = X \neq V \cup (X \smallsetminus \overline{\{x\}})$; hence there is a $y \in \overline{\{x\}}$ such that $y \notin V$. Then $\overline{\{y\}} \subseteq X \smallsetminus V$, and since $V \cup U = X$ we conclude that $\overline{\{y\}} \subseteq U$. $\square$

Note. Subfitness was introduced by Isbell in [136] and independently (as *conjunctivity*, because it is the opposite of the disjunctive property for distributive lattices) by Simmons in [245].

1.2. Before proceeding with subfitness we will discuss a slightly stronger property, the *fitness*, introduced also by Isbell in [136]. Unlike subfitness, this is first of all a very important algebraic property, with not quite so much immediate topological impact. In connection with the purposes of this section, it is important that it is a hereditary variant of subfitness (which is not generally inherited in sublocales; see 1.5 below). As a separation axiom it has also some aspects similar with regularity, see 5.5.

A locale (frame) L is said to be *fit* if

$$a \not\leq b \quad \Rightarrow \quad \exists c, \ a \vee c = 1 \ \text{ and } \ c \to b \not\leq b. \tag{Fit}$$

1.2.1. Lemma. (1) *Each fit locale is subfit.*

(2) *Each sublocale S of a fit locale L is fit.*

Proof. (1): Let L be fit and let a, b, c be as in the condition (Fit). Then

$$c \to b = (c \to b) \wedge (b \to b) = (c \vee b) \to b \neq b$$

and hence $c \vee b \neq 1$.

(2): Let $a, b \in S \subseteq L$, $a \not\leq b$. Denote by $x \vee_S y$ the join in S. Consider the $c \in L$ from the definition of fitness. Trivially $\nu_S(c) \vee_S a \geq c \vee a = 1$, and by III.5.3.1(b) $\nu_S(c) \to b = c \to b \not\leq b$. $\square$

1.3. Before presenting a series of characteristics of fitness it will be useful to introduce the following notation: for a sublocale S of a general L set

$$S' = \mathord{\downarrow}(S \smallsetminus \{1\}) \ (= \{x \in L \mid \nu_S(x) \neq 1\}).$$

1.3.1. Lemma. *For any sublocale S and any $c \in L$, $S \subseteq \mathfrak{o}(c)$ iff $\nu_S(c) = 1$.*

Proof. If $\nu_S(c) = 1$ then for any $s \in S$, by III.5.3.1(b) and (H1) in III.3.1.1, $c \to s = \nu_S(c) \to s = 1 \to s = s$. If $s = \nu_S(c) \neq 1$ then $c \leq s$ and $c \to s = 1 \neq s$, by (H2) in III.3.1.1. $\square$

1.3.2. Proposition. *The following statements about a locale L are equivalent:*

(1) L *is fit.*

(2) *For any sublocales S and T of L, $S' = T' \Rightarrow S = T$.*

(3) *For any sublocale S of L, $S = \bigcap\{\mathfrak{o}(x) \mid \nu_S(x) = 1\}$.*

(4) *Each sublocale is an intersection of open sublocales.*

(5) *Each closed sublocale is an intersection of open sublocales.*

Proof. $(1) \Rightarrow (2)$: Let $S' = T'$ and let $b \in T$, $b \neq 1$. Set $a = \nu_S(b)$. Suppose $a \vee c = 1$ and $b \vee c \leq a_1 \in S$. Then $a_1 \geq a \vee c = 1$ so that $b \vee c \notin S' = T'$. By property (H8) (III.3.1.1), however, $(b \vee c) \wedge (c \to b) \leq b$ and hence $b \vee c \leq (c \to b) \to b \in T$ so that $(c \to b) \to b = 1$ and $c \to b = b$ by (H2) and (H3). Therefore, by (Fit), $a \leq b$ and hence $b = \nu_S(b) \in S$.

$(2) \Rightarrow (3)$: The inclusion $\subseteq$ follows from 1.3.1. On the other hand, if

$$a \in T = \bigcap\{\mathfrak{o}(x) \mid \nu_S(x) = 1\}$$

we have $x \to a = a$ whenever $\nu_S(x) = 1$. Thus, if $\nu_S(a) = 1$ we have $a = a \to a = 1$. Hence $T \smallsetminus \{1\} \subseteq S'$, consequently $T' \subseteq S' \subseteq T'$, and by (2), $S = T$.

$(3) \Rightarrow (4) \Rightarrow (5)$ is trivial.

$(5) \Rightarrow (1)$: By III.6.2.1, if $\uparrow a$ is an intersection of some open sublocales then it is the intersection of all open sublocales $\mathfrak{o}(c)$ with $\nu_{\uparrow a}(c) = 1$. As

$$\nu_{\uparrow a}(c) = \bigwedge\{s \mid a \leq s, \ c \leq s\} = a \vee c,$$

we have $\mathfrak{c}(a) = \uparrow a = \bigcap\{\mathfrak{o}(c) \mid a \vee c = 1\}$ and hence, if $c \to b = b$ for all c such that $a \vee c = 1$ then $a \leq b$. $\square$

Note. The original Isbell's definition in [136] was the property (3).

1.3.3. The equivalence $(1) \Leftrightarrow (2)$ above has an interesting algebraic interpretation.

Corollary. *A frame L is fit iff for the frame congruences on L one has the implication*

$$E_1 1 = E_2 1 \quad \Rightarrow \quad E_1 = E_2$$

(that is, if the congruence classes of the top coincide then the congruences coincide). $\square$

(Indeed, recall that for the congruence E corresponding to a sublocale S one has $x E y$ iff $(\forall s \in S,\ x \le s \Leftrightarrow y \le s)$. Thus, $x \in S'$ iff $(\exists s \ne 1,\ s \in S$ such that $x \le s)$ iff $(x, 1) \notin E$ and hence $E_1 = E_2$ iff $S_1 = S_2$ iff $S_1' = S_2'$ iff $E_1 1 = E_2 1$.)

1.4. Subfit locales have similar characteristics like those in 1.3.2.

Proposition. *The following statements about a locale L are equivalent:*

(1) *L is subfit.*

(2) *For a sublocale $S \subseteq L$, $S \smallsetminus \{1\}$ is cofinal in $L \smallsetminus \{1\}$ only if $S = L$.*

(3) *If $S \ne L$ for a sublocale $S \subseteq L$ then there is a closed $\mathfrak{c}(x) \ne \mathsf{O}$ such that $S \cap \mathfrak{c}(x) = \mathsf{O}$.*

(4) *For each open sublocale $\mathfrak{o}(a)$, $\mathfrak{o}(a) = \bigvee \{\mathfrak{c}(x) \mid x \vee a = 1\}$.*

(5) *Each open sublocale is a join of closed sublocales.*

Proof. (1)$\Rightarrow$(2): Let $b \in L$ and $a = \nu_S(b)$. If $a \vee c = 1$ we have $\nu_S(b \vee c) \ge a \vee c = 1$ and hence $b \vee c = 1$. Thus, $a \le b$, that is, $b \in S$.

(2)$\Leftrightarrow$(3): the latter is just an immediate reformulation of the former.

(4)$\Leftrightarrow$(5) follows immediately from the fact that $\mathfrak{c}(x) \subseteq \mathfrak{o}(a)$ iff $x \vee a = 1$ (indeed $\mathfrak{c}(x) \subseteq \mathfrak{o}(a)$ iff $\mathfrak{c}(x) \cap \mathfrak{c}(a) = \mathsf{O}$ iff $\mathfrak{c}(x \vee a) = \mathsf{O}$ iff $x \vee a = 1$).

(3)$\Rightarrow$(4): Set $S = \bigvee \{\mathfrak{c}(x) \mid x \vee a = 1\}$ and suppose $\mathfrak{c}(y) \cap (\mathfrak{c}(a) \vee S) = \mathsf{O}$. Then, $\mathfrak{c}(y) \cap \mathfrak{c}(a) = \mathsf{O}$ and, by complementation, $\mathfrak{c}(y) \subseteq \mathfrak{o}(a)$. Therefore, $y \vee a = 1$, and we conclude that $\mathfrak{c}(y) \subseteq S$ and finally $\mathfrak{c}(y) = \mathfrak{c}(y) \cap (\mathfrak{c}(a) \vee S) = \mathsf{O}$. Thus by (3), $\mathfrak{c}(a) \vee S = L$ and by complementation again $\mathfrak{o}(a) \subseteq S$ (the reverse inclusion is obvious).

(4)$\Rightarrow$(1): If $a \not\le b$ we have $\mathfrak{c}(b) \not\subseteq \mathfrak{c}(a)$ (as $b \in \mathfrak{c}(b) \smallsetminus \mathfrak{c}(a)$) and hence $\mathfrak{o}(a) \not\subseteq \mathfrak{o}(b)$. Thus there is a c such that $c \vee a = 1$ and $\mathfrak{c}(c) \not\subseteq \mathfrak{o}(b)$, that is, $c \vee b \ne 1$. $\qquad\square$

1.4.1. Using the same procedure as in 1.3.3 we obtain the following interpretation of the equivalence (1)$\Leftrightarrow$(2).

Corollary. *A frame L is subfit iff for the frame congruences on L one has the implication*

$$E1 = \{1\} \quad \Rightarrow \quad E = \Delta = \{(x, x) \mid x \in L\}$$

(that is, if the congruence class of the top is trivial than the whole of the congruence is trivial). $\qquad\square$

1.5. Proposition. *A locale is fit iff each of its sublocales is subfit.*

Proof. The implication $\Rightarrow$ follows from 1.2.1.

$\Leftarrow$: Let each sublocale of L be subfit. We will prove that L satisfies the property (2) from 1.3.2. Let S, T be sublocales of L and let $S' = T'$. Consider the sublocale $S \vee T$. By the assumption it is subfit. Consider an element $s \wedge t \in S \vee T$, $s \wedge t < 1$. If $s < 1$ we have $s \wedge t \leq s < 1$ with $s \in S$. If $s = 1$ then $t < 1$ and we have an $s' \in S$ such that $t \leq s' < 1$. Thus, $s \wedge t \leq s' < 1$ with $s' \in S$ again and $S \smallsetminus \{1\}$ is cofinal in $S \vee T \smallsetminus \{1\}$ and we have, by (2) in 1.4, $S = S \vee T$. Similarly $T = S \vee T$ and we conclude that $S = T$. $\qquad\square$

Note. If a *space* X is T_1 then each of its *subspaces* is T_1 and hence subfit. That does not mean that the locale $\mathsf{Lc}(X)$ is then necessarily fit, though. There can be non-induced sublocales of $\mathsf{Lc}(X)$ that are not subfit.

1.6. A frame homomorphism $h\colon M \to L$ is *co-dense* if

$$h(a) = 1 \quad \Rightarrow \quad a = 1$$

(compare with the density from III.8.2 amounting to $h(a) = 0 \;\Rightarrow\; a = 0$). In the language of localic maps (called co-dense if its adjoint is) this is expressed as follows.

Proposition. *A localic map is co-dense if $f[L \smallsetminus \{1\}]\ (= f[L] \smallsetminus \{1\})$ is cofinal in $M \smallsetminus \{1\}$.*

Proof. As for the equality recall II.2.3: $f(x) \geq 1$ iff $x \geq f^*(1) = 1$.

Now if $f[L \smallsetminus \{1\}]$ is cofinal in $M \smallsetminus \{1\}$ and $b < 1$ then there is an $a < 1$ such that $b \leq f(a)$ and hence $f^*(b) \leq a < 1$. Conversely, if f^* is co-dense and $b < 1$ in M then $f^*(b) \leq a < 1$, and $b \leq f(a)$. $\qquad\square$

From 1.4.1 we immediately obtain

1.6.1. Corollary. *A frame L is subfit iff each co-dense frame homomorphism $h\colon L \to M$ is one-one.* $\qquad\square$

(Consider the congruence $E = \{(x, y) \mid h(x) = h(y)\}$ of L. If h is co-dense then $E1 = \{1\}$ and hence E is trivial; conversely, if the condition holds and $E1 = \{1\}$ consider the quotient map $L \to L/E$.)

In the language of localic maps this has the following formulation:

1.6.2. Corollary. *A locale L is subfit iff each co-dense localic map $f\colon M \to L$ is onto.* $\qquad\square$

1.7. Proposition. *Every complemented sublocale of a subfit locale is subfit.*

Proof. Let T be the complement of a sublocale $S \subseteq L$, that is, $S \cap T = \mathsf{O}$ and $S \vee T = L$. Let $S_0 \subseteq S$ be such that $S_0 \smallsetminus \{1\}$ is cofinal in S. Consider the join $S_0 \vee T$. Let $x \in L$. Then $x = s \wedge t$ for some $s \in S$ and $t \in T$. If $x < 1$ then either $t < 1$ and $x = s \wedge t \leq t \in S_0 \vee T$ or $t = 1$ and then $s < 1$ and there is an $s_0 \in S_0$, $s_0 < 1$, such that $x = s \leq s_0$. Thus, $S_0 \vee T \smallsetminus \{1\}$ is cofinal in $L \smallsetminus \{1\}$ and $S_0 \vee T = L$. Consequently $S = S \cap (S_0 \vee T) = S \cap S_0 = S_0$. $\qquad\square$

1.8. Recall from III.7.2 that open localic maps are precisely the localic maps $f \colon L \to M$ for which the left adjoint f^* is a complete Heyting homomorphism. There we saw that open frame homomorphisms $h \colon M \to L$ are characterized by the existence of a map ϕ such that

$$x \wedge \phi(a) = y \wedge \phi(a) \quad \text{iff} \quad h(x) \wedge a = h(y) \wedge a. \tag{1.8.1}$$

If we set

$$E = \{(x,y) \mid h(x) \wedge a = h(y) \wedge a\} \quad \text{and} \quad E' = \{(x,y) \mid x \wedge \phi(a) = y \wedge \phi(a)\}$$

we can rewrite (1.8.1) as $E = E'$ and if M is fit this reduces, by 1.3.3, to $E1 = E'1$, that is, to

$$\phi(a) \leq x \quad \text{iff} \quad x \wedge \phi(a) = \phi(a) \quad \text{iff} \quad h(x) \wedge a \quad \text{iff} \quad a \leq h(x)$$

so that (1.8.1) reduces to h being also a right adjoint; hence

> *if M is a fit frame then a frame homomorphism $h \colon M \to L$ is open iff it is a complete lattice homomorphism.*

(Thus, the Heyting condition comes for free.)

In fact, this holds already for subfit frames. We do not have it as a direct corollary of the characterization theorem, but the proof is easy.

Proposition. *If M is subfit then a frame homomorphism $h \colon M \to L$ is open iff it is a complete lattice homomorphism.*

Proof. Let $h \colon M \to L$ be an open frame homomorphism. By III.7.2, h is, in particular, a complete lattice homomorphism.

Conversely, let h be a complete lattice homomorphism, denote by ϕ the corresponding left adjoint and assume M is subfit. Then $\phi(a \wedge h(b)) = \phi(a) \wedge b$ (and by III.7.2 h is open). Indeed:

We have always

$$\phi(a \wedge h(b)) \leq \phi(a) \wedge \phi h(b) \leq \phi(a) \wedge b.$$

On the other hand, if $\phi(a) \wedge b \not\leq \phi(a \wedge h(b))$ then, by subfitness, there exists $c \in L$ for which

$$(\phi(a) \wedge b) \vee c = 1 \neq \phi(a \wedge h(b)) \vee c.$$

But

$$h(\phi(a \wedge h(b)) \vee c) = h(\phi(a \wedge h(b))) \vee h(c) = h\phi(a \wedge h(b)) \vee h(c)$$
$$\geq (a \wedge h(b)) \vee h(c) = (a \vee h(c)) \wedge (h(b) \vee h(c))$$
$$= (a \vee h(c)) \wedge h(b \vee c) = a \vee h(c) \geq a.$$

Thus $\phi(a \wedge h(b)) \vee c \geq \phi(a)$ and consequently $\phi(a \wedge h(b)) \vee c \geq \phi(a) \vee c = 1$, a contradiction. $\qquad\square$

2. Mimicking the Hausdorff axiom

2.1. Isbell's approach. A topological space X is Hausdorff (or T_2) if and only if the diagonal $\{(x,x) \mid x \in X\}$ is closed in $X \times X$. This was imitated by Isbell in [136]: a frame L is called Hausdorff whenever the diagonal

$$\Delta \colon L \to L \oplus L$$

is a closed localic map.

Recall from IV.5.5.2 that the diagonal Δ is given by the formula

$$\Delta(a) = \{(x,y) \mid x \wedge y \leq a\},$$

and its left adjoint Δ^* (the codiagonal homomorphism – see IV.5.2) is given by

$$\Delta^*(U) = \bigvee\{x \wedge y \mid (x,y) \in U\} = \bigvee\{x \mid (x,x) \in U\}$$

(in particular, $\Delta^*(x \oplus y) = x \wedge y$). Thus, this amounts to requiring that $\Delta[L] \subseteq L \oplus L$ is a closed sublocale, and since $\bigwedge \Delta[L] = \Delta(0)$, to

$$\Delta[L] = {\uparrow}\Delta(0).$$

The element $\Delta(0) = \{(x,y) \mid x \wedge y = 0\}$ of $L \oplus L$ will play, repeatedly, a basic role. We will denote it by

$$d_L.$$

Since we will also consider another Hausdorff type property we will call a frame (locale) satisfying the just mentioned one *I-Hausdorff* (short for *Isbell-Hausdorff*; sometimes one speaks of a *strongly Hausdorff* frame [136]).

Note that the I-Hausdorff property is only an imitation of the classical one, not its exact extension: we use a square that is not an exact extension of the classical square of spaces (since $\mathsf{Lc}(X \times Y)$ is not generally isomorphic to $\mathsf{Lc}(X) \oplus \mathsf{Lc}(Y)$). We have that

if $\mathsf{Lc}(X)$ is I-Hausdorff then X is Hausdorff,

but the reverse implication generally does not hold ([145]).

2.1.1. Obviously Δ^* is onto (for instance, $\Delta^*(a \oplus a) = a$). Thus, as $\Delta^*\Delta\Delta^* = \Delta^*$, we have

$$\Delta^*\Delta = \mathrm{id}. \tag{2.1.1}$$

From the adjunction we further have that

$$U \subseteq \Delta\Delta^*(U). \tag{2.1.2}$$

2.2. Here is a useful characterisation.

Lemma. *The following statements about a locale L are equivalent:*

(1) *L is I-Hausdorff.*
(2) *For all saturated $U \supseteq d_L$, $\Delta\Delta^*(U) = U$; in other words, the restrictions of Δ and Δ^* to $L \to \uparrow d_L$ and $\uparrow d_L \to L$ respectively are mutually inverse isomorphisms.*
(3) *There exists a mapping $\alpha \colon L \to \uparrow d_L$ such that $\alpha\Delta^* = \gamma = (U \mapsto U \vee d_L)$.*

Furthermore, an α satisfying (3) is necessarily an isomorphism inverse to the restriction of Δ^*, hence, the restriction of Δ to $L \to \uparrow d_L$.

Proof. (1)$\Rightarrow$(2): For the $U \supseteq d_L$ choose $a = \beta(U)$ such that $U = \Delta(a)$. Then, $U = \Delta\beta(U)$ and, by (2.1.1), $\Delta^*(U) = \Delta^*\Delta\beta(U) = \beta(U)$. Thus, $U = \Delta\Delta^*(U)$.

(2)$\Rightarrow$(3): Set

$$\alpha = (a \mapsto \Delta(a)) \colon L \to \uparrow d_L.$$

Then $U \vee d_L = \Delta\Delta^*(U \vee d_L) = \Delta(\Delta^*(U) \vee \Delta^* d_L) = \Delta(\Delta^*(U) \vee 0) = \Delta(\Delta^*(U)) = \alpha\Delta^*(U)$.

(3)$\Rightarrow$(1): Since γ and Δ^* are homomorphisms and Δ^* is onto, α is a homomorphism, and it is onto since γ is. Finally,

$$\Delta^*\alpha\Delta^*(U) = \Delta^*\gamma(U) = \Delta^*(U) \vee \Delta^*\Delta(0) = \Delta^*(U)$$

since $\Delta^*\Delta(0) = 0$; thus, α is one-one, hence an isomorphism, and (1) follows.

Moreover, we have seen that α is necessarily the inverse to the restriction of Δ^*. $\square$

2.3. Proposition. *A locale L is I-Hausdorff iff for every saturated $U \supseteq d_L$ one has the implication*

$$(a \wedge b, a \wedge b) \in U \quad \Rightarrow \quad (a, b) \in U.$$

Consequently, in any I-Hausdorff locale we have

$$(a \oplus b) \vee d_L = (b \oplus a) \vee d_L.$$

Proof. ⇒: Let L be an I-Hausdorff locale. Then by 2.2, $\Delta\Delta^*(U) = U$. If $(a \wedge b, a \wedge b) \in U$ then $a \wedge b \leq \Delta^*(U)$ and $(a, b) \in \Delta(\Delta^*(U)) = U$.

⇐: Let the implication hold and let x_i, $i \in J$, be in L. Then for a saturated $U \supseteq d_L$ we have the implication

$$(x_i, x_i) \in U \quad \Rightarrow \quad (x_i, x_j) \in U \text{ for any } j \tag{2.3.1}$$

(since if $(x_i, x_i) \in U$ then $(x_i \wedge x_j, x_i \wedge x_j) \in U$).

We will use 2.2(3). Set $\alpha(a) = (a \oplus a) \vee d_L$. Obviously α preserves meets, and by (2.3.1) and IV.5.2 also

$$\alpha\Big(\bigvee_{i \in J} a_i\Big) = \Big(\Big(\bigvee_{i \in J} a_i\Big) \oplus \Big(\bigvee_{i \in J} a_i\Big)\Big) \vee d_L$$

$$= \Big(\bigvee_{i,j \in J} (a_i \oplus a_j)\Big) \vee d_L = \bigvee_{i \in J} ((a_i \oplus a_i) \vee d_L) = \bigvee_{i \in J} \alpha(a_i).$$

Thus, α is a frame homomorphism and hence so is also $\alpha\Delta^*$; we have

$$\alpha\Delta^*(a \oplus b) = \alpha(a \wedge b) = ((a \wedge b) \oplus (a \wedge b)) \vee d_L = (a \oplus b) \vee d_L,$$

the last equality by the assumed implication again. Since the elements $a \oplus b$ generate $L \oplus L$ (recall IV.5.2(1)), the statement follows. □

2.4. Proposition. *Each sublocale of an I-Hausdorff locale is I-Hausdorff.*

Proof. Let $S \subseteq L$ be a sublocale and let U be saturated in $S \times S$; let

$$U \supseteq \Delta_S(0_S) = \{(x, y) \in S \times S \mid x \wedge y = \nu_S(0)\}.$$

Then $\downarrow U$ (with $\downarrow$ taken in $L \times L$) is saturated in $L \times L$: if $(x_i, y) \in \downarrow U$ there are $x_i' \geq x_i$, $y' \geq y$ with $(x_i', y') \in U$; hence $(\bigvee_{i \in J}^S x_i', y') \in U$ and

$$\bigvee_{i \in J} x_i \leq \bigvee_{i \in J} x_i' \leq \nu_S\Big(\bigvee_{i \in J} x_i'\Big) = \bigvee_{i \in J}^S x_i',$$

so that $(\bigvee_{i \in J} x_i, y) \in \downarrow U$. Further, $\downarrow U \supseteq d_L$: in fact, if $x \wedge y = 0$ then $\nu_S(x) \wedge \nu_S(y) = \nu_S(0)$ and hence

$$(x, y) \leq (\nu_S(x), \nu_S(y)) \in \Delta_S(0_S) \subseteq U.$$

Now let $(a \wedge b, a \wedge b) \in U$ with $(a, b) \in S \times S$. Then $(a, b) \in \downarrow U$ and since U is a down-set in $S \times S$, $(a, b) \in U$. □

2.5. Coequalizers of homomorphisms with I-Hausdorff source. One of the important properties of Hausdorff spaces is that if $f_1, f_2 \colon X \to Y$ are continuous maps and Y is Hausdorff then the equalizer set $\{x \mid f_1(x) = f_2(x)\}$ is closed. The following proposition is a point-free counterpart of this statement, proved first by Banaschewski for regular frames [13]. It holds (more generally, as we will see below in 5.4.2) for I-Hausdorff frames as well (this case is due to Chen [61]).

2.5.1. Proposition. *Let L be I-Hausdorff and $h_1, h_2 \colon L \to M$ frame homomorphisms. Set $c = \bigvee\{h_1(x) \wedge h_2(y) \mid x \wedge y = 0\}$. Then*

$$\check{c} = (a \mapsto a \vee c) \colon M \to \uparrow c = \mathfrak{c}(c)$$

is the coequalizer of h_1 and h_2.

Proof. Obviously, $c = \Delta^*((h_1 \oplus h_2)(d_L))$. Consider the homomorphism α from 2.2. By 2.3 we have $(a \oplus 1) \vee d_L = (1 \oplus a) \vee d_L$ and hence

$$\check{c}h_1(a) = h_1(a) \vee c = \Delta^*(h_1 \oplus h_2)((a \oplus 1) \vee d_L)$$
$$= \Delta^*(h_1 \oplus h_2)((1 \oplus a) \vee d_L) = h_2(a) \vee c = \check{c}h_2(a).$$

On the other hand, if $\phi h_1 = \phi h_2 = h$ for a $\phi \colon M \to K$, we have

$$\phi(c) = \bigvee\{h(x \wedge y) \mid x \wedge y = 0\} = 0$$

and hence we can define $\overline{\phi} \colon \uparrow c \to K$ by $\overline{\phi}(x) = \phi(x)$ to obtain $\overline{\phi} \cdot \check{c} = \phi$.   □

In the language of localic morphisms we then have

2.5.2. Proposition. *Let $f_1, f_2 \colon M \to L$ be localic morphisms with L a I-Hausdorff locale. Then their equalizer is*

$$j_{\uparrow c} \colon \uparrow c \subseteq M \quad \text{where} \quad c = \bigvee\{f_1^*(x) \wedge f_2^*(y) \mid x \wedge y = 0\}.$$ □

(The inclusion map is the right adjoint of $\check{c}$.)

2.5.3. Proposition. *Let $g \colon L \to M$ be a localic map such that $\overline{g[L]} = M$. Then for any localic maps $f_1, f_2 \colon M \to N$ with N I-Hausdorff we have the implication*

$$f_1 g = f_2 g \quad \Rightarrow \quad f_1 = f_2.$$

In the frame language: if $k \colon M \to L$ is dense then we have the implication

$$kh_1 = kh_2 \quad \Rightarrow \quad h_1 = h_2$$

for any $h_1, h_2 \colon N \to M$ with N I-Hausdorff. Thus, in the (full) subcategory of I-Hausdorff frames, each dense homomorphism is a monomorphism.

Proof. It will be done in the frame language, but it is a useful exercise to prove the first statement using 2.5.2 and the formula for equalizer from IV.3.3.1.

If $kh_1 = kh_2$ then there is a ϕ such that $k = \phi\check{c}$. Then $k(c) = \phi\check{c}(c) = 0$ and if k is dense, $c = 0$, $\check{c}$ is the identity isomorphism, and $h_1 = h_2$.   □

2.6. The axiom of Dowker and Strauss. In 1972 ([69]), Dowker and Strauss suggested a number of point-free conditions as counterparts of separation axioms. As a Hausdorff type axiom one has there the following requirement (called (S_2')):

> If $a \vee b = 1$, $a, b \neq 1$ then there exist $u \not\leq a$, $v \not\leq b$ with $u \wedge v = 0$.

We will adopt its weakly hereditary variant:

> If $a \vee b \neq a, b$ then there exist $u \not\leq a$, $v \not\leq b$ with $u \wedge v = 0$

and speak of the *DS-Hausdorff property* (short for *Dowker-Strauss-Hausdorff*; sometimes one speaks of *weakly Hausdorff* frames).

2.6.1. Lemma. *For any $a, b \in L$,*

$$U = {\downarrow}(a, a \wedge b) \cup {\downarrow}(a \wedge b, b) \cup \mathbf{n}$$

is a saturated subset of $L \times L$.

Proof. Let (x_i, y) be in U, $i \in J$. If $y = 0$ then $(\bigvee_{i \in J} x_i, y) \in U$ trivially. If $0 \neq y \leq a \wedge b$ then $x_i \leq a$ and $(\bigvee_{i \in J} x_i, y) \in {\downarrow}(a, a \wedge b) \subseteq U$; if $y \not\leq a \wedge b$ then $y \leq b$ and $x_i \leq a \wedge b$, and $(\bigvee_{i \in J} x_i, y) \in {\downarrow}(a \wedge b, b) \subseteq U$. Symmetrically for $(x, \bigvee_{i \in J} y_i)$. $\qquad\square$

2.6.2. Proposition. *Each I-Hausdorff locale is DS-Hausdorff.*

Proof. Suppose L is not DS-Hausdorff. Then there are $a, b \neq a \vee b$ such that whenever $u \wedge v = 0$ and $(u, v) \leq (a, b)$ then either $u \leq b$ or $v \leq a$, that is, for the set U from the lemma, $d_L \cap (a \oplus b) \subseteq U$.

Now let L be I-Hausdorff. Then by 2.3, $(a, b) \in (a \wedge b) \oplus (a \wedge b) \vee d_L$ and hence

$$(a, b) \in ((a \wedge b) \oplus (a \wedge b) \vee d_L) \cap (a \oplus b) \subseteq (a \wedge b) \oplus (a \wedge b) \vee U = U$$

(as $(a \wedge b) \oplus (a \wedge b) \subseteq U$) which is a contradiction. $\qquad\square$

3. I-Hausdorff frames and regular monomorphisms

In Chapter IV (starting already in III.1) we discussed special epimorphisms in **Frm** (extremal, strong and regular ones) and plain epimorphisms (the hard case). As the reader certainly observed, we evaded any kind of special monomorphisms (the plain ones are, for trivial reasons, precisely the one-one homomorphisms). This subject is in general a rather difficult one. However, in the Hausdorff case, at least the regular monomorphisms are fairly transparent.

3.1. Consider a subframe L of a frame M and denote by $j\colon L \to M$ the embedding homomorphism. Further, let us denote by $\iota_1, \iota_2\colon M \to M \oplus M$ the coproduct injections and use the notation

$$\check{c} = (x \mapsto x \vee c)\colon K \to {\uparrow}c$$

as in 2.5.

Recall the d_L from 2.1, set

$$d_L^M = (j \oplus j)(d_L) = \bigvee\{x \oplus y \in M \oplus M \mid x, y \in L,\ x \wedge y = 0\}$$

and, finally,

$$\Gamma_M(L) = \{a \in M \mid (a \oplus 1) \vee d_L^M = (1 \oplus a) \vee d_L^M\}.$$

3.1.1. Observations. (1) $\Gamma_M(L)$ *is a subframe of* M, *and* $\Gamma_M(\mathbf{2}) = \mathbf{2}$.

(2) *If* $L \subseteq K \subseteq M$ *then* $d_L^M \leq d_K^M$ *and hence* $\Gamma_M(L) \subseteq \Gamma_M(K)$.

(The embedding $\Gamma_M(L) \to M$ is the equalizer of $\check{d}_L^M \iota_1$ and $\check{d}_L^M \iota_2$; $d_{\mathbf{2}}^M = \{1 \oplus 0, 0 \oplus 1\} = 0$, and $a \oplus 1 = 1 \oplus a$ only if $a = 0$ or $a = 1$. The last statement is trivial.)

3.1.2. Lemma. *Let* L *be I-Hausdorff. Then*

$$L \subseteq \Gamma_M(L) \quad \text{and} \quad \Gamma_M(\Gamma_M(L)) = \Gamma_M(L).$$

Proof. The first inequality immediately follows from 2.3. For the equality set $K = \Gamma_M(L)$. By 3.1.1(2) it suffices to prove that $d_K^M \leq d_L^M$. If $x, y \in K$ we have $(x \oplus 1) \vee d_L^M = (1 \oplus x) \vee d_L^M$ and similarly with y so that, if $x \wedge y = 0$,

$$(x \oplus 1) \vee d_L^M = ((x \oplus 1) \wedge (1 \oplus y)) \vee d_L^M$$
$$= ((x \oplus 1) \wedge (y \oplus 1)) \vee d_L^M = ((x \wedge y) \oplus 1) \vee d_L^M = d_L^M.$$

Thus, $x \oplus y \leq d_L^M$. $\square$

3.1.3. Proposition. *Let* L *be an I-Hausdorff frame. Then a monomorphism* $h\colon L \to M$ *is regular iff* $\Gamma_M(h[L]) = h[L]$.

Proof. We can assume $L \subseteq M$ and $h = j\colon L \subseteq M$. Suppose j is the equalizer of $f, g\colon M \to K$ and consider the coequalizer γ of $\iota_1 j$ and $\iota_2 j$. By 2.5.1 $\gamma = \check{d}_L^M\colon M \oplus M \to {\uparrow}d_L^M$. By the definition of coproducts, there is $f \oplus g\colon M \oplus M \to K$ such that $(f \oplus g)\iota_1 = f$ and $(f \oplus g)\iota_2 = g$. Then, since $(f \oplus g)\iota_1 j = (f \oplus g)\iota_2 j$, there is a $k\colon {\uparrow}d_L^M \to K$ such that $k\gamma = f \oplus g$:

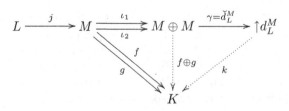

Now if $\gamma\iota_1\phi = \gamma\iota_2\phi$ then $f\phi = g\phi$ and hence there is precisely one $\overline{\phi}$ such that $j\overline{\phi} = \phi$. Thus j is the equalizer of $\breve{d}_l^M \iota_1$ and $\breve{d}_l^M \iota_2$, that is,

$$L = \{a \in M \mid (a \oplus 1) \vee d_L^M = (1 \oplus a) \vee d_L^M\} = \Gamma_M(L).$$

Conversely, if $L = \Gamma_M(L) = \{a \in M \mid (a \oplus 1) \vee d_L^M = (1 \oplus a) \vee d_L^M\}$ then j is an equalizer by definition.   □

3.2. In fact the operation Γ_M also makes in the Hausdorff case the epimorphisms more transparent.

3.2.1. Lemma. *Let L be a subframe of M. Then any two frame homomorphisms $f, g\colon M \to N$ coinciding on L coincide on all of the $\Gamma_M(L)$.*

Proof. Consider the codiagonal $\nabla\colon N \oplus N \to N$. We have

$$\nabla(f \oplus g)(d_L^M) = \bigvee\{f(x \wedge y) \mid x, y \in L, \ x \wedge y = 0\} = 0,$$

$\nabla(f \oplus g)(a \oplus 1) = f(a)$ and $\nabla(f \oplus g)(1 \oplus a) = g(a)$. Now if $(a \oplus 1) \vee d_L^M = (1 \oplus a) \vee d_L^M$ then

$$f(a) = \nabla(f \oplus g)((a \oplus 1) \vee d_L^M) = \nabla(f \oplus g)((1 \oplus a) \vee d_L^M) = g(a). \qquad \square$$

3.2.2. Proposition. *Let L be an I-Hausdorff frame. Then a frame homomorphism $h\colon L \to M$ is an epimorphism iff $\Gamma_M(L) = M$.*

Proof. If $\Gamma_M(L) = M$ then h is an epimorphism by 3.2.1. Now suppose $\Gamma_M(L) \neq M$. The embedding $j\colon \Gamma_M(L) \subset M$ is the equalizer of

$$\alpha = \breve{d}_{h[L]}^M \cdot \iota_1 \quad \text{and} \quad \beta = \breve{d}_{h[L]}^M \cdot \iota_2,$$

by definition, and hence $\alpha \neq \beta$. But $h\alpha = h\beta$ so that h is not an epimorphism.   □

3.3. Notes. (1) In **Frm** the regular monomorphisms do not coincide with the extremal ones. By [212] they are not even closed under composition.

(2) One would like some special monomorphisms in **Frm** to model quotient continuous maps. A partial characterization is obtained as follows. Define $E_M(L)$ as the smallest subframe of M containing L that is closed under the solutions x of the equations

$$a \vee x = b,$$
$$a \wedge x = c.$$

By [226], for a wide class of spaces (including all the metrizable ones) a continuous $p\colon X \to Y$ is a quotient iff $E_M(h[L]) = h[L]$ for $h = \Omega(p)$ and $L = \Omega(Y)$.

The two mentioned operators are related by the inclusion $E_M(L) \subseteq \Gamma_M(L)$ (see [214]).

4.   Aside: Raney identity

In this section we will see that under the DS-Hausdorff condition, spatiality can be warranted by strengthening the frame distribution law.

4.1. The system of *all* subsets of a set X satisfies the complete distributivity law

$$\bigcup_{i \in J} \bigcap \mathcal{A}_i = \bigcap \left\{ \bigcup_{i \in J} \phi(i) \ \middle|\ \phi \in \prod_{i \in J} \mathcal{A}_i \right\} \tag{4.1.1}$$

for any system $\mathcal{A}_i$, $i \in J$, of sets of subsets of X.

 Now consider the frame $\Omega(X)$ of open sets of a topological space X. Let $\mathcal{A}_i$, $i \in J$, be a system of *finite* subsets of $\Omega(X)$. Then the left-hand side of (4.1.1) is open, and hence so is the intersection on the right-hand side, which then coincides with the meet in the lattice $\Omega(X)$. Thus, in $\Omega(X)$ one has for any system $\mathcal{F}_i$, $i \in J$, of finite $\mathcal{F}_i \subseteq \Omega(X)$,

$$\bigcup_{i \in J} \bigcap \mathcal{F}_i = \bigwedge \left\{ \bigcup_{i \in J} \phi(i) \ \middle|\ \phi \in \prod_{i \in J} \mathcal{F}_i \right\}.$$

4.2. Quasitopologies. A complete lattice L is said to be a *quasitopology* if it satisfies the distributivity law

$$\bigvee_{i \in J} \bigwedge F_i = \bigwedge \left\{ \bigvee_{i \in J} \phi(i) \ \middle|\ \phi \in \prod_{i \in J} F_i \right\} \tag{RD}$$

for any system F_i, $i \in J$, of finite $F_i \subseteq L$. (Note that each quasitopology is a frame: set $F_i = \{a, b_i\}$ to obtain $a \wedge \bigvee_{i \in J} b_i$ on the right-hand side realizing that if $\phi(k) = a$ for some k then $\bigvee_{i \in J} \phi(i) \geq a$.)

 We have seen above that every topology is a quasitopology. In 1953 ([231]), Raney proved that

  every quasitopology is a complete homomorphic image of a topology

and formulated the problem whether every quasitopology is (isomorphic to) a topology. In our terminology,

  each frame satisfying (RD) is a complete sublocale of a spatial frame

(by a *complete sublocale* we understand such one that the associated frame homomorphism is complete), and the problem is whether each frame satisfying (RD) is spatial.

 This was answered in the negative in [168]. However, the answer was shown to be positive for DS-Hausdorff frames, as we will see shortly.

4.3. Recall that an element $p \neq 1$ of a distributive lattice L is prime (or meet-irreducible) if $x_1 \wedge x_2 \leq p$ implies that $x_1 \leq p$ or $x_2 \leq p$ (II.3.3). Equivalently, if $x_1 \wedge x_2 = p$ implies that $x_1 \leq p$ or $x_2 \leq p$.

It is said to be *semiprime* if $x_1 \wedge x_2 = 0$ implies that $x_1 \leq p$ or $x_2 \leq p$, and *a-semiprime* for an $a \in L$ if for any $x_1, \ldots, x_n \in L$,

$$x_1 \wedge \cdots \wedge x_n \leq a \quad \Rightarrow \quad \exists i, \; x_i \leq p.$$

Obviously we have the implication

$$a\text{-semiprime} \quad \Rightarrow \quad \text{semiprime}.$$

4.4. Lemma. *In a DS-Hausdorff frame, each semiprime element is prime.*

Proof. If p is not prime we have $a, b \not\leq p$ such that $a \wedge b = p$. Hence $a \vee b \neq a, b$ and there are u, v such that $u \wedge v = 0$, $u \not\leq a$ and $v \not\leq b$. Then $u, v \not\leq a \wedge b = p$ and hence p is not semiprime. $\qquad\square$

4.4.1. Note. It may be of some interest to consider (semiprime $\Rightarrow$ prime) as a sort of separation axiom.

4.5. Lemma *A frame L satisfies* (RD) *iff each $a \in L$ is a meet of a-semiprime elements.*

Proof. Let L satisfy (RD). For $a \in L$ take the system F_i, $i \in J$, of all finite $F \subseteq L$ such that $\bigwedge F \leq a$. Since in particular $\{a\}$ is in the system we have

$$a = \bigvee_{i \in J} \bigwedge F_i = \bigwedge \left\{ \bigvee_{i \in J} \phi(i) \;\middle|\; \phi \in \prod_{i \in J} F_i \right\}.$$

Now each $p = \bigvee_{i \in J} \phi(i)$ with $\phi \in \prod_{i \in J} F_i$ is a-semiprime: if $x_1 \wedge \cdots \wedge x_n \leq a$ then $\{x_1, \ldots, x_n\} = F_k$ for some k and we have an $x_j = \phi(k) \leq \bigvee_{i \in J} \phi(i) = p$.

On the other hand let each $a \in L$ be a meet of a-semiprime elements. Take a system F_i, $i \in J$, of finite subsets of L and set

$$a = \bigvee_{i \in J} \bigwedge F_i.$$

The element a is $\bigwedge_{m \in M} p_m$ with semiprime p_m. Take a fixed m. Then, since $\bigwedge F_i \leq a$ there is a $\phi_m(k) \in F_k$ that is $\leq p_m$. Hence, for thus defined ϕ_m, $\bigvee_{i \in J} \phi_m(i) \leq p_m$. Hence $\bigwedge \left\{ \bigvee_{i \in J} \phi(i) \;\middle|\; \phi \in \prod_{i \in J} F_i \right\} \leq p_m$ for all M and we have

$$\bigwedge \{ \bigvee_{i \in J} \phi(i) \mid \phi \in \prod_{i \in J} F_i \} \leq \bigwedge_{m \in M} p_m = a = \bigvee_{i \in J} \bigwedge F_j.$$

The other inequality is trivial. $\qquad\square$

4.6. Proposition. *Let L be DS-Hausdorff. Then it is spatial if and only if it satisfies the distributivity rule* (RD).

Proof. If L is spatial then it satisfies (RD) by 4.1. Now let L satisfy (RD). Then by 4.5 and 4.4 each $a \in L$ is a meet of prime elements and hence L is spatial by II.5.3. $\qquad\square$

5. Quite like the classical case: Regular, completely regular and normal

5.1. Classical regularity. A space X is regular if for any closed $F \subseteq X$ and any $x \notin F$ there are open V_1, V_2 such that $V_1 \cap V_2 = \emptyset$, $x \in V_1$ and $F \subseteq V_2$. Equivalently, if $x \in U$ and U is open there is an open V such that $x \in V \subseteq \overline{V} \subseteq U$ (take $F = X \smallsetminus U$ and set $V = V_1$). Thus we see that

$$X \text{ is regular iff for every open } U \subseteq X,\ U = \bigcup\{V \in \Omega(X) \mid \overline{V} \subseteq U\}.$$

The relation $\overline{V} \subseteq U$ can be expressed as

$$V \prec U \quad \equiv \quad \exists W \text{ open such that } V \cap W = \emptyset \text{ and } W \cup U = X$$

(or, considering the pseudocomplement V^* in $\Omega(X)$, that is, $X \smallsetminus \overline{V}$,

$$V \prec U \quad \equiv \quad V^* \cup U = X. \text{)}$$

5.2. The "rather below" relation. In the setting of general frames define

$$a \prec b \quad \equiv \quad a^* \vee b = 1, \tag{rb1}$$

or equivalently

$$a \prec b \quad \equiv \quad \exists c,\ a \wedge c = 0 \text{ and } c \vee b = 1. \tag{rb2}$$

The first formulation is more concise, but the latter is sometimes more handy.
The fact that $a \prec b$ is usually referred to by saying that

$$a \text{ is } rather\ below\ b.$$

5.2.1. Lemma. (1) $a \prec b \Rightarrow a \leq b$.

(2) $0 \prec a \prec 1$ for every $a \in L$.

(3) $x \leq a \prec b \leq y \Rightarrow x \prec y$.

(4) If $a \prec b$ then $b^* \prec a^*$.

(5) If $a \prec b$ then $a^{**} \prec b$.

(6) If $a_i \prec b_i$ for $i = 1, 2$ then $a_1 \vee a_2 \prec b_1 \vee b_2$ and $a_1 \wedge a_2 \prec b_1 \wedge b_2$.

Proof. We use the standard properties of pseudocomplements.

(1): $a = a \wedge 1 = a \wedge (a^* \vee b) = a \wedge b$.

(2) is trivial and (3) follows from the implication $x \leq a \Rightarrow a^* \leq x^*$.

(4) follows from $b \leq b^{**}$ and (5) from $a^{***} = a^*$.

(6): $(a_1 \vee a_2)^* \vee (b_1 \vee b_2) = (a_1^* \wedge a_2^*) \vee (b_1 \vee b_2) \geq (a_1^* \vee b_1) \wedge (a_2^* \vee b_2) = 1$ and $(a_1 \wedge a_2)^* \vee (b_1 \wedge b_2) \geq (a_1^* \vee a_2^*) \vee (b_1 \wedge b_2) \geq (a_1^* \vee b_1) \wedge (a_2^* \vee b_2) = 1$. $\square$

5.3. Regularity. A locale L is said to be *regular* if

$$a = \bigvee \{x \mid x \prec a\} \text{ for every } a \in L. \tag{reg}$$

From 5.1 we see that this is an exact extension of the homonymous classical property, that is, we have

Observation. *A space X is regular iff the locale $\mathsf{Lc}(X)$ is regular.*

5.4. Although the I-Hausdorff property is not a precise equivalent of the classical property, it stands in an analogous relation to regularity.

5.4.1. Lemma. *Let L be an arbitrary locale and let $U \subseteq L \times L$ be saturated and such that $d_L \subseteq U$. Let $(a \wedge b, a \wedge b) \in U$. Then $(x, y) \in U$ for every $x \prec a$, $y \prec b$.*

(Compare this statement with 2.3.)

Proof. Let $x^* \vee a = y^* \vee b = 1$. Then by IV.5.2(3)

$$
\begin{aligned}
(x, y) \in\ x \oplus y &= (x \wedge (y^* \vee b)) \oplus (y \wedge (x^* \vee a)) \\
&= ((x \wedge y^*) \vee (x \wedge b)) \oplus ((y \wedge x^*) \vee (y \wedge a)) \\
&= ((x \wedge y^*) \oplus (y \wedge x^*)) \vee ((x \wedge y^*) \oplus (y \wedge a)) \vee \\
&\quad \vee ((x \wedge b) \oplus (y \wedge x^*)) \vee ((x \wedge b) \oplus (a \wedge y)) \\
&\leq (x \oplus x^*) \vee (y^* \oplus y) \vee (x \oplus x^*) \vee ((a \wedge b) \oplus (a \wedge b)) \subseteq U
\end{aligned}
$$

since $x \oplus x^*$, $y^* \oplus y$ and $(a \wedge b) \oplus (a \wedge b)$ are subsets of the saturated U. $\square$

5.4.2. Proposition. *Each regular locale is I-Hausdorff (and hence also DS-Hausdorff).*

Proof. If the locale is regular and if $(a \wedge b, a \wedge b) \in U$ we have, in the notation of the previous proof, first,

$$(a, y) = (\bigvee \{x \mid x \prec a\}, y) \in U$$

for $y \prec b$, and then $(a, b) = (a, \bigvee \{y \mid y \prec b\}) \in U$. $\square$

5.5. Regularity compared with fitness. Compare the formulas

$$
\begin{array}{llll}
a \not\leq b & \Rightarrow & \exists c,\ a \vee c = 1 \neq b \vee c, & \text{(subfit)} \\
a \not\leq b & \Rightarrow & \exists c,\ a \vee c = 1 \text{ and } c \to b \not\leq b, & \text{(fit)} \\
a \not\leq b & \Rightarrow & \exists c,\ a \vee c = 1 \text{ and } c \to 0 \not\leq b. & \text{(reg)}
\end{array}
$$

(The last is indeed equivalent with the formula in 5.3. If that holds and if $a \not\leq b$, there is an x such that $x^* \vee a = 1$ and $x \not\leq b$. Set $c = x^*$. Then $x \leq c^* = c \to 0$ and hence $c \to 0 \not\leq b$. Conversely, if the new implication holds and $a \not\leq \bigvee \{b \mid b \prec a\}$ then we have a c such that $c \vee a = 1$ and $c^* \not\leq \bigvee \{b \mid b \prec a\}$, contradicting $c^* \prec a$.)

Thus we see that regularity is related to fitness and obviously (since $c \to 0 \le c \to b$)

$$(\text{reg}) \quad \Rightarrow \quad (\text{fit}) \quad \Rightarrow \quad (\text{subfit}).$$

The affinity is deeper. For $a \in L$ and ordinals α set

$$a_0 = 0, \quad a_{\alpha+1} = \bigvee\{b \mid \exists x, \ x \vee a = 1 \text{ and } b \wedge x \le a_\alpha\}, \text{ and}$$

$$a_\lambda = \bigvee_{\alpha < \lambda} a_\alpha \quad \text{for limit ordinals.}$$

Thus, L is regular iff $a = a_1$ for all $a \in L$; it can be proved that L is fit iff for every $a \in L$ there is an α such that $a = a_\alpha$ (see [130]).

5.6. Since for obvious reasons regularity is used more often than the weaker axioms, it may be useful to list a few facts following from 5.4 and 5.5 (and 1.3.2, 1.6, 1.8 and 2.5).

Facts. *Let L be a regular frame (locale). Then:*

(1) *If E, E' are frame congruences on L and $E1 = E'1$ then $E = E'$.*

(2) *Each sublocale of L is an intersection of open sublocales.*

(3) *Each complete frame homomorphism $h \colon L \to M$ is Heyting.*

(4) *Each co-dense frame homomorphism $h \colon M \to L$ is one-one.*

(5) *The coequalizer of frame homomorphisms $h_1, h_2 \colon L \to M$ is given by*

$$\check{c} = (a \mapsto a \vee c) \colon M \to {\uparrow}c$$

where $c = \bigvee\{h_1(x) \wedge h_2(y) \mid x \wedge y = 0\}$.

(6) *Each dense frame homomorphism $h \colon M \to L$ is a monomorphism in the category of regular frames.* □

Note. In fact, in the category of regular frames, the monomorphisms are precisely the dense homomorphisms – see 6.5.

5.7. Complete regularity. Let a and b be elements of L. We say that a is *completely below b* in L and write

$$a \lll b$$

if there are $a_r \in L$ (r rational, $0 \le r \le 1$) such that

$$a_0 = a, \quad a_1 = b \quad \text{and} \quad a_r \prec a_s \quad \text{for } r < s.$$

Note that the relation $\lll$ is the largest interpolative relation contained in $\prec$ (a relation R is *interpolative* if $aRb \Rightarrow aRcRb$ for some c). From 5.2.1 we immediately obtain

Lemma. (1) $a \lll b \Rightarrow a \leq b$.

(2) $0 \lll a \lll 1$ *for every* $a \in L$.

(3) $x \leq a \lll b \leq y \Rightarrow x \lll y$.

(4) *If* $a \lll b$ *then* $b^* \lll a^*$.

(5) *If* $a \lll b$ *then* $a^{**} \lll b$.

(6) *If* $a_i \lll b_i$ *for* $i = 1, 2$ *then* $a_1 \vee a_2 \lll b_1 \vee b_2$ *and* $a_1 \wedge a_2 \lll b_1 \wedge b_2$.   □

5.7.1. A locale L is said to be *completely regular* if

$$a = \bigvee \{x \mid x \lll a\} \text{ for every } a \in L. \tag{creg}$$

This, too, is an extension of the homonymous property from classical topology, that is

Observation. *A space* X *is completely regular iff the locale* $\mathsf{Lc}(X)$ *is completely regular.*

> (The implication $\Rightarrow$ is immediate while $\Leftarrow$ is a bit more involved, requiring the construction of suitable real functions, imitating the standard proof of Urysohn's Lemma – see XIV.6.2.2.)

5.8. Using (rb2) from 5.2 we immediately obtain (note that it is for this purpose handier than (rb1))

Lemma. *For any frame homomorphism* $h \colon L \to M$, $a \prec b \Rightarrow h(a) \prec h(b)$, *and consequently* $a \lll b \Rightarrow h(a) \lll h(b)$.   □

We immediately infer

Corollary. *Each sublocale of a regular (resp. completely regular) locale is regular (resp. completely regular).*   □

5.9. Normality. This is also an immediate extension of the homonymous classical property: a locale L is *normal* if $a \vee b = 1$ for $a, b \in L$ implies the existence of $u, v \in L$ such that $u \wedge v = 0$ and $a \vee v = 1 = b \vee u$.
(Equivalently, if $a \vee b = 1$ for $a, b \in L$ implies the existence of $u \in L$ such that $a \vee u^* = 1 = b \vee u$.)

5.9.1. Lemma. *In any normal locale the relation* $\prec$ *is interpolative. Consequently, in normal locales* $\lll$ *coincides with* $\prec$ *and regularity coincides with complete regularity.*

Proof. Let $a \prec b$ in a normal locale L. Then there are $u, v \in L$ such that $u \leq v^*$ and $a^* \vee v = 1 = b \vee u$. Thus $a \prec v$ and $v^* \vee b \geq u \vee b = 1$ so that $v \prec b$.   □

5.9.2. Proposition. *Any normal and subfit locale is completely regular.*

Proof. Since $\twoheadrightarrow\!\!\prec \,=\, \prec$ it suffices to show that L is regular. For $a \in L$ set $b = \bigvee\{x \mid x \prec a\}$. Let $a \vee c = 1$. By normality there is $u \in L$ such that $a \vee u^* = 1 = c \vee u$. Thus $u \leq b$ and $b \vee c = 1$. Hence we have proved that $a \vee c = 1$ always implies $b \vee c = 1$. Thus by subfitness $a \leq b$.   □

5.9.3. Note. In classical topology one usually thinks of the hierarchy of separation axioms as

$$T_4 \& T_1 \quad \Rightarrow \quad T_3 \& T_1 \quad \Rightarrow \quad T_2.$$

Under closer scrutiny one sees that for the latter implication one does not have to assume T_1 and that one has

$$T_3 \& T_0 \quad \Rightarrow \quad T_2$$

while $T_4 \& T_0$ does not imply T_1. Hence one has to have something stronger than T_0; not necessarily T_1, the subfitness suffices.

    The reader is certainly not surprised that here the regularity implies the Hausdorff property without any extra assumption: in the point-free context, T_0 is irrelevant.

6.   The categories RegLoc, CRegLoc, HausLoc and FitLoc

6.1. Lemma. *For any locale L,*

$$L_\prec = \{a \mid a = \bigvee\{x \mid x \prec a\}\}$$

is a subframe of L.

Proof. Clearly $1 \in L_\prec$ and $L_\prec$ is closed under joins. Moreover, if $a, b \in L_\prec$ we have by property (6) in 5.2.1 that

$$a \wedge b = \bigvee\{x \wedge y \mid x \prec a, y \prec b\} \leq \bigvee\{z \mid z \prec a \wedge b\} \leq a \wedge b. \qquad \square$$

6.2. For any subframe S of a frame L the map $p_S \colon L \to S$ given by $p_S(x) = \bigvee\{y \in S \mid y \leq x\}$ is a localic map (it is the right adjoint to the frame embedding $S \subseteq L$).

Lemma. *Let $f \colon L \to M$ be a localic map and let S be a subframe of L. If $f^*(a) \in S$ for every $a \in M$ then $f|_S \colon S \to M$ is a localic map such that $f|_S \cdot p_S = f$.*

Proof. It is obvious: by the condition on the f^* this is a frame homomorphism factorizing through the embedding $S \subseteq L$ and $f|_S$ is the right adjoint of $f^* \colon M \to S$.
                                                                                           □

6.3. The categories RegLoc and CRegLoc. Let **RegLoc** (resp. **CRegLoc**) denote the full subcategory of **Loc** generated by regular (resp. completely regular) locales. (Accordingly, we will denote by **RegFrm** and **CRegFrm** the corresponding categories of frames.)

Let L be a general locale and define $R_\alpha(L)$ for ordinals α by setting

$$R_0(L) = L, \quad R_{\alpha+1}(L) = (R_\alpha(L))_{\prec}, \quad \text{and}$$

$$R_\alpha(L) = \bigcap_{\beta < \alpha} R_\beta(L) \quad \text{for a limit ordinal } \alpha.$$

Denote $\bigcap_{\beta \in Ord} R_\beta(L)$ by $R_\infty(L)$. Since $(R_\infty(L))_{\prec} = R_\infty(L)$, this is a regular locale. We have (see AII.8.1.1)

6.3.1. Proposition. **RegLoc** *is epireflective in* **Loc**, *with epireflection maps*

$$\rho_L \colon L \to R_\infty(L)$$

given by $\rho_L(a) = \bigvee\{x \in R_\infty(L) \mid x \le a\}$.

Proof. Let M be a regular locale and let $f \colon L \to M$ be a localic map. By Lemma 5.8, for any $a \in M$,

$$f^*(a) = \bigvee\{f^*(x) \mid x \prec a\} \le \bigvee\{y \mid y \prec f^*(a)\} \le f^*(a)$$

and thus $f^*(a) \in R_1(L)$. Hence by 6.1 we have a localic $f|_{R_1(L)} \colon R_1(L) \to M$ such that $f|_{R_1(L)} \cdot p_{R_1(L)} = f$. Proceeding by induction we eventually get $f|_{R_\infty(L)} \colon R_\infty(L) \to M$ such that $f|_{R_\infty(L)} \cdot \rho_L = f$. □

Then, by AII.8.3 we conclude that

6.3.2. Corollary. *The category* **RegLoc** *is complete and cocomplete, and the limits in it coincide with those in* **Loc**. *In particular, sublocales and products of regular locales, taken in* **Loc**, *are regular.* □

6.3.3. In a similar way, now considering the subframe

$$L_{\prec\prec} = \{a \mid a = \bigvee\{x \mid x \prec\prec a\}\}$$

of L and defining $\widetilde{R}_\alpha(L)$ for ordinals α by setting

$$\widetilde{R}_0(L) = L, \quad \widetilde{R}_{\alpha+1}(L) = (R_\alpha(L))_{\prec\prec},$$

$$\widetilde{R}_\alpha(L) = \bigcap_{\beta < \alpha} \widetilde{R}_\beta(L) \quad \text{for a limit ordinal } \alpha$$

and

$$\widetilde{R}_\infty(L) = \bigcap_{\beta \in Ord} \widetilde{R}_\beta(L),$$

one gets the following

Proposition. CRegLoc *is epireflective in* **Loc,** *with epireflection maps*

$$\widetilde{\rho}_L \colon L \to \widetilde{R}_\infty(L)$$

given by $\widetilde{\rho}_L(a) = \bigvee \{x \in \widetilde{R}_\infty(L) \mid x \leq a\}.$ □

Corollary. *The category* **CRegLoc** *is complete and cocomplete, and the limits in it coincide with those in* **Loc.** *In particular, sublocales and products of completely regular locales, taken in* **Loc,** *are regular.*   □

6.4. The categories HausLoc and FitLoc. Let **HausLoc** (resp. **FitLoc**) denote the full subcategory of **Loc** generated by I-Hausdorff (resp. fit) locales.

6.4.1. Lemma. *Let L be a locale. Then:*

(a) *The subframe generated by any family of I-Hausdorff subframes of L is I-Hausdorff.*

(b) *The subframe generated by any family of fit subframes of L is fit.*

Proof. (a): Suppose that $\{M_i \mid i \in J\}$ is a family of I-Hausdorff subframes of L and M is the frame they generate. Each $x \in M$ is obtained as a join of finite meets of the form $a_1 \wedge \cdots \wedge a_k$ with $a_j \in M_{i_j}$. By 2.3, I-Hausdorff locales L are characterized by the condition

$$a \oplus b \leq (a \wedge b \oplus a \wedge b) \vee d_L \text{ for every } a, b \in L. \qquad (6.4.1)$$

Thus it suffices to check it for M. Let $a = \bigvee_i (a_1^i \wedge \cdots \wedge a_{k_i}^i)$ with $a_j^i \in M_{i_j}$ and $b = \bigvee_r (b_1^r \wedge \cdots \wedge b_{t_r}^r)$ with $b_s^r \in M_{r_s}$. Using IV.5.2(3) we obtain

$$a \oplus b = \bigvee_{i,r} (a_1^i \wedge \cdots \wedge a_{k_i}^i \oplus b_1^r \wedge \cdots \wedge b_{t_r}^r)$$

thus it suffices to show that for each i, r

$$a_1^i \wedge \cdots \wedge a_{k_i}^i \oplus b_1^r \wedge \cdots \wedge b_{t_r}^r \leq (a \wedge b \oplus a \wedge b) \vee d_M.$$

First note that

$$a_1^i \wedge \cdots \wedge a_{k_i}^i \oplus b_1^r \wedge \cdots \wedge b_{t_r}^r \leq \bigcap_{j=1}^{k_i} (a_j^i \oplus 1) \cap \bigcap_{s=1}^{t_r} (1 \oplus b_s^r).$$

Then, by hypothesis, each $a_j^i \oplus 1 \leq (a_j^i \oplus a_j^i) \vee d_{M_{i_j}}$ and each $1 \oplus b_s^r \leq (b_s^r \oplus b_s^r) \vee d_{M_{r_s}}$. Hence $a_1^i \wedge \cdots \wedge a_{k_i}^i \oplus b_1^r \wedge \cdots \wedge b_{t_r}^r$ is included in

$$\bigcap_{j=1}^{k_i} \left((a_j^i \oplus a_j^i) \vee d_{M_{i_j}} \right) \cap \bigcap_{s=1}^{t_r} \left((b_s^r \oplus b_s^r) \vee d_{M_{r_s}} \right)$$

$$\leq (a_1^i \wedge \cdots \wedge a_{k_i}^i \wedge b_1^r \wedge \cdots \wedge b_{t_r}^r \oplus a_1^i \wedge \cdots \wedge a_{k_i}^i \wedge b_1^r \wedge \cdots \wedge b_{t_r}^r) \vee d_M.$$

(b): Suppose that $\{M_i \mid i \in J\}$ is a family of fit subframes of L and M is the frame they generate. By 1.3.2, it suffices to show that each closed sublocale of M is an intersection of open sublocales. So let $\mathfrak{c}_M(m)$ be a closed sublocale of M. We may write $m = \bigvee_i (a_1^i \wedge \cdots \wedge a_{k_i}^i)$ with $a_j^i \in M_{i_j}$. Then

$$\mathfrak{c}_L(m) = \mathfrak{c}_L(\bigvee_i (a_1^i \wedge \cdots \wedge a_{k_i}^i)) = \bigcap_i (\mathfrak{c}_L(a_1^i) \vee \cdots \vee \mathfrak{c}_L(a_{k_i}^i))$$

and therefore $\mathfrak{c}_M(m) = \mathfrak{c}_L(m) \cap M = \bigcap_i (\mathfrak{c}_M(a_1^i) \vee \cdots \vee \mathfrak{c}_M(a_{k_i}^i))$. By hypothesis each M_{i_j} is fit so $\mathfrak{c}_M(a_j^i) = \bigcap_{\alpha \in \Lambda_{i_j}} \mathfrak{o}_M(x_\alpha^{i_j})$ with $x_\alpha^{i_j} \in M_{i_j}$. Finally,

$$\mathfrak{c}_M(m) = \bigcap_i \left(\bigcap_{\alpha \in \Lambda_{i_1}} \mathfrak{o}_M(x_\alpha^{i_1}) \vee \cdots \vee \bigcap_{\alpha \in \Lambda_{i_{k_i}}} \mathfrak{o}_M(x_\alpha^{i_{k_i}}) \right)$$

$$= \bigcap_i \bigcap_{\alpha_{i_1} \in \Lambda_{i_1}, \ldots, \alpha_{k_i} \in \Lambda_{i_{k_i}}} \left(\mathfrak{o}_M(x_{\alpha_{i_1}}^{i_1}) \vee \cdots \vee \mathfrak{o}_M(x_{\alpha_{k_i}}^{i_{k_i}}) \right)$$

$$= \bigcap_i \mathfrak{o}_M \left(\bigwedge_{\alpha_{i_1} \in \Lambda_{i_1}, \ldots, \alpha_{k_i} \in \Lambda_{i_{k_i}}} (x_{\alpha_{i_1}}^{i_1} \vee \cdots \vee x_{\alpha_{k_i}}^{i_{k_i}}) \right)$$

and $\bigwedge_{\alpha_{i_1} \in \Lambda_{i_1}, \ldots, \alpha_{k_i} \in \Lambda_{i_{k_i}}} (x_{\alpha_{i_1}}^{i_1} \vee \cdots \vee x_{\alpha_{k_i}}^{i_{k_i}}) \in M$. $\square$

6.4.2. Proposition. **HausLoc** *and* **FitLoc** *are epireflective in* **Loc**.

Proof. Let L be a locale and let $\mathrm{Haus}(L)$ be the subframe of L generated by all I-Hausdorff subframes of L. We have then the localic map (recall 6.2)

$$p_{\mathrm{Haus}(L)} \colon L \to \mathrm{Haus}(L)$$

and by 6.4.1 $\mathrm{Haus}(L)$ is I-Hausdorff. Finally, given a localic map $f \colon L \to M$ with M I-Hausdorff, by 2.4 $f[L]$ is I-Hausdorff which means that $f^*(a) \in \mathrm{Haus}(L)$ for every $a \in M$. Hence by 6.2 the map $f|_{\mathrm{Haus}(L)}$ is the map we want. Uniqueness follows from the fact that $p_{\mathrm{Haus}(L)}$ is onto.

A similar reasoning can be made for the fit case. $\square$

6.4.3. Corollary. *The categories* **HausLoc** *and* **FitLoc** *are complete and cocomplete, and the limits in it coincide with those in* **Loc**. *In particular, sublocales and products of Hausdorff (resp. fit) locales, taken in* **Loc**, *are Hausdorff (resp. fit).*
          $\square$

6.5. By 2.5 equalizers in **HausLoc** are closed sublocales, given by a remarkable simple formula: the equalizer of $f_1, f_2 \colon L \to M$ is just the closed sublocale $j \colon {\uparrow}c \subseteq L$, where c stands for

$$\bigvee \{ f_1^*(x) \wedge f_2^*(y) \mid x \wedge y = 0 \}.$$

We have seen (2.5.3) that in **HausLoc** (and, in particular, in **RegLoc** and **CRegLoc**) each dense morphism is an epimorphism. The converse is also true:

Proposition. *The epimorphisms in* **RegLoc** *and* **CRegLoc** *are precisely the dense localic maps.*

Proof. It will be done in frame language, and for the regular case. The completely regular case follows by similar reasoning (we only have to adjust $\prec$ to $\prec\!\!\prec$).

We already know that each dense morphism is a monomorphism in **RegFrm**. Now suppose a homomorphism $h\colon L \to M$ between regular frames is not dense. Thus, there is an $a \neq 0$ such that $h(a) = 0$. Set

$$N = \{(x,y) \in L \times L \mid x \vee a = y \vee a\}.$$

Obviously N is a subframe of $L \times L$.

For $(u,v) \in L \times L$ define

$$(uv)_1 = u \wedge (v \vee a) \quad \text{and} \quad (uv)_2 = v \wedge (u \vee a).$$

Then $(uv)_1 \vee a = (u \vee a) \wedge (v \vee a) = (uv)_2 \vee a$ so that $((uv)_1, (uv)_2) \in N$. Now let $(x,y) \in N$, $u \prec x$ and $v \prec y$. We have elements s, t such that

$$s \wedge u = 0, s \vee x = 1, t \wedge u = 0 \text{ and } t \vee y = 1.$$

Then obviously $(uv)_i \wedge (st)_i = 0$, $i = 1, 2$, and furthermore

$$(st)_1 \vee x = (s \wedge (t \vee a)) \vee x = (s \vee x) \wedge (t \vee a \vee x) = (s \vee x) \wedge (t \vee y \vee a) = 1$$

and similarly $(st)_2 \vee y = 1$ so that we have $((uv)_1, (uv)_2) \wedge ((st)_1, (st)_2) = 0$ and $((st)_1, (st)_2) \vee (x,y) = 1$, that is, $((uv)_1, (uv)_2) \prec (x,y)$ in N. Now we have

$$
\begin{aligned}
\bigvee\{((uv)_1, &(uv)_2) \mid u \prec x, v \prec y\} \\
&= (\bigvee_{u \prec x, v \prec y} u \wedge (v \vee a), \bigvee_{u \prec x, v \prec y} v \wedge (u \vee a)) \\
&= (\bigvee_{u \prec x} u \wedge (\bigvee_{v \prec y} v \vee a), \bigvee_{v \prec y} v \wedge (\bigvee_{u \prec x} u \vee a)) \\
&= (x \wedge (y \vee a), y \wedge (x \vee a)) = (x, y)
\end{aligned}
$$

(since $x \vee a = y \vee a$). Thus, N is regular.

Now define $g_i\colon N \to L$ ($i = 1, 2$) by setting $g_i(x_1, x_2) = x_i$. Then $g_1(a, 0) = a \neq 0 = g_2(a, 0)$ while

$$hg_i(x_1, x_2) = h(x_i) = h(x_i) \vee h(a) = h(x_i \vee a) = h(x_1 \vee a)$$

and hence h is not a monomorphism. □

6.5.1. Note. Proposition 6.5 can be extended to I-Hausdorff locales (frames) as well. One uses the same subframe N: only proving that N is I-Hausdorff is somewhat less transparent than the regularity above.

6.6. Well-poweredness. Consider the set

$$\mathfrak{J}(L)$$

of all the ideals of a frame L, ordered by inclusion (this is a frame – see VII4.1.1 for a proof).

6.6.1. Lemma. *Let* $h \colon L \to M$ *be a dense frame homomorphism with* L *regular. Then the mapping*

$$\phi \colon L \to \mathfrak{J}(M)$$

defined by $\phi(a) = \downarrow \{h(x) \mid x \prec a\}$ *is one-one.*

Proof. If h is dense then $h_*(0) = 0$ for the right adjoint h_*. For each ideal J of M set

$$\psi(J) = \bigvee\{h_*(b) \mid b \in J\}.$$

We will show that $\psi\phi = \mathrm{id}$. Since

$$\psi(\phi(a)) = \bigvee\{h_* h(x) \mid x \prec a\}$$

and $h_* h(x) \geq x$ it suffices to show that if $x \prec a$ then $h_* h(x) \prec a$.

If $x \prec a$ then there is a v such that $x \wedge v = 0$ and $a \vee v = 1$. Then of course $a \vee h_* h(v) = 1$; on the other hand, since h and h_* preserve finite meets we have

$$h_* h(x) \wedge h_* h(v) = h_* h(x \wedge v) = h_*(0) = 0$$

and hence $h_* h(x) \prec a$. □

6.6.2. Corollary. *The category* **RegFrm** *is well powered.*

Proof. By the lemma, if $h \colon L \to M$ is a monomorphism then the cardinality $|L|$ of L is at most $|\mathfrak{J}M|$. □

Chapter VI

More on Sublocales

1. Subspaces and sublocales of spaces

1.1. Let X be a $(T_0\text{-})$space and let $A \subseteq X$ be a subspace. We have the embedding

$$j_A \colon A \subseteq X$$

and the sublocale homomorphism (i.e., frame quotient)

$$\Omega(j_A) = (U \mapsto U \cap A) \colon \Omega(X) \to \Omega(A).$$

The associated frame congruence on $\Omega(X)$ is, of course,

$$E_A = \{(U,V) \mid U \cap A = V \cap A\}. \tag{1.1.1}$$

The homomorphism $\Omega(j_A)$ has the right adjoint $V \mapsto \bigcup\{U \mid U \cap A = V\}$ and consequently (see III.5) our subspace is represented by the sublocale

$$\widetilde{A} = \{U \mid U \text{ maximal in the class } E_A U\} \subseteq \mathsf{Lc}(X)$$

(corresponding to E_A in the isomorphism $\mathcal{Sl}(\mathsf{Lc}(X))^{\mathrm{op}} \cong \mathfrak{C}(\mathsf{Lc}(X))$ from III.5.2).

1.1.1. Observation. *For any $A, B \subseteq X$ we have*

$$E_{A \cup B} = E_A \cap E_B.$$

Consequently (use the isomorphism $\mathcal{Sl}(\mathsf{Lc}(X))^{\mathrm{op}} \cong \mathfrak{C}(\mathsf{Lc}(X))$ again),

$$\widetilde{A \cup B} = \widetilde{A} \vee \widetilde{B}.$$

1.2. How well the sublocale $\widetilde{A}$ represents the subspace A. Recall I.4. The congruences E_A on $\Omega(X)$ (see I.4.1.1) – and hence the sublocales $\widetilde{A}$ of $\mathsf{Lc}(X)$ – uniquely represent the subsets (subspaces) of X if and only if X is T_D. More precisely one generally has

$$A \subseteq B \quad \Rightarrow \quad \widetilde{A} \subseteq \widetilde{B} \tag{1.2.1}$$

but only under T_D

$$A \subseteq B \quad \text{iff} \quad \widetilde{A} \subseteq \widetilde{B} \tag{1.2.2}$$

(while already the assumption that distinct subsets induce distinct sublocales implies T_D).

1.2.1. Remark. If X is a T_D-space and $\widetilde{A}$ and $\widetilde{B}$ are mutual complements in $\mathsf{Lc}(X)$ (or, E_A and E_B are mutual complements in $\mathfrak{C}(\mathsf{Lc}(X))$) then $B = X \smallsetminus A$.

(Indeed, $E_A \cap E_B = E_{A \cup B}$ is the smallest congruence, that is, E_X; hence $A \cap B = X$. Since $A \cap B \subseteq A, B$ we have $E_{A \cap B} \supseteq E_A, E_B$ and hence $E_{A \cap B} \supseteq E_\emptyset = E_A \vee E_B$ and finally $A \cap B = \emptyset$.)

But the converse does not hold, that is, $B = X \smallsetminus A$ does not make generally $\widetilde{B}$ a complement of $\widetilde{A}$. Recall III.8.3: if both A and B are dense they meet at least in B_L.

1.3. Recall II.2.4 and the meet-irreducibles

$$\widetilde{x} = X \smallsetminus \overline{\{x\}}$$

representing in $\mathsf{Lc}(X)$ the points $x \in X$. We have

1.3.1. Proposition. *Let X be a T_D-space. Then*

$$x \in A \quad \text{iff} \quad \widetilde{x} \in \widetilde{A}.$$

Proof. Let $x \in A$ and let $U E_A(X \smallsetminus \overline{\{x\}})$, that is, $U \cap A = A \smallsetminus \overline{\{x\}}$. Then $x \notin U$ (else $x \in U \cap A$) and consequently $X \smallsetminus U \ni x$, hence $X \smallsetminus U \supseteq \overline{\{x\}}$ and finally $U \subseteq X \smallsetminus \overline{\{x\}}$; thus, the latter is maximal in the class.

Conversely, assume $x \notin A$, and let X be a T_D-space. Then $X \smallsetminus \overline{\{x\}}$ is not maximal in its E_A-class since this class also contains $W = (X \smallsetminus \overline{\{x\}}) \cup \{x\}$. We just have to prove that W is open, for which it suffices to show that it is a neighbourhood of x. There is a $U \ni x$ such that $U \smallsetminus \{x\}$ is open, that is, $U \smallsetminus \{x\} = U \smallsetminus \overline{\{x\}}$. Thus,

$$U = (U \smallsetminus \{x\}) \cup \{x\} = (U \smallsetminus \overline{\{x\}}) \cup \{x\} \subseteq (X \smallsetminus \overline{\{x\}}) \cup \{x\}. \qquad \square$$

Remark. Note that the implication $\Rightarrow$ holds without any assumption on the space while for the equivalence, in view of the statement (1.2.2) in 1.2, the assumption T_D is necessary.

1.4. In fact, every frame congruence can be expressed in a form extending the formula (1.1.1). We have

1.4.1. Proposition. *Let S be a general sublocale of a locale L. Then we have for the corresponding congruence E_S*

$$aE_Sb \quad \text{iff} \quad \mathfrak{o}(a) \cap S = \mathfrak{o}(b) \cap S$$

(where $\mathfrak{o}(x)$ is the open sublocale associated with x).

Proof. First note that in any distributive lattice, if u, v are complemented with complements $\overline{u}, \overline{v}$, then for any w

$$u \wedge w = v \wedge w \quad \text{iff} \quad \overline{u} \wedge w = \overline{v} \wedge w$$

(if we have the first equation then $\overline{u} \wedge w = \overline{u} \wedge (v \vee \overline{v}) \wedge w = \overline{u} \wedge ((u \wedge w) \vee (\overline{v} \wedge w)) = \overline{u} \wedge \overline{v} \wedge w$, and by symmetry $\overline{v} \wedge w = \overline{u} \wedge \overline{v} \wedge w$). Thus, by III.6.1 it suffices to prove that aE_Sb iff $\uparrow a \cap S = \uparrow b \cap S$.

Recall III.5.3. We have

$$aE_Sb \text{ iff } \nu_S(a) = \nu_S(b) \text{ iff } \bigwedge \{s \in S \mid a \leq s\} = \bigwedge \{s \in S \mid b \leq s\}$$

$$\text{iff for every } s \in S, a \leq s \equiv b \leq s.$$

In other words, iff $\uparrow a \cap S = \uparrow b \cap S$. $\qquad\qquad\square$

2. Spatial and induced sublocales

2.1. In the previous section we have seen that, under T_D, subspaces of spaces are well represented by the corresponding sublocales. That does not mean, however, that all the sublocales of a space (more precisely, of a spatial locale) should be (induced by) subspaces. Typically there are many sublocales that are not.

Consider, say, the real line (or, for that matter, any open subspace X of a Euclidean space). The rational points constitute a dense subspace R while the complement set $C = X \smallsetminus R$ is dense in X as well. Now this makes $\widetilde{R}$ and $\widetilde{C}$ dense in $L = \mathsf{Lc}(X)$ (see III.8.2) and by the Isbell's Density Theorem they contain B_L, the smallest dense sublocale of L. Thus, the sublocale $\widetilde{R} \cap \widetilde{C}$ can not be induced by a subspace $Y \subseteq X$: by 1.3.1 such Y would have to be empty. This example shows that

- there are sublocales that are not subspaces (this, of course, is seen already on the $B_{\mathsf{Lc}(X)}$, without reference to R and C since it is in our case non-atomic Boolean – recall II.5.4 – and hence not spatial at all), and

- unlike the join, the intersection of induced sublocales is not necessarily induced; in particular, we do not necessarily have $\widetilde{A} \cap \widetilde{B} = \widetilde{A \cap B}$ (compare with 1.1.1).

2.2. The sublocale $\widetilde{A}$ in 1.1 is isomorphic with $\Omega(A)$ and hence spatial. Sometimes one speaks of the $\widetilde{A}$ as of the spatial sublocales of $\mathsf{Lc}(X)$ but it is imprecise and misleading. We will use, rather, the term *induced* sublocale (or, a *sublocale induced by a subspace*; *spatially induced* in [246]) while the adjective *spatial* will be used for sublocales $S \subseteq \mathsf{Lc}(X)$ isomorphic with an $\mathsf{Lc}(Y)$ where Y is not necessary in any relation with X.

In general, a sublocale $S \subseteq \mathsf{Lc}(X)$ can be spatial without being induced. However, we have

2.2.1. Proposition. *Let X be a sober space. Then each spatial sublocale $S \subseteq L = \mathsf{Lc}(X)$ is induced by a subspace.*

Proof. Let $\phi\colon S \to \Omega(Y)$ be an isomorphism and let j^* be the left adjoint to the embedding $j\colon S \subseteq L = \Omega(X)$. By I.1.6.2 we have a continuous mapping $f\colon Y \to X$ such that $\Omega(f) = \phi \cdot j^*$. Since $\Omega(f)$ is onto and Y is T_0 (we work with T_0-spaces only, but Y could have been replaced by a T_0-space with the same open set lattice anyway), f is one-to-one (if $f(x) = f(y)$ with $x \neq y$ choose a U such that $x \notin U \ni y$ and a V such that $U = f^{-1}[V]$ to obtain a contradiction $f(x) \notin V \ni f(y)$). Thus we have a decomposition

$$f = (Y \xrightarrow{g=(x \mapsto f(x))} A = f[X] \xrightarrow{k = \subseteq} X)$$

with k a subspace embedding and g a homeomorphism. Thus (recall III.5.1), the sublocales associated with f and k coincide, and the one corresponding to $\Omega(f)$ is

$$\Omega(f)_*[\Omega(Y)] = (\phi j^*)_*[\Omega(Y)] = j[\phi[\Omega(Y)]] = j[S] = S.$$

Now recall 1.1: $\widetilde{A}$ is the sublocale induced by the embedding $k\colon A \subseteq X$, and hence $\widetilde{A} = S$. $\qquad\square$

2.3. Let X be a general (T_0-)space. Recall the sobrification $\lambda_X\colon X \to \mathsf{SpLc}(X)$ from II.6.3. We will represent the space $\mathsf{sob}X = \mathsf{SpLc}(X)$, and the sobrification as the embedding

$$\lambda_X\colon X \subseteq \mathsf{sob}X = X \cup \{\mathcal{F} \mid \mathcal{F} \text{ free completely prime filter in } \mathsf{Lc}(X)\}.$$

Then the open sets of $\mathsf{sob}X = \mathsf{SpLc}(X)$, the original $\Sigma_U = \{\mathcal{F} \mid U \in \mathcal{F}\}$, will appear as

$$U^{\mathsf{s}} = U \cup \{\mathcal{F} \text{ free} \mid U \in \mathcal{F}\}.$$

Recall III.4.6.1 and III.4.6.4. We have the isomorphism $\phi_{\mathsf{Lc}(X)}\colon \mathsf{Lc}(X) \to \mathsf{Lc}(\mathsf{sob}X)$ inverse to $\Omega(\lambda_X)$. In our notation, we have the mutually inverse isomorphisms of frames

$$\phi_{\mathsf{Lc}(X)}\colon \Omega(X) \to \Omega(\mathsf{sob}X), \qquad U \mapsto U^{\mathsf{s}},$$
$$\Omega(\lambda_X)\colon \Omega(\mathsf{sob}X) \to \Omega(X), \qquad V \mapsto V \cup X. \tag{2.3.1}$$

2.3.1. Proposition. *Let* $S \subseteq \mathsf{Lc}(X) = \Omega(X)$ *be a spatial sublocale and let* $q \colon \Omega(X) \to S$ *be the corresponding sublocale homomorphism. Then there is a subset* $A \subseteq \mathsf{sob}X$ *such that*

$$q(U) = q(V) \quad \text{iff} \quad U^{\mathsf{s}} \cap A = V^{\mathsf{s}} \cap A.$$

Proof. We have a sublocale homomorphism

$$\Omega(\mathsf{sob}X) \xrightarrow{\ \Omega(\lambda_X)\ } \Omega(X) \xrightarrow{\ q\ } S$$

with the sober $\mathsf{sob}X$ so that by 2.2.1 there is a subspace A such that $U E_A V$, that is,

$$U \cap A = V \cap A \text{ iff } q(U \cap X) = q(V \cap X).$$

Using the isomorphisms (2.3.1) we obtain the stated equivalence. □

2.3.2. Note. This formula seems to be an easy source of examples of sublocales that are spatial but not induced. The problem is, however, more difficult and cannot be solved immediately.

Let us look more closely at the situation:

– The induced sublocales correspond to the congruences E_A with $A \subseteq X$ (note that it does not matter whether we think of them as congruences on $\Omega(X)$ or on $\Omega(\mathsf{sob}X)$: as $A \subseteq X$, $U \cap A = V \cap A$ iff $U^{\mathsf{s}} \cap A = V^{\mathsf{s}} \cap A$.

– The spatial ones correspond to the congruences E_A with $A \subseteq \mathsf{sob}X$, and we have more such A's.

But:

$$A \text{ non-trivial sobrification is never } T_D.$$

(Indeed, let $\mathcal{F}$ be a free completely prime filter and let $U^{\mathsf{s}} \smallsetminus \{\mathcal{F}\}$ be open for some $U^{\mathsf{s}} \ni \mathcal{F}$; then $U^{\mathsf{s}} \smallsetminus \{\mathcal{F}\} = V^{\mathsf{s}}$ for some V, and $U = U^{\mathsf{s}} \cap X = V^{\mathsf{s}} \cap X = V$, a contradiction.)

Thus (recall 1.2), for all what we know, if $A \not\subseteq X$ there still can exist a $B \subseteq X$ such that $E_A = E_B$.

Non-induced spatial sublocales do exist, though (see [91]).

3. Complemented sublocales of spaces are spatial

3.1. For a subset A of a locale L define

$$\mathsf{m}(A) = \{\textstyle\bigwedge B \mid B \subseteq A\}.$$

Extending the definition

$$A \vee B = \{a \wedge b \mid a \in A, b \in B\}$$

from III.3.2 to general subsets (not only for sublocales; this is introduced just for the purpose of this section) we immediately see that

$$m(A \cup B) = m(A) \vee m(B). \tag{3.1.1}$$

We have seen (II.5.1) that a locale is spatial if and only if the frame homomorphism

$$\sigma_L^* = (a \mapsto \Sigma_a) \colon L \to \mathsf{LcSp}(L)$$

is one-to-one, that is, if for $a \not\leq b$, $\Sigma_a \not\subseteq \Sigma_b$.

When representing the spectrum points by meet-irreducible elements (see $\mathsf{Sp}'L$ and $\Sigma_a' = \{p \mid a \not\leq p\}$ in II.4.5) we can rewrite this criterion as follows:

L is spatial

iff $(a \not\leq b \Rightarrow \exists p$ meet-irreducible and such that $a \not\leq p$ and $b \leq p)$

iff $\mathsf{mSp}'L = L$.

3.1.1. Lemma. *Let $S \subseteq L$ be a sublocale. Then an element $p \in S$ is meet-irreducible in S iff it is meet-irreducible in L. In other words,*

$$\mathsf{Sp}'(S) = \mathsf{Sp}'(L) \cap S.$$

Proof. Obviously if p is meet-irreducible in L then it is such also in S. Now let p be meet-irreducible in S and let $a, b \in L$ and $a \wedge b \leq p$. Then (by III.5.3) $\nu_S(a) \wedge \nu_S(b) \leq \nu_S(p) = p$ and hence, say, $\nu_S(a) \leq p$ and consequently $a \leq \nu_S(a) \leq p$. $\qquad\square$

3.1.2. Corollary. *For any sublocale S, $\mathsf{mSp}'S$ is a sublocale.* $\qquad\square$

(Indeed, by III.10.1.1 and 3.1.1, if $a \in \mathsf{Sp}'S$ then $x \to a$ is either 1 or a itself; hence if $A \subseteq \mathsf{Sp}'S$ then by ($\bigwedge$-distr) in III.3.1, $x \to \bigwedge A = \bigwedge_{a \in A}(x \to a) \in \mathsf{mSp}'S$.)

3.2. Proposition. *For any two sublocales $S, T \subseteq L$ one has*

$$\mathsf{Sp}'(S \vee T) = \mathsf{Sp}'S \cup \mathsf{Sp}'T \quad \text{and hence} \quad \mathsf{mSp}'(S \vee T) = \mathsf{mSp}'S \vee \mathsf{mSp}'T.$$

Consequently, if S, T are spatial sublocales of a locale then $S \vee T$ is also spatial.

Proof. $\mathsf{Sp}'(S \vee T) \supseteq \mathsf{Sp}'S \cup \mathsf{Sp}'T$ by 3.1.1. Now if p is meet-irreducible in $S \vee T$ then, first of all, it is an element of $S \vee T$ and hence $p = a \wedge b$ with $a \in S$ and $b \in T$. By meet-irreducibility either $p = a$ or $p = b$. The second equality follows from (3.1.1).

The last statement follows from the second equality, which yields, for spatial S, T, $\mathsf{mSp}'(S \vee T) = \mathsf{mSp}'S \vee \mathsf{mSp}'T = S \vee T$. $\qquad\square$

3.2.1. Corollary. *Let X be a sober space and $A, B \subseteq X$ subspaces. Then $\widetilde{A} \vee \widetilde{B}$ is an induced sublocale.* $\qquad\square$

(As for the formula: by 1.3.1 we have $\widetilde{x} \in \widetilde{A \cup B}$ iff either $\widetilde{x} \in \widetilde{A}$ or $\widetilde{x} \in \widetilde{B}$, that is, either $x \in A$ or $x \in B$. Now recall (I.1.1) that in a sober X, the $\widetilde{x}$ are precisely the meet-irreducible elements.)

A sublocale S need not be spatial, even if L is spatial. For example,

$$B_{\Omega(\mathbb{R})} = \{U^* \mid U \in \Omega(\mathbb{R})\} = \{U \in \Omega(\mathbb{R}) \mid U = U^{**}\}$$

has no meet-irreducibles, since meet-irreducible elements of $\Omega(\mathbb{R})$ are of the form $\mathbb{R} \smallsetminus \{x\}$, $x \in \mathbb{R}$, and $U^* = \mathrm{int}(\mathbb{R} \smallsetminus U)$. However, for complemented sublocales, we have

3.3. Proposition. *Let L be a spatial frame. Then every complemented sublocale of L is spatial.*

Proof. Let S, T be sublocales and let $S \vee T = L$ and $S \cap T = \mathsf{O}$. By 3.2 we have $\mathsf{mSp}'(S) \vee \mathsf{mSp}'(T) = \mathsf{mSp}'(S \vee T) = \mathsf{mSp}'(L) = L$, and since $\mathsf{mSp}'(U) \leq U$ for any sublocale we obtain $S = S \cap L = (S \cap \mathsf{mSp}(S)) \vee \mathsf{O} = \mathsf{mSp}'(S)$. $\qquad\square$

3.3.1. Corollary. *Let X be sober. Then every complemented sublocale S of $\mathsf{Lc}(X)$ is induced. If, moreover, the space is also T_D and S is induced by $A \subseteq X$ then the complement of S is induced by $X \smallsetminus A$.* $\qquad\square$

3.3.2. Remark. The system of all subspaces of a space is, for trivial reasons, always a Boolean algebra. Now by 3.3 we see that (in the sober case) the system of all sublocales of the corresponding locales is a Boolean algebra only if there are no sublocales but subspaces (see Section 7 below).

4. The zero-dimensionality of $\mathcal{S\!l}(L)^{\mathrm{op}}$ and a few consequences

Before we start let us mention that most of the facts on complementariness of sublocales in this and in the two following sections go back to Isbell.

4.1. The lattice (in fact, co-frame) $\mathcal{S\!l}(L)$ of sublocales is typically not Boolean. The complemented elements, nevertheless, abound, and by III.6.5 each $S \in \mathcal{S\!l}(L)$ can be expressed as $S = \bigcap\{\mathfrak{c}(x) \vee \mathfrak{o}(y) \mid \nu_S(x) = \nu_S(y)\}$ and since each $\mathfrak{c}(x) \vee \mathfrak{o}(y)$ is complemented we have that

$$S = \bigwedge\{C \mid C \text{ complemented, } S \subseteq C\}. \qquad (4.1.1)$$

In the opposite frame $\mathcal{S\!l}(L)^{\mathrm{op}}$ this means that each element is a join of complemented ones, which is usually expressed, extending the homonymous concept from spaces, by saying that

$$\mathcal{S\!l}(L)^{\mathrm{op}} \text{ is zero-dimensional.}$$

4.1.1. Notation. The complement of an element a will be denoted by

$$a^c.$$

4.2. Proposition. *Let L^{op} be zero-dimensional. Then*

$$a \leq b \quad \text{iff} \quad (\forall c \text{ complemented in } L, \ a \wedge c \neq 0 \ \Rightarrow \ b \wedge c \neq 0).$$

Proof. If $a \leq b$ then the formula obviously holds. Now let $a \nleq b$. We have

$$b = \bigwedge \{d \mid d \text{ complemented}, \ b \leq d\}$$

and hence there is a complemented d such that $b \leq d$ and $a \nleq d$. Set $c = d^c$. Then $a \wedge c \neq 0$ (since else $a = a \wedge (c \vee d) \leq d$), and $b \wedge c = 0$. □

4.3. Corollary. *Let L be a distributive complete lattice such that L^{op} is zero-dimensional. Then L is subfit.*

Proof. (Recall V.1.1.) Applying 4.2 for the opposite L^{op} we obtain that if $a \nleq b$ then there is a (complemented, but we do not need that) c such that $a \vee c = 1$ while $b \vee c \neq 1$. □

4.4. Now we will present a simple criterion of complementariness in distributive complete lattices. An element a in such a lattice L is said to be *linear* (resp. *co-linear*) if for each system b_i, $i \in I$, of elements of L

$$a \wedge \bigvee_{i \in J} b_i = \bigvee_{i \in J} (a \wedge b_i) \quad (\text{resp.} \quad a \vee \bigwedge_{i \in J} b_i = \bigwedge_{i \in J} (a \vee b_i) \ )$$

(thus for instance in a frame each element is linear but not necessarily co-linear).

4.4.1. Lemma. *Each complemented element a in a distributive complete lattice is linear and co-linear.*

Proof. Since the assumption is self-dual it suffices to prove that a is linear. Let c be the complement of a. Obviously

$$(a \wedge \bigvee_{i \in J} b_i) \vee a = a = (\bigvee_{i \in J} (a \wedge b_i)) \vee a$$

and by distributivity $(a \wedge \bigvee_{i \in J} b_i) \vee c = (\bigvee_{i \in J} b_i) \vee c$ and also

$$(\bigvee_{i \in J} (a \wedge b_i)) \vee c = \bigvee_{i \in J} ((a \wedge b_i) \vee c) = \bigvee_{i \in J} ((a \vee c) \wedge (b_i \vee c)) = \bigvee_{i \in J} b_i \vee c.$$

Thus

$$a \wedge \bigvee_{i \in J} b_i = (a \wedge \bigvee_{i \in J} b_i) \vee (a \wedge c) = (\bigvee_{i \in J} (a \wedge b_i) \vee a) \wedge (\bigvee_{i \in J} b_i \vee c)$$

$$= (\bigvee_{i \in J} (a \wedge b_i)) \vee (a \wedge c) = \bigvee_{i \in J} (a \wedge b_i). \qquad \square$$

4.4.2. Theorem. *Let a distributive complete lattice be subfit. Then an element $a \in L$ is complemented iff it is co-linear.*

Proof. Let a be co-linear. Set

$$a^{\#} = \bigwedge\{x \mid a \vee x = 1\}.$$

By co-linearity, $a \vee a^{\#} = 1$. Suppose $a \wedge a^{\#} > 0$. Then by subfitness there is a $c \neq 1$ such that $(a \vee c) \wedge (a^{\#} \vee c) = (a \wedge a^{\#}) \vee c = 1$. Thus $a \vee c = 1$ and hence $c \geq a^{\#}$. But then $c \geq a^{\#} \vee c = 1$, a contradiction. $\qquad\square$

4.4.3. From 4.4.2 and 4.3 we immediately obtain

Corollary. *A sublocale S is complemented in $\mathcal{S\ell}(L)$ iff it is linear.* $\qquad\square$

4.5. The lattice $\mathcal{S\ell}(L)$, as a co-frame, is a co-Heyting algebra. That is, there is an operation $S \smallsetminus T$ such that

$$S \smallsetminus T \subseteq U \quad \text{iff} \quad S \subseteq T \vee U.$$

We will speak of $S \smallsetminus T$ as of the *pseudodifference*.

The pseudodifference models the difference of subsets of a set; we will see soon that it relates to a natural concept of difference analogously as the pseudocomplement relates to the complement.

In a complete Heyting algebra we obviously have

$$a \to b = \bigvee\{x \mid x \wedge a \leq b\}$$

(and in a complete co-Heyting algebra,

$$a \smallsetminus b = \bigwedge\{x \mid a \leq b \vee x\} \,).$$

In the zero-dimensional case we have a perhaps somewhat surprising formula.

4.5.1. Proposition. *Let L be a zero-dimensional frame. Then*

$$a \to b = \bigwedge\{x \mid x \geq b \text{ and } x \vee a = 1\}.$$

Proof. If $x \geq b$ and $x \vee a = 1$ then

$$a \to b = (a \to b) \wedge (x \vee a) = ((a \to b) \wedge a) \vee ((a \to b) \wedge x)$$
$$\leq b \vee ((a \to b) \wedge x) \leq (a \to b) \wedge x$$

and hence $a \to b \leq x$. Thus,

$$a \to b \leq \bigwedge\{x \mid x \geq b \text{ and } x \vee a = 1\}.$$

On the other hand, let u be complemented and $u \leq \bigwedge\{x \mid x \geq b \text{ and } x \vee a = 1\}$. By zero-dimensionality, it suffices to check that $u \leq a \to b$. We have the implication

$$x \geq b \text{ and } x \vee a = 1 \quad \Rightarrow \quad u \leq x$$

and hence

$$x \geq b \text{ and } x \vee a = 1 \quad \Rightarrow \quad x \vee u^c = 1.$$

Now let v be complemented and $v \leq a$. Since $b \vee v^c \geq b$ and $b \vee v^c \vee a = 1$, we have $b \vee v^c \vee u^c = 1$ and $v = v \wedge (b \vee u^c)$, that is, $v \leq b \vee u^c$, and taking the join of all such v, $a \leq b \vee u^c$ and finally $b = b \vee (u \wedge u^c) \geq (b \vee u) \wedge a$ and $u \leq b \vee u \leq a \to b$. $\square$

4.5.2. Corollary. *In $\mathcal{S}\ell(L)$ we have the formula*

$$S \smallsetminus T = \bigvee\{U \mid U \leq S \text{ and } U \cap T = \mathsf{O}\}. \qquad \square$$

5. Difference and pseudodifference, residua

In the first part of this section we will discuss pseudodifference and difference in complete co-Heyting algebras (co-frames). Later we will concentrate on the co-frames $\mathcal{S}\ell(L)$ and, using their special properties (zero-dimensionality of $\mathcal{S}\ell(L)^{\mathrm{op}}$, existence of smallest dense sublocales), derive a sufficient and a necessary condition for a sublocale to be complemented. This will prepare us for the Isbell's Development Theorem that will be the highlight of the following section.

5.1. We have already observed, and exploited, the fact that a frame is a general complete Heyting algebra. Similarly, a co-frame is a complete co-Heyting algebra, that is, it possesses an extra operation, the *pseudodifference* (often called *relative complement*) $a \smallsetminus b$ characterized by the formula

$$a \smallsetminus b \leq c \quad \text{iff} \quad a \leq b \vee c.$$

This operation mimics the difference $A \smallsetminus B$ of subsets of a set and satisfies the formula dual to those discussed in III.3. Thus for instance ($\bigwedge$-distr) appears as

$$\left(\bigvee_{i \in J} a_i\right) \smallsetminus b = \bigvee_{i \in J}(a_i \smallsetminus b), \qquad (\text{$\bigwedge$-distr}^{\mathrm{op}})$$

(H5) transforms to

$$b \vee (a \smallsetminus b) = b \vee a, \qquad (\text{H5}^{\mathrm{op}})$$

(H6) to

$$a \vee c = b \vee c \quad \text{iff} \quad a \smallsetminus c = b \smallsetminus c, \qquad (\text{H6}^{\mathrm{op}})$$

or (H7) to

$$(a \smallsetminus b) \smallsetminus c = (a \smallsetminus c) \smallsetminus b = a \smallsetminus (b \vee c), \qquad (\text{H7}^{\mathrm{op}})$$

and so on. Note how intuitively satisfactory such rules are if we think of the operation as a sort of difference of subsets.

5.2. The *difference* $a - b$ of two elements of a co-Heyting algebra is the (unique – see AI.6.3.5) solution x of the equations

$$(a \wedge b) \vee x = a$$
$$(a \wedge b) \wedge x = 0, \tag{diff1}$$

if it exists, which it does not have to, of course.

One customarily uses the equivalent system

$$(a \wedge b) \vee x = a$$
$$b \wedge x = 0, \tag{diff}$$

(by the first of the equations, $x \leq a$ and hence if $a \wedge b \wedge x = 0$ then $b \wedge x = 0$).

5.2.1. Observations. (1) *The complement a^c is the difference $1 - a$.*

(2) $a - b$ *exists iff* $a - (b \wedge a)$ *exists, and these differences are equal.*

(3) *If b is complemented then $a - b$ always exists and equals $a \wedge b^c$. Consequently, if b is an intersection of a with a complemented c then $a - b$ exists (and equals $a \wedge c^c$).*

(Only (3) needs a comment, if at all: $(a \wedge b) \vee (a \wedge b^c) = a \wedge (b \vee b^c) = a$, $b \wedge (a \wedge b^c) = 0$; the consequence follows from (2).)

5.2.2. The reader probably expects that $a - b$, if it exists, coincides with $a \smallsetminus b$. This is indeed so, and we have a bit more.

Lemma. *Let $a_1 - b$ exist for some $a_1 \geq a$. Then $a - b$ exists and we have*

$$a \smallsetminus b = a - b = a \wedge (a_1 - b).$$

Proof. For $x = a_1 - b$ we have $(a_1 \wedge b) \vee x = a_1$ and $b \wedge x = 0$. Hence $b \wedge (a \wedge x) = 0$ and $(a \wedge b) \vee (a \wedge x) = a \wedge (b \vee x) = a$, the last equality since $b \vee x \geq (a \wedge b) \vee a$.

Now let $y = a - b$. Then $(a \wedge b) \vee y = a$ and $b \wedge y = 0$. Thus if $y \leq c$ then $a = (a \wedge b) \vee y \leq (a \wedge b) \vee c \leq b \vee c$, and if $a \leq b \vee c$ then

$$y = y \wedge a \leq (y \wedge b) \vee (y \vee c) = 0 \vee (y \wedge c) \leq c. \qquad \square$$

5.2.3. Lemma. *Let $a \leq b \leq c$ and let $b - a$ and $c - b$ exist. Then $c - a$ exists and $c - a = (c - b) \vee (b - a)$.*

Proof. Set $x = b - a$ and $y = c - b$. Thus $a \vee x = b$, $b \vee y = c$ and $a \wedge x = b \wedge y = 0$, and hence $a \vee x \vee y = b \vee y = c$ and $a \wedge (x \vee y) = a \wedge y \leq b \wedge y = 0$. $\qquad \square$

5.2.4. Lemma. *Let $a - b$ exist and let $b \leq a$. Then $a - b$ is complemented with complement c iff c satisfies the equations*

$$a \vee c = 1 \quad \text{and} \quad a \wedge c = b. \tag{$*$}$$

Proof. Let $(*)$ hold. Hence $c = b \vee c$ and (by (H5^{op}))

$$(a - b) \vee c = (a - b) \vee b \vee c = a \vee c = 1,$$

and $(a - b) \wedge c \leq a \wedge c = b$ and hence $(a - b) \wedge c = (a - b) \wedge b \wedge c = 0$.

Conversely, let c be the complement of $a - b$. Then trivially $a \vee c \geq (a-b) \vee c = 1$, and $a \wedge c = ((a - b) \vee b) \wedge c = b \wedge c = b \wedge (c \vee (a - b)) = b$. □

5.3. In the sequel we will work exclusively in the co-frame $\mathcal{S\!\ell}(L)$. We will use the following specific properties of this lattice.

- $\mathcal{S\!\ell}(L)^{\mathrm{op}}$ is zero-dimensional; hence we will be able to use the formulas from Section 4.

- There are special sublocales, the open $\mathfrak{o}(a)$ and the closed $\mathfrak{c}(a) = {\uparrow}a$ complementing each other.

- We have the closure $\overline{S}$, the smallest closed sublocale containing S.

- For each $S \in \mathcal{S\!\ell}(L)$ there exists the smallest dense sublocale $\partial S \subseteq S$, namely $\partial S = \mathfrak{b}(\bigwedge S)$ (recall III.10.3) (smallest dense in the sense that $\overline{\partial S} \supseteq S$ and whenever $\overline{T} \supseteq S$ then $T \supseteq \partial S$).

- The order in $\mathcal{S\!\ell}(L)$ is $\subseteq$ and the meets are the intersections; we will use the symbols $\cap, \bigcap$ to keep this in mind, and also to avoid confusion with the $\bigwedge$ used inside L.

5.3.1. Lemma. *For any two sublocales S, T we have*

$$T \subseteq \overline{S \cap T} \quad \text{iff} \quad \partial T \subseteq S.$$

Proof. $\Rightarrow$: If $T \subseteq \overline{S \cap T}$ then $\bigwedge T \geq \bigwedge(S \cap T) \geq \bigwedge T$, hence $\bigwedge T = \bigwedge(S \cap T) \in S$, and $\partial T = \mathfrak{b}(\bigwedge T) \subseteq S$.

$\Leftarrow$: If $\partial T \subseteq S$ then $\bigwedge T \in S$, hence $\bigwedge T \in T \cap S$, and $T \subseteq \overline{T} \subseteq \overline{S \cap T}$. □

5.4. The α-residua. For a sublocale S set

$$r(S) = S \cap \overline{(\overline{S} \smallsetminus S)}$$

and further define for ordinals α

$$r_0(S) = S, \quad r_{\alpha+1}(S) = r(r_\alpha(S)) \quad \text{and}$$

$$r_\lambda(S) = \bigcap_{\alpha < \lambda} r_\alpha(S) \quad \text{for limit ordinals } \lambda.$$

5.4.1. Lemma. *For every α there exists a closed, and hence complemented, A_α such that $r_\alpha(S) = S \cap A_\alpha$. Consequently, the difference $S - r_\alpha(S)$ always exists.*

Proof. $r_0(S) = S \cap L$. Let $r_\alpha(S)$ be $S \cap A$ with A closed. Then $r_{\alpha+1}(S) = S \cap A \cap \overline{B}$ (for $B = (\overline{S \cap A} \smallsetminus S \cap A)$), hence a meet of S and a closed sublocale. Finally let $r_\alpha(S) = S \cap A_\alpha$ for $\alpha < \lambda$ limit. Then

$$r_\lambda(S) = \bigcap_{\alpha < \lambda} r_\alpha(S) = \bigcap_{\alpha < \lambda} (S \cap A_\alpha) = S \cap \bigcap_{\alpha < \lambda} A_\alpha.$$

For the consequence see 5.2.1(3). □

5.4.2. Proposition. *Each $S - r_\alpha(S)$ is complemented. Consequently, if there is an α such that $r_\alpha(S) = 0$ then S is complemented.*

Proof. First, for any sublocale T we have

$$T - r(T) = T - \overline{(\overline{T} \smallsetminus T)} = \overline{T} - \overline{(\overline{T} \smallsetminus T)}. \qquad (*)$$

Indeed, by (H5^{op}) we have $T \vee (\overline{T} \smallsetminus T) = \overline{T}$. Thus,

$$\overline{T} - \overline{\overline{T} \smallsetminus T} = (T \vee (\overline{T} \smallsetminus T)) \cap \overline{(\overline{T} \smallsetminus T)}^c = T \cap \overline{(\overline{T} \smallsetminus T)}^c$$

$$= T - \overline{(\overline{T} \smallsetminus T)} = T - T \cap \overline{(\overline{T} \smallsetminus T)} = T \smallsetminus r(T).$$

Thus, $T - r(T) = \overline{T} \cap \overline{(\overline{T} \smallsetminus T)}^c$, an intersection of two complemented sublocales, is complemented. In particular, $S \smallsetminus r_1(S)$ and each $r_\alpha(S) \smallsetminus r_{\alpha+1}(S)$ are complemented. Hence if we already know that $S - S_\alpha$ is complemented then also (recall 5.2.3)

$$S - r_{\alpha+1}(S) = (S - r_\alpha(S)) \vee (r_\alpha(S) - r_{\alpha+1}(S))$$

is. Now let λ be a limit ordinal and let C_α be the complements of $S - r_\alpha(S)$, $\alpha < \lambda$. Set $C = \bigwedge_{\alpha < \lambda} C_\alpha$. We have

$$C_\alpha \vee r_\alpha(S) = (C_\alpha \vee r_\alpha(S)) \cap ((S - r_\alpha(S)) \vee C_\alpha) = 0 \vee C_\alpha \vee 0 \vee (r_\alpha(S) \cap C_\alpha) = C_\alpha,$$

hence $r_\alpha(S) \leq C_\alpha$ and consequently

$$C \cap S = (\bigwedge_{\alpha < \lambda} C_\alpha) \cap S = \bigcap_{\alpha < \lambda} (C_\alpha \cap S) = \bigcap_{\alpha < \lambda} C_\alpha \cap (r_\alpha(S) \vee (S - r_\alpha(S)))$$

$$= \bigcap_{\alpha < \lambda} C_\alpha \cap r_\alpha(S) = \bigcap_{\alpha < \lambda} r_\alpha(S) = r_\lambda(S).$$

Finally, using the co-frame distributivity we obtain $C \vee S = \bigwedge_{\alpha < \lambda} (C_\alpha \vee S) = L$ and hence C is the complement of $S - r_\lambda(S)$ by 5.2.4. □

5.5. We say that a sublocale S is *doubly dense* in a sublocale T if both $S \cap T$ and $T \smallsetminus S$ are dense in T. Obviously

if S is doubly dense in T, it is doubly dense in $\overline{T}$.

5.5.1. Lemma. *A complemented S is doubly dense in no non-empty T.*

Proof. If C is the complement of S then $T \smallsetminus S = T \cap C$. Let T be non-empty. Then $\partial T \neq 0$ and hence we cannot have both $\partial T \subseteq S$ and $\partial T \subseteq C$, that is, by 5.3.1, we cannot have both $T \subseteq \overline{S \cap T}$ and $T \subseteq \overline{C \cap T} = \overline{T \smallsetminus S}$. □

6.   Isbell's Development Theorem

6.1. Let κ be a limit ordinal and

$$(A_\alpha)_{\alpha < \kappa}$$

a (transfinite) decreasing sequence of closed sublocales of L. Since the A_α are complemented we have the differences

$$S - A_\alpha \ (= S \wedge A_\alpha^c) \text{ for any sublocale } S,$$

and we can define, for *even* α (that is, $\alpha = 2n$ or $\alpha = \lambda + 2n$ where λ is a limit ordinal; otherwise we speak of an *odd* ordinal)

$$B_0 = A_0,$$
$$B_{\alpha+1} = B_\alpha - A_{\alpha+1},$$
$$B_{\alpha+2} = B_{\alpha+1} \vee A_{\alpha+2},$$

and for limit λ, $0 < \lambda < \kappa$,

$$B_\lambda = \bigcap \{B_\alpha \mid \alpha < \lambda, \ \alpha \text{ even}\}.$$

Note that

$$B_\alpha \supseteq A_\beta \ \text{ for any } \ \beta \geq \alpha.$$

6.2. Proposition. *The sequence*

$$B_0, B_2, \ldots, B_\alpha, \ldots \quad (\alpha \text{ even})$$

decreases and the sequence

$$B_1, B_3, \ldots, B_\alpha, \ldots \quad (\alpha \text{ odd})$$

increases. Further, for any odd α and any even β, $B_\alpha \leq B_\beta$.

Proof. Let α be even. Then by 5.2.1 and (H5^{op}),

$$B_{\alpha+2} = (B_\alpha - A_{\alpha+1}) \vee A_{\alpha+2} \leq (B_\alpha - A_{\alpha+1}) \vee A_{\alpha+1} \leq B_\alpha \vee A_{\alpha+1} = B_\alpha.$$

If α is odd we have

$$B_{\alpha+2} = (B_\alpha \vee A_{\alpha+1}) - A_{\alpha+2} = ((B_{\alpha-1} \cap A_\alpha^c) \vee A_{\alpha+1}) \cap A_{\alpha+2}^c$$
$$= (B_{\alpha-1} \cap A_\alpha^c \cap A_{\alpha+2}^c) \vee (A_{\alpha+1} - A_{\alpha+2})$$
$$= (B_{\alpha-1} \cap A_\alpha^c) \vee (A_{\alpha+1} - A_{\alpha+2}) = B_\alpha \vee (A_{\alpha+1} - A_{\alpha+2}) \geq B_\alpha.$$

Now we know that the even sequence decreases because the definition of the B_λ, λ limit, makes sure that the descent continues. For the odd sequence, however, we have to be somewhat more careful. Let α be odd and let λ be the first next limit ordinal. For an even ordinal β with $\alpha < \beta < \lambda$ we have $B_\alpha \leq B_{\beta+1} \leq B_\beta$ and hence $B_\alpha \leq B_\lambda$. Further, $B_\alpha = C \cap A_\alpha^c \leq A_{\lambda+1}^c$ and hence $B_\alpha \leq B_\lambda \cap A_{\lambda+1}^c = B_{\lambda+1}$.

   The last statement easily follows from $B_{\alpha+1} \leq B_\alpha$ for even α, and from the decreasing resp. increasing even resp. odd sequences. $\square$

6.3. Proposition. *Set*

$$M = \bigcap\{B_\alpha \mid \alpha \text{ even}\} \quad \text{and} \quad J = \bigvee\{B_\alpha \mid \alpha \text{ odd}\}.$$

Then

$$M = J \vee \bigcap_\alpha A_\alpha.$$

Proof. By 6.2 $J \leq M$, and since $A_{\alpha+2} \leq B_{\alpha+2}$ we have

$$\bigcap_\alpha A_\alpha = \bigcap\{A_{\alpha+2} \mid \alpha \text{ even}\} \leq M$$

as well. Thus, we have to prove that $M \leq J \vee \bigcap_\alpha A_\alpha$. Recall 4.2: it suffices to prove that if $C \cap M \neq 0$ then either $C \cap J \neq 0$ or $C \cap \bigcap_\alpha A_\alpha \neq 0$, for every complemented C.

Thus, suppose that $C \cap M \neq 0$ and $C \cap J = 0$. Then for α even,

$$C \cap M \leq C \cap B_{\alpha+2} = C \cap (B_{\alpha+1} \vee A_{\alpha+2}) = C \cap A_{\alpha+2} \leq A_{\alpha+2}$$

and since A_α decreases, $0 \neq C \cap M \leq \bigcap_\alpha A_\alpha$. $\qquad\square$

6.4. Consequently, if one has $\bigcap_\alpha A_\alpha = 0$ then

$$S = \bigcap\{B_\alpha \mid \alpha \text{ even}\} = \bigvee\{B_\alpha \mid \alpha \text{ odd}\}.$$

The sequence $(A_\alpha)_{\alpha<\kappa}$ is then said to be a *development* of this common value S.
We will assume, furthermore, that for each limit ordinal λ

$$\bigcap_{\alpha<\kappa} A_\alpha = A_\lambda \tag{6.4.1}$$

(any development can be easily so adjusted: it suffices to insert the meet at the limit step, repeat it at the next, and proceed with the original A_λ at the $(\lambda+2)$th place).

6.4.1. Observation. *If $(A_\alpha)_{\alpha<\kappa}$ is a development of S and if C is closed then $(A_\alpha \cap C)_{\alpha<\kappa}$ is a development of $S \cap C$.*

(The associated $\widetilde{B}_\alpha$ are $B_\alpha \cap C$.)

6.5. Lemma. *Each S that is doubly dense in no non-empty T has a development.*

Proof. If A is non-void closed, S is not doubly dense in A and hence some of the $\overline{A \cap S}, \overline{A \setminus S}$ is a proper closed sublocale of A. Thus, for any closed A',

$$\text{if } A = \overline{A' \cap S} \text{ then } \overline{A \setminus S} \neq A, \text{ and if } A = \overline{A' \setminus S} \text{ then } \overline{A \wedge S} \neq A \tag{6.5.1}$$

(in the first case,

$$\overline{A \cap S} = \overline{\overline{(A' \cap S)} \cap S} \supseteq \overline{A' \cap S \cap S} = \overline{A' \cap S} = A$$

and in the second one,

$$\overline{A \smallsetminus S} = \overline{\overline{(A' \smallsetminus S)} \smallsetminus S} \supseteq \overline{(A' \smallsetminus S) \smallsetminus S} = \overline{A' \smallsetminus S} = A$$

– here we use (H7^{op})). Thus, if we set

$$A_0 = \overline{S}, \quad A_{\alpha+1} = \overline{A_\alpha \smallsetminus S}, \quad \text{and} \quad A_{\alpha+2} = \overline{A_{\alpha+1} \cap S} \text{ for } \alpha \text{ even,}$$

$$\text{and } A_\lambda = \bigcap_{\alpha < \lambda} A_\alpha \text{ for } \lambda \text{ limit}$$

we obtain a transfinite sequence $(A_\alpha)_\alpha$ which by (6.5.1) strictly decreases, with the possible exception of the limit values where one can have $A_{\lambda+1} = A_\lambda$, until we obtain 0 for a limit ordinal κ. Take $(A_\alpha)_{\alpha < \kappa}$.

For the corresponding sequence $(B_\alpha)_\alpha$ we have

$$B_\alpha \geq S \text{ for even } \alpha, \quad \text{and} \quad B_\alpha \leq S \text{ for odd } \alpha$$

(by induction, if $\beta = \alpha + 1$ is non-limit even we have

$$B_\beta = B_\alpha \vee A_\beta = (B_{\alpha-1} \smallsetminus A_\alpha) \vee A_\beta \geq (B_{\alpha-1} \smallsetminus A_\alpha) \vee (A_\alpha \cap S)$$
$$\geq ((B_{\alpha-1} \smallsetminus A_\alpha) \vee A_\alpha) \cap S \geq B_{\alpha-1} \cap S = S,$$

and

$$B_{\beta+1} = (B_\beta \smallsetminus A_{\beta+1}) = (B_\alpha \vee A_\beta) \smallsetminus A_{\beta+1}$$
$$\leq (B_\alpha \vee A_\beta) \smallsetminus (A_\beta \smallsetminus S) \leq (S \vee A_\beta) \smallsetminus (A_\beta \smallsetminus S)$$
$$= (S \smallsetminus (A_\beta \smallsetminus S)) \vee (A_\beta \smallsetminus (A_\beta \smallsetminus S)) \leq S,$$

and if λ is a limit ordinal

$$B_{\lambda+1} = B_\lambda \smallsetminus A_{\lambda+1} = \bigcap \{B_\alpha \mid \alpha \text{ even}, \ \alpha < \lambda\} \smallsetminus (A_\lambda \smallsetminus S)$$
$$= \bigcap \{B_{\alpha+1} \vee A_{\alpha+2} \mid \alpha \text{ even}, \ \alpha < \lambda\} \smallsetminus (A_\lambda \smallsetminus S)$$
$$\leq \bigcap \{S \vee A_{\alpha+2} \mid \alpha \text{ even}, \ \alpha < \lambda\} \smallsetminus (A_\lambda \smallsetminus S)$$
$$= (S \vee \bigcap \{A_{\alpha+2} \mid \alpha \text{ even}, \ \alpha < \lambda\}) \smallsetminus (A_\lambda \smallsetminus S)$$
$$= (S \vee A_\lambda) \smallsetminus (A_\lambda \smallsetminus S) = (S \smallsetminus (A_\lambda \smallsetminus S)) \vee (A_\lambda \smallsetminus (A_\lambda \smallsetminus S)) \leq S$$

and trivially $B_\lambda = \bigcap \{B_\alpha \mid \alpha \text{ even}, \ \alpha < \lambda\} \geq S$).

Thus (since $\bigcap_{\alpha < \kappa} A_\alpha = 0$) we have

$$S = \bigcap \{B_\alpha \mid \alpha \text{ even}\} = \bigvee \{B_\alpha \mid \alpha \text{ odd}\}. \qquad \square$$

6.6. Theorem. *The following statements about a sublocale S are equivalent.*

(1) *S is complemented.*

(2) *S is doubly dense in no non-empty sublocale.*

(3) *S has a development.*

(4) $r_\alpha(S) = 0$ *for sufficiently large α.*

Proof. We have (4)$\Rightarrow$(1) by 5.4.2, (1)$\Rightarrow$(2) by 5.5.1, and (2)$\Rightarrow$(3) by 6.4.1.

(3)$\Rightarrow$(4): First we will slightly extend our definition of development. For simplicity in the notation we have considered sequences $(A_\alpha)_{\alpha<\kappa}$ starting with A_0. All that has been said holds equally well for a modified definition in which we consider sequences $(A_\alpha)_{\beta\leq\alpha<\kappa}$ starting with an ordinal $\beta < \kappa$.

Note, however, that if we have a development $(A_\alpha)_{\alpha<\kappa}$ of S then a restricted development $(A_\alpha)_{\beta\leq\alpha<\kappa}$ does not necessarily result in the same S. In particular realize that

$$\text{if } (A_\alpha)_{\beta\leq\alpha<\kappa} \text{ is a development of } S \text{ then } S \subseteq A_\beta.$$

Let $(A'_\alpha)_{\alpha<\kappa}$ be a development of a sublocale T; consider the development

$$(A'_\alpha \cap (\overline{\overline{T} \smallsetminus T}))_{\alpha<\kappa}$$

of the residuum $r(T)$ (recall 6.4.1).

We have $A'_0 - A'_1 \leq T \leq \overline{T} \leq A'_0$, hence $\overline{T} \smallsetminus T \leq A'_0 \smallsetminus (A'_0 \smallsetminus A'_1) \leq A'_1$ (because $A'_0 \leq (A'_0 \smallsetminus A'_1) \vee A'_1$) so that

$$A'_0 \cap (\overline{\overline{T} \smallsetminus T}) - A'_1 \cap (\overline{\overline{T} \smallsetminus T}) = (A'_0 \cap (A'_1)^c) \cap (\overline{\overline{T} \smallsetminus T}) = 0$$

and we have $B_2 = A_0$ and hence we have for $r(T)$ also the development

$$(A'_\alpha \cap r(T))_{2\leq\alpha<\kappa} \quad \text{of} \quad r(T).$$

By induction we see that $r_\beta(S)$ has developments

$$(A_\alpha \cap \overline{\overline{r}_{\beta-1}(S) \smallsetminus r_{\beta-1}(S)})_{\beta'\leq\alpha<\kappa}$$

if β is non-limit, and

$$(A_\alpha \cap \bigcap_{\gamma<\beta} \overline{\overline{r}_\gamma \smallsetminus r_\gamma(S)})_{\beta'\leq\alpha<\kappa}$$

if β is limit, *with β' increasing with β.* Since the original development $(A_\alpha)_{\alpha<\kappa}$ can be, without any effect on the resulting S, extended by a string of zeros between κ and the first next limit ordinal we see that $r_\gamma(S)$ will be 0 for sufficiently large γ. $\qquad\square$

7. Locales with no non-spatial sublocales

We already know that sublocales of a space are not necessarily induced, not even spatial. In the following two sections we will discuss the cases when they are (cf. [189, 246]).

7.1. The set of all primes (meet-irreducibles) in L will be denoted by $\mathrm{Pr}L$, and the set of all $p \in \mathrm{Pr}L \cap {\uparrow}a$ by $\mathrm{Pr}(a)$. Thus, L is spatial iff for each $a \in L$, $a = \bigwedge \mathrm{Pr}(a)$. We will use the representation of the spectrum from II.4.5, that is,

$$\mathrm{Sp}'L = (\mathrm{Pr}L, \sigma L) \ \text{with} \ \sigma L = \{\Sigma'_a \mid a \in L\}, \ \Sigma'_a = \{p \mid a \not\leq p\}.$$

7.1.1. Lemma. *For every* $p \in \mathrm{Pr}(a)$ *there is a minimal* $q \in \mathrm{Pr}(a)$ *such that* $q \leq p$.

Proof. Let $C \subseteq \mathrm{Pr}(a)$ be a chain. Set $r = \bigwedge C$. Then r is a prime. Indeed, let $a \wedge b \leq r$. Then for each $p \in C$ either $a \leq p$ or $b \leq p$. Suppose $b \not\leq q$ for some individual $q \in C$. Then $b \not\leq p$, and hence $a \leq p$, for all $p \leq q$ and finally $a \leq r$. Now the statement follows from Zorn's Lemma (AI.2). $\square$

Consequently, if we denote by $\mathrm{MinPr}(a)$ the set of all minimal elements of $\mathrm{Pr}(a)$,

$$\bigwedge \mathrm{MinPr}(a) = \bigwedge \mathrm{Pr}(a), \ \text{and}$$

$$L \ \text{is spatial iff for each} \ a \in L, a = \bigwedge \mathrm{MinPr}(a).$$

7.2. Recall from III.10.2

$$\mathfrak{b}(a) = \{x \to a \mid x \in L\} = \{x \mid x = (x \to a) \to a\} = \{(x \to a) \to a \mid x \in L\}.$$

It is the smallest sublocale of L containing a, it is Boolean, and for each prime p, $\mathfrak{b}(p) = \{p, 1\}$ (III.10.1.1).

7.2.1. Proposition. *For any sublocale* S *of* L *we have:*

(a) S *is prime in* $\mathcal{Sl}(L)^{\mathrm{op}}$ *iff* $S = \mathfrak{b}(p)$ *for some* $p \in \mathrm{Pr}L$.

(b) $\mathrm{mSp}'(S) = \bigvee\{\mathfrak{b}(p) \mid p \in \mathrm{Sp}'(S)\}$ (mM *is as in* 3.1).

(c) S *is spatial iff* $S = \bigvee\{\mathfrak{b}(p) \mid p \in \mathrm{Sp}'(S)\}$.

Proof. (a): If $S = \mathfrak{b}(p) = \{p, 1\}$ then $S \neq \mathsf{O}$, and $p \in S \subseteq S_1 \vee S_2$ implies $p = a \wedge b$ for some $a \in S_1$ and $b \in S_2$. Hence, say, $p = a \in S_1$, and thus $\mathfrak{b}(p) \subseteq S_1$.

Conversely, if S is prime in $\mathcal{Sl}(L)^{\mathrm{op}}$ then $S \supseteq \{a, 1\}$ for $a = \bigwedge S \neq 1$ (because $S \neq \mathsf{O}$). Let $b \in S$ be such that $a < b$. Then $S \not\subseteq \mathfrak{c}(b) = {\uparrow}a$. But $S \subseteq \mathfrak{c}(b) \vee \mathfrak{o}(b)$ and hence by primeness $S \subseteq \mathfrak{o}(b)$. In particular, $b \in S \subseteq \mathfrak{o}(b)$ and $b \in \mathfrak{o}(b) \cap \mathfrak{c}(b) = \mathsf{O}$, that is, $b = 1$. Hence $S = \{a, 1\}$. By III.10.1.2, a is necessarily prime in L.

(b): The inclusion $\supseteq$ is obvious since

$$\bigcup\{\mathfrak{b}(p) \mid p \in \mathrm{Pr}(S)\} = \bigcup\{\{p, 1\} \mid p \in \mathrm{Pr}(S)\} = \mathrm{Pr}(S).$$

The reverse inclusion is easy as well: for any $\bigwedge Y \in \mathrm{mSp}'(S)$,

$$Y \subseteq \mathrm{Pr}(S) \subseteq \bigcup\{\mathfrak{b}(p) \mid p \in \mathrm{Pr}(S)\}$$

and hence $\bigwedge Y \in \bigvee_{p \in \mathrm{Pr}(S)} \mathfrak{b}(p)$.

(c) is an immediate consequence of (b) and 3.1.   □

7.2.2. Corollary. *Every sublocale of L is spatial iff $\mathscr{Sl}(L)^{\mathrm{op}}$ is spatial.*
   In particular, if $\mathscr{Sl}(L)^{\mathrm{op}}$ is spatial, then L is spatial.

Proof. By II.5.3, $\mathscr{Sl}(L)^{\mathrm{op}}$ is spatial iff each sublocale S of L is a join in $\mathscr{Sl}(L)$ of primes of $\mathscr{Sl}(L)^{\mathrm{op}}$. Use (a) and (c) above.   □

7.3. By 7.2.1(a), $(p \mapsto \mathfrak{b}(p))$ is a one-one mapping of $\mathrm{Pr}L$ onto $\mathrm{Pr}(\mathscr{Sl}(L)^{\mathrm{op}})$. Since obviously $\bigwedge \mathfrak{b}(p) = p$ we have the inverse pair of bijections

$$\mathrm{Pr}L \xrightleftharpoons[g=(S \mapsto \bigwedge S)]{f=(p \mapsto \mathfrak{b}(p))} \mathrm{Pr}(\mathscr{Sl}(L)^{\mathrm{op}}). \tag{7.3.1}$$

This shows that the spectrum of $\mathscr{Sl}(L)^{\mathrm{op}}$ (often referred to as the *assembly of L*, see [246]) has essentially the same points as the spectrum of L. It has, however, a different topology.
   Recall the *Skula topology* from I.4.1.2. We have

7.3.1. Proposition. *The maps from (7.3.1) constitute a homeomorphic pair*

$$(\mathrm{Pr}L, \mathrm{Sk}(\sigma L)) \xrightleftharpoons[g]{f} \mathrm{Sp}'(\mathscr{Sl}(L)^{\mathrm{op}}).$$

In particular, if $\mathrm{Sp}'L$ is T_D then $\mathrm{Sp}'(\mathscr{Sl}(L)^{\mathrm{op}})$ is discrete.

Proof. Let Σ'_S be in $\sigma(\mathscr{Sl}(L)^{\mathrm{op}})$. Since $S = \bigvee_{i \in J}(\mathfrak{c}(a_i) \wedge \mathfrak{o}(b_i))$ for suitable open and closed sublocales we have $\Sigma'_S = \bigcup_{i \in J}(\Sigma'_{\mathfrak{c}(a_i)} \cap \Sigma'_{\mathfrak{o}(b_i)})$ and hence the sets $\Sigma'_{\mathfrak{c}(a)}$ and $\Sigma'_{\mathfrak{o}(b)}$ constitute a subbase for the topology. We have

$$f^{-1}[\Sigma'_{\mathfrak{c}(a)}] = f^{-1}[\{\mathfrak{b}(p) \mid \mathfrak{c}(a) \not\leq \mathfrak{b}(p)\}] = \{p \mid \mathfrak{c}(a) \not\sqsupseteq \mathfrak{b}(p)\} = \Sigma'_a$$

and by complementarity $f^{-1}[\Sigma'_{\mathfrak{o}(a)}] = \mathrm{Pr}L \smallsetminus \Sigma'_a$.
   The continuity of g follows similarly: for every $a \in L$,

$$g^{-1}[\Sigma'_a] = \{\mathfrak{b}(p) \mid \mathfrak{c}(a) \not\sqsupseteq \mathfrak{b}(p)\}$$

and by complementarity $g^{-1}[\mathrm{Pr}L \smallsetminus \Sigma'_a] = \Sigma'_{\mathfrak{o}(a)}$.   □

7.4. Lemma. *Let $a \in L$ satisfy $a = \bigwedge \mathsf{MinPr}(a)$. For each $p \in \mathsf{MinPr}(a)$,*

$$p \to a = \bigwedge \{q \in \mathsf{MinPr}(a) \mid q \neq p\}.$$

Proof. Let $b = \bigwedge \{q \in \mathsf{MinPr}(a) \mid q \neq p\}$. Since $p \wedge b = a$, then $b \leq p \to a$. On the other hand, if $q \in \mathsf{MinPr}(a)$ and $q \neq p$, since $p \nleq q$, by the minimality of q, and $(p \to a) \wedge p = p \wedge a \leq a \leq q$ by (H5), we have $p \to a \leq q$. Hence, $p \to a \leq b$. $\square$

7.5. Essential prime elements. A minimal prime p of a is called *essential* [189] if

$$a = \bigwedge \mathsf{MinPr}(a) \quad \text{and} \quad a \neq \bigwedge \{q \in \mathsf{MinPr}(a) \mid q \neq p\}.$$

Let $\mathsf{EssPr}(a)$ denote the set of essential primes of a.

7.5.1. Lemma. *The following are equivalent for $a \in L$ satisfying $a = \bigwedge \mathsf{MinPr}(a)$.*

(1) $p \in \mathsf{EssPr}(a)$.
(2) $p \in \mathsf{Pr}L$ and there exists $b \in L$ such that $a = b \wedge p$ and $b \nleq p$.
(3) $p \in \mathsf{Pr}L$ and $(p \to a) \to a = p$; in other words, $p \in \mathsf{Pr}(\mathfrak{b}(a))$.

Proof. (1)$\Rightarrow$(2): Take $b = \bigwedge \{q \in \mathsf{MinPr}(a) \mid q \neq p\}$. Then $a = b \wedge p$ and $b \nleq p$ (because $b \neq a$).

(2)$\Rightarrow$(3): Let $p \in \mathsf{Pr}(a)$ and consider the b from (2). Then $b \leq p \to a$ and consequently, using (H5) and (H3),

$$((p \to a) \to a) \wedge b \leq ((p \to a) \to a) \wedge (p \to a) = (p \to a) \wedge a = a \leq p.$$

Since $b \nleq p$ then $(p \to a) \to a \leq p$. By (H9), $(p \to a) \to a \geq p$.

(3)$\Rightarrow$(1): Assume p is a prime of L and $(p \to a) \to a = p$. Then immediately $p \in \mathsf{Pr}(a)$. Let us show that $p \in \mathsf{MinPr}(a)$. Assume $a \leq q \leq p$ with q prime. If $q \neq p$ we have $p \to a \leq q$ (because $(p \to a) \wedge p \leq a \leq q$ and $p \nleq q$). Thus

$$p \to a = (p \to a) \wedge q \leq (p \to a) \wedge p \leq a,$$

and hence $p = (p \to a) \to a = 1$, contradicting the fact that p is prime. Hence $p \in \mathsf{MinPr}(a)$.

Finally, let $b = \bigwedge \{q \in \mathsf{MinPr}(a) \mid q \neq p\}$. By 7.4, $p \to a = b$. Consequently,

$$p = (p \to a) \to a = b \to a = b \to (b \wedge p) = b \to p,$$

and therefore $b \nleq p$ (as $p \neq 1$). Hence, $a = b \wedge p \neq b$. $\square$

7.5.2. Remark. It follows from the implication (1)$\Rightarrow$(3) above that for any $a, b \in L$ such that $b \to a \neq 1$, if $p \in \mathsf{EssPr}(b \to a)$ then $b \nleq p$. Indeed, if $b \leq p$ we would have

$$p = (p \to (b \to a)) \to (b \to a)$$
$$= ((p \wedge b) \to a)) \to (b \to a) = (b \to a) \to (b \to a) = 1,$$

contradicting the assumption that p is prime.

7.6. Other characterizations. Here are two more characterizations of the locales whose sublocales are all spatial ([189]).

7.6.1. Proposition. *The following are equivalent for a locale L.*

(1) *Every sublocale of L is spatial.*

(2) *For every $a \neq 1$, $\mathsf{EssPr}(a) \neq \emptyset$.*

(3) *For every $a \in L$, $a = \bigwedge \mathsf{EssPr}(a)$.*

Proof. $(1) \Rightarrow (2)$: Suppose $a \neq 1$. Since, by 7.2.2, $\mathcal{Sl}(L)^{\mathrm{op}}$ is spatial, we have by II.5.3 and 7.2.1(a),

$$\mathfrak{b}(a) = \bigvee \mathsf{Sp}'(\mathfrak{b}(a)) = \bigvee \{\mathfrak{b}(p) \mid p \in \mathsf{Sp}'(L), \mathfrak{b}(p) \leq \mathfrak{b}(a)\}.$$

But $\mathfrak{b}(p) \leq \mathfrak{b}(a)$ means that $p \in \mathfrak{b}(a)$, that is, $p = (x \to a) \to a$ for some $x \in L$ and hence by (H10) $p = (p \to a) \to a$. Therefore

$$\mathfrak{b}(a) = \bigvee \{\mathfrak{b}(p) \mid p \in \mathsf{Sp}'(L), p = (p \to a) \to a\}.$$

Since $a \in \mathfrak{b}(a)$, $a = \bigwedge Y$ for a $Y \subseteq \bigcup \{\mathfrak{b}(p) \mid p \in \mathsf{Sp}'(L), p = (p \to a) \to a\}$ and we can write

$$a \geq \bigwedge \{p \mid p \in \mathsf{Sp}'(L), p = (p \to a) \to a\} \geq \bigwedge \{p \mid p \in \mathsf{Sp}'(L), p \geq a\} \geq a.$$

Thus $a = \bigwedge \mathsf{Pr}(a) = \bigwedge \mathsf{MinPr}(a)$. Since $a \neq 1$, the meets above are non-empty so that there is a $p \in \mathsf{Sp}'(L)$ such that $p = (p \to a) \to a$. By 7.5.1, $p \in \mathsf{EssPr}(a)$.

$(2) \Rightarrow (3)$: Let $a \in L$ and $b = \bigwedge \mathsf{EssPr}(a)$. Suppose $a < b$. Then $b \to a \neq 1$ and hence there is a $p \in \mathsf{EssPr}(b \to a)$. By III.3.1.1(H6), we have

$$(p \to a) \to a \leq (((p \wedge b) \to a) \wedge b) \to a$$
$$= ((p \wedge b) \to a) \to (b \to a) = (p \to (b \to a)) \to (b \to a).$$

By 7.5.1, essential primeness of p and (H9),

$$(p \to (b \to a)) \to (b \to a) = p \leq (p \to a) \to a.$$

Thus $(p \to a) \to a = p$ and $p \in \mathsf{EssPr}(a)$. Hence $p \geq b$ which contradicts 7.5.2.

$(3) \Rightarrow (1)$: Let S be a sublocale of L and let $a \in S$. We have $a = \bigwedge \mathsf{EssPr}(a)$ in L. Now if $p \in \bigwedge \mathsf{EssPr}(a)$ we have by 7.5.1(3) $p = (p \to a) \to a$ and hence $p \in S$ so that $a = \bigwedge \{p \mid p \in \mathsf{MinPr}(a), p \in S\}$.   □

7.6.2. Corollary. *The locale $\mathsf{Lc}(X)$ has a non-spatial sublocale iff X has a closed subset F for which $B_{\mathsf{Lc}(F)}$ has no points.*

Proof. By 7.6.1 and 7.5.1, $\mathsf{Lc}(X)$ has a non-spatial sublocale iff there is a proper $A \in \mathsf{Lc}(X)$ such that $\mathfrak{b}(A)$ has no prime element. Since $\mathfrak{b}(A) = B_{\mathfrak{c}(A)} = B_{\mathsf{Lc}(X \smallsetminus A)}$ this means that there is a non-void closed subset F of X such that $B_{\mathsf{Lc}(F)}$ has no points.   □

8. Spaces with no non-induced sublocales

In this final section we will discuss the spaces in which all sublocales are not only spatial but even induced. This is simpler in the T_D case than in the general one. Therefore, after a brief discussion of some necessary concepts we will prove the theorem for this special case; then we will analyze what happens after dropping the T_D hypothesis.

8.1. Scattered and weakly scattered spaces. Recall from I.4.1 that a point x of a space X is *isolated* (resp. *weakly isolated*) in $S \subseteq X$ if there is an open U such that $S \cap U = \{x\}$ (resp. $x \in S \cap U \subseteq \overline{\{x\}}$). A space X is *scattered* (resp. *weakly scattered – corrupt* in [246]) if every non-void closed $S \subseteq X$ contains an isolated point (resp. a weakly isolated point). Note that, since each isolated point of $\overline{S}$ is an isolated point of S, in the former case it is the same as assuming an isolated point in every $S \subseteq X$.

Equivalently, X is a (weakly) scattered space if each closed subset is the closure of the set of its (weakly) isolated points.

8.1.1. Proposition. *A space is scattered iff it is T_D and weakly scattered.*

Proof. By I.4.2 it suffices to prove that a scattered space is T_D. By hypothesis, for each $x \in X$ the closure $\overline{\{x\}}$ has an isolated point y. Then $y = x$ (otherwise, $x \notin \{y\} = \overline{\{x\}} \cap U$ for some open U would imply $x \in X \smallsetminus U$ and thus $\overline{\{x\}} \subseteq X \smallsetminus U$, that is, $\overline{\{x\}} \cap U = \emptyset$). Hence x is an isolated point of $\overline{\{x\}}$ so that for a suitable open set $U \ni x$, $U \smallsetminus \{x\} = U \smallsetminus \overline{\{x\}}$. $\square$

8.2. The T_D case. Recall 1.1 and notice that the right adjoint of the homomorphism $\Omega(j_A)$ is given by

$$U \cap A \mapsto \bigcup \{V \in \Omega(X) \mid V \cap A = U \cap A\} = \mathrm{int}((X \smallsetminus A) \cup U).$$

Therefore, for each $A \subseteq X$, the induced sublocale $\widetilde{A}$ is just

$$\widetilde{A} = \{\mathrm{int}((X \smallsetminus A) \cup U) \mid U \in \Omega(X)\}.$$

By 1.2 we already know that for a T_D-space X, the map

$$\rho_X \colon \mathfrak{P}(X) \to \mathcal{S}\ell(\mathsf{Lc}(X))^{\mathrm{op}}$$

given by $A \subseteq X \mapsto \widetilde{A}$ is one-one. For which spaces X is ρ_X surjective? In order to answer this we consider, for each sublocale S of $\mathsf{Lc}(X)$, the set

$$\sigma_X(S) = \bigcup \{j_S^*(U) \smallsetminus U \mid U \in \Omega(X)\} \tag{8.2.1}$$

where j_S is the localic embedding $S \to \mathsf{Lc}(X)$.

(That is, $j_S^*(U) = \bigwedge \{V \in S \mid U \leq V\}$.)

It is easy to check that

$$x \in \sigma_X(S) \text{ iff } x \in j_S^*(X \smallsetminus \overline{\{x\}}). \tag{8.2.2}$$

(Indeed: the implication $\Leftarrow$ is trivial since $x \notin X \smallsetminus \overline{\{x\}}$; conversely, if $x \in j_S^*(U) \smallsetminus U$ for some open U, then $x \in \overline{\{x\}} \subseteq X \smallsetminus U$, so that $U \subseteq X \smallsetminus \overline{\{x\}}$ and thus $x \in j_S^*(U) \subseteq j_S^*(X \smallsetminus \overline{\{x\}})$.)

Now until 8.3, where we will start to discuss the general case, the space X will always be T_D. Let $A \subseteq X$. From I.4.2 it immediately follows that

$$x \in \text{int}(A \cup X \smallsetminus \overline{\{x\}}) \text{ iff } x \in A. \tag{8.2.3}$$

8.2.1. Lemma. *For every $A \subseteq X$ and every sublocale S of $\mathsf{Lc}(X)$ we have:*

(a) $\sigma_X(\widetilde{A}) = X \smallsetminus A$.

(b) $\sigma_X(S) \subseteq X \smallsetminus A \Rightarrow \widetilde{A} \subseteq S$.

Proof. (a): It is a straightforward exercise to check that $j_{\widetilde{A}}^*(X \smallsetminus \overline{\{x\}}) = \text{int}(X \smallsetminus A \cup X \smallsetminus \overline{\{x\}})$. Thus, by (8.2.3) above, $x \in j_{\widetilde{A}}^*(X \smallsetminus \overline{\{x\}})$ iff $x \in X \smallsetminus A$.

(b): If $\sigma_X(S) \subseteq X \smallsetminus A$ then, for each $U \in \Omega(X)$, $j_S^*(U) \subseteq X \smallsetminus A \cup U$ so that $j_S^*(U) \subseteq \text{int}(X \smallsetminus A \cup U)$. In particular,

$$j_S^*(\text{int}(X \smallsetminus A \cup U)) \subseteq \text{int}(X \smallsetminus A \cup \text{int}(X \smallsetminus A \cup U)) \subseteq \text{int}(X \smallsetminus A \cup U),$$

and hence $j_S^*(\text{int}(X \smallsetminus A \cup U)) = \text{int}(X \smallsetminus A \cup U)$, that is, $\text{int}(X \smallsetminus A \cup U) \in S$. This means that $\widetilde{A} \subseteq S$, as stated. $\qquad\square$

8.2.2. Lemma. *Let F be a closed subset of X and let W be the set of all isolated points of F. For $A = X \smallsetminus F$ we have:*

(a) $\sigma_X(\mathfrak{b}(A)) = X \smallsetminus W$.

(b) *If $\mathfrak{b}(A)$ is induced then $\mathfrak{b}(A) = \widetilde{W}$.*

(c) *If X is scattered then $\mathfrak{b}(A) = \widetilde{W}$.*

Proof. (a): First we observe that, by (8.2.2),

$$x \in \sigma_X(\mathfrak{b}(A)) \quad \text{iff} \quad x \in X \smallsetminus \text{cl}(F \cap \text{int}(A \cup \overline{\{x\}})). \tag{$*$}$$

In fact,

$$
\begin{aligned}
j_{\mathfrak{b}(A)}^*(X \smallsetminus \overline{\{x\}}) &= ((X \smallsetminus \overline{\{x\}}) \to A) \to A = \text{int}(A \cup \overline{\{x\}}) \to A \\
&= (X \smallsetminus \text{cl}(F \cap X \smallsetminus \overline{\{x\}})) \to A = \text{int}(A \cup \text{cl}(F \cap X \smallsetminus \overline{\{x\}})) \\
&= X \smallsetminus \text{cl}(X \smallsetminus (F \cap X \smallsetminus \overline{\{x\}}))) = X \smallsetminus \text{cl}(F \cap \text{int}(A \cup \overline{\{x\}})).
\end{aligned}
$$

Now, if $x \in W$ then $x \in F \cap U \subseteq \overline{\{x\}}$ for an open U. But

$$U = (U \cap F) \cup (U \cap A) \subseteq \overline{\{x\}} \cup A.$$

Thus $x \in U \subseteq \operatorname{int}(A \cup \overline{\{x\}})$ and therefore $x \in F \cap \operatorname{int}(A \cup \overline{\{x\}})$. By $(*)$, this means that $x \notin \sigma_X(\mathfrak{b}(A))$.

Conversely, if $x \notin \sigma_X(\mathfrak{b}(A))$ that is $x \in \operatorname{cl}(F \cap \operatorname{int}(A \cup \overline{\{x\}})) \subseteq F$ then $F \cap \operatorname{int}(A \cup \overline{\{x\}}) \neq \emptyset$. Hence there is a

$$y \in F \cap \operatorname{int}(A \cup \overline{\{x\}}) \subseteq F \cap (A \cup \overline{\{x\}}) = F \cap \overline{\{x\}} \subseteq \overline{\{x\}}.$$

Thus $x \in \operatorname{int}(A \cup \overline{\{x\}})$ (otherwise we would have

$$y \in \overline{\{y\}} \subseteq \overline{\{x\}} \subseteq F \cap X \smallsetminus \operatorname{int}(A \cup \overline{\{x\}}),$$

a contradiction). Hence $x \in F \cap \operatorname{int}(A \cup \overline{\{x\}}) \subseteq \overline{\{x\}}$ which shows that x is weakly isolated. By T_D it is isolated.

(b): If $\mathfrak{b}(A) = \widetilde{B}$ for some $B \subseteq X$ then, using 8.2.1 and (a), we get

$$\mathfrak{b}(A) = \widetilde{B} = (X \widetilde{\smallsetminus \sigma_X}(\widetilde{B})) = (X \widetilde{\smallsetminus \sigma_X}(\mathfrak{b}(A))) = \widetilde{W}.$$

(c): The inclusion $\widetilde{W} \subseteq \mathfrak{b}(A)$ is always true (from 8.2.2(a) and 8.2.1(b)). Now let X be scattered. Then $\overline{W} = F$. Since $\mathfrak{b}(A) \subseteq \mathfrak{c}(A)$ and

$$A = X \smallsetminus \overline{W} = \operatorname{int}(X \smallsetminus W) = \widetilde{W}(\emptyset) = \bigwedge \widetilde{W},$$

we have $\mathfrak{b}(A) \subseteq \uparrow (\bigwedge \widetilde{W}) = \overline{\widetilde{W}} \subseteq \widetilde{W}$. □

8.2.3. Theorem. *For a T_D-space X, all sublocales of $\mathsf{Lc}(X)$ are induced if and only if X is scattered.*

Proof. Let F be a closed subset of X and let $A = X \smallsetminus F$. By hypothesis, $\mathfrak{b}(A)$ is induced so, by the preceding lemma, $\mathfrak{b}(A) = \widetilde{W}$. Then

$$F = X \smallsetminus A = X \smallsetminus ((\emptyset \to A) \to A) = X \smallsetminus \widetilde{W}(\emptyset) = X \smallsetminus \operatorname{int}(X \smallsetminus W) = \overline{W}$$

and X is scattered.

Conversely, let S be a sublocale of $\mathsf{Lc}(X)$ and $A = X \smallsetminus \sigma_X(S)$. By 8.2.1, $\sigma_X(\widetilde{A}) = \sigma_X(S)$. Now it suffices to check that by scatteredness, σ_X is injective. So let $S, T \in \mathcal{Sl}(\mathsf{Lc}(X))^{\mathrm{op}}$, $S \neq T$. We may assume that there is some $A \in \mathsf{Lc}(X)$ such that $A \in S$ and $A \notin T$. Then $\mathfrak{b}(A) \subseteq S$ and $\mathfrak{b}(A) \not\subseteq T$ and, by 8.2.2, $\mathfrak{b}(A) = \widetilde{W}$ for some W. Then $X \smallsetminus W = \sigma_X(\mathfrak{b}(A)) \supseteq S$ and $X \smallsetminus W \not\supseteq T$, which shows that σ_X is one-one. □

8.3. The general case. Since we will not assume T_D any more, ρ_X will not be necessarily injective (distinct subsets may induce equal sublocales). Recall 2.2 and also I.4.2: the Skula topology is now not necessarily trivial and will come handy in the following analysis.

We will write $\mathrm{int}_{Sk}(\cdot)$ for the Skula interior operator and $\mathrm{cl}_{Sk}(\cdot)$ for the corresponding closure operator. We cannot use (8.2.3). Instead, we have:

Fact. *For each $A \subseteq X$,*

$$x \in \mathrm{int}(A \cup X \smallsetminus \overline{\{x\}}) \quad \text{iff} \quad x \in \mathrm{int}_{Sk}(A). \tag{8.3.1}$$

Proof. First notice that for each $x \in X$, the family

$$U \cap \overline{\{x\}} \quad (x \in U \in \Omega(X))$$

forms a base for the Sk-open neighbourhoods of x (because

$$U \cap (X \smallsetminus V) = \bigcup \{U \cap \overline{\{x\}} \mid x \in U \cap (X \smallsetminus V)\}$$

for any $U, V \in \Omega(X)$). So, for any $x \in \mathrm{int}_{Sk}(A)$ we may conclude that $x \in U \cap \overline{\{x\}} \subseteq A$ for some $U \in \Omega(X)$. But now $U \subseteq A \cup X \smallsetminus \overline{\{x\}}$ so that $x \in U \subseteq \mathrm{int}(A \cup X \smallsetminus \overline{\{x\}})$. Conversely, if $x \in U = \mathrm{int}(A \cup X \smallsetminus \overline{\{x\}})$ then $x \in U \subseteq A \cup X \smallsetminus \overline{\{x\}}$. Hence $x \in U \cap \overline{\{x\}} \subseteq A$. $\qquad\square$

By manipulating complements we then get

$$x \in \mathrm{cl}_{Sk}(A) \text{ iff } x \in \mathrm{cl}(A \cap \overline{\{x\}}). \tag{8.3.2}$$

Replacing (8.2.3) by (8.3.1) and using (8.3.2) we adapt the proofs of 8.2.1, 8.2.2 and 8.2.3 to obtain

8.3.1. Lemma. *For every $A \subseteq X$ and every sublocale S of $\mathsf{Lc}(X)$ we have:*

(a) $\sigma_X(\widetilde{A}) = X \smallsetminus \mathrm{cl}_{Sk}(A)$.

(b) *If A is Sk-closed then*

$$\sigma_X(S) \subseteq X \smallsetminus A \Rightarrow \widetilde{A} \subseteq S. \qquad\square$$

8.3.2. Lemma. *Let F be a closed subset of X and let W be the set of all weakly isolated points of F. For $A = X \smallsetminus F$ we have:*

(a) $\sigma_X(\mathfrak{b}(A)) = X \smallsetminus W$.

(b) *If $\mathfrak{b}(A)$ is induced then $\mathfrak{b}(A) = \widetilde{W}$.*

(c) *If X is weakly scattered then $\mathfrak{b}(A) = \widetilde{W}$.* $\qquad\square$

8.3.3. Theorem. *For a space X, all sublocales of $\mathsf{Lc}(X)$ are induced if and only if X is weakly scattered.* $\qquad\square$

Chapter VII

Compactness and Local Compactness

The cover definition of compactness is basically point-free; therefore there is no surprise that the basic facts are very much like in the classical case. But a surprise does come: the point-free variant of Stone-Čech compactification is fully constructive (no choice principle and no use of the excluded middle). Thus in particular, the fact that products of compact (regular) locales are compact is constructive, unlike the Tychonoff Theorem of classical topology (see Section 4 – in particular 4.5 and 6.5).

In the latter parts of this chapter we present (a.o.) the famous Hofmann-Lawson duality between locally compact frames and locally compact spaces. Thus, for this very important class of spaces the point-free theory is quite parallel with the classical one (here, however, the axiom of choice is necessary).

1. Basics, and a technical lemma

1.1. A *cover* of a frame L is a subset $A \subseteq L$ such that $\bigvee A = 1$. A *subcover* of a cover A is a subset $B \subseteq A$ such that (still) $\bigvee B = 1$.

> **Note.** In classical topology this corresponds to the notion of *open cover*, that is, of a cover consisting of open sets. This is what we will almost exclusively be concerned with. A brief encounter with another type of covers will be duly emphasized.

Quite like in classical topology, a locale (frame) is said to be *compact* if each cover has a finite subcover.

More generally, a frame is α-compact if each cover has a subcover of cardinality less than α. Thus, "compact" is the same as "$\aleph_0$-compact". We will be also interested in the $\aleph_1$-compact frames, that is, in frames such that each cover has an *at most countable* subcover. These are referred to as *Lindelöf frames*.

1.2. Observation. *Each subframe and each closed sublocale of a compact (more generally, α-compact) frame L is compact (resp. α-compact).*

Proof. Both statements hold for the same reason: the joins (non-void, in the latter case) – unlike for instance in non-closed sublocales – coincide with those in L (III.6.4(1)).   □

1.3. Proposition. *An image of a compact (more generally, α-compact) sublocale $S \subseteq L$ under a localic map $f \colon L \to M$ is compact (resp. α-compact).*

Proof. Let $j \colon S \subseteq L$ be the embedding. Consider the onto - (one-one) decomposition of $f \cdot j$ from IV.1.4

Since g is onto we have for the associated frame homomorphism $g^* \colon f[S] \to S$, $gg^* = \mathrm{id}$. Now if $\bigvee_{i \in J} a_i = 1$ in $f[S]$ we have $g^*(\bigvee_{i \in J} a_i) = \bigvee_{i \in J} g^*(a_i) = 1$ in S and hence $g^*(\bigvee_{i \in K} a_i) = \bigvee_{i \in K} g^*(a_i) = 1$ for a finite K (or K of cardinality $< \alpha$ in the more general case), and hence $\bigvee_{i \in K} a_i = gg^*(\bigvee_{i \in K} a_i) = 1$.   □

1.4. The $\pi_2\pi_1$-Lemma. The following lemma will be useful more than once.
  Recall from IV.5.6 the nucleus μ generated by the prenucleus $\nu = \pi_2\pi_1$ where

$$\pi_1(U) = \{\bigvee A, b) \mid A \times \{b\} \subseteq U\}, \quad \text{and}$$

$$\pi_2(U) = \{a, \bigvee B) \mid \{a\} \times B \subseteq U\}.$$

We have

Lemma. *Let L_1 be compact and let $(1, b) \in \mu(U)$. Then $(1, b) \in \nu(U) = \pi_2\pi_1(U)$.*

Proof. Let α be the first ordinal such that $(1, b) \in \nu_\alpha(U)$. Then, unless $\alpha = 0$, it cannot be a limit ordinal. We will show that $\alpha = 1$. Indeed, otherwise $(1, b) \in \nu(\nu_\beta(U))$ for some non-limit $\beta < \alpha$ and hence $(1, b) \in \nu(\nu(W)) = \pi_2(\pi_1\pi_2\pi_1(W))$ (where $W = \nu_{\beta-1}(U)$). Then $b = \bigvee B$ for some B such that $\{1\} \times B \subseteq \pi_1\pi_2\pi_1(V)$ and for each $y \in B$ we have $(1, y) \in \pi_1\pi_2\pi_1(V)$ and hence there is an A_y such that $\bigvee A_y = 1$ and $A_y \times \{y\} \subseteq \pi_2\pi_1(V)$. By compactness there are finite $C_y \subseteq A_y$ such that $\bigvee C_y = 1$, and for every $x \in C_y$, $(x, y) \in \pi_2\pi_1(V)$. By Lemma IV.5.6, $(\bigvee C_y, y) = (1, y) \in \pi_2\pi_1(V)$ and hence

$$(1, b) = (1, \bigvee B) \in \pi_2\pi_2\pi_1(V) = \pi_2\pi_1(V) = \nu(V) = \nu_\beta(U)$$

contradicting the minimality of α.   □

2. Compactness and separation

We will see that similar to classical spaces, compact I-Hausdorff locales are normal. But the Hausdorff property is not quite the counterpart of the classical homonymous one, while regularity, complete regularity and normality are. Therefore we will later consider rather the compact regular locales in contexts analogous to those in which one classically considers the compact Hausdorff spaces. Nevertheless, we will start with proving that a compact I-Hausdorff locale is regular.

2.1. Recall the $d_L = \{(x, y) \mid x \wedge y = 0\}$ from IV.2.1.

2.1.1. Lemma. *For $U = (a \oplus 1) \cup d_L = \downarrow(a, 1) \cup d_L$ we have*

$$\pi_2\pi_1(U) = \Big\{(x, y) \mid y \leq \bigvee\{b \mid x \leq a \vee b^*\}\Big\}.$$

Proof. I. Let $y \leq \bigvee\{b \mid x \leq a \vee b^*\}$. Set $B = \{b \mid x \leq a \vee b^*\}$ and for $b \in B$ define $x_1(b) = x \wedge a$ and $x_2(b) = x \wedge b^*$. Then $(x_1(b), b), (x_2(b), b) \in \pi_1(U)$ and hence $(x, b) = (x_1(b) \vee x_2(b), b) \in \pi_1(U)$. Thus, $\{x\} \times B \subseteq \pi_1(U)$ and finally $(x, \bigvee B) \in \pi_2\pi_1(U)$.

II. Let $(x, y) \in \pi_2\pi_1(U)$. Then $\{x\} \times B \subseteq \pi_1(U)$ for some B such that $\bigvee B = y$. Thus, for every $b \in B$ there is an $A(b)$ such that $x = \bigvee A(b)$ and $A(b) \times \{b\} \subseteq \downarrow(a, 1) \cup d_L$. Then for a $z \in A(b)$ either $z \leq a$ or $z \leq b^*$ and hence $x \leq a \vee b^*$, and $y \leq \bigvee\{b \mid x \leq a \vee b^*\}$. □

2.1.2. Proposition. *A compact I-Hausdorff frame is regular.*

Proof. We have $(1, a) \in 1 \oplus a$. Our frame is I-Hausdorff and hence, by V.2.3, $(1 \oplus a) \vee d_L = (a \oplus 1) \vee d_L$. Now by 1.4, $(1, a) \in \pi_2\pi_1(\downarrow(a, 1) \cup d_L)$ and, by the lemma, $a \leq \bigvee\{b \mid 1 = a \vee b^*\} = \bigvee\{b \mid b \prec a\}$. □

2.2. The following statement (including the proof) is quite like in classical spaces.

Proposition. *Every regular Lindelöf locale (in particular, every compact regular locale) is normal; consequently, the relation $\prec$ in such a locale interpolates, and the locale is completely regular.*

Proof. Let $a \vee b = 1$ in L. By regularity, $a = \bigvee\{x \mid x \prec a\}$ and $b = \bigvee\{y \mid y \prec b\}$. Thus, $\bigvee\{x \mid x \prec a\} \vee b = 1 = a \vee \bigvee\{y \mid y \prec b\}$ and by the Lindelöf property, there are

$$x_1, x_2, \cdots \prec a \quad \text{and} \quad y_1, y_2, \cdots \prec b$$

satisfying

$$\bigvee_{i=1}^{\infty} x_i \vee b = 1 = a \vee \bigvee_{i=1}^{\infty} y_i.$$

We may assume that $x_1 \leq x_2 \leq \cdots$ and $y_1 \leq y_2 \leq \cdots$. Set $u_i = x_i \wedge y_i^*$ and $v_i = x_i^* \wedge y_i$. We have

$$a \vee v_i = a \vee (x_i^* \wedge y_i) = (a \vee x_i^*) \wedge (a \vee y_i) = a \vee y_i$$

and similarly $b \vee u_i = b \vee x_i$. Consequently, taking $u = \bigvee_{i=1}^{\infty} u_i$ and $v = \bigvee_{i=1}^{\infty} v_i$ we have

$$a \vee v = \bigvee_{i=1}^{\infty} (a \vee y_i) = a \vee \bigvee_{i=1}^{\infty} y_i = 1$$

and similarly $b \vee u = 1$. Finally, $u_i \wedge v_j = x_i \wedge y_i^* \wedge y_j \wedge x_j^* = 0$ for every i, j (if $i \leq j$ then $x_i \wedge x_j^* = 0$ and otherwise if $i \geq j$ then $y_j \wedge y_i^* = 0$). Hence $u \wedge v = \bigvee_{i,j=1}^{\infty} (u_i \wedge v_j) = 0$. $\qquad\square$

2.2.1. Corollary. *A compact I-Hausdorff locale is normal.* $\qquad\square$

2.2.2. Proposition. *Let L be compact and let M be regular. Then each dense localic map $f: L \to M$ is onto.*

In frame setting: each dense frame homomorphism $h: M \to L$ is one-one.

Proof. We will prove the statement on frame homomorphisms. By V.5.6 it suffices to prove h is co-dense.

Suppose $h(a) = 1$. Since $a = \bigvee\{x \mid x \prec a\}$, the set $\{h(x) \mid x \prec a\}$ is a cover of L and hence there are $x_1, \dots, x_n$ such that $\bigvee_{i=1}^{n} h(x_i) = 1$. By V.5.2.1, $x = x_1 \vee \cdots \vee v_n \prec a$ and we have $x^* \vee a = 1$ and $h(x) = 1$. Now $h(x^*) \leq h(x)^* = 0$, hence by density $x^* = 0$, and finally $a = 1$. $\qquad\square$

2.2.3. Corollary. *A compact sublocale S of a regular locale L is closed.*

Proof. Decompose the embedding $j: S \subseteq L$ into

$$S \xrightarrow{\;j'=\subseteq\;} \overline{S} \xrightarrow{\;j''=\subseteq\;} L .$$

By V.4.8, $\overline{S}$ is regular. Since j' is dense, it is onto by 2.2.2, and hence it is the identity map. $\qquad\square$

Note. In the classical setting the analogous statement holds with a Hausdorff space in place of our regular locale. Here, the regularity is essential.

3. Kuratowski-Mrówka characterization

3.1. The famous Kuratowski-Mrówka Theorem characterizes compact spaces as those spaces X for which the natural projection

$$p_Y : X \times Y \to Y$$

is closed for every space Y. We will prove the counterpart of this characteristic in the point-free setting.

In this section we will use the symbols

$$\iota_L = (x \mapsto x \oplus 1): L \to L \oplus M, \quad \iota_M = (x \mapsto 1 \oplus x): M \to L \oplus M$$

for the coproduct injections, but since we will be mostly interested in the latter, we will write just ι for ι_M if there will be no danger of confusion, and

$$p: L \oplus M \to M$$

for the right adjoint of ι (the "localic projection" of the product in **Loc**). Thus,

$$1 \oplus x \leq u \quad \text{iff} \quad x \leq p(u).$$

3.2. Recall the closed localic maps and frame homomorphisms, and their characteristics in III.7.3, in particular the one in (4) stating that a localic map f is closed iff

$$x \leq f(u) \vee y \iff f^*(x) \leq u \vee f^*(y).$$

Since the implication $\Rightarrow$ holds always (if $x \leq f(u) \vee y$ then $f^*(x) \leq f^* f(u) \vee f^*(y) \leq u \vee f^*(y)$), it is the $\Leftarrow$ what is of the essence. Thus in our case, asking whether the projection p above is closed amounts to the question whether

$$1 \oplus x \leq u \vee (1 \oplus y) \quad \Rightarrow \quad x \leq p(u) \vee y. \tag{3.2.1}$$

3.3. Lemma. *Let L be compact and let $1 \oplus x \leq \bigvee_{i \in J}(a_i \oplus b_i) \vee (1 \oplus y)$. Then*

$$x \leq \bigvee\{ \bigwedge_{i \in K} b_i \mid K \subseteq J \text{ such that } \bigvee_{i \in K} a_i = 1\} \vee y.$$

Proof. By 1.4, $(1, x) \in \pi_2\pi_1(U)$ where

$$U = \bigcup_{i \in J} {\downarrow}(a_i, b_i) \cup {\downarrow}(1, y).$$

Thus there is a B such that $\{x\} \times B \subseteq \pi_1(U)$ and $\bigvee B = x$ so that for each $b \in B$ there is an $A(b)$ such that

$$\bigvee A(b) = 1 \quad \text{and} \quad A(b) \times \{b\} \subseteq U.$$

Set

$$B' = \{b \in B \mid b \not\leq y\} \quad \text{and} \quad B'' = \{b \in B \mid b \leq y\},$$

and for $b \in B'$ set $K(b) = \{i \mid b \leq b_i\}$ so that

$$b \leq \bigwedge_{i \in K(b)} b_i \quad \text{and} \quad A(b) \times \{b\} \subseteq \bigcup_{i \in K(b)} {\downarrow}(a_i, b_i).$$

Thus

(1) for each $a \in A(b)$ there is an $i \in K(b)$ with $a \leq a_i$ and hence $\bigvee_{i \in K(b)} a_i = 1$, and

(2) $\bigvee\{\bigwedge_{i \in K(b)} b_i \mid b \in B'\} \vee y \geq \bigvee B = x$,

and the statement follows. $\qquad\square$

3.4. Now suppose L is not compact. Choose a cover A such that no finite $K \subseteq A$ is a cover, and define

$$\widetilde{L} = \{U \subseteq L \mid 1 \in U \;\Rightarrow\; \exists \text{ finite } K \subseteq A, \; \uparrow(\bigvee K) \subseteq U\}.$$

It is easy to check that $\widetilde{L}$ is a topology on L and hence a frame, with unions for joins and finite intersections for finite meets.

Lemma. *Set*

$$c = \bigvee\{a \oplus \uparrow a \mid a \in A\} \quad (\in L \times \widetilde{L}).$$

Then $c \vee (1 \oplus (L \smallsetminus \{1\})) = 1$ *and* $p(c) = 0$.

Proof. Put $x = c \vee (1 \oplus (L \smallsetminus \{1\}))$. For $a \in A$ we have

$$a \oplus 1_{\widetilde{L}} = a \oplus L = (a \oplus \uparrow a) \vee (a \oplus (L \smallsetminus \{1\})) \leq c \vee (1 \oplus (L \smallsetminus \{1\})) = z$$

and since A is a cover we see that $1 = (\bigvee A) \oplus 1 \leq z$.

Now let $p(c) \neq 0$. Then there is an $x \in p(c)$ such that $x \neq 1$. Consider the frame homomorphism $\xi \colon \widetilde{L} \to \mathbf{2}$ defined by $\xi(U) = 1$ iff $x \in U$ and define

$$\phi \colon L \oplus \widetilde{L} \to L \quad \text{by} \quad \phi \cdot \iota_L = \mathrm{id} \text{ and } \phi \cdot \iota_{\widetilde{L}} = \xi$$

(thus $\phi = \mathrm{id} \oplus \xi$, if we consider $L \oplus \mathbf{2}$ as L with the first injection the identity and the second one the trivial embedding $\mathbf{2} \subseteq L$). Then

$$\phi(a \oplus \uparrow a) = \phi((a \oplus 1) \wedge (1 \oplus \uparrow a)) = a \wedge \xi(\uparrow a) = \begin{cases} a \text{ if } a \leq x, \\[4pt] 0 \text{ if } a \not\leq x. \end{cases}$$

Hence $\phi(c) \leq x$ while $\phi(1 \oplus p(c)) = 1 \wedge 1 = 1$ and hence $1 \oplus p(c) \not\leq c$ and finally $p(c) \not\leq p(c)$, a contradiction. $\qquad\square$

3.5. Proposition (Kuratowski-Mrówka Theorem for locales). *A locale L is compact iff the product projection $p \colon L \oplus M \to M$ (the coproduct injection $\iota \colon M \to L \oplus M$, in frame language) is closed for every locale M.*

Proof. I. Let L be compact. Let $1 \oplus x \leq u \vee (1 \oplus y)$ with $u = \bigvee\{a_i \oplus b_i \mid i \in J\}$. Set

$$\bigvee\{\bigwedge_{i \in K} b_i \mid K \subseteq J \text{ such that } \bigvee_{i \in K} a_i = 1\}.$$

We have $1 \oplus b \leq u$ since if $\bigvee_{i \in K} a_i = 1$ then

$$(1, \bigwedge_{i \in K} b_i) \in \bigvee_{i \in K}(a_i \oplus b_i) \subseteq \bigvee_{i \in J}(a_i \oplus b_i)$$

and hence $b \leq p(u)$. Now by 3.3, $x \leq b \vee y \leq p(u) \vee y$, and the implication in (3.2.1) holds.

II. Let L not be compact. Use 3.4. As $1 \oplus 1 \leq c \vee (1 \oplus (L \smallsetminus \{1\}))$ while $1 \not\leq L \smallsetminus \{1\}$, (3.2.1) does not hold. $\qquad\square$

4. Compactification

In this section we will present a point-free counterpart of the Stone-Čech compactification of completely regular spaces, as constructed by Banaschewski and Mulvey in [32].

4.1. A not very satisfactory but instructive compactification. Recall that an *ideal* in a frame L (more generally, in a distributive lattice, but here we are not interested in the general case) is a non-empty subset $I \subseteq L$ such that

(I1) $a, b \in I \quad \Rightarrow \quad a \vee b \in I$, and

(I2) $b \leq a \in I \quad \Rightarrow \quad b \in I$.

Note. We assume the ideals non-empty, but allow for the (improper) ideal $I = L$.

4.1.1. Proposition. *The set*

$$\mathfrak{J}(L)$$

of all the ideals in L, ordered by inclusion, is a compact frame.

Proof. First note that any intersection of ideals is an ideal.

Second, any system I_j, $j \in J$, of ideals has a supremum in $\mathfrak{J}(L)$, namely

$$\bigvee_{j \in J} I_j = \{ \bigvee F \mid F \text{ finite } \subseteq \bigcup_{j \in J} I_j \}.$$

Indeed, this set is an ideal (trivially we have (I1), and if $x \leq \bigvee F$ then by distributivity $x = \{a \wedge x \mid a \in F\}$ with $\{a \wedge x \mid a \in F\} \subseteq \bigcup_{j \in J} I_j$) containing all the I_j, and that if K is an ideal containing all the I_j then necessarily $\bigvee F \in K$ for any finite $F \subseteq \bigcup_{j \in J} I_j$.

Now if K, I_j are ideals then trivially $K \cap (\bigvee_{j \in J} I_j) \supseteq \bigvee_{j \in J} (K \cap I_j)$, and if $F \subseteq \bigcup_{j \in J} I_j$ and $x = \bigvee F$ is in K then $F \subseteq \bigcup_{j \in J} (I_j \cap K)$ and $x \in \bigvee_{j \in J} (K \cap I_j)$.

Finally, $\mathfrak{J}(L)$ is compact: indeed, if $\bigvee I_j = L = 1_{\mathfrak{J}(L)}$ then in particular $1 \in \bigvee I_j$ and hence $1 = \bigvee F$ for some finite $F \subseteq \bigcup I_j$. But then $F \subseteq \bigcup_{j \in J_0} I_j$ for some finite $J_0 \subseteq J$, hence $1 \in \bigvee_{j \in J_0} I_j$ and $L = \bigvee_{j \in J_0} I_j$. $\square$

4.1.2. The dense embedding $\alpha\colon L \to \mathfrak{J}(L)$. Consider the mappings

$$v = (I \mapsto \bigvee I)\colon \mathfrak{J}(L) \to L \quad \text{and} \quad \alpha = (a \mapsto {\downarrow}a)\colon L \to \mathfrak{J}(L).$$

We obviously have

$$v\alpha(a) = a \quad \text{and} \quad I \subseteq \alpha v(I)$$

and hence these maps are adjoint, v to the left and α to the right; hence

v preserves joins.

It preserves finite meets as well: we have

$$v(I_1) \wedge v(I_2) = \bigvee I_1 \wedge \bigvee I_2 = \bigvee \{a_1 \wedge a_2 \mid a_j \in I_j\}$$
$$\leq \bigvee \{a \mid a \in I_1 \cap I_2\} = v(I_1 \cap I_2) \leq v(I_1) \wedge v(I_2)$$

(the first inequality because of (I2): $a_1 \wedge a_2 \leq a_j$ and hence it belongs to both the I_j's; the second one is trivial). Thus, v is a frame homomorphism, and α is a localic map.

4.1.3. Observation. α *is a dense localic embedding.*

(Indeed, it is obviously one-one, and if $I \neq 0$, that is, $I \neq \{0\}$, then $v(I) = \bigvee I \neq 0$.)

Thus, we have for each locale a dense embedding into a compact one – if we wish we can view it as a compact extension of L in which L is dense. But it is not very satisfactory: $\mathfrak{J}(L)$ is considerably larger even if the L we started with was already compact itself. In general we have to accept it as it is, but for completely regular locales this drawback can be mended.

4.2. Regular ideals. An ideal I is said to be *regular* if

(R) for every $a \in I$ there is a $b \in I$ such that $a \prec\!\!\prec b$.

4.2.1. Proposition. *The set*

$$\mathfrak{R}(L)$$

of regular ideals in L is a subframe of $\mathfrak{J}(L)$. Hence it is a compact locale.

Proof. Let I_1, I_2 be regular ideals and let $a \in I_1 \cap I_2$. There are $b_j \in I_j$ such that $a \prec\!\!\prec b_j$. Then by V.5.7 $a \prec\!\!\prec b_1 \wedge b_2 \in I_1 \cap I_2$.

Now, let I_j, $j \in J$, be regular, $F \subseteq \bigcup_{j \in J} I_j$ finite. For $x \in F$ choose $j(x)$ such that $x \in I_{j(x)}$ and an $x' \in I_{j(x)}$ such that $x \prec\!\!\prec x'$. Set $F' = \{x' \mid x \in F\}$. Then $F' \subseteq \bigcup_{j \in J} I_j$ and by V.5.7, $\bigvee F \prec\!\!\prec \bigvee F'$. $\qquad\square$

4.2.2. The regular ideals $\sigma(a)$. For an $a \in L$ set

$$\sigma(a) = \{x \mid x \prec\!\!\prec a\}.$$

Obviously

$$\text{each } \sigma(a) \text{ is a regular ideal.}$$

Lemma. *If $a \prec\!\!\prec b$ then $\sigma(a) \prec \sigma(b)$.*

Proof. Interpolate $a \prec\!\!\prec x \prec y \prec\!\!\prec b$. Then $y \in \sigma(b)$ and, by V.5.7, $x^* \in \sigma(a^*)$. Thus, $1 = x^* \vee y \in \sigma(a^*) \vee \sigma(b)$ and $\sigma(a^*) \vee \sigma(b) = L = 1_{\mathfrak{J}(L)}$. If $x \in \sigma(a^*) \cap \sigma(a)$ then $x \leq a^* \wedge a = 0$ and hence $x = 0$, and $\sigma(a^*) \cap \sigma(a) = \{0\} = 0_{\mathfrak{J}(L)}$. Use (rb2) in V.5.2. $\qquad\square$

4.2.3. Proposition. *If L is completely regular then $\mathfrak{R}(L)$ is completely regular.*

Proof. As $\mathfrak{R}(L)$ is compact it suffices to prove it is regular. We have, for any ideal I, by interpolativity of $\lhd\!\lhd$,

$$I = \bigcup\{\sigma(a) \mid a \in I\} = \bigvee\{\sigma(a) \mid a \in I\}$$

and since L is completely regular, each a is the join $\bigvee\{b \mid b \lhd\!\lhd a\}$ and hence

$$\sigma(a) = \bigcup\{\sigma(b) \mid b \lhd\!\lhd a\} = \bigvee\{\sigma(b) \mid b \lhd\!\lhd a\}.$$

By the Lemma in 4.2.2, $I = \bigcup\{\sigma(b) \mid b \lhd\!\lhd a,\ a \in I\} \supseteq \bigvee\{K \mid K \lhd\!\lhd I\}$.　　□

4.3. $\mathfrak{R}(L)$ as a compactification. Let L be a completely regular frame. Then, writing again $v(I) = \bigvee I$ we have

$$v\sigma(a) \quad \text{and} \quad I \subseteq \sigma v(I). \tag{4.3.1}$$

The mapping $v\colon \mathfrak{R}(L) \to L$ is a frame homomorphism for precisely the same reason as the v in 4.1.2 and hence σ is a localic embedding. This time we have

Proposition. *Let L be compact. Then v and σ are mutually inverse isomorphisms.*

Proof. Let $x \in \sigma v(I)$. Thus in particular $x \prec \bigvee I$, and $x^* \vee \bigvee I = 1$. By compactness there are $y_1, \dots, y_n \in I$ such that $x^* \vee y_1 \vee \cdots \vee y_n = 1$. But $y = y_1 \vee \cdots \vee y_n \in I$ and we have $x \prec y$ and hence $x \in I$. Thus, $\sigma v(I) \subseteq I$ and the statement follows from (4.3.1).　　□

4.4. Functoriality. Let $h\colon L \to M$ be a frame homomorphism. Then for a regular ideal I, $h[I]$ has the properties (I1) and (R), and if we extend this set to

$$\mathfrak{R}(h)(I) = {\downarrow}h[I]$$

we have a regular ideal and we easily see that

$$(\mathfrak{R}(g)\mathfrak{R}(h))(I) = \mathfrak{R}(gh)(I) \quad \text{and} \quad \mathfrak{R}(\mathrm{id})(I) = I.$$

Thus, we have obtained a functor

$$\mathfrak{R}\colon \mathbf{CRegFrm} \to \mathbf{RegKFrm}$$

(the latter stands for the category of regular compact frames). Furthermore, if we specify the L in the definition above writing $v_L\colon \mathfrak{R}(L) \to L$ we obtain a natural transformation (indeed, the diagrams

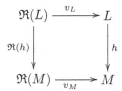

commute: $v_H \Re(h)(I) = \bigvee \downarrow h[I] = \bigvee h[I] = h(\bigvee I) = h v_L(I)$). Hence taking into account that v_L is onto, that is, a sublocale homomorphism, we conclude that we have got a coreflection of **CRegFrm** onto **RegKFrm** resp. a reflection of **CRegLoc** onto **RegKLoc**.

4.5. The last includes a rather surprising fact. If one checks carefully the construction and the proofs one learns that everything has been done constructively (meaning: without any choice principle, and even without the double negation). Furthermore, the fact that a reflective subcategory is closed under products (see AII.8) is also a fully constructive matter. See also [46].

Now in classical topology the fact that a product of compact regular spaces is compact is not constructive (it is equivalent to the Boolean Ultrafilter Theorem). Hence we have here a constructive proof of something that has no such in the classical setting.

The fact that a product of general compact spaces is compact (Tychonoff Theorem) is even equivalent with the (unrestricted) Axiom of Choice. And even here there is a constructive proof of the point-free counterpart (Johnstone [144], see also Kříž [163]) albeit by no means so simple as the reflectivity presented here. Constructive procedures like this are one of pleasant features of point-free topology.

5. Well below and rather below.
Continuous completely regular frames

5.1. Let us briefly recapitulate the relation $\ll$ and its properties. We say that an element a is *well below* b in a poset L if

$$\text{for every directed } D, \quad b \leq \sup D \;\Rightarrow\; \exists d \in D, \; a \leq d.$$

We will work in lattices. There this property can be reformulated to

$$b \leq \bigvee A \quad \Rightarrow \quad \exists \text{ finite } F \subseteq A, \; a \leq \bigvee F.$$

One obviously has

$$a \leq a' \ll b' \leq b \quad \Rightarrow \quad a \ll b \tag{5.1.1}$$

and in a lattice

$$a_i \ll b, \; i = 1, 2 \quad \Rightarrow \quad a_1 \vee a_2 \ll b \tag{5.1.2}$$

(but NOT as one would like to expect $a \ll b_i \Rightarrow a \ll b_1 \wedge b_2$). Thus,

$$\text{in a lattice, the sets } \{x \mid x \ll a\} \text{ are directed.}$$

We will consider the relation $\ll$ in frames, but it should be noted that this relation is a concept playing a very important role in a much more general context, in particular in not necessarily distributive lattices [112].

Recall that a lattice L is said to be *continuous* if for every $a \in L$,

$$a = \bigvee \{x \mid x \ll a\}$$

(in the general case of continuous poset one has to assume that the sets $\{x \mid x \ll a\}$ are directed; here it is automatic). An easy but important fact is that

in a continuous lattice the relation $\ll$ interpolates.

(Indeed, let $a \ll b$; we have $b = \bigvee \{x \mid x \ll b\}$ and applying the rule for each individual $x \ll b$ we obtain $b = \bigvee \{y \mid y \ll x \ll b$ for some $x\}$; the last join is obviously directed and hence there are x, y such that $a \leq y \ll x \ll b$. In particular, if $a \ll b$ and $b \leq \bigvee D$ with a directed D, one has a $d \in D$ such that $a \ll d$).

Since we will be interested in continuous frames it is perhaps useful to realize that in complete continuous lattices the frame distributivity follows from the plain one: if $x \ll (\bigvee_{i \in J} a_i) \wedge b$ then by (5.1.1) $x \ll \bigvee_{i \in J} a_i$ and $x \ll b$; the former yields $x \ll \bigvee_F a_i$ with a finite $F \subseteq J$, and hence by distributivity $x \ll (\bigvee_{i \in F} a_i) \wedge b = \bigvee_{i \in F} (a_i \wedge b) \leq \bigvee_{i \in J} (a_i \wedge b)$. Thus, $(\bigvee_{i \in J} a_i) \wedge b \leq \bigvee_{i \in J} (a_i \wedge b)$ and the other inequality is trivial.

5.1.1. Let X be a locally compact space. Then the frame $\Omega(X)$ is continuous. Indeed, if U is open and $x \in U$ choose an open $V(x)$ and a compact K such that $x \in V(x) \subseteq K \subseteq U$. Then $V(x) \ll U$ and $U = \bigcup \{V(x) \mid x \in U\}$.

In fact, we will see in the next section that such frames $\Omega(X)$ are, up to isomorphism, precisely the continuous ones.

Note that an $\Omega(X)$ with X compact is in general not necessarily continuous (compactness does not imply local compactness). But under the regularity condition it is, and we have more. See below.

5.2. The relations "well below", "rather below" and "completely below" are closely connected. We have

5.2.1. Lemma. (a) *In any frame,*

$$a \prec b \ll 1 \implies a \ll b.$$

(b) *If the frame is regular (resp. completely regular) then*

$$a \ll b \implies a \prec b$$

(*resp. $a \prec\!\prec b$*).

Proof. (a): If $a \prec b$ then $a^* \vee b = 1$ and hence if $b \leq \bigvee D$ then $1 \leq a^* \vee \bigvee D$, and $b \leq a^* \vee d$ for some $d \in D$. Then $a = a \wedge b \leq a \wedge d \leq d$.

(b): Let $a \ll b$. Since $b = \bigvee \{x \mid x \prec b\}$ and the join is directed, there is an $x \prec b$ such that $a \leq x \prec b$, and hence $a \prec b$. Similarly for $\prec\!\prec$. $\square$

5.2.2. Proposition. *Every open sublocale of a regular compact locale L is continuous.*

Proof. Let $a \in L$. The associated open sublocale $\mathfrak{o}(a)$ is isomorphic with the subposet $\downarrow a$ (see III.6.1.1). Now $\downarrow a$ is not a sublocale nor a subframe of L. Nevertheless it has some pleasant features:

(1) the lattice operations, with the exception of the void meet 1 (which is now a) coincide with those in L;

(2) the way below relation of $\downarrow a$ coincides with that of L;

(3) although the relation rather below $\prec_a$ of $\downarrow a$ does not necessarily coincide with the $\prec$ of L (we only have $x \prec y \Rightarrow x \prec_a y$), if L is regular we have in $\downarrow a$, $x = \bigvee\{y \mid y \prec x\}$ with the stronger $\prec$.

Now let L be compact regular and let $b \in \downarrow a$. We have $b \leq a \ll 1$ in L (compactness of L is the same as $1_L \ll 1_L$) and hence if $x \prec b$ then $x \ll b$ (in L, and by (2) in $\downarrow a$ as well). Since (see (3)), $b = \bigvee\{x \mid x \prec b\}$, $\prec$ as in L, we have by 5.2.1(a) $b = \bigvee\{x \mid x \ll b\}$. $\qquad\square$

5.3. Proposition. *A completely regular frame is continuous iff it is open in its Stone-Čech compactification.*

Proof. The implication $\Leftarrow$ is in 5.2.2. Thus, we have to prove that if L is continuous then the homomorphism $v \colon \mathfrak{R}L \to L$ from 4.3 is open. That is, we need a regular ideal A in L such that

$$v(I_1) = v(I_2) \ (\text{that is}, \ \bigvee I_1 = \bigvee I_2) \quad \text{iff} \quad I_1 \cap A = I_2 \cap A.$$

Set

$$A = \{x \mid x \ll 1\}.$$

It is obviously an ideal (by (5.1.1) and (5.1.2)). Furthermore, if $x \in A$, that is, $x \ll 1$, interpolate $x \ll y \ll 1$; then $y \in A$ and $x \lll y$ by 5.2.1(b).

     We have $v(A) = \bigvee\{x \mid x \ll a\} = 1$ by continuity, hence $I_1 \cap A = I_2 \cap A$ implies $v(I_1) = v(I_2)$.

     Finally let $\bigvee I_1 = \bigvee I_2$ and $a \in I_1 \cap A$. The set $I_1 \cap A$ is a regular ideal and hence there is a $b \in I_1 \cap A$ such that $a \lll b$; as $b \ll 1$, we have $a \ll b$ by 5.2.1(a). Since ideals are directed and $b \leq \bigvee I_2 = \bigvee I_1$ there is an $x \in I_2$ such that $a \leq x$. Hence $a \in I_2$. $\qquad\square$

5.4. Furthermore we have

5.4.1. Lemma. *Let L_1, L_2 be frames and $\iota_i \colon L_i \to L_1 \oplus L_2$ the coproduct injections. Let $a_i \in L_i$. Then the subposet $\downarrow(a_1 \oplus a_2) \subseteq L_1 \oplus L_2$, together with the homomorphisms*

$$\iota_i' = (x \mapsto \iota_i(x) \wedge (a_1 \oplus a_2)) \colon \downarrow a_1 \to \downarrow(a_1 \oplus a_2),$$

constitutes the coproduct of $\downarrow a_1$ and $\downarrow a_2$.

Proof. Let $h_i \colon \mathop{\downarrow}a_i \to M$ be frame homomorphisms. Take the homomorphisms $\widehat{a}_i = (x \mapsto a_i \wedge x) \colon L \to \mathop{\downarrow}a_i$ and the $g \colon L_1 \oplus L_2 \to M$ such that $g\iota_i = h_i\widehat{a}_i$. Then $g(x_1 \oplus x_2) = h(x_1 \wedge a_1) \wedge h(x_2 \wedge a_2)$ and hence if $(x_1 \oplus x_2) \wedge (a_1 \oplus a_2) = (y_1 \oplus y_2) \wedge (a_1 \oplus a_2)$ then $g(x_1 \oplus x_2) = g(y_1 \oplus y_2)$ and hence there is a homomorphism $h \colon \mathop{\downarrow}(a_1 \oplus a_2) \to M$ such that $h((x_1 \oplus x_2) \wedge (a_1 \oplus a_2)) = g(x_1 \oplus x_2)$. Thus, for $x \in \mathop{\downarrow}a_1$ we have $h(\iota_i'(x)) = h(\iota_i(x) \wedge (a_1 \oplus a_2)) = g\iota_i(x) = h_i(x)$. $\square$

5.4.2. Proposition. *Finite coproducts of completely regular continuous frames are continuous.*

Proof. Coproducts of regular compact frames are regular compact (see 4.5). Thus, the statement follows from 5.3, 5.4.1, and 5.2.2. $\square$

6. Continuous is the same as locally compact. Hofmann-Lawson duality

6.1. An unusual use of Scott topology. We have already encountered the famous Scott topology in II.6.4. There it constituted the (lattice of open sets of the) space we were interested in. In this section, however, we will consider the Scott topology on a frame that is already a point-free view of a space itself. It is as if we had a topological space (in fact, as we will see shortly, for an important class of frames not just "as if"), and then took the set of its open sets for points of a new space, introducing a topology on that.

Thus, we have a frame L and declare a subset $U \subseteq L$ (*Scott*) *open* if

(1) $\mathop{\uparrow}U = U$, and

(2) if $D \subseteq L$ and $\bigvee D \in U$ then $D \cap U \neq \emptyset$ (equivalently, if $\bigvee A \in U$ for an $A \subseteq L$ then $\bigvee B \in U$ for a finite $B \subseteq A$).

Checking that this constitutes a topology is an easy exercise.

6.2. Proposition (Modified Birkhoff's Theorem). *Let F be a Scott open filter in a frame L and let $b \notin F$. Then there exists a Scott open prime filter G such that $b \notin G \supseteq F$.*

Proof. Recall AI.6.4.3; we will do almost exactly the same, we only have to watch the openness.

Let $\mathcal{F}$ be a chain of open filters in L such that none of its elements contains b. Then $\bigcup \mathcal{F}$ is an *open* filter again and it does not contain b. Thus, by Zorn's Lemma we can infer that in the poset of open filters (ordered by inclusion) there is a $G \supseteq F$ maximal with the property that $b \notin G$. We will prove that it is prime.

Let $u \vee v \in G$ and $u, v \notin G$. Set

$$H = \{x \mid x \vee v \in G\}.$$

Checking that H is a filter is straightforward, and we also easily see that it is Scott open: if D is directed and $\bigvee D \in H$ we have $\bigvee D \vee v \in G$, and since $\{d \vee v \mid d \in D\}$ is also directed, there is a $d \in D$ such that $d \vee v \in G$, and d is in H. Obviously $G \subseteq H$, and because $u \in H \smallsetminus G$, H is strictly larger. Thus, necessarily $b \in H$, that is, $b \vee v \in G$; again, $b, v \notin G$ and we can repeat the procedure with

$$H' = \{x \mid b \vee x \in G\}$$

to obtain $b \in H$ and a contradiction $b = b \vee b \in G$. $\qquad\square$

6.3. The following observation, together with showing that there are sufficiently many open filters is one of the main moments of the duality theorem below: completely prime filters in a frame may not exist at all (for instance see II.5.4) while prime filters abound (every filter can be majorized by such).

6.3.1. Lemma. *A filter F in a frame L is completely prime iff it is prime and Scott open.*

Proof. Let F be prime and Scott open and let $\bigvee A \in F$. By the openness there is a finite $B \subseteq A$ such that $\bigvee B \in F$, and by the primeness there is an $a \in B$ such that $a \in F$.

    If F is completely prime and if $\bigvee D \in F$ then there is, trivially, a $d \in D$ with $d \in F$. $\qquad\square$

6.3.2. Lemma. *Let $c \ll a$ in a continuous frame L. Then there is an open filter F such that $a \in F \subseteq {\uparrow}c$.*

Proof. Since L is continuous we can interpolate (see 5.1) inductively

$$a = a_0 \gg a_1 \gg a_2 \gg \cdots \gg a_n \gg \cdots \gg c.$$

Set
$$F = \{x \mid x \geq a_n \ \text{ for some } \ n\}.$$

Obviously F is a filter contained in ${\uparrow}c$, and if $\bigvee D \geq a_n$ for a directed D, there is a $d \in D$ such that $d \geq a_{n+1}$. $\qquad\square$

6.3.3. Proposition. *Each continuous lattice is spatial.*

Proof. We will prove that if $a \nleq b$ there is a completely prime filter P such that $a \in P$ and $b \notin P$.

    Since L is continuous there is a $c \ll a$ such that $c \nleq b$. Consider the F from 6.3.2. Since $F \subseteq {\uparrow}c$, $b \notin F$ and hence there is by 6.2 an open prime filter P such that $b \notin P \ni a$. By 6.3.1, P is completely prime. $\qquad\square$

6.3.4. In other words,

   each locally compact frame is spatial.

   Consequently, each compact regular frame is spatial.

6.4. We will use the representation of the spectrum as in II.4.2, that is, as the set of completely prime (now ≡ Scott open prime) filters with the open sets $\Sigma_a = \{F \mid a \in F\}$.

6.4.1. Lemma. *A subset $K \subseteq \Sigma L$ is compact iff $\bigcap\{P \mid P \in K\}$ is a Scott open subset of L.*

Proof. Let K be compact and let $\bigvee A \in \bigcap\{P \mid P \in K\}$. Thus, $\bigvee A \in P$, that is, $P \in \Sigma_{\bigvee A}$, for every $P \in K$ and $K \subseteq \Sigma_{\bigvee A} = \bigcup\{\Sigma_a \mid a \in A\}$ and we have a finite $B \subseteq A$ such that $K \subseteq \bigcup\{\Sigma_a \mid a \in B\}$. Thus for each $P \in K$, $\bigvee B \in P$, and hence $\bigvee B \in \bigcap\{P \mid P \in K\}$ so that $\bigcap\{P \mid P \in K\}$ is open.

On the other hand let $\bigcap\{P \mid P \in K\}$ be open and let $K \subseteq \bigcup\{\Sigma_a \mid a \in A\}$. Then for every $P \in K$, $P \in \Sigma_{\bigvee A}$ hence $\bigvee A \in \bigcap\{P \mid P \in K\}$. As this is open, $\bigvee B \in \bigcap\{P \mid P \in K\}$ for some finite $B \subseteq A$ and we infer that $K \subseteq \bigcup\{\Sigma_a \mid a \in B\}$. $\square$

6.4.2. Lemma. *Let a frame L be continuous. Then the spectrum ΣL is locally compact.*

Proof. Let $P \in \Sigma_a$, that is, $a = \bigvee\{c \mid c \ll a\} \in P$, and since P is completely prime there is a $c \ll a$ such that $c \in P$. Consider an open filter F such that $a \in F \subseteq \uparrow c$, and set

$$K = \{Q \in \Sigma L \mid F \subseteq Q\}.$$

By 6.2, $\bigcap\{Q \mid Q \in K\} = F$ (if $b \notin F$, there is a $Q \in K$ such that $b \notin Q$), hence it is open and hence, by 6.4.1, K is compact. Now we have $P \in \Sigma_c$, further if $Q \in \Sigma_c$ then $F \subseteq \uparrow c \subseteq Q$ and hence $Q \in K$, and finally if $Q \in K$ then $a \in Q$. Thus,

$$P \in \Sigma_c \subseteq K \subseteq \Sigma_a$$

and we have found a compact neighbourhood of P included in Σ_a. $\square$

6.4.3. Theorem (Hofmann-Lawson's Duality). *Denote by* **ContFrm** *the category of continuous frames and frame homomorphisms, and by* **LKSob** *the category of sober locally compact spaces and continuous mappings. Then the restriction of the functors Σ (that is,* Sp *viewed as a functor defined on* **Frm***) and Ω constitutes a dual isomorphism of categories.*

Proof. By 6.3.3 the functors restrict to a dual isomorphism of **ContFrm** with the subcategory of **Sob** generated by the spaces of the form ΣL, L continuous. By 5.1.1 and 6.4.2 this is precisely **LKSob**. $\square$

6.5. Notes on coproducts. (1) Restricting the duality further to

CR continuous frames   vs.   CR locally compact spaces

("CR" stands for "completely regular") and to

regular compact frames   vs.   regular compact spaces

(for the latter recall 5.2.2) we see that in particular by 5.4.2,

> *finite products of completely regular compact spaces correspond precisely to finite coproducts of completely regular continuous frames,*

and from the latter,

> *general products of regular compact spaces correspond precisely to the coproducts of regular compact frames.*

(2) Another phenomenon worth noting has to do with the last mentioned fact. We have seen in 4.5 that the existence of products in the category of compact regular locales (coproducts in that of compact regular frames) is a constructive fact needing, in particular, no choice principle. Now the duality translates it to the existence of products in the category of compact regular spaces, in other words to the fact that products of compact regular spaces are compact. The last is not choice-free (as it was already mentioned sooner, it is equivalent with the Boolean Ultrafilter Theorem). There is no mystery: the choice is hidden in the proof of spatiality. What happens is that a product of compact regular spaces is compact unconditionally, but without a choice principle it might not have enough points.

7. One more spatiality theorem

7.1. Lemma (a variant of Baire's Theorem). *Let X be a compact regular space and let $U_n \subseteq X$, $n \in \mathbb{N}$, be dense open subsets. Then $\bigcup_{n=1}^{\infty} U_n$ is dense in X.*

Proof. Let V be non-void open. Since X is regular we can find an open non-void V_1 such that $\overline{V_1} \subseteq V \cap U_1$. Suppose we have already found non-void open V_i, $i = 1, \ldots, n$, such that

$$\overline{V_i} \subseteq V_{i-1} \cap U_{i-1} \quad \text{and} \quad V_1 \supseteq \overline{V_2} \supseteq V_2 \supseteq \overline{V_3} \supseteq \cdots .$$

Then choose a non-void open V_{n+1} such that $\overline{V_{n+1}} \subseteq V_n \cap U_n$. For thus chosen sequence $(V_n)_n$ we have by compactness $A = \bigcup_{n=1}^{\infty} \overline{V_n} \neq \emptyset$ and $A \subseteq V \cap \bigcup_{n=1}^{\infty} U_n$. $\square$

7.2. Proposition. *Let X be a compact Hausdorff space and let $(U_n)_n$ be a sequence of dense open subspaces. Then the intersection of open sublocales*

$$\bigcap_{n=1}^{\infty} \mathfrak{o}(U_n)$$

of $\Omega(X)$ (meet in $\mathcal{Sl}(\Omega(X))$) is isomorphic to $\Omega(Y)$ where $Y = \bigcap_{n=1}^{\infty} U_n$.

Proof. By III.11.4.1,

$$\bigcap_{n=1}^{\infty} \mathfrak{o}(U_n) = \Omega(X)/R \ \text{ where } R = \{(V \cap U_n, V) \mid V \in \Omega(X), \ n \in \mathbb{N}\},$$

and the elements of $\Omega(X)/R$ (the R-saturated ones) are the open $W \subseteq X$ for which

$$(\exists n, \ V \cap U_n \subseteq W) \quad \Rightarrow \quad V \subseteq W. \tag{7.2.1}$$

Let $j \colon Y \to X$ be the embedding map. Since we have

$$\Omega(j)(V \cap U_n) = (V \cap U_n) \cap Y = V \cap Y = \Omega(j)(V)$$

there is a homomorphism

$$h \colon \Omega(X)/R \to \Omega(Y)$$

such that $h\mu = \Omega(j)$ for the sublocale homomorphism $\mu \colon \Omega(X) \to \Omega(X)/R$ and we have by III.11.3.1, $h(W) = \Omega(j)(W) = W \cap Y$.

Obviously h is onto. We have to prove it is also one-one. Since the frames in question are regular it suffices to prove that h is codense (see V.5.6), that is, that

$$h(W) = Y \quad \Rightarrow \quad W = X.$$

Consider a $W \in \Omega(X)/R$ such that $W \subset X$. Denote by K the (non-void) compact subspace $X \smallsetminus W$. For every V such that $V \cap K \neq \emptyset$, that is, $V \nsubseteq W$, we have by (7.2.1)

$$\forall n, \ V \cap U_n \nsubseteq W, \quad \text{that is, } V \cap U_n \cap K \neq \emptyset.$$

Thus each $U_n \cap K$ is dense in K and hence $Y \cap K = \bigcup_n (U_n \cap K) \neq \emptyset$ by 7.1. Thus, $Y \nsubseteq W$ and hence, finally, $h(W) = Y \cap W \neq Y$. $\qquad\square$

7.3. From 7.2 and 6.3.4 we now immediately obtain

Corollary. *A meet of countably many open sublocales ("a G_δ-sublocale") of a compact regular frame is spatial.* $\qquad\square$

8. Supercompactness. Algebraic, superalgebraic and supercontinuous frames

8.1. A frame L is *supercompact if* each cover of L contains 1_L, that is, the only subcover consisting of one element. This may be not a most interesting concept: it is a sort of $\bigvee$-primeness, or $\bigvee$-irreducibility of the top.

More interesting is the more general notion of a *supercompact element* $a \in L$, namely one for which $a \leq \bigvee_{i \in J} x_i$ implies that $a \leq x_j$ for some $j \in J$ (of course, again, it is the $\bigvee$-primeness, or $\bigvee$-irreducibility of a).

8.1.1. Lemma. *An element $a \in L$ is supercompact iff there is a $b \in L$ such that*

$$L \smallsetminus {\uparrow}a = {\downarrow}b.$$

Proof. If a is supercompact consider $b = \bigvee\{x \mid a \not\leq x\}$. Then $a \not\leq b$, by supercompactness, and hence if $x \leq b$ then $a \not\leq x$. If $a \not\leq x$ then $x \leq b$ by the definition.

On the other hand, if such b exists and $a \leq \bigvee_{i \in J} x_i$ then $\bigvee_{i \in J} x_i \not\leq b$, hence $x_j \not\leq b$ for some j, and $a \leq x_j$.   □

8.1.2. Note. Thus, the supercompactness of the whole of L simply means that 1 is an immediate successor of an element majorizing all the other elements. This was what we meant by indicating that the supercompactness of a frame is a much less interesting concept than that of the individual elements.

8.2. The super-way-below relation. This is the relation

$$a \lll b$$

defined by the formula

$$b \leq \bigvee_{i \in J} x_i \quad \Rightarrow \quad \exists j \in J, \ a \leq x_j.$$

Thus

$$a \text{ is supercompact iff } a \lll a.$$

A frame L is *supercontinuous* if

$$\forall a \in L, \quad a = \bigvee\{x \mid x \lll a\}.$$

8.2.1. A frame L is said to be *algebraic* (resp. *superalgebraic*) if

$$\forall a \in L, \quad a = \bigvee\{x \mid x \ll x \leq a\}$$

$$(\text{resp. } \forall a \in L, \quad a = \bigvee\{x \mid x \lll x \leq a\}).$$

8.2.2. Remarks. (a) More generally, algebraic lattices (or just posets, where we have to assume the join in the definition directed) play a fundamental role in theoretical computer science (in particular, theory of domains). They are of course also of interest in topology, but it will be the superalgebraic ones in which we will encounter an old acquaintance shortly.

(b) In other words, algebraic (superalgebraic) frames are those in which the set of compact (supercompact) elements constitutes a (join) basis.

8.3. Proposition. *A frame L is superalgebraic iff it is isomorphic to the frame $\mathfrak{D}(X, \leq)$ of down-sets of a poset $(X, \leq)$.*

Proof. Obviously $\mathfrak{D}(X, \leq)$ is superalgebraic: each individual $\downarrow x$ is supercompact (if $\downarrow x \subseteq \bigcup_{i \in J} U_i$ with U_i down-sets then $x \in U_j$, and hence $\downarrow x \subseteq U_j$ for some j), and a down-set U is the union $\bigcup\{\downarrow x \mid x \in U\}$.

Now let L be superalgebraic. Set

$$X = \{x \in L \mid x \text{ supercompact}\}$$

and endow X with the original order from L. Consider the maps

$$\alpha = (a \mapsto {\downarrow}a = \{x \in X \mid x \le a\}) \colon L \to \mathfrak{D}(X, \le),$$

$$\beta = (U \mapsto \bigvee U) \colon \mathfrak{D}(X, \le) \to L.$$

Obviously both the maps are monotone. Now $\beta\alpha(a) = a$ by superalgebraicity and trivially $\alpha\beta(U) \supseteq U$. Finally, if $x \in \alpha\beta(U)$, that is, $x \le \bigvee U$, and x is supercompact then $x \le u$ for some $u \in U$; but U is a down-set, and hence $x \in U$. Thus, also $\alpha\beta(U) = U$ and α, β are mutually inverse isomorphisms. □

(For more about superalgebraic lattices and related notions see [86], and other papers of M. Erné ([85, 87, 88]).)

8.4. Proposition. *Let L be supercontinuous, and let $h\colon L \to M$ be a sublocale homomorphism (quotient frame homomorphism) that preserves all meets. Then M is supercontinuous.*

(This is an instance of a more general fact from [86].)

Proof. For $b \in M$ consider

$$a = \bigwedge\{x \in L \mid b \le h(x)\}.$$

Then $h(a) = \bigwedge\{h(x) \mid b \le h(x)\} \ge b \ge h(a)$ (the latter inequality since $b = h(x)$ for some x and hence $x \ge a$).

Now let $x \lll a$ in L and let $b \le \bigvee_{i \in J} y_i$ in M; let $y_i = h(x_i)$. Then $b \le h(\bigvee_{i \in J} x_i)$ and hence $a \le \bigvee_{i \in J} x_i$ so that $x \le x_j$ for some j, and $h(x) \le y_j$. Thus,

$$x \lll a \quad \Rightarrow \quad h(x) \lll b.$$

Since $a = \bigvee\{x \mid x \lll a\}$ we have $b = h(a) = \bigvee\{h(x) \mid x \lll a\} \le \bigvee\{y \mid y \lll b\}$ in M. □

8.5. Recall V.4. We will now use the following characterization of completely distributive lattices; L is completely distributive iff

$$\text{for every } \mathcal{Y} \subseteq \mathfrak{D}(L), \quad \bigwedge\{\bigvee A \mid A \in \mathcal{Y}\} = \bigvee(\bigcap\mathcal{Y}).$$

Further, we will use the concept of separatedness: a poset $(X, \le)$ is *separated* if

$$\text{for any } a \nleq b \text{ there are } c, d \text{ such that } a \nleq d, \ c \nleq b \text{ and } {\uparrow}c \cup {\downarrow}d = X.$$

Both are from Raney [231], and also the following fact is, in essence, due to this author.

8.5.1. Proposition. *For a complete lattice the following conditions are equivalent.*

(1) *L is supercontinuous.*

(2) *L is separated.*

(3) *L is completely distributive.*

(4) *L is a complete homomorphic image of a down-set frame.*

Proof. (1)$\Rightarrow$(2): Let L be supercontinuous and let $a \not\leq b$. Then there is a $c \lll a$ such that $c \not\leq b$. Set $d = \bigvee(L \smallsetminus \uparrow c)$. Then we cannot have $a \leq d$ since else $c \leq x$ for some $x \in L \smallsetminus \uparrow c$, a contradiction. Trivially, $\downarrow d \supseteq L \smallsetminus \uparrow c$, that is, $\uparrow c \cup \downarrow d = L$.

(2)$\Rightarrow$(3): Let L be separated. Set

$$a = \bigwedge\{\bigvee A \mid A \in \mathcal{Y}\} \quad \text{and} \quad b = \bigvee(\bigcap \mathcal{Y}).$$

We trivially have $b \leq a$.

Suppose $a \not\leq b$. Choose $c, d \in L$ such that $a \not\leq d$, $c \not\leq b$ and $L = \uparrow c \cup \downarrow d$. Now for an arbitrary $A \in \mathcal{Y}$, $\bigvee A \not\leq d$ and hence there is an $a \in A$, $a \not\leq d$, making $a \geq c$ and $c \in A$ (because it is a down-set). Thus, $c \in \bigcap \mathcal{Y}$ while it was assumed that $c \not\leq \bigvee(\bigcap \mathcal{Y})$.

(3)$\Rightarrow$(4): Consider the down-set frame $\mathfrak{D}(L)$ and the mapping $v \colon \mathfrak{D}(L) \to L$ defined by $v(A) = \bigvee A$. Since for the $\alpha = (a \mapsto \downarrow a) \colon L \to \mathfrak{D}(L)$, $v\alpha = \text{id}$ and $\alpha v(A) \supseteq A$, v is a left Galois adjoint and hence it preserves all joins. Finally, for any $\mathcal{Y} \subseteq \mathfrak{D}(L)$ we have

$$v(\bigcap \mathcal{Y}) = \bigvee(\bigcap \mathcal{Y}) = \bigwedge\{\bigvee A \mid A \in \mathcal{Y}\} = \bigwedge\{v(A) \mid A \in \mathcal{Y}\},$$

that is, v preserves also all meets.

(4)$\Rightarrow$(1): Follows from 8.3 and 8.4. $\square$

Chapter VIII

(Symmetric) Uniformity and Nearness

Uniformity, one of the most natural enrichments of the classical topology, is a fundamental concept in frames and locales as well. It will feature in bigger or smaller roles in almost all of the following chapters. Here we will discuss the (from the point-free view, particularly natural) approach via systems of special covers.

1. Background

1.1. In classical topology there are, in principle, two ways of defining uniformity, that is a structure capturing the idea of uniform continuity (and related phenomena, like generalizations of Cauchy sequences and mappings preserving them).

One of them (going back to A. Weil, [272]) defines a uniformity on a set X as a non-empty set $\mathcal{U}$ of subsets of $X \times X$ (the "entourages" or "neighbourhoods of the diagonal") containing the diagonal $\Delta = \{(x, x) \mid x \in X\}$ such that

(E1) $V \supseteq U \in \mathcal{U} \;\Rightarrow\; V \in \mathcal{U}$,

(E2) $U, V \in \mathcal{U} \;\Rightarrow\; U \cap V \in \mathcal{U}$,

(E3) $U \in \mathcal{U} \;\Rightarrow\; U^{-1} = \{(x, y) \mid (y, x) \in U\} \in \mathcal{U}$,

(E4) and for every $U \in \mathcal{U}$ there is a $V \in \mathcal{U}$ satisfying

$$V \circ V = \{(x, y) \mid \exists z, \ (x, z), (z, y) \in V\} \subseteq U.$$

One then has an associated topology with $M \subseteq X$ open iff for each $x \in M$ there is a $U \in \mathcal{U}$ such that $Ux = \{y \mid (y, x) \in U\} \subseteq M$.

For the other approach, that appeared first in Tukey's monograph [259] and much popularized by Isbell's book [135], let us first recall a few concepts. A *cover* $\mathcal{U}$ of a set X is a set of subsets of X such that $\bigcup \mathcal{U} = X$. We say that a cover $\mathcal{U}$

is a *refinement* of a cover $\mathcal{V}$, and write $\mathcal{U} \leq \mathcal{V}$, if for each $U \in \mathcal{U}$ there is a $V \in \mathcal{V}$ such that $U \subseteq V$. Finally, we write for a cover $\mathcal{U}$ and a subset $A \subseteq X$ (resp. an element $x \in X$)

$$\mathcal{U} * A = \bigcup\{U \in \mathcal{U} \mid U \cap A \neq \emptyset\} \quad (\text{resp. } \mathcal{U} * x = \mathcal{U} * \{x\} = \bigcup\{U \in \mathcal{U} \mid U \ni x\}).$$

Now a uniformity on X is a non-empty system $\mathfrak{U}$ of covers of X such that

(C1) $\mathcal{U} \in \mathfrak{U}$ and $\mathcal{U} \leq \mathcal{V}$ implies $\mathcal{V} \in \mathfrak{U}$,

(C2) if $\mathcal{U}, \mathcal{V} \in \mathfrak{U}$ then $\mathcal{U} \wedge \mathcal{V} = \{U \cap V \mid U \in \mathcal{U}, V \in \mathcal{V}\} \in \mathfrak{U}$, and

(C3) for every $\mathcal{U} \in \mathfrak{U}$ there is a $\mathcal{V} \in \mathfrak{U}$ such that $\{V * V \mid V \in \mathcal{V}\} \leq \mathcal{U}$.

In the associated topology, a set $M \subseteq X$ is open iff for each $x \in M$ there is a $\mathcal{U} \in \mathfrak{U}$ such that $\mathcal{U} * x \subseteq M$.

The concept of uniform continuity is expressed by declaring a mapping $f \colon (X, \mathfrak{U}) \to (Y, \mathfrak{V})$ *uniformly continuous* if

$$\text{for each } \mathcal{V} \in \mathfrak{V}, \quad \{f^{-1}[V] \mid V \in \mathcal{V}\} \in \mathfrak{U}.$$

The resulting category of uniform spaces and uniformly continuous maps will be denoted

$$\textbf{UniSp.}$$

We have

1.1.1. Fact *A uniformly continuous mapping is continuous in the associated topologies.*

Proof. Let M be open in $(Y, \mathfrak{V})$ and $x \in f^{-1}[M]$. There is a $\mathcal{V} \in \mathfrak{V}$ such that $\mathcal{V} * f(x) \subseteq M$. We have $\{f^{-1}[V] \mid V \in \mathcal{V}\} \in \mathfrak{U}$; now $x \in f^{-1}[V]$ iff $f(x) \in V$, and then $V \subseteq \mathcal{V} * f(x) \subseteq M$, and $f^{-1}[V] \subseteq f^{-1}[M]$. Thus, $\{f^{-1}[V] \mid V \in \mathcal{V}\} * x \subseteq f^{-1}[M]$. $\qquad\square$

Note. In the definition of uniformity above one often uses, instead of (C3), a formally weaker condition

(C3′) for every $\mathcal{U} \in \mathfrak{U}$ there is a $\mathcal{V} \in \mathfrak{U}$ such that $\{V * x \mid x \in X\} \leq \mathcal{U}$.

This, however, is in fact equivalent. Indeed, choose for $\mathcal{U}$ covers $\mathcal{V}, \mathcal{W} \in \mathfrak{U}$ such that

$$\{V * x \mid x \in X\} \leq \mathcal{W} \quad \text{and} \quad \{W * x \mid x \in X\} \leq \mathcal{U}.$$

For $V_0 \in \mathcal{V}$ and $x \in V_0$ choose $W_x \in \mathcal{W}$ such that $\mathcal{V} * x \subseteq W_x$. Then $V_0 \subseteq W_x$ for all x, and if we choose $x_0 \in V_0$ we have $W_x \subseteq \mathcal{W} * x_0 \subseteq U$ for some $U \in \mathcal{U}$, and finally

$$V * V \subseteq \bigcup\{\mathcal{V} * x \mid x \in V\} \subseteq \bigcup\{W_x \mid x \in V\} \subseteq \mathcal{W} * x \subseteq U.$$

We have chosen the formula (C3) because the equivalence with the former (entourage) definition is more obvious, and also because it will be handier in the point-free context.

1.2. The two definitions of uniformity are easily seen to be equivalent. The first one, however, can be relaxed by dropping the symmetry requirement (E3) which cannot be imitated in the cover description. This useful generalization will be discussed in Chapter XII below.

There is a third way of describing uniformities in terms of certain families of pseudometrics (called *gauges*), due to Bourbaki (1948) and used in Gillman and Jerison's book [113], which can be also treated in the point-free setting (see [202]).

1.3. Whatever definition we chose, we have to realize that the situation we will have in the point-free context will radically differ in the following: in the classical situation one has, first, a uniformity defined on an unstructured underlying set, and a topology is derived from this structure; in the point-free context we start with the space structure already given, and expect the uniformity to be a genuine enrichment.

Thus, we have to answer the following question:

Take a uniformity $\mathfrak{U}$ on a set defined, say, as a system of covers, and a topology τ. What makes $\mathfrak{U}$ to induce τ as the associated topology (in which V is a neighbourhood of a point x iff there is a $\mathcal{U} \in \mathfrak{U}$ such that $\mathcal{U} * x \subseteq V$)?

It is not hard. First we will see that one can restrict the uniformity $\mathfrak{U}$ to a system of covers open in τ. More precisely, we have

1.3.1. Proposition. *For each $\mathcal{U} \in \mathfrak{U}$ define $\mathcal{U}^\circ$ as the system of all interiors $\mathrm{int}(U)$ of the sets $U \in \mathcal{U}$. Set $\mathfrak{U}^\circ = \{\mathcal{U}^\circ \mid \mathcal{U} \in \mathfrak{U}\}$. Then:*

(1) *each $\mathcal{U}^\circ$ is in $\mathfrak{U}$ and (of course) $\mathcal{U}^\circ \leq \mathcal{U}$;*

(2) *a mapping $f \colon (X, \mathfrak{U}) \to (Y, \mathfrak{V})$ is uniformly continuous iff*

$$\text{for each } \mathcal{V} \in \mathfrak{V}^\circ, \quad \{f^{-1}[V] \mid V \in \mathcal{V}\} \in \mathfrak{U}^\circ; \qquad (*)$$

(3) *$\mathfrak{U}$ satisfies the conditions (C1)–(C3) above, with the proviso that in (C1) we consider only the open covers $\mathcal{V} \geq \mathcal{U}$.*

Proof. (1): For $\mathcal{U} \in \mathfrak{U}$ consider the $\mathcal{V}$ from (C3). Now for each $V \in \mathcal{V}$ we have $V \subseteq \mathrm{int}(U)$ for the U such that $V * V \subseteq U$. Thus, $\mathcal{V} \leq \mathcal{U}^\circ$ which makes $\mathcal{U}^\circ$ a cover and an element of $\mathfrak{U}$.

(2) follows immediately from (1) and 1.1.1 (for the second implication, if $(*)$ holds and $\mathcal{U} \in \mathfrak{U}$ then $\{f^{-1}[V] \mid V \in \mathcal{V}\} \geq \{f^{-1}[V] \mid V \in \mathcal{V}^\circ\}$).

(3) is an immediate consequence of (1): if $\mathcal{U}, \mathcal{V} \in \mathfrak{U}$ are open covers then $\mathcal{U} \wedge \mathcal{V}$ is an open cover, and if $\{\mathcal{V} * V \mid V \in \mathcal{V}\} \leq \mathcal{U}$ then $\{\mathcal{V}^\circ * V \mid V \in \mathcal{V}^\circ\} \leq \mathcal{U}$. □

Second, we have the following property of the open sets.

1.3.2. Proposition. *$U \in \tau$ iff*

$$U = \bigcup \{V \in \tau \mid \exists \mathcal{U} \in \mathfrak{U}^\circ \text{ such that } \mathcal{U} * V \subseteq U\}.$$

Proof. The union is trivially open. Now let U be open and let $x \in U$. Hence there is a $W \in \mathfrak{U}$ such that $W * x \subseteq U$. There is an open cover $\mathcal{V} \in \mathfrak{U}$ such that $\{\mathcal{V} * V \mid V \in \mathcal{V}\} \leq \mathcal{W}$. Take a $V_0 \in \mathcal{V}$ such that $x \in V_0$, and a $W \in \mathcal{W}$ such that $\mathcal{V} * V_0 \subseteq W$. Then $x \in W$ and hence $W \subseteq U$. Thus, for each $x \in U$ we have a $\mathcal{V} \in \mathfrak{U}$ and an open V_0 such that $x \in V_0 \subseteq \mathcal{V} * V_0 \subseteq W \subseteq U$. $\qquad\square$

The facts in 1.3.1 and 1.3.2 allow us to define the uniformity and related concepts in the point-free context. This will be exploited in the following sections.

2. Uniformity and nearness in the point-free context

2.1. Recall that a *cover* of a frame L is a subset $A \subseteq L$ such that $\bigvee A = 1$. We say that a cover A *refines* a cover B (or that A is a *refinement* of B) and write

$$A \leq B$$

if

$$\forall a \in A \; \exists b \in B \;\; \text{such that} \;\; a \leq b$$

(this notion and notation is sometimes used for arbitrary subsets A, B).

Further, set
$$A \wedge B = \{a \wedge b \mid a \in A, b \in B\}.$$

Note that $A \wedge B$ is a common refinement of A and B, maximal in the preorder $\leq$ of covers.

2.2. For a cover A of L and an element $x \in L$ set

$$Ax = \bigvee\{a \mid a \in A, \; a \wedge x \neq 0\}.$$

This corresponds to the $\mathcal{U} * V$ from the preceding section. Since the notation in a frame is more transparent, we can omit the asterisk without any danger of confusion.

Furthermore, if A, B are two covers we will write

$$AB = \{Ab \mid b \in B\}.$$

The following formulas are very easy to check.

2.2.1. Facts. (1) *For any cover A, $x \leq Ax$,*

(2) $A \leq B$ *and* $x \leq y \;\Rightarrow\; Ax \leq By$,

(3) $A(Bx) \leq (AB)x = A(B(Ax))$, *and*

(4) $(A_1 \wedge \cdots \wedge A_n)(B_1 \wedge \cdots \wedge B_n) \leq (A_1 B_1) \wedge \cdots \wedge (A_n B_n).$ $\qquad\square$

Also the following is an immediate

2.2.2. Fact. *We have*

$$A\left(\bigvee_{i\in J} x_i\right) = \bigvee_{i\in J} Ax_i.$$

In other words, the correspondence $(x \mapsto Ax)\colon L \to L$ *preserves suprema; hence it has a right adjoint* $(x \mapsto x/A)\colon L \to L$, *for which we have the obvious formula* $x/A = \bigvee\{y \mid Ay \leq x\}$. $\square$

2.2.3. Proposition. *Let* $h\colon L \to M$ *be a frame homomorphism. Then for any* $A \subseteq L$ *and* $x \in L$, $h[A]h(x) \leq h[Ax]$. *Consequently, for* $A, B \subseteq L$, $h[A]h[B] \leq h[AB]$.

(Note that the "$\leq$" in the latter is something else than the same symbol in the former inequality.)

Proof. If $h(a) \wedge h(x) = h(a \wedge x) \neq 0$ then $a \wedge x \neq 0$. Now for $b \in B$, $h[A]h(b) \leq h[Ab]$ and hence $h[A]h[B]$ refines $h[AB]$. $\square$

2.3. The relation "uniformly below".

Let $\mathcal{A}$ be a system of covers on L. The relation $\triangleleft_{\mathcal{A}}$ on L (the subscript will be sometimes omitted) is defined by setting

$$x \triangleleft_{\mathcal{A}} y \quad \equiv_{\mathrm{df}} \quad \exists A \in \mathcal{A},\ Ax \leq y.$$

It follows immediately from 2.2.3 that

2.3.1. Corollary. *Let* $h\colon L \to M$ *be a frame homomorphism. Let* $\mathcal{A}, \mathcal{B}$ *be sets of covers of* L *resp.* M, *and let for every* $A \in \mathcal{A}$ *there exist a* $B \in \mathcal{B}$ *such that* $h[A] \geq B$. *Then*

$$x \triangleleft_{\mathcal{A}} y \quad \Rightarrow \quad h(x) \triangleleft_{\mathcal{B}} h(y). \qquad \square$$

The uniformly below relation has the following properties.

2.3.2. Lemma. (1) $x' \leq x \triangleleft_{\mathcal{A}} y \leq y' \quad \Rightarrow \quad x' \triangleleft_{\mathcal{A}} y'$.

(2) *If for any* $A_1, A_2 \in \mathcal{A}$ *there is a common refinement* $B \in \mathcal{A}$ *then*

$$x \triangleleft_{\mathcal{A}} y_1, y_2 \Rightarrow x \triangleleft_{\mathcal{A}} (y_1 \wedge y_2) \quad \text{and}$$
$$x_1, x_2 \triangleleft_{\mathcal{A}} y \Rightarrow (x_1 \vee x_2) \triangleleft_{\mathcal{A}} y.$$

(3) $x \triangleleft_{\mathcal{A}} y \quad \Rightarrow \quad x \prec y$.

(4) $x \triangleleft_{\mathcal{A}} y \quad \Rightarrow \quad x^{**} \triangleleft_{\mathcal{A}} y$.

Proof. (1) follows from 2.2.1(1).

(2): If $A_i x \leq y_i$ $(i = 1, 2)$ then $Bx \leq y_1 \wedge y_2$; if $A_i x_i \leq y$ $(i = 1, 2)$ then $Bx_i \leq y$ and hence $B(x_1 \vee x_2) = Bx_1 \vee Bx_2 \leq y$ by 2.2.2.

(3): If $Ax \leq y$ then $1 = \bigvee A = \bigvee\{a \mid a \wedge x = 0\} \vee \bigvee\{a \mid a \wedge x \neq 0\} = \bigvee\{a \mid a \leq x^*\} \vee Ax \leq x^* \vee y$. Now recall V.5.2.

(4): For the double pseudocomplement x^{**} one has $a \wedge x^{**} = 0$ iff $a \wedge x = 0$ (see $(*)$ in AI.7.1.2). Hence $Ax = Ax^{**}$. $\square$

2.3.3. Note. If $x \prec y$ we have the cover $\{x^*, y\}$. Obviously $\{x^*, y\}x \leq y$. Thus,

> *whenever a system $\mathcal{A}$ contains all finite covers, $\lhd_{\mathcal{A}}$ coincides with $\prec$.*

2.4. Admissible systems. A system of covers $\mathcal{A}$ of L is said to be *admissible* if

$$\forall x \in L, \quad x = \bigvee \{y \mid y \lhd_{\mathcal{A}} x\}. \tag{2.4.1}$$

Using the right adjoint from 2.2.2 this can be expressed as

$$\forall x \in L, \quad x = \bigvee \{x/A \mid A \in \mathcal{A}\}.$$

2.4.1. Proposition. (1) *L is regular iff there is an admissible system of covers $\mathcal{A}$ on L iff the system of all covers of L is admissible.*

(2) *If $\mathcal{A}$ is admissible and $\mathcal{B} \supseteq \mathcal{A}$ then $\mathcal{B}$ is admissible.*

(3) *If $\mathcal{A}$ is admissible on L then L is (join-) generated by $\bigcup \mathcal{A}$.*

Proof. (1): If $\mathcal{A}$ is admissible then L is regular by 2.3.2(3). If L is regular then the system of all covers is admissible by 2.3.3.

(2) is obvious.

(3): Set $\mathcal{C}(\mathcal{A}, x) = \{a \mid \exists A \in \mathcal{A}, \exists y \in L, \ a \in A \text{ and } Ay \leq x\}$. Then by the admissibility of $\mathcal{A}$, $x = \bigvee \mathcal{C}(\mathcal{A}, x)$. $\qquad\square$

2.5. Uniformity on a frame. A *uniformity* on a frame L is a non-empty admissible system of covers $\mathcal{A}$ such that

(U1) $A \in \mathcal{A}$ and $A \leq B \ \Rightarrow \ B \in \mathcal{A}$,

(U2) $A, B \in \mathcal{A} \ \Rightarrow \ A \wedge B \in \mathcal{A}$,

(U3) for every $A \in \mathcal{A}$ there is a $B \in \mathcal{A}$ such that $BB \leq A$.

(A cover B such that $BB \leq A$ is often called a *star-refinement* of A; thus, (U3) is often expressed by saying that each $A \in \mathcal{A}$ has a star-refinement in $\mathcal{A}$.)

> Recall the (C1), (C2) and (C3) from 1.1, and the propositions 1.3.1 and 1.3.2 (note that the latter motivates the admissibility condition). We see that our concept of uniformity is a correct extension of the homonymous classical notion; that is, uniformities on a topological space $X = (X, \tau)$ coincide with the just defined uniformities on $\Omega(X)$.

If $\mathcal{A}$ is a uniformity on L we will call the pair $(L, \mathcal{A})$ a *uniform frame* (or *uniform locale*, according to the perspective chosen).

2.6. Nearness. An admissible system of covers of L satisfying (U1) and (U2) will be called *nearness*, and the $(L, \mathcal{A})$ is then referred to as a *nearness frame*. This is a point-free extension of the concept of nearness as known in the classical context ([126], [128]) but in fact the admissibility condition is for a nearness more restrictive than necessary. A weaker one will be discussed in Section 4 below.

One speaks of a *strong nearness* if, besides (U1) and (U2) one has, moreover,

(SN) for each $A \in \mathcal{A}$ the cover $\{b \mid b \lhd_A a \in A\}$ is in $\mathcal{A}$.

(Note that (SN) holds true for a uniformity: if $BB \leq A$ then obviously $B \leq \{b \mid b \lhd_A a \in A\}$.)

From 2.4.1 we immediately obtain

2.6.1. Corollary. *A frame admits a nearness iff it admits a strong nearness iff it is regular.* □

2.7. Bases and subbases. Often one works with a *basis of nearness*, an admissible system of covers $\mathcal{A}$ satisfying

(U2′) For any $A, B \in \mathcal{A}$ there is a $C \in \mathcal{A}$ such that $C \leq A, B$.

Then one has the least nearness containing $\mathcal{A}$, the system of covers

$$\mathcal{B} = \{B \mid \exists A \in \mathcal{A}, \ A \leq B\}.$$

Note that the relation $\lhd_{\mathcal{B}}$ coincides with $\lhd_{\mathcal{A}}$.

If $\mathcal{A}$ satisfies

(SN′) for each $A \in \mathcal{A}$ there is a $B \in \mathcal{A}$ such that $B \leq \{b \mid b \lhd_A a \in A\}$

then $\mathcal{B}$ is a strong nearness, and if $\mathcal{A}$ satisfies (U3) then $\mathcal{B}$ is a uniformity (then one speaks of a *basis of uniformity*).

A *subbasis of nearness* is a system of covers $\mathcal{A}$ such that

$$\mathcal{B} = \{B \mid \exists A_1, \ldots, A_n \in \mathcal{A} \text{ such that } A_1 \wedge \cdots \wedge A_n \leq B\}$$

is admissible. Then, $\mathcal{B}$ is the smallest nearness containing $\mathcal{A}$ and, again, if $\mathcal{A}$ satisfies (SN′) resp. (U3) then $\mathcal{B}$ is a strong nearness resp. uniformity.

(Here we do not necessarily have $\lhd_{\mathcal{B}}=\lhd_{\mathcal{A}}$. Therefore one has to be careful when constructing nearnesses or uniformities by means of subbases. Sometimes $\mathcal{A}$ itself is admissible and then, since $\mathcal{A} \subseteq \mathcal{B}$, $\mathcal{B}$ is admissible automatically; but it is not always the case.)

2.8. Uniformizability and complete regularity. This is similar to the classical case.

2.8.1. Lemma. *If $\mathcal{A}$ is a (basis of) uniformity then $\lhd_{\mathcal{A}}$ interpolates. Consequently,*

$$x \lhd_A y \quad \Rightarrow \quad x \lll y.$$

Proof. Let $x \lhd_A y$. Then $Ax \leq y$ for some $A \in \mathcal{A}$ and if we choose a $B \in \mathcal{A}$ such that $BB \leq A$ we have $x \lhd_A Bx$, and by 2.2.1 $B(Bx) \leq (BB)x \leq Ax \leq y$ and hence $Bx \lhd_A y$.

For the second statement recall 2.3.2(3) and V.5.7. □

2.8.2. Theorem. *A frame L admits a uniformity iff it is completely regular.*

Proof. If L admits a uniformity then it is completely regular by 2.8.1. Now let L be completely regular. For a sequence $a_1 \lll a_2 \lll \cdots \lll a_n$ $(n > 1)$ define a cover

$$A(a_1, \ldots, a_n) = \{a_2, a_1^* \wedge a_3, a_2^* \wedge a_4, \ldots, a_{n-2}^* \wedge a_n, a_{n-1}^*\}$$

(as before, x^* is the pseudocomplement of x),

If we interpolate $a_1 \lll u_1 \lll v_1 \lll a_2 \lll u_2 \cdots \lll v_{n-1} \lll a_n$ and set $B = A(a_1, u_1, v_1, a_2, u_2, \cdots, v_{n-1}, a_n)$ we easily check that $BB \leq A(a_1, a_2, \ldots, a_n)$. Thus, the system $\mathcal{S}$ of all the $A(a_1, a_2, \ldots, a_n)$ is a subbasis of a uniformity $\mathcal{A}$. Now $\mathcal{S}$ is admissible since if $x \lll a$ then $A(x, a)x \leq a$; since $\mathcal{S} \subseteq \mathcal{A}$, $\mathcal{A}$ is admissible as well. □

2.9. Fine uniformity. A union of uniformities is obviously a uniformity. Thus,

each completely regular frame L has a largest uniformity.

This uniformity is called the *fine uniformity of L*.

One has trivially the largest (strong) nearness on any regular L, the system of all covers of L: in general a cover does not have to have a star-refinement.

A cover A of a general L is said to be *normal* if there are covers A_n, $n \in \mathbb{N}$, such that

$$A = A_1 \quad \text{and} \quad A_n \geq A_{n+1}A_{n+1} \quad \text{for all} \quad n.$$

Thus, the fine uniformity of (a completely regular) L consists of all the normal covers of L.

We will soon see that for instance in a compact regular frame each cover is normal (Section 5 below); spaces in which this happen (the paracompact ones) will be discussed in Chapter IX.

3. Uniform homomorphisms. Modelling embeddings. Products

In this and the following sections it will be convenient to use the frame point of view. In particular we will represent embeddings as a special type of surjective homomorphisms.

3.1. Let $(L, \mathcal{A}), (M, \mathcal{B})$ be uniform (or just nearness) frames. A *uniform homomorphism* $h \colon (L, \mathcal{A}) \to (M, \mathcal{B})$ is a frame homomorphism $h \colon L \to M$ such that

$$\forall A \in \mathcal{A}, \quad h[A] \in \mathcal{B}.$$

If $\mathcal{A}, \mathcal{B}$ are bases of uniformities (resp. nearnesses) this requirement transforms to

$$\forall A \in \mathcal{A},\ \exists B \in \mathcal{B},\ \ h[A] \geq B.$$

Set

$$h[[\mathcal{A}]] = \{h[A] \mid A \in \mathcal{A}\}.$$

Then the requirement (for uniformities resp. nearnesses resp. strong nearnesses, not for the bases) can be written as

$$h[[\mathcal{A}]] \subseteq \mathcal{B}. \tag{3.1.1}$$

3.1.1. The following is an easy

Observation. *Let $h\colon L \to M$ be a frame homomorphism, let $\mathcal{S}$ be a subbasis of a nearness $\mathcal{A}$ on L, and let $\mathcal{B}'$ be a basis of a nearness $\mathcal{B}$ on M. Then h is a uniform homomorphism $(L, \mathcal{A}) \to (M, \mathcal{B})$ iff for each $A \in \mathcal{S}$ there exists a $B \in \mathcal{B}'$ such that $B \leq h[A]$.*

(Indeed, we obviously have $h[A_1] \wedge \cdots \wedge h[A_n] = h[A_1 \wedge \cdots \wedge A_n]$.)

3.2. The resulting categories will be denoted by

UniFrm resp. **NearFrm** resp. **SNearFrm.**

We will often use the fact that

3.2.1. *All dense uniform homomorphisms are monomorphisms in* **RegFrm** *resp.* **NearFrm.**

(Since the frames carrying the objects of these categories are – at least – regular – recall V.5.6.)

3.3. Ue-surjections (embeddings of uniform and nearness sublocales).

3.3.1. Lemma. (1) *Let $h\colon L \to M$ be a surjective frame homomorphism and let $\mathcal{A}$ be a nearness on L. Then $h[[\mathcal{A}]]$ is a nearness on M. If $\mathcal{A}$ is a strong nearness resp. uniformity then so is $h[[\mathcal{A}]]$.*

(2) *If $g\colon (L, \mathcal{A}) \to (K, \mathcal{B})$ is uniform and if $f\colon M \to K$ is a frame homomorphism such that the diagram*

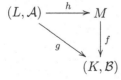

commutes, then f is a uniform homomorphism $(M, h[[\mathcal{A}]]) \to (K, \mathcal{B})$.

Proof. (1): Let $h[A] \leq B$. If $h(a) \leq b$ choose, first, an $a' \in L$ such that $h(a') = b$, and then set $a_b = a \vee a'$; if b does not majorize any of the $a \in A$ choose a_b arbitrarily. Then $A \leq A' = \{a_b \mid b \in B\}$ and $B = h[A]$. Thus we have (U1).

If $A, B \in \mathcal{A}$ then as we have already observed

$$h[A] \wedge h[B] = \{h(a) \wedge h(b) = h(a \wedge b) \mid a \in A, b \in B\} = h[A \wedge B].$$

If $b = h(a) \in M$ we have

$$b = h(\bigvee\{x \mid x \vartriangleleft_{\mathcal{A}} a\}) = \bigvee\{h(x) \mid x \vartriangleleft_{\mathcal{A}} a\}) \leq \bigvee\{y \mid y \vartriangleleft_{\mathcal{B}} b\} \leq b$$

by 2.3.1; hence $h[[\mathcal{A}]]$ is admissible.

Let $\mathcal{A}$ be a strong nearness. Consider an $h[A] \in h[[\mathcal{A}]]$. If $b \vartriangleleft_{\mathcal{A}} a$ then $h(b) \vartriangleleft_{h[[\mathcal{A}]]} h(a)$ so that

$$h[\{b \mid b \vartriangleleft_{\mathcal{A}} a \in A\}] \leq \{c \mid c \vartriangleleft_{h[[\mathcal{A}]]} h(a)\}$$

and hence $\{c \mid c \vartriangleleft_{h[[\mathcal{A}]]} h(a)\}$ is in $h[[\mathcal{A}]]$. Finally, if $\mathcal{A}$ is a uniformity, $A \in \mathcal{A}$, and if $B \in \mathcal{A}$ is such that $BB \leq A$ then $h[B]h[B] \leq h[BB] \leq h[A]$ and we have (U3).

(2) is obvious. $\square$

We will speak of $h: (L, \mathcal{A}) \to (M, h[[\mathcal{A}]])$ as a *ue-surjection* (short for *uniform embedding surjection*). In other words, $h: (L, \mathcal{A}) \to (M, \mathcal{B})$ is a ue-surjection iff it is a surjective uniform homomorphism and if each $B \in \mathcal{B}$ is $h[A]$ for some $A \in \mathcal{A}$.

(This concept models an embedding of a generalized uniform subspace into a larger one as an induced subobject.)

Technically, we do not have to check it all. We have

3.3.2. Lemma. *A surjective uniform homomorphism $h: (L, \mathcal{A}) \to (M, \mathcal{B})$ is a ue-surjection iff $\{h[A] \mid A \in \mathcal{A}\}$ is a basis of $\mathcal{B}$.*

Proof. For the right adjoint h_* of h we have $x \leq h_* h(x)$ and, as h is onto, $h h_* = \mathrm{id}$. Thus, if $B \geq h[A]$ we have $A' = h_*[B] \geq h_* h[A] \geq A$ and hence $A' \in \mathcal{A}$ and $h[A'] = h h_*[B] = B$. $\square$

It should be noted that the definition of a ue-surjection is in a way less subtle than that of sublocale: the sublocale in question is simply endowed with the smallest uniformity under which the localic embedding is still uniform. It is not necessarily an extremal epimorphism in the category of uniform or nearness frames. The monomorphisms in these categories are not necessarily one-one (which played a role in identifying the extremal epimorphisms in **Frm** with natural subspace embedding) – they are just dense. In Section 6 we will encounter very important dense ue-surjections that are not one-one (and hence not isomorphisms, as an extremal epi-mono-morphism should be).

3.4. Since for a ue-surjection $h\colon (L, \mathcal{A}) \to (M, \mathcal{B})$ each $B \in \mathcal{B}$ is an image $h[A]$ of an $A \in \mathcal{A}$ we have $A \leq h_* h[A] = h_*[B]$ and hence $h_*[B] \in \mathcal{A}$. In particular,

$$\text{each } h_*[B] \text{ is a cover.} \tag{3.4.1}$$

Later we will need the following useful criterion.

Proposition. *Let* $h\colon (L, \mathcal{A}) \to (M, \mathcal{B})$ *be a dense surjective uniform homomorphism and let* $\mathcal{A}$ *be a strong nearness. Let* $h_*\colon M \to L$ *be the right adjoint of* h *(that is, the corresponding localic embedding). Then* h *is a ue-surjection iff* $\{h_*[B] \mid B \in \mathcal{B}\}$ *is a basis of* $\mathcal{A}$.

Proof. Let h be a ue-surjection. We have $h(h_* h(x) \wedge x^*) = h(x) \wedge h(x^*) = 0$ and since h is dense, $h_* h(x) \wedge x^* = 0$, that is, $h_* h(x) \leq x^{**}$. For $A \in \mathcal{A}$ and $C = \{c \mid c \triangleleft a \in A\}$ $(\in \mathcal{A})$ we have $C \leq h_* h[C] \leq \{c^{**} \mid c \triangleleft a \in A\} \leq A$ by 2.3.2(4). Thus $\{h_*[B] \mid B \in \mathcal{B}\}$ is a basis of $\mathcal{A}$. On the other hand, if $\{h_*[B] \mid B \in \mathcal{B}\}$ is a basis of $\mathcal{A}$, each $B \in \mathcal{B}$ is $h[h_*[B]]$ with $h_*[B] \in \mathcal{A}$. $\qquad\square$

3.5. Coproducts. Let $(L_i, \mathcal{A}_i)$ be nearness frames $(i \in J)$. Consider the coproduct

$$\iota_i \colon L_i \to L = \bigoplus_{j \in J} L_j$$

in the category of frames (recall IV.4).

Let $A_i \in \mathcal{A}_i$ be a system of uniform covers such that $A_i = \{1\}$ for all but finitely many i. Set

$$\oplus_{i \in J} A_i = \{\oplus_{i \in J} a_i \mid a_i \in A_i\}.$$

3.5.1. Lemma. (1) *Each* $\oplus_{i \in J} A_i$ *is a cover of* L.

(2) *The system* $\mathcal{A} = \{\oplus_{i \in J} A_i \mid A_i \in \mathcal{A}_i, A_i = \{1\}$ *for all but finitely many* $i\}$ *is an admissible system of covers of* L.

(3) $\mathcal{A}$ *is a basis of nearness generated by the subbasis* $\bigcup\{\iota_i[[A_i]] \mid i \in J\}$.

Proof. (1): Let $\{i_1, \dots, i_n\}$ be all the indices for which $A_i \neq \{1\}$. Then

$$\bigvee(\oplus_{i \in J} A_i) = \bigvee\{a_{i_1} \oplus \cdots \oplus a_{i_n} \mid a_{i_j} \in A_{i_j}\}$$
$$= \bigvee A_{i_1} \oplus \bigvee A_{i_2} \oplus \cdots \oplus \bigvee A_{i_n} = 1 \oplus 1 \oplus \cdots \oplus 1 = 1$$

by distributivity.

(2): We have $\oplus a_i \wedge \oplus x_i = \oplus(a_i \wedge x_i) \neq 0$ iff for all i, $a_i \wedge x_i \neq 0$. Thus,

$$(\oplus_{i \in J} A_i)(\oplus_{i \in J} x_i) = \oplus_{i \in J}(A_i x_i) \tag{3.5.1}$$

and $\oplus b_i \triangleleft_\mathcal{A} \oplus c_i$ iff $b_i \triangleleft_{\mathcal{A}_i} c_i$ for all i. Consequently $\oplus c_i = \bigvee\{\oplus b_i \mid \oplus b_i \triangleleft_\mathcal{A} \oplus c_i\}$, and since each $U \in L$ is the join $U = \bigvee\{\oplus c_i \mid \oplus c_i \leq U\}$ the statement follows.

(3): Obviously $\oplus_{i \in J} A_i = \iota_{i_1}[A_{i_1}] \wedge \cdots \wedge \iota_{i_n}[A_{i_n}]$ where $i_1, \dots, i_n$ are all the indices with $A_i \neq \{1\}$. $\qquad\square$

3.5.2. Proposition. $(\iota_i \colon (L_i, \mathcal{A}_i) \to (\bigoplus_{j \in J} L_j, \mathcal{A}))_{i \in J}$ *is the coproduct of the system* $((L_i, \mathcal{A}_i))_{i \in J}$ *in the category* **NearFrm**.

If all the $\mathcal{A}_i$ are strong nearnesses then $\mathcal{A}$ is a strong nearness, and if they are uniformities then $\mathcal{A}$ is a uniformity.

Proof. Obviously ι_i are uniform homomorphisms (as $\iota_i[A] = \bigoplus_{j \in J} A_j$ where $A_i = A$ and $A_j = \{1\}$ otherwise). Now let $h_i \colon (L_i, \mathcal{A}_i) \to (M, \mathcal{B})$ be uniform homomorphisms and let $h \colon L \to M$ be the frame homomorphism such that $h\iota_i = h_i$ for all i. Then for $A \in \mathcal{A}_i$, $h[\iota_i[A]] = h_i[A] \in \mathcal{B}$ and h is a uniform homomorphism by 3.1.1 and 3.5.1(3).

The statement on strong nearness and uniformity easily follows from the formula (3.5.1) in the proof of 3.5.1(2).   □

3.6. By 3.3.1(2), if $h_1, h_2 \colon (K, \mathcal{B}) \to (L, \mathcal{A})$ are uniform homomorphism we can construct a coequalizer in **NearFrm**, **SNearFrm** or **UniFrm** by taking the coequalizer $h \colon L \to M$ in **Frm** and then the uniform homomorphism $h \colon (L, \mathcal{A}) \to (M, h[[\mathcal{A}]])$. By 3.5.2 we have coproducts. Thus (recall AII.5.3.2) we obtain

Corollary. *The categories* **NearFrm**, **SNearFrm** *and* **UniFrm** *are cocomplete.*   □

3.7. Spectrum. The adjunction from II.4 can be easily extended to uniform and nearness frames. We will do it for the uniform ones, and will use the contravariant description as in II.4.4.

3.7.1. The functor Ω. For a uniform space $(X, \mathfrak{U})$ set

$$\Omega(X, \mathfrak{U}) = (\Omega(X), \mathfrak{U}^\circ)$$

where $\mathfrak{U}^\circ$ is the system of all covers from $\mathfrak{U}$ containing only open sets. By 1.3.1 and 1.3.2, $\Omega(X, \mathfrak{U})$ is a uniform frame. Now if $f \colon (X, \mathfrak{U}) \to (Y, \mathfrak{V})$ is a uniform mapping, $\Omega(f)$ is a uniform homomorphism (1.3.1 again), and we have a contravariant functor

$$\Omega \colon \mathbf{UniSp} \to \mathbf{UniFrm}.$$

3.7.2. The functor Σ. Let $(L, \mathcal{A})$ be a uniform frame. On the spectrum ΣL consider

$$\Sigma_{\mathcal{A}} = \{\Sigma_A \mid A \in \mathcal{A}\} \quad \text{where} \quad \Sigma_A = \{\Sigma_a \mid a \in A\}.$$

Using the obvious fact that $\Sigma_a \neq \emptyset$ implies $a \neq 0$ we see that (in the notation from Section 1)

$$\Sigma_A * \Sigma_b = \bigcup\{\Sigma_a \mid a \in A, \ \Sigma_a \cap \Sigma_b = \Sigma_{a \wedge b} \neq \emptyset\}$$
$$\subseteq \bigcup\{\Sigma_a \mid a \in A, \ a \wedge b \neq 0\} = \Sigma_{Ab}. \tag{$*$}$$

We immediately infer that

$$c \lhd_{\mathcal{A}} b \quad \Rightarrow \quad \Sigma_c \lhd_{\Sigma_{\mathcal{A}}} \Sigma_b$$

so that, by II.4.1,

$$\Sigma_b = \Sigma_{\bigvee\{c \mid c \lhd_{\mathcal{A}} b\}} = \bigcup\{\Sigma_c \mid c \lhd_{\mathcal{A}} b\} \leq \bigcup\{\Sigma_c \mid \Sigma_c \lhd_{\Sigma_{\mathcal{A}}} \Sigma_b\}.$$

Now by (∗) again $\Sigma_A * \Sigma_B \leq \Sigma_{AB}$ and obviously $\Sigma_A \wedge \Sigma_B = \{\Sigma_a \cap \Sigma_b = \Sigma_{a \wedge b} \mid a \in A, \ b \in B\} = \Sigma_{A \wedge B}$ and hence $\Sigma_{\mathcal{A}}$ is a basis of uniformity which is by 1.3.2 in agreement with the topology of ΣL.

If $h\colon (L, \mathcal{A}) \to (M, \mathcal{B})$ is a uniform homomorphism and $\Sigma_B \in \Sigma_{\mathcal{B}}$ then by II.4.3.1 (translated as $(\Sigma h)^{-1}[\Sigma_a] = \Sigma_{h(a)}$),

$$\{(\Sigma h)^{-1}[\Sigma_a] \mid a \in A\} = \{\Sigma_{h(a)} \mid a \in A\} = \Sigma h[A]$$

and hence Σh is a uniform mapping $(\Sigma M, \Sigma_{\mathcal{B}}) \to (\Sigma L, \Sigma_{\mathcal{A}})$ and we have a contravariant functor

$$\Sigma\colon \mathbf{UniFrm} \to \mathbf{UniSp}.$$

3.7.3. The adjunction. The formulas for ϕ and λ in II.4.6.1 provide us with adjunction units again. We only have to show that thus defined maps are uniform. We have $\phi_L[A] = \{\Sigma_a \mid a \in A\}$, and since

$$\lambda_X^{-1}[\Sigma_U] = \{x \in X \mid \lambda_X(x) \in \Sigma_U\} = \{x \in X \mid U \in \lambda_X(x)\} = U,$$

we have $\{\lambda_X^{-1}[\Sigma_U] \mid U \in \mathcal{U}\} = \mathcal{U}$ for any uniform cover $\mathcal{U}$.

Note that in the covariant interpretation,

Ω is the left adjoint and the spectrum is the right one,

like in the plain spectrum adjunction.

4. Aside: admitting nearness in a weaker sense

4.1. We have seen in 1.3.2 that the condition of generating the associated topology τ on X by a uniformity $\mathfrak{U}$, namely,

$$\forall U \in \tau, \quad U = \{x \in X \mid \exists \mathcal{U} \in \mathfrak{U}^\circ \text{ such that } \mathcal{U} * x \subseteq U\} \qquad (4.1.1)$$

can be replaced by the formally stronger

$$\forall U \in \tau, \quad U = \bigcup\{V \in \tau \mid \exists \mathcal{U} \in \mathfrak{U}^\circ \text{ such that } \mathcal{U} * V \subseteq U\}. \qquad (4.1.2)$$

Therefore we have adopted this latter formula for the definition of admissibility in the point-free context.

We have used the same formula for the admissibility of a nearness as well. However, here the two definitions are not equivalent; thus, in fact the requirement is too strong and we can think of relaxing it.

4.2. The condition (4.1.1) can be equivalently formulated with subsets instead of points:

$$\forall U \in \tau, \quad U = \bigcup\{A \subseteq X \mid \exists \mathcal{U} \in \mathfrak{U}^\circ \text{ such that } \mathcal{U} * A \subseteq U\}.$$

and this can be easily mimicked in the point-free context.

Recall Chapter III, in particular Sections 2 and 5. For a cover A of L set

$$\widetilde{A} = \{\mathfrak{o}(a) \mid a \in A\}.$$

By III.6.1.5, $\widetilde{A}$ is a cover of $\mathcal{S\ell}(L)$, that is, $\bigvee \widetilde{A} = L$. For any sublocale $S \subseteq L$ set, like in 1.1,

$$\widetilde{A} * S = \bigvee\{\mathfrak{o}(a) \mid a \in A, \mathfrak{o}(a) \cap S \neq \mathsf{O}\} = \mathfrak{o}(\bigvee\{a \mid a \in A, \mathfrak{o}(a) \cap S \neq \mathsf{O}\}).$$

Now we will say that a system $\mathcal{A}$ of covers of L satisfying (U1) and (U2) is a *weak nearness* on L if

$$\forall u \in L, \quad \mathfrak{o}(u) = \bigcup\{S \subseteq L \text{ a sublocale} \mid \exists A \in \mathcal{A} \text{ such that } \widetilde{A} * S \subseteq \mathfrak{o}(U)\}. \quad (4.2.1)$$

4.2.1. Lemma. (1) $\mathfrak{o}(a) \cap S \neq \mathsf{O}$ *iff* $\mathfrak{o}(a) \cap \overline{S} \neq \mathsf{O}$. *Thus,* $\widetilde{A} * S = \widetilde{A} * \overline{S}$.

(2) $\mathfrak{o}(a) \cap \mathfrak{c}(b) \neq \mathsf{O}$ *iff* $a \not\leq b$. *Consequently,* $\widetilde{A} * \mathfrak{c}(b) = \mathfrak{o}(\bigvee\{a \mid a \in A, \ a \not\leq b\})$.

(3) $\mathfrak{c}(b) \subseteq \mathfrak{o}(a)$ *iff* $a \vee b = 1$.

Proof. We will write S^* for the complement in $\mathcal{S\ell}(L)$, if it exists.

(1): We have $\mathfrak{o}(a) \cap S \neq \mathsf{O}$ iff $S \not\subseteq \mathfrak{o}(a)^* = \mathfrak{c}(a) = \uparrow a$ iff there is an $s \in S$, $s \not\geq a$ iff $\bigwedge S \not\leq a$.

(2): $\mathfrak{o}(a) \cap \mathfrak{c}(b) \neq \mathsf{O}$ iff $\mathfrak{c}(b) = \uparrow b \not\subseteq \mathfrak{o}(a)^* = \uparrow a$ iff $a \not\leq b$.

(3): $\mathfrak{c}(b) \subseteq \mathfrak{o}(a)$ iff $\mathfrak{c}(b) \cap \mathfrak{c}(a) = \mathsf{O}$ iff $\mathfrak{c}(b \vee a) = \mathfrak{c}(1)$ iff $a \vee b = 1$. $\qquad\square$

4.2.2. Taking into account that $\mathfrak{o}(u) \subseteq \mathfrak{o}(v)$ iff $u \leq v$ we obtain

Corollary. *A system $\mathcal{A}$ of covers of L is a weak nearness on L iff*

$$\forall u \in L, \quad \mathfrak{o}(a) = \bigvee\{\mathfrak{c}(b) \mid \exists A \in \mathcal{A}, \ \bigvee\{a \mid a \in A, \ a \not\leq b\} \leq u\}. \qquad\square$$

4.3. Proposition. *A frame admits a weak nearness iff it is subfit.*

Proof. If L admits a weak nearness then it is subfit by V.1.4(5) and Corollary 4.2.2.

Now let L be subfit. We will show that the system of all covers is a weak nearness on L. Let $\mathfrak{c}(b) \subseteq \mathfrak{o}(u)$. Then by 4.2.1(3), $b \vee u = 1$ and we have a cover $\{b, u\}$. Now

$$\widetilde{\{b, u\}} * \mathfrak{c}(b) \leq \mathfrak{o}(u)$$

since $\mathfrak{o}(b) \cap \mathfrak{c}(b) = \mathsf{O}$. Thus, the join $\mathfrak{o}(u) = \bigvee\{\mathfrak{c}(b) \mid b \vee u = 1\}$ from V.1.4(4) proves (4.2.1) for the system of all covers of L. $\qquad\square$

4.4. Thus in particular admitting a weak nearness is not a hereditary property. From 4.3 and V.1.5 we obtain

Corollary. *A frame L is fit iff each of its sublocales admits a weak nearness.* $\qquad\square$

5. Compact uniform and nearness frames. Finite covers

5.1. Proposition. *A compact regular frame L admits precisely one nearness, namely the system of all covers, and this nearness is a uniformity.*

Proof. Let $\mathcal{A}$ be a nearness on L. Let C be an arbitrary cover of L. Set $B = \{b \mid b \prec c \in C\}$. Now choose a finite subcover $\{b_1, b_2, \ldots, b_n\}$ of B, and covers $A_i \in \mathcal{A}$ such that $A_i b_i \leq c_i \in C$. For $A = A_1 \wedge \cdots \wedge A_n$ we have $A b_i \leq c_i$. Now for each non-zero $a \in A$ there is a b_i with $a \wedge b_i \neq 0$, and hence $a \leq c_i$. Thus, $A \leq C$ and hence $C \in \mathcal{A}$.

Finally, since a regular compact frame is completely regular (VII.2.2), the unique nearness on L has to be a uniformity (2.8.2). $\qquad\square$

5.2. Corollary. *Let L be compact and $(M, \mathcal{A})$ an arbitrary nearness frame. Then each homomorphism $M \to L$ is uniform.* $\qquad\square$

5.3. Recall 2.8. The system of all covers of a compact frame is, as we have just observed, a uniformity. This happens for a much broader class of frames, the *paracompact* ones (including, e.g., all the completely regular frames with a countable admissible system of covers, to mention a class very different from the compact frames discussed here). It is characterized by a number of interesting properties: see the next chapter.

6. Completeness and completion

6.1. In a classical uniform space one has a very transparent concept of a *Cauchy point*, roughly speaking, a filter of open sets containing arbitrarily small ones (where the smallness is determined by the uniformity in question), representing the neighbourhood system of an ideal point. For a precise definition see Chapter X; at this moment we just need a rough idea. A uniform *space* $(X, \mathfrak{U})$ is complete if each Cauchy point P represents an $x \in X$, namely, the intersection $\bigcap P$ (the *center* of P; one sometimes uses the formulation "a uniform space X is complete if every Cauchy point in X has a center").

For a point-free treatment, another property equivalent with completeness of spaces will be more suitable. Namely,

   - a uniform space X is complete iff for each uniform embedding $f \colon X \to Y$ the image $f[X]$ is closed in Y,

in other words,

   - a uniform space X is complete iff each dense uniform embedding $f \colon X \to Y$ is a uniform isomorphism.

This will be adopted as the definition of completeness in this chapter. The point of view of Cauchy points will, however, not be neglected. It will be discussed in Chapter X.

6.2. Recall the definition of a ue-surjection (mimicking embeddings of uniform subspaces) in our context (3.3). A uniform frame (more generally, a strong nearness frame) $(L, \mathcal{A})$ is *complete* if each dense ue-surjection $h \colon (M, \mathcal{B}) \to (L, \mathcal{A})$ is an isomorphism.

Note that in particular

$$\textit{every compact regular frame is complete.} \tag{6.2.1}$$

Indeed: for a compact regular L, any dense onto homomorphism $M \to L$ with regular M is an isomorphism (VII.2.3.1); this makes M compact, and hence there is only one uniformity (also) on M, and hence this isomorphism is a uniform one.

6.3. A *completion* of a uniform frame $(L, \mathcal{A})$ is a dense ue-surjection $(M, \mathcal{B}) \to (L, \mathcal{A})$ such that $(M, \mathcal{B})$ is a complete uniform frame.

6.4. Regular Cauchy maps. Let $(L, \mathcal{A})$ be a uniform frame (more generally, a strong nearness frame) and M a frame. A mapping $\phi \colon L \to M$ (not necessarily a homomorphism) is said to be a *Cauchy map* (with respect to $\mathcal{A}$) if

$$\phi(0) = 0, \quad \phi(1) = 1, \quad \phi(a \wedge b) = \phi(a) \wedge \phi(b), \quad \text{and}$$

for every $A \in \mathcal{A}$, $\phi[A]$ is a cover of M.

It is said to be *regular* if

$$\forall a \in L, \quad \phi(a) = \bigvee \{\phi(x) \mid x \triangleleft_{\mathcal{A}} a\}.$$

6.4.1. Lemma. *Let $\mathcal{A}$ be a strong nearness on L and let $h \colon (M, \mathcal{B}) \to (L, \mathcal{A})$ be a dense ue-surjection. Then the right adjoint $h_* \colon L \to M$ is a regular Cauchy map with respect to $\mathcal{A}$.*

Proof. By density, $h_*(0) = 0$, and as h_* is a right adjoint it preserves all meets (we need the finite ones only, though); by (3.4.1), the $h_*[A]$ with $A \in \mathcal{A}$ are covers. Thus, h_* is a Cauchy map and we have to prove the regularity formula.

For any $z \in M$ we have the implication

$$z \triangleleft_{\mathcal{B}} h_*(a) \quad \Rightarrow \quad h(z) \triangleleft_{\mathcal{A}} a.$$

(Indeed: $h(z) \triangleleft_{\mathcal{A}} hh_*(a) \le a$.) Consequently,

$$\begin{aligned}
h_*(a) = \bigvee \{z \mid z \triangleleft_{\mathcal{B}} h_*(a)\} &\le \bigvee \{z \mid h(z) \triangleleft_{\mathcal{A}} a\} \\
&\le \bigvee \{h_* h(z) \mid h(z) \triangleleft_{\mathcal{A}} a\} \le \bigvee \{h_*(x) \mid x \triangleleft_{\mathcal{A}} a\} \le h_*(a). \qquad \square
\end{aligned}$$

6.5. The frame $\mathbf{C}(L, \mathcal{A})$. Let $(L, \mathcal{A})$ be a nearness frame. For $a \in L$ set

$$\mathfrak{k}(a) = \{x \mid x \lhd_{\mathcal{A}} a\}.$$

Recall the down-set functor $\mathfrak{D}$ from IV.2 and regard L (similarly as we have already done) for a moment, as its underlying meet-semilattice. On the frame $\mathfrak{D}L$ consider the relation R consisting of all the pairs

$$(\downarrow a, \mathfrak{k}(a)), \ a \in L \quad \text{and} \quad (L, \downarrow C), \ C \in \mathcal{A}$$

and set

$$\mathbf{C}(L, \mathcal{A}) = (\mathfrak{D}L)/R.$$

As in III.11, it will be viewed as the frame of the *saturated* (more precisely, R-saturated, but we will omit the prefix) down-sets $U \in \mathfrak{D}L$.

6.5.1. The saturation conditions. Consider the following two conditions:

(R1) $\mathfrak{k}(a) \subseteq U \ \Rightarrow \ a \in U$, and
(R2) $\{a\} \wedge C \subseteq U$ for a $C \in \mathcal{A} \ \Rightarrow \ a \in U$.

We have

Lemma. *A down-set $U \in \mathfrak{D}L$ is now saturated in the R above iff it satisfies* (R1) *and* (R2).

Proof. We have $W = \bigcup\{\downarrow b \mid b \in W\}$, $\mathfrak{k}(a) \subseteq \downarrow a$ and $\downarrow C \subseteq L$. Thus, we can rewrite the saturation conditions to

$$\forall a, b \in L, \quad (\mathfrak{k}(a) \cap \downarrow b \subseteq U \ \Rightarrow \ \downarrow a \cap \downarrow b = \downarrow(a \wedge b) \subseteq U),$$

$$\forall C \in \mathcal{A}, \forall a \in L, \quad (\downarrow C \cap \downarrow a \subseteq U \ \Rightarrow \ \downarrow a \subseteq U).$$

Now take into account that $\downarrow x \subseteq U$ iff $x \in U$, $\downarrow C \cap \downarrow a \subseteq U$ iff $\{a\} \wedge C \subseteq U$, and $\mathfrak{k}(a \wedge b) \subseteq \mathfrak{k}(a) \cap \downarrow b$, and the statement easily follows. $\qquad\square$

The following is an immediate

6.5.2. Observation. *Each $\downarrow a$ is saturated.*

((R1) follows from the admissibility, and if $A \wedge \{b\} \subseteq \downarrow a$ then for all $c \in A$ we have $c \wedge b \leq a$, and hence $b = (\bigvee A) \wedge b = \bigvee\{c \wedge b \mid c \in A\} \leq a \in \downarrow a$.)

6.6. The nearness resp. uniformity $\mathcal{A}^{\downarrow}$. For a cover $A \in \mathcal{A}$ define

$$A^{\downarrow} = \{\downarrow a \mid a \in A\} \subseteq \mathbf{C}(L, \mathcal{A}).$$

6.6.1. Lemma. (1) *Each $A^{\downarrow}$ is a cover of $\mathbf{C}(L, \mathcal{A})$.*
(2) $A^{\downarrow} \wedge B^{\downarrow} = (A \wedge B)^{\downarrow}$.
(3) $A^{\downarrow}\downarrow x \leq \downarrow(Ax)$. *Consequently,* $A^{\downarrow}B^{\downarrow} \leq (AB)^{\downarrow}$.

Proof. (1): $\bigvee\{\downarrow a \mid a \in A\} \supseteq \bigcup\{\downarrow a \mid a \in A\} = \downarrow a \supseteq A \wedge \{1\}$, hence, by (R1), $1 \in \bigvee A^{\downarrow}$, and $L \subseteq \bigvee A^{\downarrow}$.

(2): In $\mathbf{C}(L, \mathcal{A})$ the meet coincides with the intersection and we have $\downarrow a \cap \downarrow b = \downarrow(a \wedge b)$.

(3): If $\downarrow a \cap \downarrow x \neq \{0\}$ then $a \wedge b \neq 0$ and $a \leq Ax$; hence $\downarrow a \subseteq \downarrow(Ax)$. $\qquad\square$

Now set
$$A^{\downarrow} = \{A^{\downarrow} \mid A \in \mathcal{A}\}.$$

6.6.2. Lemma. *We have the implication*

$$x \vartriangleleft_A y \quad \Rightarrow \quad \downarrow x \vartriangleleft_{A^{\downarrow}} \downarrow y.$$

Proof. Use 6.6.1(3). If $Ax \leq y$ then $A^{\downarrow}\downarrow x \subseteq \downarrow(Ax) \subseteq \downarrow y$. $\qquad\square$

6.6.3. Proposition. $A^{\downarrow}$ *is a basis of nearness on* $\mathbf{C}(L, \mathcal{A})$. *If* $\mathcal{A}$ *is a strong nearness resp. uniformity then* $A^{\downarrow}$ *is a basis of strong nearness resp. uniformity.*

Proof. First, by 6.6.1(1), $A^{\downarrow}$ is a system of covers. Since each $U \in \mathbf{C}(L, \mathcal{A})$ is a down-set, we have $U = \bigcup\{\downarrow u \mid u \in U\} = \bigvee\{\downarrow u \mid u \in U\}$, and by 6.6.2 each $\downarrow u$ is the join $\bigvee\{\downarrow v \mid \downarrow v \vartriangleleft_{A^{\downarrow}} \downarrow u\}$. Thus, $A^{\downarrow}$ is admissible.

$A^{\downarrow}$ is a basis of nearness by 6.6.1(2).

If $\mathcal{A}$ is a strong nearness we have for an $A^{\downarrow} \in \mathcal{A}^{\downarrow}$

$$\{\downarrow b \mid \downarrow b \vartriangleleft_{A^{\downarrow}} \downarrow a \in A^{\downarrow}\} \geq \{\downarrow b \mid b \vartriangleleft_A a\} = \{b \mid b \vartriangleleft_A a\}^{\downarrow} \in A^{\downarrow}.$$

Finally, if $\mathcal{A}$ is a uniformity and $A \in \mathcal{A}$ take a $B \in \mathcal{A}$ with $BB \leq A$; then, by 6.6.1(3), $B^{\downarrow}B^{\downarrow} \leq (BB)^{\downarrow} \leq A^{\downarrow}$. $\qquad\square$

6.7. The dense ue-surjection $v_{(L,\mathcal{A})} \colon \mathbf{C}(L, \mathcal{A}) \to (L, \mathcal{A})$. We have the maps

$$v_L = (U \mapsto \bigvee U) \colon \mathfrak{D}L \to L, \quad \lambda_L = (x \mapsto \downarrow x) \colon L \to \mathfrak{D}L$$

in the Galois adjunction

$$v_L \lambda_L(x) = x, \quad \lambda_L v_L(U) \supseteq U,$$

hence v_L on the left and λ_L on the right. Since

$$v_L(U) \wedge v_L(V) = \bigvee\{x \wedge y \mid x \in U, \, y \in V\}$$
$$\leq \bigvee\{z \mid z \in U \cap V\} = v_L(U \cap V) \leq v_L(U) \wedge v_L(V),$$

v_L is a frame homomorphism.

Now, $v_L(\ell(a)) = a = v_L(\downarrow a)$ and for a cover A, $v_L(\downarrow A) = \bigvee A = 1 = v_L(L)$ so that v_L equalizes the relation R above and we have, by III.11, a homomorphism

$$v_{(L,\mathcal{A})} \colon \mathbf{C}(L, \mathcal{A}) \to L.$$

Since $v_L(\downarrow x) = x$, the map is onto, and obviously it is dense ($\bigvee U = 0$ only if $U = \{0\}$). Further recall that $\lambda_L(x) = \downarrow x$ is always in $\mathbf{C}(L, \mathcal{A})$ and hence $v_{(L,\mathcal{A})}$

has a right adjoint

$$\lambda_{(L,\mathcal{A})} = (x \to \downarrow x) \colon (L, \mathcal{A}) \to \mathbf{C}(L, \mathcal{A}).$$

Now observe that $v_{(L,\mathcal{A})}[A^{\downarrow}] = A$; hence $v_{(L,\mathcal{A})}$ is uniform. Finally, since $\lambda_{(L,\mathcal{A})}[A] = A^{\downarrow}$, we can conclude using Proposition 3.4 that

$$v_{(L,\mathcal{A})} \colon \mathbf{C}(L, \mathcal{A}) \to L \text{ is a dense ue-surjection.}$$

6.7.1. Note. Recall the proof of Proposition 3.4. The condition with the right adjoint implies that h is a ue-surjection in any case. One needed the strong nearness property for the other implication. Thus, for a plain nearness we are mimicking something more than just a dense uniform embedding. The difference between this and the standard embedding will play a role.

6.8. $v_{(L,\mathcal{A})} \colon \mathbf{C}(L, \mathcal{A}) \to (L, \mathcal{A})$ **is a completion.**

6.8.1. Lemma. *Let* $(L, \mathcal{A})$ *be a strong nearness frame and let* $\phi \colon L \to M$ *be a regular Cauchy map with respect to* $\mathcal{A}$. *Then there is a frame homomorphism* $f \colon \mathbf{C}(L, \mathcal{A}) \to M$, *given by* $f[U] = \bigvee \phi[U]$, *such that the diagram*

$$(L, \mathcal{A}) \xrightarrow{\ \lambda_L\ } \mathbf{C}(L, \mathcal{A})$$

$$\searrow{\phi} \qquad \downarrow{f}$$

$$M$$

commutes.

Proof. By IV.2.3 we have a frame homomorphism $f \colon \mathfrak{D}L \to M$ such that $f(\downarrow a) = \phi(a)$. Thus, $f(U) = \bigvee \phi[U]$. We have

$$f(\mathfrak{k}(a)) = \bigvee \phi[\mathfrak{k}(a)] = \bigvee\{\phi(x) \mid x \lhd a\} = \phi(a) = f(\downarrow a)$$

and

$$f(\downarrow C) = f(\bigcup\{\downarrow c \mid c \in C\}) = \bigvee\{f(\downarrow c) \mid c \in C\} = \bigvee \phi[C] = 1,$$

hence f preserves the relation R, and by III.11.3.1 can be restricted to the desired frame homomorphism f. $\qquad\square$

6.8.2. Lemma. *Let* $(L, \mathcal{A}), (M, \mathcal{B})$ *be strong nearness frames and let* $h \colon (M, \mathcal{B}) \to (L, \mathcal{A})$ *be a dense ue-surjection. Then there is a dense ue-surjection* $f \colon \mathbf{C}(L, \mathcal{A}) \to (M, \mathcal{B})$ *such that the diagram*

$$\mathbf{C}(L, \mathcal{A}) \xrightarrow{\ f\ } (M, \mathcal{B})$$

$$\searrow{v_{(L,\mathcal{A})}} \qquad \downarrow{h}$$

$$(L, \mathcal{A})$$

commutes.

Proof. By 6.4.1, $h_* \colon L \to M$ is a regular Cauchy map with respect to $\mathcal{A}$. Applying 6.8.1 we obtain a frame homomorphism $f \colon \mathbf{C}(L, \mathcal{A}) \to M$ such that $f(U) = \bigvee h_*[U]$. Then

$$hf(U) = h(\bigvee h_*[U]) = \bigvee hh_*[U] = \bigvee U = v(U).$$

Now by 3.4 $\{h_*[C] \mid C \in \mathcal{A}\}$ is an admissible system of covers and hence $h_*[L] \supseteq \bigcup\{h_*[C] \mid C \in \mathcal{A}\}$ generates M (recall 2.4.1(3)) and since $f[\mathbf{C}(L, \mathcal{A})] \supseteq f[\lambda[L]] = h_*[L]$ we have $f[\mathbf{C}(L, \mathcal{A})] = M$.

If $f(U) = \bigvee h_*[U] = 0$ we have $h_*(u) = 0$ for all $u \in U$ and since $hh_*(u) = u$, $U = \{0\}$.

Finally, from $hf = v$ we obtain for the right adjoints that $f_*h_* = \lambda$. Thus, $f_*[h_*[C]] = C^\downarrow$ and since the covers $h_*[C]$ generate $\mathcal{B}$ we see that f is a ue-surjection. $\qquad\square$

6.8.3. Theorem. *For every strong nearness frame $(L, \mathcal{A})$ there exists a strong nearness completion $v_{(L, \mathcal{A})} \colon \mathbf{C}(L, \mathcal{A}) \to (L, \mathcal{A})$, unique up to isomorphism. If $(L, \mathcal{A})$ is uniform then $\mathbf{C}(L, \mathcal{A})$ is uniform as well.*

Proof. I. $\mathbf{C}(L, \mathcal{A})$ *is complete*: Let $g \colon (M, \mathcal{B}) \to \mathbf{C}(L, \mathcal{A})$ be a dense ue-surjection, Consider the dense ue-surjection

$$h = v_{(L, \mathcal{A})} \cdot g \colon (M, \mathcal{B}) \to (L, \mathcal{A})$$

and the dense ue-surjection $f \colon \mathbf{C}(L, \mathcal{A}) \to (M, \mathcal{B})$ such that $hf = v$ from 6.8.2. We have, then, $vgf = v$ and since v is dense and hence monomorphic, $gf = \mathrm{id}$. Finally, $fgf = f$ and f is onto, hence also $fg = \mathrm{id}$, and f, g are mutually inverse uniform homomorphisms.

II. *Unicity:* Let $g \colon (M, \mathcal{B}) \to (L, \mathcal{A})$ be another completion. By 6.8.2 there is a dense ue-surjection $f \colon \mathbf{C}(L, \mathcal{A}) \to (M, \mathcal{B})$. Since $(M, \mathcal{B})$ is complete, f is an isomorphism. $\qquad\square$

7.   Functoriality. CUniFrm is coreflective in UniFrm

7.1. Recall the definitions from 6.4. Now we will show how plain Cauchy mappings can be approximated by regular ones.

For a Cauchy map $\phi \colon L \to M$ with respect to a strong nearness $\mathcal{A}$ on L set

$$\phi^\circ(a) = \bigvee\{\phi(x) \mid x \lhd_{\mathcal{A}} a\}.$$

We have

Lemma. (1) *If $\phi, \psi \colon L \to M$ are Cauchy mappings with respect to a strong nearness $\mathcal{A}$ and if $\psi \leq \phi$ then $\phi^\circ \leq \psi$.*

(2) *ϕ° is a regular Cauchy mapping.*

> (Note the inequalities in (1): ϕ° is, hence, the smallest Cauchy map ψ such that $\psi \leq \phi$.)

Proof. (1): If $x \vartriangleleft_{\mathcal{A}} a$ in L then $Ax \le a$ for an $A \in \mathcal{A}$. Thus, for $b \in A$ either $b \le a$ or $b \wedge x = 0$, that is, $b \le x^*$; in other words, $A \le \{a, x^*\}$ and hence $\{a, x^*\} \in \mathcal{A}$. Consequently, ψ being Cauchy with respect to $\mathcal{A}$, $\psi(a) \vee \psi(x^*) = 1$. Since $\phi(x) \wedge \psi(x^*) \le \phi(x) \wedge \phi(x^*) = 0$ we have $\phi(x) = \phi(x) \wedge (\psi(x^*) \vee \psi(a)) \le \psi(a)$ and conclude that $\phi^\circ(a) \le \psi(a)$.

(2): For $A \in \mathcal{A}$ take a $B \in \mathcal{A}$ such that $B \le \{x \mid x \vartriangleleft_{\mathcal{A}} a \in A\}$ (in the uniform case we can take, of course, a B such that $BB \le A$)). Then $\bigvee \phi^\circ[A] \ge \bigvee \phi[B] = 1$. Obviously $\phi^\circ(0) = 0$ and by distributivity

$$\phi^\circ(a) \wedge \phi^\circ(b) = \bigvee\{\phi(x) \mid x \vartriangleleft_{\mathcal{A}} a\} \wedge \bigvee\{\phi(y) \mid y \vartriangleleft_{\mathcal{A}} b\}$$
$$= \bigvee\{\phi(x \wedge y) \mid x \vartriangleleft_{\mathcal{A}} a, y \vartriangleleft_{\mathcal{A}} b\} \le \bigvee\{\phi(z) \mid z \vartriangleleft_{\mathcal{A}} a \wedge b\}$$
$$= \phi^\circ(a \wedge b) \le \phi^\circ(a) \wedge \phi^\circ(b).$$

Now we know that $\phi^{\circ\circ}$ is a Cauchy mapping, obviously $\le \phi$. Thus, by (1), $\phi^\circ \le \phi^{\circ\circ}(\le \phi^\circ)$ and by the definition of ϕ°, $\phi^\circ(a) = \phi^{\circ\circ}(a) = \bigvee\{\phi^\circ(x) \mid x \vartriangleleft a\}$ so that ϕ° is regular. $\qquad\square$

7.2. Now let $h \colon (L, \mathcal{A}) \to (M, \mathcal{B})$ be a uniform homomorphism. Then

$$\phi = \lambda_{(M,\mathcal{B})} \cdot h \colon L \to \mathbf{C}(M, \mathcal{B})$$

is a Cauchy map with respect to $\mathcal{A}$ ($\lambda_{(M,\mathcal{B})}$ is Cauchy with respect to $\mathcal{B}$; the uniform property of h transfers it to $\mathcal{A}$: check it as an easy exercise). Now by 7.1 we have

$$\phi^\circ \colon L \to \mathbf{C}(M, \mathcal{B})$$

regular Cauchy with respect to $\mathcal{A}$. Then, by 6.8.1, we have a frame homomorphism $f \colon \mathbf{C}(L, \mathcal{A}) \to \mathbf{C}(M, \mathcal{B})$ such that $f\lambda_{(L,\mathcal{A})} = \phi^\circ$. Set

$$\mathbf{C}h = f.$$

Theorem. *Denote by* **CUniFrm** *(resp.* **CSNearFrm***) the full subcategories of* **UniFrm** *(resp.* **SNearFrm***) generated by the complete objects. Then the construction above yields a functor*

$$\mathbf{C} \colon \mathbf{UniFrm} \to \mathbf{CUniFrm} \quad (\textit{resp.} \quad \mathbf{C} \colon \mathbf{SNearFrm} \to \mathbf{CSNearFrm}).$$

Furthermore, the dense ue-surjections $v_{(L,\mathcal{A})}$ constitute a natural transformation

$$v \colon J \cdot \mathbf{C} \overset{\cdot}{\to} \mathrm{Id}$$

(where J is the embedding functor) providing a coreflection of **UniFrm** *(resp.* **SNearFrm***) onto* **CUniFrm** *(resp.* **CSNearFrm***).*

Proof. We have $\mathbf{C}h[A^{\downarrow}] = \phi^\circ[A] = \{\bigvee\{\downarrow h(x) \mid x \vartriangleleft a\} \mid a \in A\}$; if we take a $B \in \mathcal{A}$ such that $B \le \{x \mid x \vartriangleleft a \in A\}$, the cover $\phi^\circ[A]$ is refined by $h[B]^{\downarrow}$. Thus, $\mathbf{C}h$ is a uniform homomorphism.

We have

$$v_{(M,\mathcal{B})}\mathbf{Ch}(\downarrow a) = v\phi^\circ(a) = \bigvee\{v\lambda h(x) \mid x \lhd_\mathcal{A} a\}$$
$$= \bigvee\{h(x) \mid x \lhd_\mathcal{A} a\} = h(a) = h v_{(L,\mathcal{A})}(\downarrow a).$$

Since the $\downarrow a$'s generate $\mathbf{C}(L,\mathcal{A})$ we see that the diagram

$$
\begin{array}{ccc}
\mathbf{C}(L,\mathcal{A}) & \xrightarrow{v_{(L,\mathcal{A})}} & (L,\mathcal{A}) \\
{\scriptstyle Ch}\downarrow & & \downarrow{\scriptstyle h} \\
\mathbf{C}(M,\mathcal{B}) & \xrightarrow[v_{(M,\mathcal{B})}]{} & (M,\mathcal{B})
\end{array}
$$

commutes.

Since $v_{(M,\mathcal{B})}$ is dense and hence a monomorphism, there is only one g satisfying $v_{(M,\mathcal{B})} \cdot g = h \cdot v_{(L,\mathcal{A})}$. Consequently we now easily infer that $\mathbf{C}(hg) = \mathbf{Ch} \cdot \mathbf{Cg}$ and $\mathbf{C}(\mathrm{id}) = \mathrm{id}$ and that $v\colon J \cdot \mathbf{C} \xrightarrow{\cdot} \mathrm{Id}$ is a natural transformation.

Finally, since the v's are dense ue-surjections, if $(L,\mathcal{A})$ is complete, $v_{(L,\mathcal{A})}$ is an isomorphism; hence v is a coreflection. $\qquad\square$

7.3. General nearness frames. The reader has observed that the proofs in this section were dependent on the nearnesses in question being strong. This is not just a technical issue. In fact, completion of general nearness frames is not functorial (see [31]).

Nevertheless we have a unique completion of general nearness frames. Define in this context a *strong ue-surjection* as a dense ue-surjection $h\colon (L,\mathcal{A}) \to (M,\mathcal{B})$ such that

$$\{h_*[B] \mid B \in \mathcal{B}\} \text{ is a basis of } \mathcal{A}$$

(the property from 3.4), and modify the definition of completeness and completion using strong dense ue-surjections. For strong nearnesses (in particular for uniformities) everything remains the same as before but for plain nearnesses the new condition is stronger; it is, however, still satisfied by the $v_{(L,\mathcal{A})}$ (see 6.7.1). A nearness frame has now a unique (up to isomorphism) completion; but the modification does not make it functorial.

Unlike in the strong and in particular uniform case, the subcategory of complete nearness frames is not coreflective in the category of all nearness frames. But still it behaves well in the following respect. One has that

all coproducts of complete frames are complete

(see [228]).

7.4. (Fully) Cauchy homomorphisms. In classical topology one also considers mappings between uniform spaces that are not necessarily uniform, but preserve Cauchy sequences (in metric spaces; in the more general context preserving Cauchy filters is assumed) and speaks of *Cauchy maps*. This concept makes sense in the point-free setting as well; it will be discussed in Chapter X together with other properties of the Cauchy spectrum. Here let us introduce a relevant concept that does not need the Cauchy filters at all.

A frame homomorphism $h: L \to M$ is a *fully Cauchy map* $(L, \mathcal{A}) \to (M, \mathcal{B})$ if there exists a frame homomorphism $g: \mathbf{C}(L, \mathcal{A}) \to \mathbf{C}(M, \mathcal{B})$ such that the diagram

$$
\begin{array}{ccc}
\mathbf{C}(L, \mathcal{A}) & \xrightarrow{v_{(L, \mathcal{A})}} & (L, \mathcal{A}) \\
{\scriptstyle g}\downarrow & & \downarrow{\scriptstyle h} \\
\mathbf{C}(M, \mathcal{B}) & \xrightarrow[v_{(M, \mathcal{B})}]{} & (M, \mathcal{B})
\end{array}
$$

commutes. This will be also discussed in Chapter X, together with Cauchy points.

8. An easy completeness criterion

In this very short final section we will introduce a simple completeness criterion.

8.1. A frame admitting a uniformity is regular and we know that for regular frames, co-dense homomorphisms are one-one (V.5.6). Since $v_{(L, \mathcal{A})}$ is onto we obtain that

  – $(L, \mathcal{A})$ *is complete iff* $\bigvee U = 1$ *for a saturated* U *implies that* $U = L$.

8.2. As an easy application we see that

  – *the fine (strong) nearness (that is the nearness consisting of all covers) is always complete.*

Indeed: if $\bigvee U = 1$ for a saturated U then, first of all, U is a cover and hence an element of the nearness; we have $\{1\} \wedge U = U \subseteq U$ and hence, by (R2), $1 \in U$ and $U = L$.

8.3. Here is a version of the criterion that can be sometimes expedient.

Proposition. *A uniform frame* $(L, \mathcal{A})$ *is complete iff* L *is the only* $U \in \mathfrak{D}L$ *satisfying* (R2) *and* $\bigvee U = 1$.

Proof. Let U satisfy (R2). Then $\widetilde{U} = \{a \mid \mathfrak{k}(a) \subseteq U\}$ satisfies (R2) as well:

Let $\{a\} \wedge C \subseteq \widetilde{U}$. Choose $D \in \mathcal{A}$ such that $DD \le C$. Fix an $x \in \mathfrak{k}(a)$. For $d \in D$ we have a $c \in C$ such that $Dd \le c$, and hence $x \wedge d \in \mathfrak{k}(a \wedge c) \subseteq U$, as $a \wedge c \in \widetilde{U}$. Thus, $\{x\} \wedge D \subseteq U$ and hence $x \in U$. Hence $\mathfrak{k}(a) \subseteq U$ and $a \in \widetilde{U}$.

Furthermore, $\widetilde{U}$ also satisfies (R1) (as we have a uniformity, the relation $\lhd_{\mathcal{A}}$ interpolates and $\mathfrak{k}(a) = \bigcup\{\mathfrak{k}(b) \mid b \lhd_{\mathcal{A}} a\}$) and $\bigvee \widetilde{U} \ge \bigvee U = 1$. Thus, $1 \in \widetilde{U}$, that is, $1 \in \mathfrak{k}(1) \subseteq U$, and hence $U = L$. $\qquad\square$

Chapter IX

Paracompactness

Paracompactness behaves in the point-free context better than in the classical one. It has all the properties known from spaces (full normality, various characteristics by the properties of covers, see 2.3.4), and it is implied by metrizability. But we also have a very pleasing fact that a frame is paracompact iff it admits a complete uniformity (which is not true in the classical case), and this in turn, very much in contrast with the classical situation, yields the reflectivity of paracompact locales in the general ones (making them closed under limit constructions).

1. Full normality

1.1. As we have seen in VIII.2.9, a completely regular frame has a largest uniformity, the so-called fine uniformity. Unlike the fine nearness on a regular frame, simply consisting of all the covers, the fine uniformity does not necessarily contain them all. Namely, it can very well happen that a cover of a completely regular frame has no star refinement. This led to the notion of the *normal cover* A (see VIII.2.9) such that there are covers A_n with

$$A = A_1 \text{ and } A_{n-1} \geq A_n A_n \text{ for all } n \in \mathbb{N}.$$

The fine uniformity is then the system of all normal covers.

1.2. The term "normal cover" was probably inspired by the behaviour of finite covers in normal frames.

1.2.1. Lemma. *In a normal frame there is for each cover $\{a_1, \ldots, a_n\}$ a cover $\{b_1, \ldots, b_n\}$ such that for all i, $b_i \prec a_i$.*

Proof. By the definition of normality there exist b_1, v such that

$$b_1 \wedge v = 0, \quad b_1 \vee (a_2 \vee \cdots \vee a_n) = 1 \quad \text{and} \quad a_1 \vee v = 1.$$

Thus, $v \leq b_1^*$ and $a_1 \vee b_1^* = 1$, that is, $b_1 \prec a_1$, and we have a cover $\{b_1, a_2, \ldots, a_n\}$. Repeating the procedure for $a_2 \vee (b_1 \vee a_3 \vee \cdots \vee a_n)$ we get a cover $\{b_1, b_2, a_3, \ldots, a_n\}$ with $b_2 \prec a_2$, $\{b_1, b_2, b_3, a_4, \ldots, a_n\}$ with $b_3 \prec a_3$, etc. $\qquad\square$

1.2.2. Proposition. *In a normal frame, each finite cover has a finite star-refinement. Consequently, each finite cover is normal.*

In fact, normal frames are precisely those frames in which each finite cover is normal.

Proof. I. Let $A = \{a_1, a_2, \ldots, a_n\}$ be a finite cover. By 1.2.1 choose a cover $B = \{b_1, b_2, \ldots, b_n\}$ with $b_i \prec a_i$. Now $C_i = \{b_i^*, a_i\}$ $(i = 1, 2, \ldots, n)$ are covers and we have the finite cover

$$C = B \wedge C_1 \wedge \cdots \wedge C_n.$$

We claim that $CC \leq A$. Indeed, fix an element of C,

$$c = b_i \wedge \bigwedge_{j=1}^{n} c_j \quad (c_j \text{ either } b_j^* \text{ or } a_j),$$

and pick another $x \in C$, $x = b_k \wedge \bigwedge_{j=1}^{n} x_j$, $(x_j$ either b_j^* or $a_j)$. Now if $x \wedge c \neq 0$ then $x_i \neq b_i^*$, hence $x_i = a_i$, and $x \leq a_i$.

II. Let each finite cover have a star-refinement. In particular if $a \vee b = 1$ there is a cover C with $CC \leq \{a, b\}$. Set

$$w_1 = \bigvee \{c \in C \mid Cc \leq a\}, \quad w_2 = \bigvee \{c \in C \mid Cc \leq b\} \quad \text{and} \quad u = w_2^*, \ v = u^*.$$

Then $Cw_1 \leq a$, $Cw_2 \leq b$ and hence, by VIII.2.3.2, $w_1^* \vee a = 1$ and $u \vee b = w_2^* \vee b = 1$. Further, since $w_1 \vee w_2 = 1$, $w_1^* = w_1^* \wedge (w_1 \vee w_2) \leq w_2^{**} = v$ and hence $a \vee v = 1$. This shows that the frame is normal. $\qquad\square$

> Note that in the second statement we did not have to assume that each finite cover has a *finite* star-refinement.

1.3. If not only the finite covers but *all* covers of a regular frame are normal one often speaks of a

fully normal frame

(it will come under another name, *paracompact frame* in the following sections). Note that we have formally assumed just regularity; however, such a frame is normal by 1.2.2 and hence completely regular by V.5.9.1.

The fact that full normality implies normality is a source of examples of completely regular spaces in which not every cover has a star refinement. See also 2.7.1 below.

We have already encountered an important class of fully normal frames, the compact frames (see VIII.5). In the next subsection we will discuss another interesting case.

1.4. Countable admissible systems of covers

1.4.1. Recall the operation AB with covers A, B from VIII.2. We have

Lemma. *If $\mathcal{A}$ is an admissible system of covers then so is also the system*

$$\mathcal{B} = \{AA \mid A \in \mathcal{A}\}.$$

Proof. Let $y \lhd_\mathcal{A} y_1 \lhd_\mathcal{A} y_2 \lhd_\mathcal{A} x$. Then $A_1 y \leq y_1$, $A_2 y_1 \leq y_2$ and $A_3 y_2 \leq x$ for some $A_1, A_2, A_3 \in \mathcal{A}$. Thus, if we set $A = A_1 \wedge A_2 \wedge A_3$ we obtain $A(A(Ay)) \leq x$. By VIII.2.2.1(3), $A(A(Ay)) = (AA)y$, hence $(AA)y \leq x$ and $y \lhd_\mathcal{B} x$. Now we have

$$x = \bigvee\{y_2 \mid y_2 \lhd_\mathcal{A} x\} = \bigvee\{y_1 \mid \exists y_2, \, y_1 \lhd_\mathcal{A} y_2 \lhd_\mathcal{A} x\}$$
$$= \bigvee\{y \mid \exists y_1, y_2, \, y \lhd_\mathcal{A} y_1 \lhd_\mathcal{A} y_2 \lhd_\mathcal{A} x\} \leq \bigvee\{y \mid y \lhd_\mathcal{B} x\}. \qquad \square$$

1.4.2. Recall the adjunction $Bx \leq y$ iff $x \leq y/B$ from VIII.2.2.2. For a cover A of L and $x \in L$ set

$$b(A, x) = \{a \in A \mid a \wedge (x/(AA)(AA)) \neq 0\}.$$

We have

Lemma. $\bigvee b(A, x) = A(x/(AA)(AA))$ *and* $A(\bigvee b(A, x)) \leq x/(AA)$.

Proof. The equality is in the definition of Ay.

Now, by the adjunction of Bx and y/B we have $B(y/B) \leq y$ and $x/(BB) \leq (X/B)/B$.

(The latter since $y \leq x/BB$ iff $(BB)y \leq x$ which implies by VIII.2.2.1(3) $B(By) \leq x$ and this then iff $By \leq x/B$ iff $y \leq (x/B)/B$.)

Applying that for $B = AA$ we obtain

$$A(\bigvee b(A, x)) = A(A(x/(AA)(AA)) \leq (AA)(x/(AA)(AA)))$$
$$\leq (AA)((x/(AA))/(AA)) \leq x/(AA). \qquad \square$$

1.4.3. Theorem. *Each frame L admitting a countable system of covers is fully normal.*

Proof. We can assume that L admits a system of covers

$$A_1 \geq A_2 \geq \cdots \geq A_n \geq \cdots$$

(in fact, if $(A'_n)_n$ is admissible then L also admits $A_1 = A'_1$, $A_2 = A'_1 \wedge A'_2$, $A_3 = A'_1 \wedge A'_2 \wedge A'_3$, etc. since by VIII.2.2.1(4), $A_n x \leq \bigwedge_{i=1}^n A'_i x$).

We will prove that L is fully normal. Let C be a general cover of L. Set

$$B = \bigcup \{b(A_n, c) \mid c \in C, \ n = 1, 2, \dots\}.$$

I. *B is a cover.* By 1.4.1,

$$\bigvee B = \bigvee_{c \in C} \bigvee_{n \in \mathbb{N}} \bigvee b(A_n, c) = \bigvee_{c \in C} \bigvee_{n \in \mathbb{N}} A_n(c/(A_n A_n)(A_n A_n))$$

$$\geq \bigvee_{c \in C} \bigvee_{n \in \mathbb{N}} (c/(A_n A_n)(A_n A_n)) = \bigvee_{c \in C} c = 1.$$

II. *$BB \leq C$:* Take an $a \in B$, say, $a \in b(A_k, c)$. Set

$$n = \min\{i \mid \exists x \in C \ \exists y \in b(A_i, x) \ \text{ such that } \ y \wedge a \neq 0\}.$$

Note that the set is non-void, since $a \in b(A_k, c)$, and $n \leq k$. Choose $c_0 \in C$ and $y_0 \in b(A_n, c_0)$ such that $y_0 \wedge a \neq 0$. Since $A_k \leq A_n$ we have a $b \in A_n$ such that $a \leq b$, and hence $y_0 \wedge b \neq 0$ and $b \leq A_n y_0$. Now if $u \wedge a \neq 0$ for some $u \in B$, say $u \in b(A_i, d)$, we have $A_i \leq A_n$ and hence there is a $v \in A_n$ such that $u \leq v$. Then $v \wedge b \geq u \wedge a \neq 0$, and since $b \leq A_n y_0$ we obtain, using 1.4.2,

$$u \leq v \leq A_n(A_n y_0) \leq A_n(A_n(\bigvee b_n(A_n, c_0)))$$

$$\leq A_n(c_0/A_n A_n) \leq (A_n A_n)(c_0/A_n A_n) \leq c_0$$

so that $Ba \leq c_0$. $\square$

1.5. Proposition. *Let L be a regular frame. Then the following statements are equivalent.*

(1) *L admits a countable system of covers.*

(2) *L admits a countably generated nearness.*

(3) *L admits a countably generated uniformity.*

Proof. Obviously $(3) \Rightarrow (2) \Leftrightarrow (1)$.

$(1) \Rightarrow (3)$: Assume $A_1 \geq A_2 \geq \cdots \geq A_n \geq \cdots$ (see the proof of 1.4.3). Then we can set $B_1 = A_1$ and if B_n is already chosen, we can choose, using full normality, B'_{n+1} such that $B'_{n+1} B'_{n+1} \leq B_n$ and set $B_{n+1} = B'_{n+1} \wedge A_{n+1}$. Then $\{B_1, B_2, \dots, B_n, \dots\}$ is an admissible basis of uniformity. $\square$

1.5.1. Note. Imitating the assumption from the Uniform Metrization Theorem, Isbell defined (in [136]) a frame L as metrizable if it admits a countably generated uniformity. In the next chapter we will give this notion a more "metric" sense. Anyway, accepting this definition so far on the face value, we obtain that

every metrizable frame is fully normal.

2. Paracompactness, and its various guises

2.1. The concept of full normality from the previous section is equivalent to another one, the paracompactness. The classical definition of a paracompact space is that each of its open covers admits a locally finite refinement. We will formulate it in a point-free way.

A subset X of a frame L is said to be *locally finite* (resp. *discrete*) *witnessed by a cover* W if for each $w \in W$ there are only finitely many (resp. there is at most one) $x \in X$ such that $x \wedge w \neq 0$.

Now here is the crucial definition: a frame L is said to be *paracompact* if it is regular and if each of its covers has a locally finite refinement.

2.2. Further we will need the definitions of a σ-locally finite subset X, and of a σ-discrete subset X, of a frame L. In the former one assumes that X is a countable union of locally finite sets, in the latter that X is a countable union of discrete ones.

2.3. For the main theorem we will need an auxiliary technical definition. A frame is said to be a (PC)-*frame* if

(PC) For each cover A of L there is a locally finite $X \subseteq L$ such that $X \leq A$ and $(\bigvee X)^* = 0$.

2.3.1. Lemma. *Let each cover of L have a σ-locally finite refinement. Then L is a* (PC)-*frame.*

Proof. Let A be a cover and let $B = \bigcup_{n=1}^{\infty} B_n$ refine A, with locally finite B_n witnessed by W_n. We can assume that $B_n \subseteq B_{n+1}$. For $b \in B$ set

$$n(b) = \min\{n \mid b \in B_n\}$$

and choose an antireflexive R well-ordering B such that

$$n(c) < n(b) \;\Rightarrow\; cRb.$$

Further set $V_n = W_n \wedge B_n$ and set $V = \bigcup V_n$. Since $\bigvee V_n = \bigvee W_n \wedge \bigvee B_n = \bigvee B_n$, we have $\bigvee V = \bigvee B = 1$. Finally, define for $b \in B$

$$\widetilde{b} = b \wedge (\bigvee\{c \mid cRb\})^*$$

and set $X = \{\widetilde{b} \mid b \in B\}$. Obviously $X \leq A$.

If $0 \neq x \in L$ then $x \wedge b \neq 0$ for some $b \in B$. Consider the first b, in the order R, with this property. Then $x \leq (\bigvee\{c \mid cRb\})^*$ and hence $x \wedge \widetilde{b} \neq 0$. Thus, $(\bigvee X)^* = 0$.

X is locally finite witnessed by V: Take a $v \in V$, say, $v = w \wedge c$ with $w \in W_n$ and $c \in B_n$. Whenever $n(b) > n(\geq n(c))$, we have cRb and hence $v \leq \bigvee\{u \mid uRb\}$ and $v \wedge \widetilde{b} = 0$. Thus, if $v \wedge \widetilde{b} \neq 0$ then $b \in B_n$ and as $v \leq w \in W_n$ there are only finitely many such b. $\qquad \square$

2.3.2. Lemma. *Every* (PC)-*frame is fully normal.*

Proof. For a cover B define

$$B^\times = \{\textstyle\bigvee S \mid S \subseteq B, \ \textstyle\bigwedge S \neq 0\}.$$

Since obviously $BB \leq (B^\times)^\times$ it suffices to prove that for each cover A there is a cover B such that $B^\times \leq A$.

Let $X \leq \{c \mid c \prec a \in A\}$ such that $(\bigvee X)^* = 0$ be locally finite witnessed by W. For $x \in X$ choose $a(x) \in A$ such that $x \prec a(x)$, and define

$$B = \{b \in L \mid \forall x \in X \ (b \leq a(x) \ \text{ or } \ b \leq x^*)\}.$$

Take a $w \in W$ and consider the system $x_1, \ldots, x_n$ of all the elements of X such that $x \wedge w \neq 0$. We have

$$1 = \bigwedge_{i=1}^{n} (a(x_i) \vee v_i^*) = \bigvee \left\{ \bigwedge_{i=1}^{n} \xi_i \mid \xi \in \{a(x_1), x_1^*\} \times \cdots \times \{a(x_n), x_n^*\} \right\} = s.$$

Now since $w \leq x^*$ for $x \in X \setminus \{x_1, \ldots, x_n\}$ we have

$$w \wedge \textstyle\bigvee B = w \wedge \bigvee\{b \in L \mid \forall i (b \leq a(x_i) \ \text{ or } \ b \leq x_i^*)\} = w \wedge s = w,$$

and since W is a cover we infer that $\bigvee B = 1$.

Finally let $S \subseteq B$, $\bigwedge S \neq 0$. Then there is an $x \in X$ such that $x \wedge \bigwedge S \neq 0$ and hence $s \neq x^*$ for all $s \in S$. Consequently $s \leq a(x)$ for all $s \in S$, $\bigvee S \leq a(x)$, and we conclude that $B^\times \leq A$. $\qquad\square$

2.3.3. Lemma. *Every* (PC)-*frame is paracompact.*

Proof. We already know that L is fully normal and hence we can choose, for any cover A of L, a cover B such that $BB \leq A$. Then, choose for B an $X \subseteq L$ as in (PC), locally finite witnessed by W. Further, let V be a cover such that $VV \leq W$. The set

$$D = \{(V \wedge B)x \mid x \in X\}$$

is obviously a cover, and since $V \wedge B \leq B$ and $X \leq B$ it refines A. Finally, D is locally finite witnessed by V:

If $v \wedge (V \wedge B)x \neq 0$ then obviously $Vv \wedge x \neq 0$. Choose a $w \in W$ such that $Vv \leq w$. Then $w \wedge x \neq 0$, and there are only finitely many such $x \in X$. $\qquad\square$

2.3.4. Theorem. *Let L be a regular frame. Then the following statements are equivalent.*

(1) *L is paracompact.*

(2) *Each cover of L has a σ-locally finite refinement.*

(3) *Each cover of L has a σ-discrete refinement.*

(4) *L is fully normal.*

Proof. First, let us show that $(4)\Rightarrow(3)$:

For a cover A choose A_n such that $A_0 = A$ and $A_n A_n \leq A_{n-1}$. Let A_1 be well ordered by an antireflexive R; write $b\overline{R}a$ for (bRa or $b = a$). For an $a \in A_1$ set

$$a_1 = a, \quad a_{n+1} = A_{n+1}a_n, \quad \text{and} \quad \phi_n(a) = \bigvee\{b_n \mid bRa\}.$$

Note that $A_{n+1}\phi_n(a) = \phi_{n+1}(a)$ so that $\phi_n(a) \prec \phi_{n+1}(a)$. Set

$$\widetilde{A}_n = \{\widetilde{a}_n = a_n \wedge \phi_{n+1}(a)^* \mid a \in A_1\} \quad \text{and} \quad \widetilde{A} = \bigcup \widetilde{A}_n.$$

We have

$$\bigvee_{n=1}^{\infty} (\bigvee\{\widetilde{b}_n \mid b\overline{R}a\}) = \bigvee_{n=1}^{\infty} (\bigvee\{b_n \mid b\overline{R}a\}). \tag{$*$}$$

Indeed: It obviously holds for the a first in R; now if it holds for all cRa we have

$$\bigvee_{n=1}^{\infty} \bigvee\{\widetilde{b}_n \mid b\overline{R}a\} = \bigvee_{n=1}^{\infty} (\bigvee_{cRa} \bigvee\{\widetilde{b}_n \mid b\overline{R}c\} \vee \widetilde{a}_n) = \bigvee_{n=1}^{\infty} (\bigvee\{\widetilde{b}_n \mid bRa\} \vee \widetilde{a}_n)$$

$$= \bigvee_n (\phi_n(a) \vee (a_n \wedge \phi_{n+1}(a)^*)) = \bigvee_n (\phi_{n+2}(a) \vee (a_n \wedge \phi_{n+1}(a)^*))$$

$$= \bigvee_n (\phi_{n+2}(a) \vee a_n) = \bigvee_n (\phi_n(a) \vee a_n) = \bigvee_n (\bigvee\{b_n \mid b\overline{R}a\}).$$

From $(*)$ we immediately infer that $\bigvee \widetilde{A} \geq \bigvee A_1 = 1$; Thus, $\widetilde{A}$ is a cover. For an $a \in A_1$ we have a $b \in A$ such that $A_1 a = A_1 a_1 \leq b$. Since

$$A_{n+1}a_{n+1} = A_{n+1}(A_{n+1}a_n) \leq (A_{n+1}A_{n+1})a_n \leq A_n a_n$$

we see that $A_n a_n \leq b$ for all n and hence $\widetilde{a}_n \leq a_n \leq b$. Thus, $\widetilde{A}$ refines A.

Finally, $\widetilde{A}_n$ is discrete witnessed by A_{n+1}: For $x \in A_{n+1}$ take the $a \in A_1$ first in R such that $x \wedge \widetilde{a}_n \neq 0$. If aRb we have $x \leq A_{n+1}a_n = a_{n+1} \leq \phi_{n+1}(b)$ so that $x \wedge \phi_{n+1}(b)^* = 0$ and $x \wedge \widetilde{b}_n = 0$.

Now since $(3)\Rightarrow(2)$ is trivial we have, by 2.3.1 and 2.3.2 the cycle

$$(4) \Rightarrow (3) \Rightarrow (2) \Rightarrow (PC) \Rightarrow (4).$$

Finally, since $(1)\Rightarrow(2)$ is trivial we have by 2.3.3 the cycle

$$(1) \Rightarrow (2) \Rightarrow (PC) \Rightarrow (1)$$

and the statement follows.   □

2.4. As it has been already mentioned in 1.3, a fully normal regular frame is completely regular. Hence, the regularity in the definition of paracompactness (2.1) automatically entails complete regularity.

2.5. Take a Lindelöf frame L. For each cover C of L one has a countable subcover

$$A = \{a_1, a_2, \ldots, a_n, \ldots\};$$

this being a union of discrete subsets $\{a_n\} \subseteq X$ (witnessed, of course, by any cover) makes A a σ-discrete refinement of C. Thus we obtain from 2.3.4 a substantial extension of the full normality of compact frames.

Corollary. *Each Lindelöf frame is paracompact (fully normal).*   □

2.6. From 2.3.4 and 1.5 (1.4.3) we obtain

Corollary. *Each metrizable frame (i.e., a frame with an admissible countable system of covers) is paracompact.*   □

2.6.1. Remarks. (1) Each metric space (X, ρ) has a uniformity with the countable basis $A_1, A_2, \ldots, A_n, \ldots$ where

$$A_n = \{\mathcal{B}(x; \tfrac{1}{n}) \mid x \in X\}, \quad \mathcal{B}(x; \varepsilon) = \{y \mid \rho(x, y) < \varepsilon\}.$$

Note that the substance of the metrizability theorem mentioned in 1.5.1 is in the existence of a suitable metric once one has the countably based uniformity while the just presented implication is trivial. Realize that

a space X is paracompact iff the frame $\Omega(X)$ is.

Thus, as a special case of the fact 2.6 we have the classical theorem about the paracompactness of metric spaces.

(2) Now by 1.2.2 we have that each paracompact space is normal. Thus, if we wish for an easy example of a completely regular frame that is not paracompact we can take some of the standard examples of the non-normal ones: for instance, the

$$((\omega_1 + 1) \times (\omega_0 + 1)) \smallsetminus \{(\omega_1, \omega_0)\},$$

or the square $S \times S$ of the Sörgenfrey line S (see, e.g., [161]).

There are non-paracompact spaces that are normal; such constructions are, however, very involved.

3. An elegant, specifically point-free, characterization of paracompactness

3.1. As an application of the completeness criterion VIII.8.1 we immediately learned that (VIII.8.2)

a fine nearness frame (that is, a regular frame endowed with the system of all covers) is always complete.

If a uniform frame is complete as a nearness frame then it is, of course, all the more so as a uniform frame. Consequently, we obtain from 2.3.4 an immediate

3.1.1. Corollary. *The fine uniformity on a paracompact frame is complete.*   □

3.2. Recall VIII.6.5. and write, more precisely,

$$\mathfrak{k}_{\mathcal{A}}(a) = \{x \mid x \triangleleft_{\mathcal{A}} a\}.$$

Now if $\mathcal{A} \subseteq \mathcal{B}$ then obviously $\mathfrak{k}_{A}(a) \subseteq \mathfrak{k}_{\mathcal{B}}(a)$ and hence

if (R1) *is satisfied for $\mathcal{A}$, it is satisfied for $\mathcal{B}$,*

and trivially

the same holds for (R2).

Thus we obtain from the completeness criterion VIII.8.1 the

3.2.1. Observation. *If $(L, \mathcal{A})$ is a complete uniform frame then each $(L, \mathcal{B})$ with $\mathcal{A} \subseteq \mathcal{B}$ is complete. Consequently, if a (completely regular) L admits a complete uniformity then*

$$(L, \text{fine uniformity})$$

is complete.

In general the fine uniformity is not the system of all covers; thus, admitting a complete uniformity is a non-trivial property. This will be our concern in the remainder of this section.

3.3. A technical proposition. Extrapolating the properties of the relations $\triangleleft_{\mathcal{A}}$ (and the earlier $\prec\!\!\prec$) one defines a *strong inclusion* ([15]) on a frame L as an interpolative binary relation $\triangleleft$ such that

(1) $a \triangleleft b \Rightarrow a \leq b$,

(2) $x \leq a \triangleleft b \leq y \Rightarrow x \triangleleft y$,

(3) $a \triangleleft b \Rightarrow b^* \triangleleft a^*$,

(4) $a \triangleleft b \Rightarrow a^{**} \triangleleft b$, and

(5) $a_i \triangleleft b_i$, $i = 1, 2$, $\Rightarrow a_1 \wedge a_2 \triangleleft b_1 \wedge b_2$ and $a_1 \vee a_2 \triangleleft b_1 \vee b_2$.

A strong inclusion on L is *admissible* if for each $a \in L$, $a = \bigvee\{x \mid x \triangleleft a\}$.

Let $\triangleleft$ be a strong inclusion, not necessarily admissible, on L. Define

$$L_{\triangleleft} = \{a \in L \mid a = \bigvee\{x \mid x \triangleleft a\}\}.$$

3.3.1. Lemma. *Define $a' = \bigvee\{x \mid x \triangleleft a\}$. Then:*

(a) *$a' \leq a$ and $a' \in L_{\triangleleft}$.*

(b) *If A, A_1 are covers of L such that for each $a_1 \in A_1$ there is an $a \in A$ with $a_1 \triangleleft a$ then $B = \{a' \mid a \in A\}$ is a cover.*

Proof. (a): The first is trivial. Now by interpolation,

$$a' = \bigvee\{x \mid \exists y, \ x \lhd y \lhd a\} \leq \bigvee\{x \mid \exists y \lhd a', x \lhd y\} \leq \bigvee\{x \mid x \lhd a'\}.$$

(b): If $a_1 \lhd a$ then $a_1 \leq a'$; hence $\bigvee B \geq \bigvee A_1 = 1$.   □

3.3.2. Recall V.6.1. Now we have a bit more.

Lemma. $L_\lhd$ *is a subframe of* L *and* $\lhd$ *is admissible on* $L_\lhd$.

Proof. If $a, b \in L_\lhd$ then

$$a \wedge b = \bigvee\{x \mid x \lhd a\} \wedge \bigvee\{y \mid y \lhd b\}$$

$$= \bigvee\{x \wedge y \mid x \lhd a, \ y \lhd b\} \leq \bigvee\{z \mid z \lhd a \wedge b\} \leq a \wedge b$$

(the penultimate inequality by condition (5)) and if $a_i \in L_\lhd$, $i \in J$, then

$$\bigvee_{i \in J} a_i = \bigvee_{i \in J} \bigvee\{x \mid x \lhd a_i\} \leq \bigvee\{x \mid x \lhd \bigvee_{i \in J} a_i\} \leq \bigvee_{i \in J} a_i,$$

(the penultimate inequality by condition (2)).

Now about the admissibility. By interpolation, if $a \in L_\lhd$ then by 3.3.1,

$$a = \bigvee\{x \mid x \lhd a\} = \bigvee\{x \mid \exists y, \ x \lhd y \lhd a\}$$

$$\leq \bigvee_{y \lhd a} y' \leq \bigvee\{z \mid z \in L_\lhd, \ z \lhd a\}.$$ □

3.3.3. Proposition. *In any frame* L, *each normal cover has a normal locally finite refinement witnessed by a normal cover.*

Proof. Let A be a normal cover and let $\mathcal{A} = \{A_1, A_2, \dots\}$ be such that $A_n A_n \leq A_{n-1}$. Then $\lhd = \lhd_{\mathcal{A}}$ is a strong inclusion. Since A_{n+1} is a star refinement of A_n by 3.3.1(b) we have a cover $B_n = \{a' \mid a \in A_n\}$ and it is a cover of the $L_\lhd$ above. Set $\mathcal{B} = \{B_1, B_2, \dots\}$. Since obviously

$$x \lhd y \Rightarrow x \lhd_{\mathcal{B}} y,$$

$\mathcal{B}$ is admissible, and by 2.3.3 $L_\lhd$ is paracompact. Thus, each cover of $L_\lhd$ is normal and has a locally finite refinement. Since every cover of $L_\lhd$ is also a cover of L the statement follows.   □

3.4. Lemma. *Let* A *be a locally finite normal cover and let* C_a, $a \in A$, *be normal covers. Then*

$$C = \bigcup\{\{a\} \wedge C_a \mid a \in A\}$$

is a normal cover.

Proof. If A' refines A choose for $b \in A'$ an $a \in A$ with $a \geq b$, and set $C'_b = C_a$. Then $\bigcup\{\{b\} \wedge C'_b \mid b \in A'\}$ refines $\bigcup\{\{a\} \wedge C_a \mid a \in A\}$ and consequently we can assume, using 3.3.3, that A is locally finite witnessed by a normal W. By 3.3.3 again there is a locally finite B such that

$$BB \leq A \quad \text{and} \quad B \leq W.$$

For $a \in A$ choose normal covers U_a such that $U_a U_a \leq C_a$. Since $B \leq W$, the system $\{a \in A \mid a \wedge b \neq 0\}$ is finite for any $b \in B$ and hence we can define

$$D_b = \bigwedge\{U_a \mid a \in A, a \wedge b \neq 0\} \quad \text{and} \quad D = \bigcup\{\{b\} \wedge D_b \mid b \in B\}.$$

We will show that $DD \leq C$ which will finish the proof since the structure of D allows to proceed by induction. Fix a $b \in B$ and a $d \in D_b$ and consider a general $x \in B$ and $y \in D_x$ such that $(b \wedge d) \wedge (x \wedge y) \neq 0$. If $Bb \leq a \in A$ then $x \leq a$. Further, $y = y_1 \wedge \cdots \wedge y_n$ for some $y_i \in U_{a_i}$ where $a_1, \ldots, a_n$ are the elements of A that meet b. Hence a is one of the a_k and $y \leq y_k \in U_a$. By the same reasoning we also have a $d \leq d_k \in U_a$. Now take a $c \in C_a$ such that $U_a d_k \leq c$. Since $d_k \wedge y_k \geq d \wedge y \neq 0$ we have $y_k \leq c$ and $x \wedge y \leq a \wedge y_k \leq a \wedge c$. Thus, $D(b \wedge d) \leq a \wedge c$. □

3.5. Theorem. [Isbell] *A frame L is paracompact if and only if it admits a complete uniformity.*

Proof. $\Rightarrow$ is in 3.1.1.

$\Leftarrow$: If L admits a complete uniformity then by 3.2.1 the fine uniformity $\mathcal{F}$ (that is, the system of all normal covers) of L is complete. Let A be a general cover. Set

$$U = \{u \in L \mid \exists B \in \mathcal{F}, \ \{u\} \wedge B \leq A\}.$$

Obviously $U \in \mathfrak{D}L$ and $A \subseteq U$ (since $\{a\} \wedge \{1\} = \{a\}$).

Now let $\{x\} \wedge C \subseteq U$ for some $C \in \mathcal{F}$. By 3.4 we can assume that C is locally finite. For each $c \in C$ we have a normal cover B_c such that $\{x \wedge c\} \wedge B_c \leq A$. Now

$$\{x\} \wedge \bigcup\{\{c\} \wedge B_b \mid c \in C\} = \bigcup\{\{x \wedge c\} \wedge B_c \mid c \in C\} \leq A$$

and since $\bigcup\{\{c\} \wedge B_b \mid c \in C\}$ is normal by 3.4, $x \in U$. Thus, U satisfies (R2) in $(L, \mathcal{F})$. Moreover, $\bigvee U \geq \bigvee A$ (as $A \subseteq U$) and hence, by VIII.8.3, $1 \in U$. But this means that there is a normal B such that $B = \{1\} \wedge B \leq A$, making A itself normal. □

3.5.1. Note. This characteristics of paracompactness has no classical counterpart. Each paracompact space, of course, admits a complete uniformity, the fine one consisting of all the open covers. But the converse does not hold. In fact, the other implication has a rather surprising consequence as we will see in the next section.

4. A pleasant surprise: paracompact (co)reflection

4.1. Paracompact spaces, because of the nice structure of covers (in any of the aspects in 2.3.4), and also because of being a natural generalization of metric ones, play an important role in geometry and elsewhere. But there is a drawback: they are very badly behaved in constructions. For instance a product of paracompact spaces is not necessarily paracompact; in fact even a product of a paracompact space with a metric one is not necessarily so (see [136]; in fact, constructing an example is not very difficult – certainly it is much simpler than finding a non-paracompact normal space).

The situation in the point-free context is radically different. Paracompact locales are not only closed under products in the category of all locales. They constitute there a reflective subcategory and therefore are closed under all limits (AII.8.3); and of course one has colimits as well that can be obtained modifying those from **Loc**. This will be proved in this short section, in the frame form, that is, we will construct a coreflection of the category of frames onto the subcategory of the paracompact ones. A first such construction appeared in the already cited Isbell's paper [136].

4.2. Recall from VIII.2.2.3 that for any frame homomorphism, $h\colon L \to M$ and any $A, B \subseteq L$,

$$h[A]h[B] \leq h[AB].$$

Consequently,

4.2.1. Fact. *The homomorphic image of a normal cover is a normal cover.* $\square$

Let L be a completely regular frame. Denote by $\mathcal{F}(L)$ the fine uniformity on L. Since it consists of all the normal covers, we obtain from 4.2.1

4.2.2. Fact. *Let L, M be completely regular frames. Then each frame homomorphism $h\colon L \to M$ is a uniform homomorphism*

$$h\colon (L, \mathcal{F}(L)) \to (M, \mathcal{F}(M)).$$ $\square$

4.3. Construction. For a completely regular frame L take $\mathbf{C}(L, \mathcal{F}(L))$ and define

$$\mathbf{Par}(L)$$

as its underlying frame. Further, define

$$\pi_L\colon \mathbf{Par}(L) \to L$$

as the underlying frame homomorphism of $\upsilon_{(L, \mathcal{F}(L))}\colon \mathbf{C}(L, \mathcal{F}(L)) \to (L, \mathcal{F}(L))$. Finally, denote by

$$\mathbf{Par}(h)\colon \mathbf{Par}(L) \to \mathbf{Par}(M)$$

the underlying frame homomorphism of $\mathbf{C}(h)\colon \mathbf{C}(L, \mathcal{F}(L)) \to \mathbf{C}(M, \mathcal{F}(M))$.

4.4. Theorem. *The subcategory* **ParFrm** *of paracompact frames is coreflective in the category* **CRegFrm** *of completely regular frames, with the coreflection functor* **Par** *and coreflection* $\pi = (\pi_L)_L$.

Proof. Since $\mathbf{C}(L, \mathcal{F}(L))$ is complete (and uniform), $\mathbf{Par}(L)$ is paracompact by 3.5. On the other hand, if L is paracompact, $\mathcal{F}(L)$ is the system of all covers and hence $(L, \mathcal{F}(L))$ is complete (recall 3.1.1) and π_L is an isomorphism.   □

4.4.1. By V.6.3.3 the category of completely regular frames is coreflective in **Frm**. Thus we have

Corollary. *The subcategory* **ParFrm** *of paracompact frames is coreflective in the category* **Frm** *of all frames and the corresponding category* **ParLoc** *of paracompact locales is reflective in* **Loc**.

Therefore, **ParLoc** *is complete and cocomplete and the limits in it coincide with those in* **Loc**. *In particular, products of paracompact locales, taken in* **Loc**, *are paracompact.*   □

Chapter X

More about Completion

1. A variant of the completion of uniform frames

1.1. Samuel compactification of a uniform frame. Recall the compactification from VII.4. It will be now modified for uniform frames $(L, \mathcal{A})$. Here is how it will be adapted.

(1) Instead of the regular ideals in L we will consider the $\mathcal{A}$-*regular ideals* I in $(L, \mathcal{A})$ defined by the requirement

$$\forall a \in I \; \exists b \in I, \quad a \vartriangleleft_{\mathcal{A}} b$$

(note that by VIII.2.8.1 each $\mathcal{A}$-regular ideal is regular; the converse does not necessarily hold). The set of all $\mathcal{A}$-regular ideals in $(L, \mathcal{A})$ will be denoted by

$$\widetilde{\mathfrak{R}}(L, \mathcal{A}).$$

(2) We will have again $v_{(L,\mathcal{A})} \colon \widetilde{\mathfrak{R}}(L, \mathcal{A}) \to L$ defined by $v(I) = \bigvee I$. Instead of the $\sigma(a) = \{x \mid x \lll a\}$ we will take the $\mathfrak{k}(a) = \{x \mid x \vartriangleleft_{\mathcal{A}} a\}$ from VIII.6.5.

(3) Similarly like in VII.4.4 we will define, for $h \colon (L, \mathcal{A}) \to (M, \mathcal{B})$,

$$\widetilde{\mathfrak{R}}(h)(I) = \downarrow h[I],$$

but this will be now correct for uniform h only: we need the implication $a \vartriangleleft_{\mathcal{A}} b \Rightarrow h(a) \vartriangleleft_{\mathcal{B}} h(b)$ from VIII.2.3.1.

1.1.1. Proposition. (1) $\widetilde{\mathfrak{R}}(L, \mathcal{A})$ *is a compact regular frame and* $v_{(L,\mathcal{A})} \colon \widetilde{\mathfrak{R}}(L, \mathcal{A}) \to L$ *is a dense uniform homomorphism, left adjoint to* $\mathfrak{k}_{(L,\mathcal{A})} = (a \mapsto \mathfrak{k}(a))$.

(2) $(\widetilde{\mathfrak{R}}, v)$ *is a coreflection of the category of uniform frames onto the subcategory of compact regular frames (endowed with the unique uniformities).*

Proof. (1) Proving that $\widetilde{\mathfrak{R}}(L, \mathcal{A})$ is a subframe of $\mathfrak{J}(L)$ goes along the same lines as in VII.4.2.1 (if $a_i \vartriangleleft_{\mathcal{A}} b$ then $A_i a_i \leq b$, hence $A a_i \leq b$ for $A = A_1 \wedge A_2$, and hence $a_1 \vee a_2 \lhd_{\mathcal{A}} b$, etc.), but the regularity needs a slightly modified reasoning. We have to prove that

$$a \lhd_{\mathcal{A}} b \quad \Rightarrow \quad \mathfrak{k}(a) \prec \mathfrak{k}(b).$$

For this aim interpolate $a \lhd_{\mathcal{A}} x \lhd_{\mathcal{A}} y \lhd_{\mathcal{A}} b$. Hence for some $A \in \mathcal{A}$ ($A = A_1 \wedge A_2 \wedge A_3$ where $A_1 a \le x$, $A_2 x \le y$ and $A_3 y \le b$),

$$Aa \le x, \quad Ax \le y, \quad Ay \le b.$$

Set $z = \bigvee\{t \in A \mid t \wedge x = 0\}$. Then $Ax \vee z = 1$ and hence $y \vee z = 1$, and $x \wedge z = 0$. Now $Aa \wedge z = 0$ which is the same as $a \wedge Az = 0$ (both claiming that there is no u meeting both a and z). For $c = Az$ one now has

$$\mathfrak{k}(c) \vee \mathfrak{k}(b) = L \quad \text{and} \quad \mathfrak{k}(a) \cap \mathfrak{k}(c) = \{0\},$$

the first since $y \in \mathfrak{k}(a)$ and $z \in \mathfrak{k}(c)$, the second since otherwise $a \wedge c \ne 0$. This makes $\mathfrak{k}(a) \prec \mathfrak{k}(b)$ as desired.

We obviously have $v\mathfrak{k}(a) = a$ and $\mathfrak{k}v(I) \supseteq I$, hence the adjunction, and now proving that v is a frame homomorphism goes as in VII.4.1.2. We have to prove that $v_{(L,\mathcal{A})}$ is uniform, though (it is not automatic: the source is compact, not the target). For this we have to show that for any finite cover $I_1, \dots, I_n$ of $\widetilde{\mathfrak{R}}(L, \mathcal{A})$ the cover

$$c_1 = \bigvee I_1, c_2 = \bigvee I_2, \dots, c_n = \bigvee I_n$$

of L is uniform. Since $I_1 \vee \cdots \vee I_n = L$ there are $a_j \in I_j$ such that $a_1 \vee \cdots \vee a_n = 1$. Then there is an $A \in \mathcal{A}$ such that $Aa_j \le c_j$; obviously $A \le \{c_1, \dots, c_n\}$.

(2) The proof that $(v_{(L,\mathcal{A})})_{(L,\mathcal{A})}$ is a natural transformation is the same as VII.4.4. Finally, for L compact, v_L is an isomorphism, since if $a \in \mathfrak{k}(\bigvee L)$ then $a \lhd_{\mathcal{A}} \bigvee L$ and then $a \prec \bigvee L$ and we obtain $\mathfrak{k}v(I) = I$ as in VII.4.3. $\square$

1.2. Compact frames are complete and we have the dense uniform homomorphisms $v_{(L,\mathcal{A})} \colon \widetilde{\mathfrak{R}}(L, \mathcal{A}) \to (L, \mathcal{A})$. This looks as a sort of completion of $(L, \mathcal{A})$ but it is not (which is hardly a surprise). The trouble is that it is not a ue-surjection. In close scrutiny the problem is that the subsets

$$\mathcal{K}_A = \{\mathfrak{k}(a) \mid a \in A\} \subseteq \widetilde{\mathfrak{R}}(L), \quad A \in \mathcal{A}$$

naturally reflecting the uniform covers of L do not generally cover $\widetilde{\mathfrak{R}}(L)$.

This can be mended surprisingly easily: one can simply restrict the $\widetilde{\mathfrak{R}}(L)$ to the intersection of all the open sets corresponding to the

$$K_A = \bigvee \mathcal{K}_A = \bigvee\{\mathfrak{k}(a) \mid a \in A\}, \quad A \in \mathcal{A}.$$

This intersection, that is,

$$\bigcap_{A \in \mathcal{A}} \mathfrak{o}(K_A)$$

is already covered by all the $\mathcal{K}_A$, and we can hope for the best. The rest of this section will be devoted to proving that it really works.

1.3. We will use the factorization procedure from III.11. We have

$$\mathfrak{o}(K_A) = \widetilde{\mathfrak{R}}(L, \mathcal{A})/R_A \quad \text{with} \quad R_A = \{(J \cap K_A, J) \mid J \in \widetilde{\mathfrak{R}}(L, \mathcal{A})\}$$

and the desired intersection is

$$\widetilde{\mathfrak{R}}(L, \mathcal{A})/R, \quad R = \bigcup\{R_A \mid A \in \mathcal{A}\}.$$

The result (to be endowed by more structure shortly) will be denoted by

$$\widetilde{\mathbf{C}}(L, \mathcal{A}).$$

1.3.1. An easy but important observation. Obviously, if $A \leq B$ then $K_A \leq K_B$ in $\widetilde{\mathfrak{R}}(L, \mathcal{A})$. Recalling III.11.4 (note the saturatedness conditions) we see that

the result will be the same if we take instead of $\mathcal{A}$ any of its bases.

1.4. Recall the definition of a uniform embedding surjection (ue-surjection, briefly) from VIII.3.3.

1.4.1. Lemma. *Let* $h: (L, \mathcal{A}) \to (M, \mathcal{B})$ *be a dense ue-surjection, let* h_* *be its right Galois adjoint. Then for each* $b \in M$, $\widetilde{\mathfrak{R}}(h)(\mathfrak{k}(h_*(b)) = \mathfrak{k}(b)$.

Proof. Note that $h(h_*(b)) = b$ and $h_*(b)$ is the largest a with $h(a) = b$.

If $x \in \mathfrak{k}(a)$ then $h(x) \lhd_{\mathcal{B}} h(a)$, that is, $h(x) \in \mathfrak{k}(b)$ and hence

$$\widetilde{\mathfrak{R}}(h)(\mathfrak{k}(h_*(b)) \subseteq \mathfrak{k}(c).$$

Now let $y \in \mathfrak{k}(b)$, that is, $By \leq b$ for some $B \in \mathcal{B}$. Take an $A \in \mathcal{A}$ such that $B = h[A]$ and an x such that $h(x) = y$. Then

$$h(Ax) = h(\bigvee\{a \mid a \in A, \ a \wedge x \neq 0\}) = \bigvee\{h(a) \mid a \in A, \ h(a) \wedge h(x) \neq 0\}$$

since h is dense. Thus, $h(Ax) \leq h[A]y = By \leq b$, but this means that $Ax \leq h_*(b)$ and hence $x \in \mathfrak{k}(h_*(b))$, and finally $y = h(x) \in \widetilde{\mathfrak{R}}(h)(h_*(b))$. $\qquad\square$

1.4.2. Proposition. *Let* $h: (L, \mathcal{A}) \to (M, \mathcal{B})$ *be a dense ue-surjection. Then*

$$\widetilde{\mathfrak{R}}(h): \widetilde{\mathfrak{R}}(L, \mathcal{A}) \to \widetilde{\mathfrak{R}}(M, \mathcal{B})$$

is a frame isomorphism that, furthermore, restricts to an isomorphism $\widetilde{\mathbf{C}}(L, \mathcal{A}) \to \widetilde{\mathbf{C}}(M, \mathcal{B})$.

Proof. If $\widetilde{\mathfrak{R}}(h)(I) = \{0\}$, that is, $\downarrow h[I] = \{0\}$, we have by density of h that $I = \{0\}$. Thus, $\widetilde{\mathfrak{R}}(h)(I)$ is dense and since it is a homomorphism between compact regular frames, it is one-one. Further, for each regular ideal we have

$$I = \bigcup\{\mathfrak{k}(a) \mid a \in I\} = \bigvee\{\mathfrak{k}(a) \mid a \in I\}$$

and hence, to show that $\widetilde{\mathfrak{R}}(h)$ is onto it suffices to prove that each $\ell(b)$, $b \in M$, is an image and this follows from 1.4.1.

By 1.4.1 we have $\widetilde{\mathfrak{R}}(h)(K_{h_*[B]}) = K_B$ and hence $\widetilde{\mathfrak{R}}(M,\mathcal{B})$ is the copy of $\widetilde{\mathfrak{R}}(L,\mathcal{A}')$ where $\mathcal{A}' = \{h_*[B] \mid B \in \mathcal{B}\}$. Since $\mathcal{A}'$, by VIII.3.4, is a basis of $\mathcal{A}$, we have the second statement by the Observation 1.3.1. $\qquad\square$

1.5. $\widetilde{\mathbf{C}}(L,\mathcal{A})$ as a uniform frame. Denote by

$$\mu\colon \widetilde{\mathfrak{R}}(L,\mathcal{A}) \to \widetilde{\mathbf{C}}(L,\mathcal{A})$$

the sublocale map from III.11.2 associated with $\widetilde{\mathfrak{R}}(L,\mathcal{A})$. For the

$$v = v_{(L,\mathcal{A})}\colon \widetilde{\mathfrak{R}}(L,\mathcal{A}) \to (L,\mathcal{A})$$

we have $v(K_A \cap I) = \bigvee I = v(I)$ and hence there is a

$$\overline{v} = \overline{v}_{(L,\mathcal{A})}\colon \widetilde{\mathbf{C}}(L,\mathcal{A}) \to (L,\mathcal{A}) \quad \text{such that} \quad \overline{v}\mu = v.$$

We have again $\overline{v}(I) = \bigvee I$, and we see that

$$\overline{v} \text{ is dense.}$$

Further we have $\mu(K_A) = \mu(L) = 1$ and hence all the

$$\overline{\mathcal{K}}_A = \mu[\mathcal{K}_A]$$

are covers. From the obvious fact that $\ell(a) \cap \ell(b) \neq \{0\}$ iff $a \wedge b \neq 0$ we readily deduce that

$$\widetilde{\mathcal{A}} = \{\overline{\mathcal{K}}_A \mid A \in \mathcal{A}\}$$

is a basis of uniformity on the frame $\widetilde{\mathbf{C}}(L,\mathcal{A})$, which will be from now on viewed as the resulting uniform frame.

Observation. $\overline{v}\colon \widetilde{\mathbf{C}}(L,\mathcal{A}) \to (L,\mathcal{A})$ *is a dense ue-surjection.*

(Indeed, $\overline{v}\,\mu\,\ell(a) = a$ and hence $\overline{v}[\overline{\mathcal{K}}_A] = A$.)

Finally, we can complete 1.4.2 to

1.5.1. Proposition. *Let $h\colon (L,\mathcal{A}) \to (M,\mathcal{B})$ be a dense ue-surjection. Then the restriction $\phi\colon \widetilde{\mathbf{C}}(L,\mathcal{A}) \to \widetilde{\mathbf{C}}(M,\mathcal{B})$ is a uniform isomorphism making a commutative square*

$$
\begin{array}{ccc}
\widetilde{\mathbf{C}}(L,\mathcal{A}) & \xrightarrow{\;v_{(L,\mathcal{A})}\;} & (L,\mathcal{A}) \\
{\scriptstyle\phi=\widetilde{h}}\big\downarrow & & \big\downarrow{\scriptstyle h} \\
\widetilde{\mathbf{C}}(M,\mathcal{B}) & \xrightarrow[\;v_{(M,\mathcal{B})}\;]{} & (M,\mathcal{B})
\end{array}
$$

Proof. Use 1.4.1 and VIII.3.4 again. The frame isomorphism ϕ sends a basis of the uniformity of $\widetilde{\mathbf{C}}(L,\mathcal{A})$ into a basis of the uniformity of $\widetilde{\mathbf{C}}(M,\mathcal{B})$.   □

1.6. Theorem. $\bar{v}_{(L,\mathcal{A})}\colon \widetilde{\mathbf{C}}(L,\mathcal{A}) \to (L,\mathcal{A})$ *is a completion of* $(L,\mathcal{A})$.

Proof. We will write briefly L for $(L,\mathcal{A})$ and M for $(M,\mathcal{B})$.

I. First we show that $\widetilde{\mathbf{C}}(L)$ is complete. Let $f\colon L \to \widetilde{\mathbf{C}}(M)$ be a dense ue-surjection. Consider the commuting square as in 1.5.1

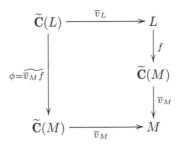

Set $g = \bar{v}_L \phi^{-1}$. Then $\bar{v}_M fg = \bar{v}_M f\bar{v}_L \phi^{-1} = v_M \phi\phi^{-1} = \bar{v}_M$ and since $\bar{v}_M$ is dense and therefore monomorphic, $fg = \mathrm{id}$. Then, as g is onto, we also have $gf = \mathrm{id}$, and f is an isomorphism.

II. Let L be complete and let $h\colon L \to M$ be a dense ue-surjection. Then in the commutative square

$$
\begin{array}{ccc}
\widetilde{\mathbf{C}}(L) & \xrightarrow{\ \bar{v}_L\ } & L \\
{\scriptstyle \phi=\tilde{h}}\downarrow & & \downarrow{\scriptstyle h} \\
\widetilde{\mathbf{C}}(M) & \xrightarrow[\ \bar{v}_M\ ]{} & M
\end{array}
$$

not only ϕ but also $\bar{v}_L$ is an isomorphism and we have the isomorphism $\psi = \phi(\bar{v}_L)^{-1}\colon L \to \widetilde{\mathbf{C}}(M)$ with $\bar{v}_M \psi = h$.   □

1.7. Thus, $\widetilde{\mathbf{C}}(L,\mathcal{A})$ is isomorphic to the completion $\mathbf{C}(L,\mathcal{A})$. But keep in mind that the construction $\mathbf{C}$ is substantially more general, since it works for general nearness frames. Here we need completely regular frames and uniformities, otherwise the compactification $\widetilde{\Re}$ would not work.

2.   Two applications

2.1. A uniformity is said to be *totally bounded* if it has a basis consisting of some of the finite covers.

2.1.1. Lemma. *Let a cover A have a finite star-refinement $B = \{b_1, \ldots, b_n\}$. Then $K_A = L$.*

Proof. Choose $a_i \in A$ with $b_i \vartriangleleft_A a_i$. Then $K_A \supseteq \bigvee_{i=1}^n \mathfrak{k}(a_i) \ni b_1 \vee \cdots \vee b_n = 1$. $\square$

2.1.2. Proposition. *A completely regular frame is compact iff it admits a totally bounded complete uniformity.*

Proof. The unique uniformity of a compact regular frame is complete (VIII.6.2) and obviously totally bounded.

   Now let a completely regular frame admit a totally bounded complete uniformity and let $\mathcal{A}$ be a basis of this uniformity consisting of some finite covers. Then each R_A is the trivial relation $\{(I, I) \mid I \in \widetilde{\mathfrak{R}}(L, \mathcal{A})\}$ and hence $\widetilde{\mathbf{C}}(L, \mathcal{A}) = \widetilde{\mathfrak{R}}(L, \mathcal{A})$ is compact. Since $(L, \mathcal{A})$ is complete, it is isomorphic with $\widetilde{\mathbf{C}}(L, \mathcal{A})$. $\square$

2.2. The following may come as a surprise.

Theorem. *The completion of any uniform frame with a countable basis of uniformity is spatial. Moreover, it is a dense sublocale of a compact Hausdorff space.*

Proof. The frame $\widetilde{\mathfrak{R}}(L, \mathcal{A})$ is compact regular and hence spatial by VII.6.3.4. Now we can choose the $\mathcal{A}$ in the construction countable, and hence the $\bigcap_{A \in \mathcal{A}} \mathfrak{o}(K_A)$ from 1.2 is a countable intersection of open sublocales. Now since L is a sublocale of $\bigcap_{A \in \mathcal{A}} \mathfrak{o}(K_A)$ and this is sublocale of $\widetilde{\mathfrak{R}}(L, \mathcal{A})$, and since L is dense in $\widetilde{\mathfrak{R}}(L, \mathcal{A})$, all the $\mathfrak{o}(K_A)$ are dense in $\widetilde{\mathfrak{R}}(L, \mathcal{A})$ and we can apply VII.7.3 to obtain that $\widetilde{\mathbf{C}}(L, \mathcal{A}) = \bigcap_{A \in \mathcal{A}} \mathfrak{o}(K_A)$ is spatial. $\square$

2.2.1. If $(L, \mathcal{A})$ has a countable basis than so has the $\widetilde{\mathbf{C}}(L, \mathcal{A})$. From the classical Uniformity Metrization Theorem (which will be also discussed later in detail in Chapter X) we now obtain

Corollary. *Each uniform frame with countable basis of uniformity can be embedded into a metric space as a dense sublocale.* $\square$

2.2.2. Notes. (1) The $(L, \mathcal{A})$ with a countable basis itself can be, however, far from spatial itself. Realize that, because of the unicity of completion, a complete uniform frame is a completion of any of its dense uniform sublocales. Thus for instance the Booleanizations (see AI.7.6) of compact metric spaces, typically lacking points.

(2) In the previous remark we mentioned compact metric spaces, about which we already know that they are complete. In fact every metric space complete in the classical sense is complete as a uniform frame; this will be another important consequence of 2.2 (see 5.3 below).

3.   Cauchy points and the resulting space

3.1. Viewing a point in generalized geometry as an entity with position but no extent was felt as somewhat unsatisfactory a long time before there was anything like a point-free topology. In fact, the pioneering paper by Carathéodory ([59], 1913), in which points are represented as systems of diminishing places, even preceded by one year Hausdorff's foundation of "pointy" topology in [125]. This idea reappeared again and again; perhaps the most relevant treatise was presented by Freundenthal ([105], 1942).

In a general frame, the first that comes to mind is the idea of a filter. Only, we lack the means to express the intuitive idea of containing arbitrarily small elements. For this purpose we can use the enrichment of the structure of frame by a uniformity $\mathcal{A}$ (or just nearness). Thinking of the individual covers $A \in \mathcal{A}$ as "measures of fineness", or "granulations", we can express the desired granular diminishing by the notion of *Cauchy filter* (more precisely, $\mathcal{A}$-*Cauchy filter*), namely a filter F such that

$$\forall A \in \mathcal{A}, \quad F \cap A \neq \emptyset \tag{Cau}$$

(that is, however fine a granulation we choose, there will be an element of F as small as the desired size of the grains).

This is still not quite satisfactory. Consider the following negative example. On the real line take the filters

$$F_+ = \{U \mid U \supseteq (0, \varepsilon) \text{ for some } \varepsilon > 0\}$$

and

$$F_- = \{U \mid U \supseteq (-\varepsilon, 0) \text{ for some } \varepsilon > 0\}.$$

Both of them are Cauchy with respect to the standard metric uniformity. But there is something wrong.

(1) We would like to think of the elements of the filter representing an abstract point as of its neighbourhoods. But some of the elements of F_+ resp. F_- do not surround 0, the only candidate for their abstract limit.

(2) One point is represented by different ends.

To avoid this we will add one more condition, the $\mathcal{A}$-*regularity*. A filter is $\mathcal{A}$-*regular* (briefly, *regular*, but there is a certain danger of confusion and one has to be careful) if

$$\forall a \in F \; \exists b \in F, \quad b \triangleleft_{\mathcal{A}} a. \tag{$\mathcal{A}$-reg}$$

3.2. Thus we have come to the concept of a

Cauchy point in $(L, \mathcal{A})$ as a regular Cauchy filter.

In a uniform frame we can approximate any Cauchy filter by a Cauchy point. Set

$$F^\circ = \{b \mid \exists a \in F, \; a \triangleleft_{\mathcal{A}} b\}.$$

We have

3.2.1. Lemma. *Let F be a Cauchy filter in a uniform frame $(L, \mathcal{A})$. Then F° is a Cauchy point.*

Proof. For $A \in \mathcal{A}$ take a $B \in \mathcal{A}$ such that $BB \leq A$. Take a $b \in F \cap B$ and an $a \in A$ with $Bb \leq a$. Then $a \in F^\circ \cap A$. Further, in the uniform case, $\lhd_{\mathcal{A}}$ interpolates, and hence F° is regular.   □

3.2.2. The condition of regularity obviously disposes with the objection (1) above. But it makes for the unicity as well. We have

Proposition. *A regular Cauchy filter is a minimal Cauchy filter. In case of a uniformity, the regular Cauchy filters are precisely the minimal ones.*

Proof. Let F be a regular Cauchy filter and let $G \subseteq F$ be a Cauchy one. For an $x \in F$ choose an $A \in \mathcal{A}$ and a $y \in F$ such that $Ay \leq x$. There is an $a \in G \cap A$; as $a \in G \subseteq F$, we have $a \wedge y \neq 0$ and hence $a \leq Ay \leq x$, and $x \in G$.

If F is a minimal Cauchy filter in a uniform frame, consider the $F^\circ \subseteq F$, which is now a Cauchy filter. Hence $F^\circ = F$ and F is regular.   □

3.2.3. Remarks. (1) We already have met with a concept of a point, namely an element of the spectrum ΣL (II.4). Note that in a regular space the spectrum points, the completely prime filters, are precisely the Cauchy points with respect to the fine nearness. Indeed: if F is completely prime and A is an arbitrary cover then $\bigvee A = 1 \in F$ and hence there is an $a \in A \cap F$; further, if $a \in F$ take the join

$$a = \bigvee \{b \mid b \prec a\} = \bigvee \{b \mid b \lhd a\}$$

to obtain by the complete primeness a $b \lhd a$ with $b \in F$. On the other hand, if F is a Cauchy point with respect to the fine nearness and if $a = \bigvee a_i \in F$ consider a $b \lhd a$ in F and the cover $1 = b^* \vee \bigvee a_i$ to obtain an $a_i \in F$ (b^* is certainly not in F since b is).

(2) Thinking of "surrounding" the limiting point would be, more precisely, expressed by *regularity* ($\forall a \in F \; \exists b \in F, \; b \prec a$). In general, the $\mathcal{A}$-regularity is a slightly stronger requirement. It is technically much more expedient for our purposes, and in view of VIII.2.8.2 the difference is marginal.

3.3. We will denote by

$$\Psi(X, \mathcal{A})$$

the set of all the Cauchy points in $(X, \mathcal{A})$. It will be endowed, first, by the topology defined by the basis

$$\{\Psi_u \mid u \in L\} \quad \text{where} \quad \Psi_u = \{P \in \Psi(L, \mathcal{A}) \mid u \in P\}$$

(it is indeed a basis of a topology: we obviously have $\Psi_u \cap \Psi_v = \Psi_{u \wedge v}$). In the following section we will make $\Psi(L, \mathcal{A})$ to a uniform space.

3.4. Enough Cauchy points. We will say that a uniform frame *has enough Cauchy points* if

$$\forall a, b \in L \ (a \not\leq b \Rightarrow \exists P \in \Psi(L, \mathcal{A}), \ b \notin P \ni a). \tag{ECP}$$

In other words, if the Cauchy points separate the elements of L; or, since $a \leq b$ implies that $\Psi_a \subseteq \Psi_b$, if

$$a \leq b \quad \equiv \quad \Psi_a \subseteq \Psi_b \tag{ECP1}$$

(note that the system $\{\Psi_u \mid u \in L\}$ is only a basis, not the whole of the topology, and that we do not generally have $\bigcup \Psi_{a_i} = \Psi_{\bigvee a_i}$; hence, (EPC1) cannot be interpreted as that the mapping $a \mapsto \Psi_a$ is an isomorphism).

3.4.1. Cauchy points can abound even when the spectral points are scarce. Here are a few facts showing how rich the class of nearness frames having enough points is.

We will start with a trivial

Observation. *Any open nearness sublocale of a nearness frame with enough Cauchy points has enough Cauchy points.*

(Represent the sublocale as the $\downarrow u$ (recall III.6.1.1) with the uniformity $\mathcal{A}' = \{A \wedge \{u\} \mid A \in \mathcal{A}\}$. Then we can separate $a \not\leq b$ by the $F \wedge \{u\}$ where F separates them in $(L, \mathcal{A})$.)

The following is more important.

3.4.2. Proposition. *Every nearness frame $(L, \mathcal{A})$ with a countable basis of $\mathcal{A}$ has enough Cauchy points.*

Proof. Let $\{A_n \mid n \in \mathbb{N}\}$ be a basis of $\mathcal{A}$. We can assume that

$$A_1 \geq A_2 \geq \cdots \geq A_n \geq \cdots .$$

Let $a \not\leq b$. Since $a = \bigvee \{a' \wedge c \mid a' \lhd_\mathcal{A} a, \ c \in A_1\}$ there is an a_1 and a $c_1 \in A_1$ such that

$$a_1 \lhd_\mathcal{A} a, \quad a_1 \leq c_1 \ \text{ and } \ a_1 \not\leq b.$$

Now let us have $a_i, c_i, \ i = 1, 2, \ldots, n$, such that

$$a_i \lhd_\mathcal{A} a_{i-1}, \quad a_i \leq c_i \in A_i \ \text{ and } \ a_i \not\leq b.$$

Since $a_n = \bigvee \{a' \wedge c \mid a' \lhd_\mathcal{A} a_n, \ c \in A_n\}$ there is an a_{n+1} and a c_{n+1} such that

$$a_{n+1} \lhd_\mathcal{A} a_n, \quad a_{n+1} \leq c_{n+1} \in A_{n+1} \ \text{ and } \ a_{n+1} \not\leq b.$$

The set

$$F = \{x \mid \exists n, \ x \geq a_n\}$$

is obviously a Cauchy point containing a but not b. $\qquad\square$

Note. Thus for instance the Booleanizations (recall AI.7.6) of metric spaces have enough Cauchy points although the spectra are often (for instance if the space is connected) void.

3.4.3. Proposition. *Let* $(L, \mathcal{A})$ *have enough Cauchy points, and let*

$$p \colon (L, \mathcal{A}) \to (M, \mathcal{B})$$

be a dense ue-surjection. Then $(M, \mathcal{B})$ *has enough Cauchy points.*

Proof. Let $\phi \colon M \to L$ be the right adjoint of p. Thus, since p is onto,

$$p(\phi(y)) = y \quad \text{and} \quad x \le \phi(p(x)).$$

In particular,

$$\phi(y) \text{ is the largest element } x \text{ of } L \text{ such that } p(x) = y. \tag{$*$}$$

Now let $a \not\le b$ in M. Then $\phi(a) \not\le \phi(b)$ (else $a = p\phi(a) \le p\phi(b) = b$), and hence there is a Cauchy point F in L such that $\phi(a) \in F$ and $\phi(b) \notin F$. Consider $G = {\uparrow}p[F]$. We have $p(x) \wedge p(y) = p(x \wedge y)$ and by density $0 \notin G$ so that G is a proper filter. If $p(x) \in G$ with $x \in F$ and $y \lhd x$ with a $y \in F$ we have $p(y) \lhd p(x)$ and G is regular. For a cover $B = p[A]$ of M, $A \in \mathcal{A}$, we have an $x \in A \cap F$ and hence $p(x) \in B \cap G$. Thus, G is a Cauchy point.

Now obviously $a = p\phi(a) \in F$. If we had $b = p(c) \in G$ for some $c \in F$ we would have by $(*)$ $c \le \phi(b)$ and $\phi(b) \in F$, a contradiction. $\qquad\square$

3.4.4. Proposition. *Let* $(L_i, \mathcal{A}_i)$, $i \in J$, *have enough Cauchy points. Then*

$$(L, \mathcal{A}) = \bigoplus_{i \in J} (L_i, \mathcal{A}_i)$$

has enough Cauchy points.

Proof. It suffices to prove that if $\oplus a_i \not\le \oplus b_i$ then there is a Cauchy point F in $(L, \mathcal{A})$ such that $\oplus a_i \in F$ and $\oplus b_i \notin F$. Since $\oplus a_i \not\le \oplus b_i$ we have a $k \in J$ such that $a_k \not\le b_k$, and $a_i \ne 0$ for all i. Now choose a Cauchy point F_k in $(L_k, \mathcal{A}_k)$ such that $b_k \notin F \ni a_k$, and for $i \ne j$ choose arbitrarily Cauchy points F_i containing a_i. Set

$$F = \{u \mid \exists x_i \in F_i,\ \oplus x_i \subseteq u\}.$$

Obviously F is a regular filter. It is Cauchy: take an $\bigoplus A_i$ as in VIII.3.5.1 with $A_i = \{1\}$ for $i \notin K$ where K is a finite subset of J. For $i \in K$ choose $x_i \in F_i \cap A_i$, otherwise set $x_i = 1$. Then $\oplus x_i \in F \cap \bigoplus A_i$.

Finally obviously $\oplus a_i \in F$ and $\oplus b_i \notin F$. $\qquad\square$

4. Cauchy spectrum

4.1. Recall the basic open sets Ψ_a of the topology of $\Psi(L, \mathcal{A})$ from 3.3. For a cover A of L set

$$\Psi_A = \{\Psi_a \mid a \in A\}.$$

Now for the whole of $\mathcal{A}$ write

$$\Psi_{\mathcal{A}} = \{\Psi_A \mid A \in \mathcal{A}\}.$$

4.1.1. Lemma. (1) *Each Ψ_A is a cover of $\Psi(L, \mathcal{A})$.*

(2) *For an $A \in \mathcal{A}$ and $b \in L$ we have*

$$\Psi_A \Psi_b \subseteq \Psi_{Ab};$$

consequently $\Psi_A \Psi_B \leq \Psi_{AB}$.

(3) *For any two $A, B \in \mathcal{A}$, $\Psi_A \wedge \Psi_B = \Psi_{A \wedge B}$.*

(4) *$\Psi_{\mathcal{A}}$ is an admissible system of covers on $\Psi(L, \mathcal{A})$.*

Proof. (1) A point $P \in \Psi(L, \mathcal{A})$ is a Cauchy filter and hence there is an $a \in P \cap A$. Then $P \in \Psi_a$.

(2) If $\Psi_a \cap \Psi_b = \Psi_{a \wedge b} \neq \emptyset$ then necessarily $a \wedge b \neq 0$ so that $a \leq Ab$ and hence $\Psi_a \subseteq \Psi_{Ab}$.

(3) $\Psi_A \wedge \Psi_B = \{\Psi_a \cap \Psi_b = \Psi_{a \wedge b} \mid a \in A, b \in B\} = \{\Psi_c \mid c \in A \wedge B\} = \Psi_{A \wedge B}$.

(4) It suffices to prove that for any $a \in L$,

$$\Psi_a = \bigcup \{\Psi_x \mid \Psi_x \lhd_{\Psi_{\mathcal{A}}} \Psi_a\}. \tag{$*$}$$

We have $\Psi_a = \{P \in \Psi(L, \mathcal{A}) \mid P \ni a\}$. Choose an $x \in P$ with $x \lhd_{\mathcal{A}} a$. Then $P \in \Psi_x$. By (2),

$$x \lhd_{\mathcal{A}} a \implies \Psi_x \lhd_{\Psi_{\mathcal{A}}} \Psi_a,$$

hence $\Psi_x \lhd_{\Psi_{\mathcal{A}}} \Psi_a$ and we conclude the equality $(*)$; note that this is *not a direct consequence* of $a = \bigvee \{x \mid x \lhd_{\mathcal{A}} a\}$: the mapping $a \mapsto \Psi_a$ in general does not preserve suprema. $\qquad\square$

4.1.2. Corollary. *Let $\mathcal{A}$ be a nearness resp. uniformity on L. Then $\Psi_{\mathcal{A}}$ is a nearness resp. uniformity on $\Psi(L, \mathcal{A})$.* $\qquad\square$

From now on,

we will regard $\Psi(L, \mathcal{A})$ as endowed by the nearness resp. uniformity $\Psi_{\mathcal{A}}$.

4.2. Completeness of spaces. By a nearness space we understand (of course) just the relaxation of the concept of a uniform space by omitting the star refinement requirement (in the standard classical literature – see [128] – one speaks of a *regular nearness*; the term nearness is sometimes used for a still more general notion [126]).

For a nearness space $(X, \mathfrak{U})$ define a mapping

$$\varepsilon_{(X, \mathfrak{U})} \colon (X, \mathfrak{U}) \to \Psi\Omega(X, \mathfrak{U})$$

by setting

$$\varepsilon_{(X, \mathfrak{U})}(x) = \mathcal{U}(x) = \{U \text{ open} \mid x \in U\}.$$

Since $\mathcal{U}(x)$ is completely prime it is indeed a Cauchy point (recall 3.2.3) but it is also directly seen from the definition of the topology induced by a uniformity (resp. nearness).

4.2.1. Proposition. $\varepsilon_{(X, \mathfrak{U})}$ *is a dense ue-surjection.*

Proof. Obviously ε is one-one (the space is T_0). We have

$$\varepsilon^{-1}[\Psi_U] = \{x \in U \mid \mathcal{U}(x) \in \Psi_U\} = \{x \mid x \in U\} = U \tag{*}$$

and hence the topology of X consists precisely of the preimages of the basis of the other space, and ε is a topological embedding.

In fact, by (*) again we have for each $\Psi_{\mathcal{U}} \in \Psi_{\mathfrak{U}}$,

$$\{\varepsilon^{-1}[\Psi_U] \mid \Psi_U \in \Psi_{\mathcal{U}}\} = \mathcal{U}$$

and hence ε is a ue-surjection.

Finally, it is dense: if $\Psi_U \neq \emptyset$ then necessarily $U \neq \emptyset$ and $\mathcal{U}(x) \in \Psi_U$ for any $x \in U$. □

4.2.2. A classical definition of the completeness of a uniform space $(X, \mathfrak{U})$ is that each Cauchy regular filter of open sets (that is, in our terminology, a Cauchy point in $\Omega(X, \mathfrak{U})$) has a center; that is, it is $\mathcal{U}(x)$ for some $x \in X$.

This leads to the following definition. A uniform (or, more generally, nearness) frame $(L, \mathcal{A})$ is said to be *Cauchy complete* if each Cauchy point is completely prime (note that for $(L, \mathcal{A}) = \Omega(X, \mathfrak{U})$ this is precisely the requirement above: such X is regular T_0 and hence sober so that each completely prime filter is some of the $\mathcal{U}(x)$).

Thus, $(L, \mathcal{A})$ is Cauchy complete if the Cauchy points are precisely the prime filters: the other implication is automatic – recall 3.2.3.

This completeness of spaces is equivalent with the statement we have imitated in VIII.6.1 to obtain a definition of a complete uniform frame.

We have

4.2.3. Proposition. *The following statements on a nearness space $(X, \mathfrak{U})$ are equivalent.*

(1) *Each dense ue-surjection $(X, \mathfrak{U}) \to (Y, \mathfrak{V})$ is an isomorphism (uniform homeomorphism).*

(2) *$\varepsilon_{(X,\mathfrak{U})}$ is an isomorphism.*

(3) *$\Omega(X, \mathfrak{U})$ is Cauchy complete.*

Proof. (1)$\Rightarrow$(2) follows from 4.2.1.

(2)$\Rightarrow$(3): If ε is an isomorphism then each Cauchy point of $(X, \mathfrak{U})$ is a $\mathcal{U}(x)$, an obviously completely prime filter.

(3)$\Rightarrow$(1): Let $(X, \mathfrak{U}) \subseteq (Y, \mathfrak{V})$ be a dense nearness subspace. Pick an arbitrary $y \in Y$ and set

$$F = \{U \cap X \mid y \in U \in \Omega(Y)\}.$$

F is a Cauchy point:

First, by density it is a proper filter. Next, for a cover $\mathcal{U} \in \mathfrak{U}$ there is a $\mathcal{V} \in \mathfrak{V}$ such that $\mathcal{U} = \{V \cap X \mid V \in \mathcal{V}\}$. Take a

$$V \in \{W \in \Omega(Y) \mid W \ni x\} \cap \mathcal{V}.$$

Then $V \cap X \in \mathcal{U} \cap F$ and we see that F is a Cauchy filter. If $y \in U \in \Omega(Y)$ we have a $\mathcal{W} \in \mathfrak{V}$ and $V \in \Omega(Y)$ such that (in the notation of VIII.1.1) $\mathcal{W} * V \subseteq U$. Now $\mathcal{W}' = \{W \cap X \mid W \in \mathcal{W}\}$ is in $\mathfrak{U}$ and we have

$$\mathcal{W}' * (V \cap X) \subseteq (\mathcal{W} * V) \cap X \subseteq U \cap X$$

and we see that F is regular.

Thus, $F = \mathcal{U}(x)$ for some $x \in X$. Since Y is in particular a T_1-space we conclude that $y = x \in X$. $\qquad\square$

Note. Thus, $(X, \mathfrak{U})$ is complete in the classical sense if $\Omega(X, \mathfrak{U})$ is Cauchy complete while $\Omega(X, \mathfrak{U})$ being complete in the sense of VIII.6 is in general a considerably stronger property (no wonder; there are much more generalized spaces to embed into to test the property). We will meet below in 5.3 an important class of spaces in which the two properties coincide, though. In his pioneering article [136], Isbell proposed to call the $(X, \mathfrak{U})$ with $\Omega(X, \mathfrak{U})$ complete as uniform frames *hypercomplete* but this terminology is not much used.

4.2.4. Theorem. *If $(L, \mathcal{A})$ is a uniform frame, or if it has enough points, then $\Psi(L, \mathcal{A})$ is Cauchy complete.*

Proof. Let $\mathcal{F}$ be a Cauchy point in $\Psi(L, \mathcal{A})$. Set

$$F = \{a \mid \Psi_a \in \mathcal{F}\}.$$

Then F is a Cauchy filter: indeed, if $A \in \mathcal{A}$ consider the Ψ_A and a $U \in \Psi_A \cap \mathcal{F}$. Then $U = \Psi_a$ for an $a \in A$ and as this Ψ_a is in $\mathcal{F}$ finally $a \in F$.

I. If $(L, \mathcal{A})$ is uniform take the Cauchy point F° (see 3.2.1). If $U \ni F^\circ$ choose a Ψ_a such that $U \supseteq \Psi_a \ni F^\circ$. Now $a \in F^\circ \subseteq F$, hence $a \in F$ and $\Psi_a \in \mathcal{F}$ and finally $U \in \mathcal{F}$. Thus, $\mathcal{U}(F^\circ) \subseteq \mathcal{F}$ and $\mathcal{U}(F^\circ) = \mathcal{F}$ by 3.2.2.

II. If $(L, \mathcal{A})$ has enough Cauchy points then F itself is already regular and hence a Cauchy point. Indeed, let $a \in F$, that is, $\Psi_a \in \mathcal{F}$. Consider a $\Psi_b \in \mathcal{F}$ such that $\Psi_A \Psi_b \subseteq \Psi_a$ for some $A \in \mathcal{A}$. If $c \in A$ and $c \wedge b \neq 0$ then there is a Cauchy point P with $c \wedge b \in P$, and hence $P \in \Psi_c \cap \Psi_b$ proving that $\Psi_c \subseteq \Psi_a$ so that by (ECP1) in 3.4, $c \leq a$ and we see that $Ab \leq a$.

Now $\mathcal{U}(F) \subseteq \mathcal{F}$ for the same reason as in I, only more directly: if $U \supseteq \Psi_a \ni F$ then $a \in F$ and $\Psi_a \in \mathcal{F}$. $\square$

4.3. Ψ as a functor. From now on we will work with uniform frames and spaces only.

4.3.1. Lemma. *Let $h \colon (L, \mathcal{A}) \to (M, \mathcal{B})$ be a uniform homomorphism. If F is a $\mathcal{B}$-Cauchy filter then $h^{-1}[F]$ is an $\mathcal{A}$-Cauchy filter.*

Proof. Obviously $h^{-1}[F]$ is a proper filter. If $A \in \mathcal{A}$ then $h[A] \in \mathcal{B}$ and hence there is a $b \in F \cap h[A]$; now for any $a \in A$ with $h(a) = b$ we have $a \in h^{-1}[F] \cap A$. $\square$

4.3.2. Consequently, using also 3.2.2, we see that we can define a mapping

$$\Psi h \colon \Psi(M, \mathcal{B}) \to \Psi(L, \mathcal{A})$$

by setting

$$\Psi h(F) = (h^{-1}[F])^\circ.$$

Proposition. *Ψh is a uniformly continuous mapping $\Psi(M, \mathcal{B}) \to \Psi(L, \mathcal{A})$. Further,*

$$\Psi(\mathrm{id}) = \mathrm{id} \quad and \quad \Psi(gh) = \Psi h \cdot \Psi g.$$

Thus we obtain a contravariant functor

$$\Psi \colon \mathbf{UniFrm} \to \mathbf{UniSp};$$

by 4.2.4 we can also view it as a functor

$$\Psi \colon \mathbf{UniFrm} \to \mathbf{CUniSp}$$

where $\mathbf{CUniSp}$ is the full subcategory of $\mathbf{UniSp}$ generated by the complete uniform spaces.

Proof. We have

$$(\Psi h)^{-1}[\Psi_a] = \{P \in \Psi(M, \mathcal{B}) \mid a \in h^{-1}[P]\}$$
$$= \{P \in \Psi(M, \mathcal{B}) \mid \exists x, \; h(x) \in P \text{ and } x \lhd_{\mathcal{A}} a\} \qquad (*)$$
$$= \bigcup\{\Psi_{h(x)} \mid x \lhd_{\mathcal{A}} a\}.$$

This union is an open set and hence Ψh is continuous.

Now let $A \in \mathcal{A}$. Choose a $C \in \mathcal{A}$ such that $CC \leq A$ and set $B = h[C]$. For a $c \in C$ there is an $a \in A$ with $c \lhd_{\mathcal{A}} a$ and, by $(*)$, $(\Psi h)^{-1}[\Psi_a] \supseteq \Psi_{h(c)}$, and we obtain

$$\{(\Psi h)^{-1}[\Psi_a] \mid a \in A\} \geq \{\Psi_b \mid b \in B\}.$$

Thus, Ψh is uniformly continuous.

Finally,

$$\Psi h(\Psi g(F)) = (h^{-1}(g^{-1}[F])^\circ)^\circ \subseteq ((gh)^{-1}[F])^\circ = \Psi(gh)(F)$$

and we infer the equality from 3.2.2. $\qquad\square$

4.4. Cauchy spectrum. Define mappings

$$\tau_{(L,\mathcal{A})} \colon \Omega\Psi(L, \mathcal{A}) \to (L, \mathcal{A})$$

by setting

$$\tau(U) = \bigvee\{x \in L \mid \Psi_x \subseteq U\}.$$

4.4.1. Proposition. *If $(L, \mathcal{A})$ has enough Cauchy points then $\tau_{(L,\mathcal{A})}$ is a uniform homomorphism.*

Proof. First, τ preserves all finite meets: obviously $\tau(\Psi(L, \mathcal{A})) = 1$, and

$$\tau(U) \wedge \tau(V) = \bigvee\{x \wedge y \mid \Psi_x \subseteq U, \Psi_y \subseteq V\}$$
$$= \bigvee\{z \mid \Psi_z \subseteq U \cap V\} = \tau(U \cap V).$$

Now define a mapping $\psi \colon (L, \mathcal{A}) \to \Omega\Psi(L, \mathcal{A})$ by setting $\psi(a) = \Psi_a$. For each open $U \subseteq \Psi(L, \mathcal{A})$ and a basic $\Psi_x \subseteq U$, $x \leq \tau(U)$ and hence $\psi(x) = \Psi_x \subseteq \psi\tau(U)$ proving that

$$U \subseteq \psi\tau(U). \qquad (\subseteq)$$

On the other hand, since we have enough Cauchy points we have by (ECP1) in 3.4

$$\tau\psi(a) = \bigvee\{x \mid \Psi_x \subseteq \Psi_a\} = \bigvee\{x \mid x \leq a\} = a$$

and hence τ is a left Galois adjoint and consequently it preserves all joins.

Now for a basic uniform cover Ψ_A of $\Psi(L, \mathcal{A})$ (with $A \in \mathcal{A}$) we have

$$\tau[\Psi_A] = \{\tau(\Psi_a) \mid a \in A\} = \{\tau\psi(a) \mid a \in A\} = A$$

and hence τ is a uniform homomorphism. $\qquad\square$

4.4.2. Denote by
$$\mathbf{UniFrm_{EC}}$$
the full subcategory of **UniFrm** generated by the uniform frames with enough Cauchy points; further, recall that for complete uniform spaces $(X, \mathfrak{U})$ the mapping $\varepsilon_{(X,\mathfrak{U})}$ is an isomorphism (4.2.3).

Theorem. *Restrict the functors Ψ and Ω to*

$$\Psi \colon \mathbf{UniFrm_{EC}} \to \mathbf{CUniSp} \quad and \quad \Omega \colon \mathbf{CUniSp} \to \mathbf{UniFrm_{EC}}.$$

Then these contravariant functors are adjoint on the left with the adjunction units $\tau = (\tau_{(L,\mathcal{A})})_{(L,\mathcal{A})}$ and $\varepsilon^{-1} = (\varepsilon^{-1}_{(X,\mathfrak{U})})_{(X,\mathfrak{U})}$.

Proof. I. First we show that τ and ε are natural transformations.

For $h \colon (L, \mathcal{A}) \to (M, \mathcal{B})$ we have, by the $(*)$ in 4.3.2,

$$\tau(\Omega\Psi h(\Psi_a)) = \tau((\Psi h)^{-1}[\Psi_a]) = \tau(\bigcup\{\Psi_{h(x)} \mid x \vartriangleleft_{\mathcal{A}} a\})$$
$$= \bigvee\{y \in M \mid \Psi_y \subseteq \bigcup\{\Psi_{h(x)} \mid x \vartriangleleft_{\mathcal{A}} a\}\} = c.$$

We will compute the join c. Since obviously $\bigcup\{\Psi_{h(x)} \mid x \vartriangleleft_{\mathcal{A}} a\} \subseteq \Psi_{h(a)}$ we obtain using the (ECP1) from 3.4

$$c \leq \bigvee\{y \mid \Psi_y \subseteq \Psi_{h(a)}\} = \bigvee\{y \mid y \leq h(a)\} = h(a).$$

On the other hand, $h(a) = \bigvee\{h(x) \mid x \vartriangleleft_{\mathcal{A}} a\}$ and each $z = h(x)$ with $x \vartriangleleft_{\mathcal{A}} a$ appears in the join defining c; hence also $h(a) \leq c$. Thus

$$\tau(\Omega\Psi h(\Psi_a)) = h(a) = h(\tau\psi(a)) = h\tau(\Psi a).$$

For a uniformly continuous $f \colon (X, \mathfrak{U}) \to (Y, \mathfrak{V})$ we have

$$\Psi\Omega(f)(\varepsilon(x)) = (\Omega(f)^{-1}[\mathcal{U}(x)])^\circ = \{V \mid f^{-1}[V] \ni x\}^\circ$$
$$= \{V \mid V \ni f(x)\}^\circ = \{V \mid V \ni f(x)\} = \varepsilon(f(x))$$

so that ε is a natural transformation, in our case (recall 4.2.3) a natural equivalence, and hence ε^{-1} is a natural equivalence as well.

II. Now consider the composition

$$\Omega(X, \mathcal{U}) \xrightarrow{\Omega(\varepsilon^{-1}_{(X,\mathcal{U})})} \Omega\Psi\Omega(X, \mathcal{U}) \xrightarrow{\tau_{\Omega(X,\mathcal{U})}} \Omega(X, \mathcal{U}).$$

Since $(X, \mathcal{U})$ is complete we have $\Psi_V = \{\mathcal{U}(x) \mid x \in V\} = \varepsilon_{(X,\mathcal{U})}[V]$. Thus,

$$\tau(\Omega(\varepsilon^{-1})(U)) = \tau(\varepsilon[U]) = \bigcup\{V \in \Omega(X) \mid \Psi_V \subseteq \varepsilon[U]\}$$
$$= \bigcup\{V \in \Omega(X) \mid \varepsilon[V] \subseteq \varepsilon[U]\} = \bigcup\{V \in \Omega(X) \mid V \subseteq U\} = U$$

and the composition is the identity.

III. Finally consider

$$\Psi(L,\mathcal{A}) \xrightarrow{\Psi(\tau_{(L,\mathcal{A})})} \Psi\Omega\Psi(L,\mathcal{A}) \xrightarrow{\varepsilon^{-1}_{\Psi(X,\mathcal{A})}} \Psi(L,\mathcal{A}).$$

We have $\tau(\Psi_a) = \tau\psi(a) = a$ and hence

$$\tau^{-1}[P] = \{\Psi_a \mid a \in P\} = \{\Psi_a \mid P \in \Psi_a\} \subseteq \{U \mid P \in U\} = \mathcal{U}(P),$$

hence by 3.2.2

$$\varepsilon^{-1}(\Psi(\tau)(P)) = \varepsilon^{-1}(\tau^{-1}[P]^\circ) = \varepsilon^{-1}(\mathcal{U}(x)) = \varepsilon^{-1}(\varepsilon(P)) = P$$

and the composition is an identity again. $\qquad\square$

4.4.3. Note. If we interpret the adjunction above localically (that is, replacing the category of uniform frames by its dual of uniform locales), the functor Ψ becomes a *left* adjoint and Ω becomes the *right* one. This contrasts with the standard spectrum adjunction where Ω is the *left* adjoint.

Not very surprisingly: the functor Σ is after all sufficiently different from the Ψ. Perhaps there is some more surprise in the fact that another restriction of the same functors, namely replacing the category **UniFrm$_{\text{EC}}$** by that of Cauchy complete uniform frames produces an adjunction with Ψ to the right and Ω to the left, with the adjunction units ε and

$$\psi = \Big(\psi_{(L,\mathcal{A})}\colon (L,\mathcal{A}) \to \Omega\Psi(L,\mathcal{A})\Big)_{(L,\mathcal{A})}$$

consisting of the maps already used in the proof above (see [42]).

This seemingly paradoxical behaviour is perhaps of some interest. It has to do with the following. The (systems of) maps τ and ψ have in general the property

$$a \leq \tau\psi(a) \quad \text{and} \quad U \leq \psi\tau U. \tag{$\leq$}$$

Now in the case of enough points the first collapses to equality. Then $(\leq)$ makes τ the left Galois adjoint and ψ the right one, while in the Cauchy complete case we have the equality $U \leq \psi\tau U$ with the converse result.

5. Cauchy completion.
The case of countably generated uniformities

5.1. Lemma. *Let* $(X,\mathfrak{U})$, $(Y,\mathfrak{V})$ *be uniform spaces. Then a uniform homomorphism* $h\colon \Omega(Y,\mathfrak{V}) \to \Omega(X,\mathfrak{U})$ *is a dense ue-surjection iff* $h = \Omega(f)$ *for a dense uniform embedding* $f\colon (X,\mathfrak{U}) \to (Y,\mathfrak{V})$ *in the classical sense.*

Proof. Since the spaces are sober, $h = \Omega(f)$ for a (unique) continuous map $f\colon (X,\mathfrak{U}) \to (Y,\mathfrak{V})$. Obviously f is one-one (consider $f(x) = f(y)$ with $x \neq y$ and an open U containing x but not y; if $U = f^{-1}[V]$ then $f(y) = f(x) \in V$

and $f(y) \notin V$, a contradiction). All the other requirements can be reformulated directly in $\Omega(f)$. □

5.2. Proposition. *Let $(X, \mathcal{A})$ have enough Cauchy points. Then the mapping*

$$\tau_{(L,\mathcal{A})} \colon \Omega\Psi(L, \mathcal{A}) \to (L, \mathcal{A})$$

is a dense ue-surjection.

Proof. We already know that it is a uniform homomorphism. Recall the mapping ψ from the proof of 4.4.1. Since $\tau\psi$ is the identity, τ is onto and since $U \subseteq \psi\tau(U)$ (formula $(\subseteq)$ in the proof) and $\psi(0) = \Psi_0 = \emptyset$, the homomorphism τ is dense.

Now recall VIII.3.4. We have $\tau_* = \psi$ and hence $\Psi_A = \tau_*[A]$. Thus the basis $\Psi_{\mathcal{A}}$ of the uniformity of $\Psi(L, \mathcal{A})$ is $\{\tau_*[A] \mid A \in \mathcal{A}\}$ and hence τ a dense ue-surjection. □

By 4.2.4, $\Psi(L, \mathcal{A})$ is Cauchy complete (because of enough Cauchy points). The dense ue-surjection

$$\tau_{(L,\mathcal{A})} \colon \Omega\Psi(L, \mathcal{A}) \to (L, \mathcal{A})$$

will be called

the *Cauchy completion* of the nearness frame $(L, \mathcal{A})$.

5.2.1. Recall VIII.6.8.2. From 5.2 we immediately obtain

Corollary. *There is a dense ue-surjection $h \colon \mathbf{C}(L, \mathcal{A}) \to \Omega\Psi(L, \mathcal{A})$ such that the diagram*

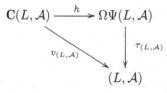

commutes. □

5.3. Theorem. *Let $\mathcal{A}$ have a countable basis. Then the Cauchy completion coincides with the completion (up to isomorphism). In particular, a Cauchy complete nearness frame with countable basis of nearness (in other words, a Cauchy complete metrizable frame) is complete.*

Proof. First, by 3.4.2 $(L, \mathcal{A})$ has enough Cauchy points and therefore the Cauchy completion is correct. By 2.2, $\mathbf{C}(L, \mathcal{A})$ is spatial, that is, there is an isomorphism $\phi \colon \Omega(X, \mathfrak{U}) \to \mathbf{C}(L, \mathcal{A})$. Take the dense ue-surjection h from 5.2.1; then by 5.1 there is a classical uniform embedding $f \colon \Psi(L, \mathcal{A}) \to (X, \mathfrak{U})$ such that $h\phi = \Omega(f)$. Since $\Psi(L, \mathcal{A})$ is complete in the classical sense (see also 4.2.3), f is an isomorphism, hence so is also $\Omega(f) = h\phi$, and since ϕ is an isomorphism, finally, h is one. □

6.   Generalized Cauchy points

6.1. In classical topology one has the two equivalent features of completeness:

  – every Cauchy point is (a neighbourhood system of) a point,
  – every dense embedding is an isomorphism.

In our context we have the completeness defined by the embedding property and so far lack the counterpart of the former characteristics. But there is one, as we will show in this short section.

6.2. A point of the spectrum of L can be viewed as a completely prime filter or equivalently as

a frame homomorphism $h\colon L \to \mathbf{2}$.

Now a Cauchy point P in $(L, \mathcal{A})$ can be similarly translated to a map $\phi\colon L \to \mathbf{2}$ with the properties that

(1) ϕ preserves finite meets and $\phi(0) = 0$,

(2) for each $A \in \mathcal{A}$, $\bigvee\{\phi(a) \mid a \in A\} = 1$, and

(3) for each $a \in L$, $\phi(a) = \bigvee\{\phi(x) \mid x \vartriangleleft_{\mathcal{A}} a\}$.

But this is a special case of a concept we already have: a Cauchy point is thus equivalently represented as

a regular Cauchy map $\phi\colon L \to \mathbf{2}$

(recall VIII.6.4).

6.3. Replacing $\mathbf{2}$ by a general frame T let us speak of frame homomorphisms $L \to T$ as of *generalized points* in L, and of regular Cauchy maps $L \to T$ as of *generalized Cauchy points* in L. Now we have

Theorem. *A nearness frame is complete if and only if each generalized Cauchy point in L is a generalized point.*

Proof. $\Rightarrow$: Let $(L, \mathcal{A})$ be complete and let $\phi\colon L \to T$ be a generalized Cauchy point. By VIII.6.8.1 there is a frame homomorphism $h\colon \mathbf{C}(L, \mathcal{A}) \to T$ such that $h\lambda_L = \phi$. Since $(L, \mathcal{A})$ is complete, λ_L is the inverse to the isomorphism $v_{(L,\mathcal{A})}$ and hence a frame homomorphism, making ϕ a frame homomorphism.

$\Leftarrow$: Let each generalized Cauchy point, that is, regular Cauchy map, be a generalized point, that is, a frame homomorphism. Take the mapping

$$\lambda = \lambda_{(L,\mathcal{A})} = (a \mapsto \downarrow a)\colon (L, \mathcal{A}) \to \mathbf{C}(L, \mathcal{A}).$$

We have $\downarrow 0 = \{0\}$ and $\downarrow 1 = L$. Further, by VIII.6.5, for $A \in \mathcal{A}$,

$$\left(\bigcup \{\downarrow a \mid a \in A\} = \downarrow A, L \right) \in R,$$

hence

$$\bigvee \{\lambda(a) \mid a \in A\} = 1 \text{ in } \mathbf{C}(L, \mathcal{A}),$$

and for each $a \in L$

$$\left(\downarrow a, \mathfrak{k}(a) = \bigcup \{\downarrow x \mid x \lhd_{\mathcal{A}} a\} \right) \in R,$$

hence

$$\lambda(a) = \bigvee \{\lambda(x) \mid x \lhd_{\mathcal{A}} a\} \text{ in } \mathbf{C}(L, \mathcal{A}).$$

Thus, λ is a regular Cauchy map and hence a frame homomorphism. Since the set $\{\downarrow a \mid a \in L\}$ generates L, λ is onto, and since $v_{(L,\mathcal{A})} \lambda = \mathrm{id}$, it is an isomorphism. $\qquad \square$

Chapter XI

Metric Frames

We have already mentioned that metrizability was defined by Isbell as the existence of a countably generated (admissible) uniformity. In this chapter we will show that in such a case there is indeed a diameter function with the properties mimicking those of the diameter in a classical metric space. We show a link of thus ensuing *metric frames* with metric spaces by a specialized spectrum adjunction, present metrization theorems extending those of the classical theory, and, finally, discuss the resulting categories.

1. Diameters and metric diameters

1.1. Denote by $\mathbb{R}_+$ the set of all non-negative reals plus $+\infty$. A *diameter* on a frame L is a mapping

$$d \colon L \to \mathbb{R}_+$$

such that

(D1) $d(0) = 0$,

(D2) $a \leq b \quad \Rightarrow \quad d(a) \leq d(b)$,

(D3) $a \wedge b \neq 0 \quad \Rightarrow \quad d(a \vee b) \leq d(a) + d(b)$, and

(D4) for every $\varepsilon > 0$, the set

$$U_\varepsilon^d = \{a \mid d(a) < \varepsilon\}$$

is a cover of L.

A diameter d on L is said to be *admissible* (or, L is said to *admit d*) if the system of covers

$$\mathcal{U}(d) = \{U_\varepsilon^d \mid \varepsilon > 0\}$$

is admissible in the sense of VIII.2.4.

The pair (L, d) is then referred to as a *metric frame*.

1.1.1. Set

$$\lhd_\varepsilon^d a = \bigvee\{x \mid U_\varepsilon^d x \le a\}.$$

In this notation the admissibility can be expressed as the requirement that

$$\forall a, \quad a = \bigvee\{\lhd_\varepsilon^d a \mid \varepsilon > 0\}. \tag{1.1.1}$$

(Note that by VIII.2.2.2, $U_\varepsilon^d a \le b$ iff $a \le \lhd_\varepsilon^d b$; thus, $\lhd_\varepsilon^d(-)$ is the right Galois adjoint of $U_\varepsilon^d(-)$.)

A frame is *metrizable* if it admits a diameter.

1.2. The definition of a diameter mimics the properties of the classical diameter

$$\operatorname{diam}_\rho(U) = \sup\{\rho(x,y) \mid x,y \in U\}$$

in a metric space (X, ρ). After the experience with the admissibility of point-free uniformities confronted with the classical relation of a topology and uniformity (VIII.1), the reader will certainly easily check that the admissibility corresponds to the classical relation of a metric with the induced topology.

We are interested in the relation of countably generated uniformities and diameters. The system of covers $\mathcal{U}(d)$, of course, creates a countably generated nearness, and this in turn implies the existence of a countably generated uniformity (recall IX.1.5). To have an immediately associated uniformity in analogy with the metric uniformity generated by a metric, our assumptions are too weak. To do better we can consider the following condition.

A diameter d is *star-additive* if

(DS) For all $a \in L$ and $S \subseteq L$ such that $a \wedge x \ne 0$ for all $x \in S$,

$$d(a \vee \bigvee S) \le d(a) + \sup\{d(x) + d(y) \mid x,y \in S\}.$$

Note that (DS) (together with (D4)) implies (D3): for each $\varepsilon > 0$, if $a \wedge b \ne 0$ choose an x such that $d(x) < \varepsilon$ and $x \wedge a \wedge b \ne 0$; then by (DS),

$$d(a \vee b) \le d(x) + d(a) + d(b) \le d(a) + d(b) + \varepsilon.$$

1.2.1. Observation. *If d is a star-additive diameter then $\mathcal{U}(d)$ is a basis of uniformity.*

(Indeed, by (DS), $U_\varepsilon^d U_\varepsilon^d \le U_{4\varepsilon}^d$.)

We will then speak of $\mathcal{U}(d)$ as of the *metric uniformity* associated with the (star-additive) diameter d.

1.3. Diameters induced by metrics have a stronger property which is expressed in the following definition.

A diameter on a frame is said to be a *metric diameter* if, moreover,

(DM)   for every $a \in L$ and $\varepsilon > 0$ there are $u, v \leq a$ such that $d(u), d(v) < \varepsilon$ and $d(a) < d(u \vee v) + \varepsilon$.

Note that this condition is obviously equivalent with

(DM′)   for every $a \in L$ and $\varepsilon > 0$ there are u, v with $u \wedge a \neq 0 \neq v \wedge a$ such that $d(u), d(v) < \varepsilon$, and $d(a) < d(u \vee v) + \varepsilon$

which is sometimes handier.

   We will see soon that by restricting ourselves to metric diameters we do not lose generality: a frame admits a diameter iff it admits a metric one, and a star-diameter can be replaced very closely by a metric one; furthermore, we gain considerable technical advantages.

1.3.1. For a diameter d define

$$\widetilde{d}(a) = \inf_{\varepsilon > 0} \sup\{d(u \vee v) \mid u \wedge a \neq 0 \neq v \wedge a, \; d(u), d(v) < \varepsilon\}$$

$$= \inf_{\varepsilon > 0} \sup\{d(u \vee v) \mid u, v \leq a, \; d(u), d(v) < \varepsilon\}$$

(the equivalence of the two formulas is obvious). We have

Lemma. $\widetilde{d}(a) \leq d(a)$, *and for any* $a, b \in L$, $\widetilde{d}(a \vee b) \geq d(a \vee b) - d(a) - d(b)$.

Proof. The first is trivial. Now the second obviously holds if some of the a, b is 0. Thus, assume that $a \neq 0 \neq b$ and choose u, v with $d(u), d(v) < \varepsilon$ and $u \wedge a \neq 0 \neq v \wedge b$. Then

$$d(a \vee b) \leq d(a \vee u \vee v \vee b) \leq d(u \vee v) + d(a) + d(b)$$

and hence

$$d(a \vee b) - d(a) - d(b) \leq d(u \vee v) \leq \sup\{d(u \vee v) \mid u \wedge a \neq 0 \neq v \wedge a, \; d(u), d(v) < \varepsilon\}.$$

$\square$

1.3.2. Proposition. (a) $\widetilde{d}$ *is a metric diameter.*

(b) *For every* $a \in L$ *and every* $\varepsilon > 0$, $U_\varepsilon^{\widetilde{d}} a = U_\varepsilon^{d} a$.

Proof. (a): The mapping $\widetilde{d}$ obviously satisfies (D1), (D2) and (D4).

(D3): Let $a \wedge b \neq 0$. For an $\varepsilon > 0$ choose an $\eta > 0$ such that

$$\alpha = \sup\{d(u \vee v) \mid u \wedge a \neq 0 \neq v \wedge a, \; d(u), d(v) < \eta\} < \widetilde{d}(a) + \varepsilon \text{ and}$$

$$\beta = \sup\{d(u \vee v) \mid u \wedge b \neq 0 \neq v \wedge b, \; d(u), d(v) < \eta\} < \widetilde{d}(b) + \varepsilon.$$

Take a $c \in U_\eta^d$ such that $0 \neq c \leq a \wedge b$ and let $x \wedge (a \vee b) \neq 0 \neq y \wedge (a \vee b)$. If $x \wedge a \neq 0 \neq y \wedge a$ or $x \wedge b \neq 0 \neq y \wedge b$ we have $d(x \vee y) \leq \alpha$ resp. $\leq \beta$ and otherwise $d(x \vee y) \leq d(x \vee c) + d(y \vee c) \leq \alpha + \beta$, hence $d(x \vee y) \leq \alpha + \beta$ in any case, and we obtain $\tilde{d}(a \vee b) \leq \alpha + \beta \leq \tilde{d}(a) + \tilde{d}(b) + 2\varepsilon$.

(DM): By 1.3.1 we have

$$\tilde{d}(a) \leq \inf_{\varepsilon > 0} \sup\{\tilde{d}(u \vee v) + 2\varepsilon \mid u, v \leq a, \ d(u), d(v) < \varepsilon\}$$

$$\leq \inf_{\varepsilon > 0} \sup\{\tilde{d}(u \vee v) \mid u, v \leq a, \ \tilde{d}(u), \tilde{d}(v) < \varepsilon\} + 2\varepsilon.$$

(b): Since $\tilde{d} \leq d$ we have $U_\varepsilon^d a \leq U_\varepsilon^{\tilde{d}} a$. On the other hand let $u \wedge a \neq 0$ and $\tilde{d}(u) < \varepsilon$. Choose an $\eta > 0$ such that $\tilde{d}(u) + 2\eta < \varepsilon$. We have

$$u = \bigvee\{x \wedge u \mid x \in U_\eta^d\}. \tag{*}$$

Fix an $x_0 \in U_\eta^d$ such that $x_0 \wedge u \wedge a \neq 0$ and take a general $x \in U_\eta^d$. Then by 1.3.1

$$d((x_0 \vee x) \wedge u) \leq \tilde{d}((x_0 \vee x) \wedge u) + d(x_0 \wedge u) + d(x \wedge u) \leq \tilde{d}(u) + 2\eta < \varepsilon$$

so that $(x_0 \vee x) \wedge u \leq U_\varepsilon^{\tilde{d}} a$, and by (*) finally $u \leq U_\varepsilon^{\tilde{d}} a$. $\qquad\square$

1.3.3. Corollary. *In the notation of VIII.2.4,* $\lhd_{\mathcal{U}(\tilde{d})} = \lhd_{\mathcal{U}(d)}$. *Consequently,* $\tilde{d}$ *is admissible iff d is.* $\qquad\square$

1.3.4. Corollary. *L is metrizable iff it admits a metric diameter.* $\qquad\square$

1.4. Proposition. (a) *A metric diameter is star-additive. Thus, it induces a uniformity (recall 1.2.1).*

(b) *A metric diameter is continuous in the sense that*

$$\text{for every up-directed } S, \quad d(\bigvee S) = \sup\{d(a) \mid a \in S\}.$$

(c) *If d is a metric diameter and $S \subseteq L$ is such that $a \leq \bigvee S$ then for each $\varepsilon > 0$ there are $x, y \in S$ such that $d(x \vee y) > d(a) - \varepsilon$.*

Proof. (a): Let $x \in S$ imply $a \wedge x \neq 0$. Then we have for u, v with $d(u), d(v) < \varepsilon$ such that $u \wedge (a \vee \bigvee S) \neq 0 \neq v \wedge (a \vee \bigvee S)$ either $u \wedge a \neq 0 \neq v \wedge a$ or $u \wedge a \neq 0 \neq v \wedge x$ for some $x \in S$, or $u \wedge x \neq 0 \neq v \wedge y$ for some $x, y \in S$. We will consider the last case, the others are only simpler. We have, then,

$$d(u \vee v) \leq d(u \vee x \vee a \vee y \vee v) \leq d(a) + d(x) + d(y) + 2\varepsilon.$$

(b): Let u, v be such that $d(u), d(v) < \varepsilon$ and $u \wedge \bigvee S \neq 0 \neq v \wedge \bigvee S$. Since S is directed there is an $a \in S$ such that $u \wedge a \neq 0 \neq v \wedge a$,

$$d(u \vee v) \leq d(u \vee a \vee v) \leq d(u) + d(a) + d(v) \leq \sup\{d(a) \mid a \in S\} + 2\varepsilon.$$

The inequality $\sup\{d(a) \mid a \in S\} \leq d(\bigvee S)$ is trivial.

(c): We can assume $d(a) \neq 0$. Then for sufficiently small ε we have $0 \neq u, v \leq a$ such that $d(u), d(v) < \frac{\varepsilon}{3}$ and $d(a) < d(u \vee v) + \frac{\varepsilon}{3}$. There are $x, y \in S$ such that $u \wedge x \neq 0$ and $y \wedge v \neq 0$. We have

$$d(u \vee v) \leq d(u \vee v \vee x \vee y) \leq d(u) + d(v) + d(x \vee y) < d(x \vee y) + 2\tfrac{\varepsilon}{3}$$

and hence $d(a) < d(x \vee y) + \varepsilon$. $\qquad\square$

Note. By 1.3.3 and 1.4(a), for any diameter whatsoever the relation $\lhd_{\mathcal{U}(d)}$ interpolates.

1.4.1. In fact, a star-additive diameter is not very far from a metric one. We have

Proposition. *Let d be a star-additive diameter. Then $\widetilde{d} \leq d \leq 2\widetilde{d}$.*

Proof. Take an $\eta > 0$, $a \neq 0$ and an $\varepsilon > 0$ such that

$$\widetilde{d}(a) > \sup\{d(u \vee v) \mid u, v \leq a, \ d(u), d(v) < \varepsilon\} - \eta. \qquad (*)$$

We can assume that $\varepsilon \leq \eta$. Set $M = \{u \in U_\eta^d \mid 0 \neq u \leq a\}$, choose a $u_0 \in M$ and consider the set $S = \{u \vee u_0 \mid u \in M\}$. Then by (DS)

$$d(a) \leq d(u_0) + \sup\{d(u \vee u_0) + d(v \vee u_0) \mid u, v \in M\}$$

and hence there are $u, v \in M$ such that

$$d(a) \leq d(u \vee u_0) + d(v \vee u_0) + 2\eta.$$

Now by $(*)$ $d(a) < 2\widetilde{d}(a) + 4\eta$. $\qquad\square$

2.   Metric spectrum

2.1. Contractive homomorphisms. Let $(X, \rho), (Y, \sigma)$ be metric spaces. A mapping $f \colon X \to Y$ is said to be *contractive* if

$$\forall x, y \in X, \quad \sigma(f(x), f(y)) \leq \rho(x, y)$$

which is obviously the same as

$$\forall x, y \in X, \forall \varepsilon > 0, \quad \rho(x, y) < \varepsilon \Rightarrow \sigma(f(x), f(y)) < \varepsilon.$$

The category of all metric spaces and contractive homomorphisms will be denoted by

$$\mathbf{Metr}.$$

Note that in **Metr** the isomorphisms are precisely the isometries, that is, the maps such that $\sigma(f(x), f(y)) = \rho(x, y)$.

2.1.1. Lemma. *A mapping $f \colon X \to Y$ is contractive with respect to ρ, σ iff in the resulting diameters for each $U \in \Omega(X)$ with $\mathrm{diam}_\rho(U) < \varepsilon$ there is a $V \in \Omega(Y)$ with $\mathrm{diam}_\sigma(V) < \varepsilon$ and $U \subseteq f^{-1}[V]$ (that is, $f[U] \subseteq V$).*

Proof. $\Rightarrow$: Let $U \in \Omega(X)$ be such that $\mathrm{diam}_\rho(U) < \varepsilon$. Choose an $\eta > 0$ such that $\mathrm{diam}_\rho(U) + 2\eta < \varepsilon$ and set

$$V = \{y \mid \exists x \in X, \ \sigma(y, f(x)) < \eta\}.$$

Then $U \subseteq f^{-1}[V]$. If $y_1, y_2 \in V$ choose $x_i \in U$ such that $\sigma(y_i, f(x_i)) < \eta$. Then $\sigma(y_1, y_2) \leq \sigma(y_1, f(x_1)) + \sigma(f(x_1), f(x_2)) + \sigma(f(x_2), y_2) \leq \eta + \rho(x_1, x_2) + \eta < \varepsilon$. Hence $\mathrm{diam}_\sigma(V) < \varepsilon$.

$\Leftarrow$: Suppose $\rho(x_1, x_2) < \sigma(f(x_1), f(x_2)) = \varepsilon$. Choose a $U \ni x_1, x_2$ such that $\mathrm{diam}_\rho(U) < \varepsilon$ (for instance, $\{x \in X \mid \rho(x, x_1), \rho(x, x_2) < \eta\}$ with sufficiently small η). Let $U \subseteq f^{-1}[V]$ with $\mathrm{diam}_\sigma(V) < \varepsilon$. Then $f(x_1), f(x_2) \in V$ and $\sigma(f(x_1), f(x_2)) < \varepsilon$, a contradiction. $\square$

2.1.2. This leads to the following definition.

Let $(L, d), (M, d')$ be metric frames. A (frame) homomorphism $h \colon L \to M$ is said to be *contractive* if for every $b \in M$ with $d'(b) < \varepsilon$ there is an $a \in L$ such that $b \leq h(a)$ and $d(a) < \varepsilon$.

The resulting category will be denoted by

$$\textbf{DFrm}$$

and its full subcategories generated by the (L, d) with metric resp. star-additive diameters by

$$\textbf{D}_\textbf{M}\textbf{Frm} \quad \text{resp.} \quad \textbf{D}_\textbf{S}\textbf{Frm}.$$

The opposite categories of metric locales will be denoted by

$$\textbf{DLoc}, \quad \textbf{D}_\textbf{M}\textbf{Loc} \quad \text{resp.} \quad \textbf{D}_\textbf{S}\textbf{Loc}.$$

By 2.1.1 we see that the formulas $\Omega(X, \rho) = (\Omega(X), \mathrm{diam}_\rho)$ and $\Omega(f)$ as in II.1.3 define contravariant functors

$$\Omega \colon \textbf{Metr} \to \textbf{DFrm} \quad \text{resp.} \quad \Omega \colon \textbf{Metr} \to \textbf{D}_\textbf{M}\textbf{Frm}$$

and in the covariant modification as in II.2.4,

$$\textsf{Lc} \colon \textbf{Metr} \to \textbf{DLoc} \quad \text{resp.} \quad \textsf{Lc} \colon \textbf{Metr} \to \textbf{D}_\textbf{M}\textbf{Loc}.$$

2.2. Lemma. *Let F be a completely prime filter in a metric frame (L, d). Then for every $a \in F$ there exists an $\varepsilon > 0$ such that for all $b \neq 0$ with $d(b) < \varepsilon$ we either have $b \leq a$ or $b \notin F$.*

Proof. By the complete primeness and (1.1.1) there is an $\varepsilon > 0$ such that $\lhd_\varepsilon^d a \in F$. Now if $d(b) < \varepsilon$ then either $b \wedge \lhd_\varepsilon^d a \neq 0$ and $b \leq a$ or $b \wedge \lhd_\varepsilon^d a = 0$ and $b \notin F$. $\square$

2.3. The spectrum. For a metric frame (L, d) consider, first the standard spectrum

$$\mathsf{Sp}L = \Sigma L = \left(\{F \mid F \text{ completely prime filter in } L\}, \{\Sigma_a \mid a \in L\}\right)$$

as in II.4.3-4.4. For $F, G \in \Sigma L$ define

$$\rho(F, G) = \rho_d(F, G) = \inf\{d(a) \mid a \in F \cap G\}.$$

2.3.1. Proposition. ρ *is a metric on* $\mathsf{Sp}L = \Sigma L$ *and the induced topology coincides with the standard topology of the spectrum.*

Proof. I. $\rho(F, F) = \inf\{d(a) \mid a \in F\} = 0$ because for any ε, $\bigvee U_\varepsilon^d = 1 \in F$ and hence we have an $a \in F$ with $d(a) < \varepsilon$.

Let $\rho(F, G) = 0$. For an $a \in F$ choose an $\varepsilon > 0$ as in 2.2. There is a $b \in F \cap G$ such that $d(b) < \varepsilon$. Since $b \in F$, $b \leq a$. Since $b \in G$, $a \in G$. Thus, $F \subseteq G$ and by symmetry $F = G$.

Trivially $\rho(F, G) = \rho(G, F)$.

If $\rho(F, G) < \alpha$ and $\rho(G, H) < \beta$, there are $a \in F \cap G$ and $b \in G \cap H$ such that $d(a) < \alpha$ and $d(b) < \beta$. Since $a, b \in G$, $a \wedge b \neq 0$ and $d(a \vee b) < \alpha + \beta$. Since $a \vee b \in F \cap H$, $\rho(F, G) < \alpha + \beta$ and we obtain the triangle inequality.

II. Let $F \in \Sigma_a$, that is, $a \in F$. Take, again, an ε from 2.2. If $\rho(F, G) < \varepsilon$ there is a $b \in F \cap G$ such that $d(b) < \varepsilon$ and it follows that $a \geq b$ is in G, and $G \in \Sigma_a$. Thus Σ_a is open in $(\Sigma L, \rho)$.

On the other hand, each $W = \{G \mid \rho(F, G) < \varepsilon\}$ is open in ΣL. Indeed, set $u = \bigvee\{a \in F \mid d(a) < \varepsilon\}$. If $G \in \Sigma_u$ then by the complete primeness there is an $a \in G$ such that $a \in F$ and $d(a) < \varepsilon$, and hence $G \in W$. If $G \in W$ then there is an $a \in G \cap F$ such that $d(a) < \varepsilon$. Hence $G \in \Sigma_u$, and we see that $W = \Sigma_u$. □

2.3.2. If $h: (L, d) \rightarrow (M, d')$ is a contractive homomorphism then the standard spectrum map
$$\Sigma h = \left(F \mapsto h^{-1}[F]: (\Sigma M, \rho_{d'}) \rightarrow (\Sigma L, \rho_d)\right)$$
is contractive. Indeed, if $\rho(F, G) < \alpha$ then there is a $b \in F \cap G$ such that $d'(b) < \alpha$ and consequently an a with $d(a) < \varepsilon$ and $h(a) \geq b$. Hence $a \in h^{-1}[F \cap G] = \Sigma h(F) \cap \Sigma h(G)$ and we see that $\rho(\Sigma h(F), \Sigma h(G)) < \alpha$.

Thus we have functors

$$\Sigma: \mathbf{DFrm} \rightarrow \mathbf{Metr} \quad \text{resp.} \quad \Sigma: \mathbf{D_M Frm} \rightarrow \mathbf{Metr}$$

and their covariant modifications

$$\mathsf{Sp}: \mathbf{DLoc} \rightarrow \mathbf{Metr} \quad \text{resp.} \quad \mathsf{Sp}: \mathbf{D_M Loc} \rightarrow \mathbf{Metr}$$

(cf. II.4) defined by $\Sigma(L, d) = \mathsf{Sp}(L, d) = (\Sigma L, \rho_d)$ and for morphisms as in the standard spectrum.

2.4. The spectrum adjunction. In the definition of the spectrum we have so far not needed any special properties of the metric frame. The spectrum adjunction, however, does not hold in full generality. We will prove it for metric diameters.

Similarly as in II.4 consider the homomorphisms

$$\phi_{(L,d)}\colon L \to \mathsf{LcSp}L(= \Omega(\Sigma L))$$

given by $\phi_{(L,d)}(a) = \Sigma_a$ and the mappings

$$\lambda_{(X,\rho)}\colon X \to \mathsf{Sp}(\mathsf{Lc}(X))$$

defined by

$$\lambda_{(X,\rho)}(x) = \mathcal{U}(x) = \{U \mid x \in U\}.$$

We have

Proposition. *The functor* $\mathsf{Sp}\colon \mathbf{D_M Loc} \to \mathbf{Metr}$ *is a right adjoint to* $\mathsf{Lc}\colon \mathbf{Metr} \to \mathbf{D_M Loc}$ *with the adjunction units* $\phi = (\phi_{(L,d)})_{(L,d)}$ *and* $\lambda = (\lambda_{(X,\rho)})_{(X,\rho)}$.

Proof. The fact that the compositions

$$\Omega(X, \rho) \xrightarrow{\phi_{\Omega(X,\rho)}} \Omega\Sigma\Omega(X, \rho) \xrightarrow{\Omega\lambda_{(X,\rho)}} \Omega(X, \rho)$$

and

$$\Sigma(L, d) \xrightarrow{\lambda_{\Sigma(L,d)}} \Sigma\Omega\Sigma(L, d) \xrightarrow{\Sigma\phi_{(L,d)}} \Sigma(L, d)$$

are identities follows precisely like in II.4.6. We have to prove, however, that the maps constituting ϕ and λ are morphisms in the respective categories.

If $\rho(x,y) < \varepsilon$ choose an η such that $\rho(x,y) + 2\eta < \varepsilon$ and open neighbourhoods U, V of x resp. y, both with diameter less than η. Then obviously $\rho_{\mathrm{diam}_\rho}(\lambda(x), \lambda(y)) < \varepsilon$.

Now let $\mathrm{diam}_\rho(\Sigma_a) < \varepsilon$. For any $F, G \in \Sigma_a$ there is a b_{FG} such that $d(b_{FG}) < \varepsilon - 3\eta$. Set

$$M = \{x \mid d(x) < \eta \text{ and } \exists F \in \Sigma L, \ x \wedge a \in F\},$$

and further $c = \bigvee M$. For $x, y \in M$ choose F, G such that $a \wedge x \in F$ and $a \wedge y \in G$. For the $b = b_{FG} \in F \cup G$ above we have $b \wedge x \neq 0 \neq b \wedge y$ and hence

$$d(x \vee y) \leq d(x \vee b \vee y) \leq d(x) + d(b) + d(y) < \varepsilon - \eta$$

and by 1.4(c)

$$d(c) < d(x \vee y) < \varepsilon.$$

Finally, if $a \in F$ there is an $x \in U_\eta^d$ such that $x \in F$, hence $x \wedge a \in F$ and hence $F \in \Sigma_c$. Thus, $\Sigma_a \subseteq \Sigma_c$. $\qquad\square$

3.   Uniform Metrization Theorem

3.1.   We will need a few auxiliary notions.

Let $x, y \in S \subseteq L$. An *xy-chain* (or simply *chain*) in S is a sequence $x_1, \ldots, x_n$ in S such that $x = x_1$, $y = x_n$ and $x_{i-1} \wedge x_i \neq 0$ for all $i > 1$. A subset $S \subseteq L$ is said to be *connected* if for every $x, y \in S$ there is an xy-chain in S.

A *prediameter* on L is a function

$$f: L \to \mathbb{R}_+$$

satisfying (D1), (D2) and (D4). It is said to be *strong* if, moreover, for every $S \subseteq L$ satisfying $a \wedge b \neq 0$ for all $a, b \in S$,

$$f(\bigvee S) \leq 2 \sup\{f(a) \mid a \in S\}.$$

3.2.   Let f be a prediameter and let $S \subseteq L$ be connected. For $x, y \in S$ set

$$\delta(f; x, y, S) = \inf\{\sum_{i=1}^{n} f(x_i) \mid (x_i)_{i=1,\ldots,n} \text{ is an } xy\text{-chain in } S\},$$

$$\delta(f; S) = \sup\{\delta(f; x, y, S) \mid x, y \in S\},$$

and define
$$d_f(a) = \inf\{\delta(f; S) \mid S \text{ connected, } a \leq \bigvee S\}.$$

We will use the obvious

3.2.1.   Observations. (a) *Let x, y_1, y_2, z be in S and let $y_1 \wedge y_2 \neq 0$. Then*

$$\delta(f; x, z, S) \leq \delta(f; x, y_1, S) + \delta(f; y_2, z, S).$$

(b) *If $x, y \in S \subseteq T$ then $\delta(f; x, y, S) \geq \delta(f; x, y, T)$.*

(c) $d_f \leq f$.

(For the last consider $S = \{a\}$.)

3.2.2.   Proposition. d_f *is a star-additive diameter.*

Proof. Obviously d_f satisfies (D1) and (D2) and since $d_f \leq f$ we also have (D4).

Now suppose there is an $a \in L$ and an S such that for every $x \in S$, $x \wedge a \neq 0$ and that

$$d_f(a \vee \bigvee S) > d_f(a) + \sup\{d_f(x) + d_f(y) \mid x, y \in S\} + 3\varepsilon.$$

For each $x \in S \cup \{a\}$ choose a connected S_x such that $x \leq \bigvee S_x$ and $\delta(f; S_x) < d_f(x) + \varepsilon$. Hence we have, for any $x, y \in S$,

$$d_f(a \vee \bigvee S) > \delta(f; S_a) + \delta(f; S_x) + \delta(f; S_y).$$

Set $T = S_a \bigcup \{ S_x \mid x \in S \}$. Then T is connected and $\bigvee T \geq a \vee \bigvee S$, and hence

$$\delta(f;T) > \delta(f;S_a) + \delta(f;S_x) + \delta(f;S_y)$$

so that there are $u, v \in T$ such that

$$\delta(f;u,v,T) > \delta(f;S_a) + \delta(f;S_x) + \delta(f;S_y).$$

We cannot have u, v both in an S_z, $z = x, y, a$: there would be

$$\delta(f;u,v,T) > \delta(f;S_z) \geq \delta(f;u,v,S_z) \geq \delta(f;u,v,T).$$

Therefore, say, either $u \in S_a$ and $v \in S_y$, or $u \in S_x$ and $v \in S_y$. In the former case choose $a_1 \in S_a$ and $v_1 \in S_y$ such that $a_1 \wedge v_1 \neq 0$ to obtain

$$\delta(f;u,v,T) > \delta(f;u,a_1,S_a) + \delta(f;v_1,v,S_y)$$
$$\geq \delta(f;u,a_1,T) + \delta(f;v_1,v,T) \geq \delta(f;u,v,T).$$

In the latter case choose $u_1 \in S_x$, $a_1, a_2 \in S_a$ and $v_1 \in S_y$ such that $u_1 \wedge a_1 \neq 0$ and $a_2 \wedge v_1 \neq 0$ and we have

$$\delta(f;u,v,T) > \delta(f;u,a_1,S_x) + \delta(f;a_1,a_2,S_a) + \delta(f;v_1,v,S_y)$$
$$\geq \delta(f;u,a_1,T) + \delta(f;a_1,a_2,T) + \delta(f;v_1,v,T) \geq \delta(f;u,v,T),$$

a contradiction again.   □

3.2.3. We will consider one more extra condition

(3W) if a, b, c are such that $a \wedge b \neq 0 \neq b \wedge c$ then

$$f(a \vee b \vee c) \leq 2 \max\{f(a), f(b), f(c)\}.$$

3.2.4. Lemma. *If a prediameter f satisfies* (3W) *then for any chain* $(x_i)_{i=1,\ldots,n}$

$$f(\bigvee_{i=1}^{n} x_i) \leq 2 \sum_{i=1}^{n} f(x_i).$$

Proof. The inequality is obvious for $n = 1$. Now let it hold for n and let us take a chain $x_1, \ldots, x_n, x_{n+1}$. Set $\alpha = \sum_{i=1}^{n+1} f(x_i)$ and consider the first k such that $\sum_{i=1}^{k} f(x_i) \geq \frac{1}{2}\alpha$. Then

$$\sum_{i=1}^{k-1} f(x_i) < \tfrac{1}{2}\alpha \quad \text{and} \quad \sum_{i=k+1}^{n+1} f(x_i) \leq \tfrac{1}{2}\alpha$$

and hence by the induction hypothesis

$$f(\bigvee_{i=1}^{k-1} x_i) < \alpha \quad \text{and} \quad f(\bigvee_{i=k+1}^{n+1} x_i) \leq \alpha,$$

and since also $f(x_k) \leq \alpha$ we obtain from (3W)

$$f(\bigvee_{i=1}^{n+1} x_i) \leq 2\alpha = 2\sum_{i=1}^{n+1} f(x_i). \qquad \square$$

3.2.5. Proposition. *For each strong prediameter f on L satisfying (3W) there is a star-additive diameter d such that for every a*

$$\tfrac{1}{4}f(a) \leq d(a) \leq f(a).$$

Proof. Let $S \subseteq L$ be connected. Fix a $u_0 \in S$ and $\varepsilon > 0$, and for a general $u \in S$ choose a $u_0 u$-chain $x_1(u), \ldots, x_n(u)$ such that

$$\sum_{i=1}^{n} f(x_i(u)) < \delta(f; u_0, u, S) + \varepsilon.$$

Set $s(u) = \bigvee_{i=1}^{n} x_i(u)$. Then any two elements of $\{s(u) \mid u \in S\}$ meet at least in $u_0 \neq 0$, and $\bigvee\{s(u) \mid u \in S\} = \bigvee S$ and we have by 3.2.4

$$f(s(u)) \leq 2\sum_{i=1}^{n} f(x_i(u)) < 2\delta(f; u_0, u, S) = 2\varepsilon \leq 2\delta(f; S) + 2\varepsilon$$

and since f is strong,

$$f(\bigvee S) \leq 2\sup\{f(s(u)) \mid u \in S\} \leq 4\delta(f; S) + 4\varepsilon.$$

Now take an $a \in L$ and a connected S such that $a \leq \bigvee S$. Since ε was arbitrary we see that $f(a) \leq 4\delta(f; S)$ and hence

$$f(a) \leq 4\inf\{\delta(f; S) \mid a \leq \bigvee S\} = 4d_f(a).$$

Use 3.2.2 and 3.2.1(c). $\qquad \square$

3.3. Theorem. *For every admissible uniformity on L with a countable basis there is a metric diameter d such that $\mathcal{A} = \mathcal{U}(d)$.*

Proof. Choose a basis $A_1, A_2, \ldots, A_n, \ldots$ of $\mathcal{A}$ such that $A_n A_n A_n \leq A_{n-1}$. Now define $f \colon L \to \mathbb{R}_+$ by setting

$$f(a) = \inf\{2^{-n} \mid a \leq x \text{ for some } x \in A_n\}$$

(typically, the infimum is a minimum, but the set can also be infinite in which case we have $f(a) = 0$, or void, in which case the value is $+\infty$). Obviously f is a prediameter.

Let $d(a) \leq 2^{-n}$ for all $a \in S$ such that $x, y \in S$ implies $x \wedge y \neq 0$. Then we have $b_a \geq a$ with b_a in A and $\bigvee\{b_a \mid a \in S\} \leq A_n b_c$ for any $c \in S$ so that there is a $b \in A_{n-1}$ such that $\bigvee S \leq b$, and

$$f(\bigvee S) \leq 2^{n-1} \leq 2 \sup\{f(a) \mid a \in S\}.$$

Thus, f is a strong prediameter.

If a, b, c are such that $a \wedge b \neq 0 \neq b \wedge c$ and if $f(a), f(b), f(c) \leq 2^{-n}$ then $a \leq x$, $b \leq y$ and $c \leq z$ for some $x, y, z \in A_n$ and $a \vee b \vee c \leq A_n(A_n z) \leq w$ for some $w \in A_{n-1}$. Thus, $f(a \vee b \vee c) \leq 2^{-(n-1)}$ and we see that f satisfies (3W).

Now take the star-additive diameter d_f from 3.2.5 and set $d = \widetilde{d_f}$. By 3.2.5 and 1.5, $\frac{1}{8}f \leq d \leq f$ and hence d generates the same uniformity as our basis $(A_n)_n$.
□

4. Metrization theorems for plain frames

4.1. We will use the concept of a *locally finite* and *discrete* subset X of a frame, witnessed by a cover W (recall IX.2.1). We have

4.1.1. Lemma. *If $X \subseteq L$ is locally finite and $x \prec a$ for each $x \in X$ then $\bigvee X \prec a$.*

Proof. Let W be the witnessing cover. For a $w \in W$ let $x_1, \ldots, x_n$ be all the elements of X such that $x_i \wedge a \neq 0$. Set $Y = X \smallsetminus \{x_1, \ldots, x_n\}$. Then $w \wedge \bigvee Y = 0$, hence $w \leq (\bigvee Y)^* = 0$ and

$$(\bigvee X)^* \vee a = \left((\bigvee Y)^* \wedge \left(\bigwedge_{j=1}^n x_j^*\right)\right) \vee a$$

$$\geq (w \vee a) \wedge \left(\bigwedge_{j=1}^n (x_j^* \vee a)\right) \geq w,$$

and since this holds for any $w \in W$ and W is a cover, $(\bigvee X)^* \vee a = 1$, that is, $\bigvee X \prec a$.
□

4.2. From IX.1.5 we know that if L admits a countable admissible system of covers then it admits a countably generated uniformity. By 3.3 we infer that, consequently, it is metrizable.

Note that whenever $\mathcal{A}$ is an admissible system of covers of L then $\bigcup \mathcal{A}$ is a (join-)basis of L; indeed,

$$b = \bigvee\{x \mid \exists A \in \mathcal{A}, \ Ax \leq a\} = \bigvee\{u \mid \exists A \in \mathcal{A}, \ u \in A, \ Ax \leq a \text{ and } u \wedge x \neq 0\}.$$

Thus, defining

a *σ-admissible* basis B of L

as a basis B that can be written as $B = \bigcup_{n=1}^{\infty} B_n$ so that $\{B_n \mid n = 1, 2, \ldots\}$ is an admissible system of covers of L, we obtain

4.2.1. Theorem. *A frame L is metrizable iff it has a σ-admissible basis.* □

4.3. This was of course a very straightforward reformulation of facts we already know. In this section we will endeavour to liberate the metrizability from the uniform connotation.

We will discuss several properties of bases (let us agree that in this section we will use the term "basis" for join-bases of frames only). Thus we will speak of

- σ-*locally finite* bases,
- σ-*discrete* bases,

and furthermore of

- *regular* bases, that is, bases B of L where for each $0 \neq a \in L$ there is a subset $C(a) \subseteq L$ such that

    (R1) $\bigvee C(a) = a$, and

    (R2) for each $c \in C(a)$, the set $\{b \in B \mid b \wedge c \neq 0 \text{ and } b \nleq a\}$ is finite.

4.4. For technical reasons we will introduce a special type of σ-admissible bases. We will say that such a basis is σ-*stratified* if in the decomposition $B = \bigcup_{n=1}^{\infty} B_n$ as above,

(S1) each B_n is locally finite, and

(S2) $B_n B_n \leq B_{n-1}$ for each $n \geq 2$.

4.4.1. Lemma. *If L has a σ-admissible basis then*

(a) *each basis of L is σ-admissible,*

(b) *L has a σ-discrete basis, and*

(c) *L has a σ-stratified basis.*

Proof. Let $\mathcal{A} = (A_n)_n$ be admissible.

(a): If B is an arbitrary basis, set $B_n = \{b \in B \mid \exists a \in A_n,\ b \leq a\}$. Then $B_n \leq A_n$ and hence for $\mathcal{B} = (B_n)_n$, $x \lhd_{\mathcal{A}} y$ implies $x \lhd_{\mathcal{B}} y$ and $\mathcal{B}$ is admissible.

(b): By IX.1.5 and IX.4.4, L is paracompact and hence we have σ-discrete refinements B_n of A_n. Obviously B is a basis again.

(c): Again by paracompactness: each cover C has a locally finite star-refinement sC; set, inductively, $B_1 = sA_1$, $B_{n+1} = s(A_n \wedge B_n)$, and consider $B = \bigcup_{n=1}^{\infty} B_n$. □

4.5. Lemma. *Every σ-stratified basis is regular.*

Proof. Let $B = \bigcup_{n=1}^{\infty} B_n$ satisfy (S1) and (S2) above with the local finiteness of B_n witnessed by W_n. These covers can be chosen so that

$$W_n \leq B_n \quad \text{and} \quad W_1 \geq W_2 \geq W_3 \geq \cdots;$$

then each W_n witnesses the local finiteness of any B_k with $k \leq n$. For $a \in L$ put

$$C(a) = \{w \mid \exists n, \ w \in W_n \text{ and } B_n w \leq a\}.$$

$(B_n)_n$ is admissible, and hence to prove that $\bigvee C(a) = a$ it suffices to prove that $\bigvee C(a) \geq x$ for any $x \lhd a$. Let $B_{n-1}x \leq a$ and let $w \wedge x$ for some $w \in W_n$. Choose $b \in B_n$ such that $w \leq b$ (recall that $W_n \leq B_n$), and a $b' \in B_{n-1}$ such that $B_n b \leq b'$. Then $b' \wedge x \neq 0$, hence $B_n w \leq B_n b \leq b' \leq a$ so that $w \in C(a)$ and we see that $x \leq W_n x \leq a$.

Take any $w \in C(a)$, say $w \in W_n$ with $B_n w \leq a$, and set

$$F = \{b \mid \exists k < n \text{ such that } b \in B_k \text{ and } b \wedge x \neq 0\}.$$

By (S1) and the choice of W_n, F is finite. Now let $b \in B$ be such that $b \wedge w \neq 0$ and $b \not\leq a$. Let $b \in B_k$. For $c \in B_j$ with $j \geq n$ and $c \wedge w \neq 0$ we have $c \leq B_j w \leq B_n w \leq a$; thus, $k < n$. Since $b \wedge w \neq 0$ we have $b \in F$ and hence the set from (R2) is a subset of F. $\square$

4.5.1. Lemma. *If L is regular and $0 \neq a \in L$ is not minimal then $a = \bigvee\{b \mid b < a\}$.*

Proof. Set $c = \bigvee\{b \mid b < a\}$. Suppose $c < a$. Take $0 \neq x \prec c$. Then $a = a \wedge (c \vee x^*) = c \vee (a \wedge x^*)$ and hence $a \wedge x^*$ has to be a. This is impossible, though, since this would mean that $a \leq x^*$, that is, $x = a \wedge x = 0$. $\square$

4.5.2. Lemma. *In a regular frame, every regular basis is locally finite.*

Proof. Let B be a regular basis of L with associated subsets $C(a)$. Then $B \cap \{x \mid x \geq a\}$ is finite for any $a \neq 0$ since each $c \in C(a)$ meets every member of this set; let $\nu(a)$ be the length of the largest chain in this intersection. Obviously $\nu(c) < \nu(a)$ whenever $0 < a < c$.

Now define

$$B_n = \{b \in B \mid \nu(b) = n \text{ or } b \text{ is minimal and } v(b) \leq n\}$$

and

$$W_n = \bigcup\{C(b) \mid b \in B_n\}.$$

We show by induction that each B_n is a cover. This is obvious for $n = 1$ since $\bigvee B = 1$ and for any $b \in B$ there are $c \geq b$ in B such that $\nu(c) = 1$. Now consider any B_n which is a cover, and a $b \in B_n$. By 4.5.1, for each non-minimal $b \in B$,

$$b = \bigvee\{c \in B \mid c < b\}$$

and hence $\nu(c) > \nu(b)$ for all c that occur. Further, replacing these c by $b_c \in B$ such that $c \le b_c$ and $\nu(b_c) = n+1$ we obtain

$$b \le \bigvee \{b_c \mid c \in B, c < b\}$$

and hence B_{n+1} is a cover.

As a result the same holds for W_n. We will now show that B_n is locally finite witnessed by W_n. For any $c \in W_n$ let $b \in B_n$ be such that $c \wedge b \ne 0$. Now $c \in C(d)$ for some $d \in B_n$. We cannot have $b < d$; this is excluded in the case of minimal d and if d is not minimal then $\nu(d) = n$ and $\nu(b)$ for $b < d$ is greater than n. Thus, $b = d$ or $b \not\le d$ and as $c \wedge b \ne 0$, there are only finitely many such $b \in B$. □

4.6. Lemma. *In a regular frame, a σ-locally finite basis is σ-admissible.*

Proof. Let $B = \bigcup_{n=1}^{\infty} B_n$ be a basis with locally finite B_n witnessed by W_n. For each $x \in L$ and $k = 1, 2, \ldots$ set

$$x_k = \bigvee \{b \in B_k \mid b \prec x\} \quad (\prec x \text{ by } 4.1.1)$$

and for each $w \in W_n$ and k set

$$\{b(w,1), b(w,2), \ldots, b(w, l_w)\} = \{b \in B_n \mid w \wedge b \ne 0\}, \quad \text{and}$$

$$S_k(w) = \{b(w,1), (b(w,1)_k)^*\} \wedge \{b(w,2), (b(w,2)_k)^*\} \wedge \cdots$$
$$\cdots \wedge \{b(w, l_w), (b(w, l_w)_k)^*\}.$$

Let $\mathcal{A}$ be the set of covers

$$A_{nk} = \{w \wedge s \mid w \in W_n, s \in S_k(w)\}, \quad n, k = 1, 2, \ldots.$$

We have $A_{nk} b_k \le b$ for any $b \in B_n$: indeed, if $w \wedge s \wedge b_k \ne 0$ for $w \in W_n$ and $s \in S_k(w)$ then also $w \wedge b \ne 0$ and hence $b = b(w,i)$ for some i, and since $s \wedge b_k \ne 0$ this implies that $s \le b$, showing that $w \wedge s \le b$. Now, $x = \bigvee x_k$ by regularity, and since $b_k \lhd_{\mathcal{A}} b$ as shown, $\mathcal{A}$ is admissible. Use 4.4.1. □

4.7. Theorem. *Let L be a regular frame. Then the following statements are equivalent (in (1) through (5) the classical metrization theorems thus generalized are indicated in brackets).*

(1) *L is metrizable.*

(2) *L has a σ-admissible basis (Moore).*

(3) *L has a σ-discrete basis (Bring).*

(4) *L has a σ-locally finite basis (Nagata-Smirnov).*

(5) *L has a regular basis (Archangelskij).*

Proof. (1)⇔(2) is in 4.2.1, (2)⇒(3) is in 4.4.1, (3)⇒(4) is trivial and (4)⇒(2) is in 4.6; (2)⇒(5) follows from 4.4.1 and 4.5, and finally (5)⇒(4) is in 4.5.2. □

5.   Categories of metric frames

5.1. Proposition. *The category* $\mathbf{D_M Frm}$ *is monocoreflective in* $\mathbf{D_S Frm}$.

Proof. If (L, d) is star-additive then by 1.5, $\frac{1}{2}d(a) \leq \widetilde{d}(a) \leq d(a)$ and hence we have contractive homomorphisms

$$\iota_{(L,d)} = (a \mapsto a) \colon (L, \widetilde{d}) \to (L, d).$$

Let (M, d') be in $\mathbf{D_M Frm}$ and let $h \colon (M, d') \to (L, d)$ be a contractive homomorphism. We will prove that for every $b \in L$ with $\widetilde{d}(b) < \varepsilon$ there is an $a \in M$ such that $d'(a) < \varepsilon$ and $h(a) \geq b$. Thus we will have a contractive

$$\widetilde{h} = (a \mapsto h(a)) \colon (M, d') \to (L, \widetilde{d}),$$

unique such that $\iota_{(L,d)} \widetilde{h} = h$.

Thus let $\widetilde{d}(b) < \varepsilon$; we can assume that $b \neq 0$. Choose an $\eta_1 > 0$ such that $\widetilde{d}(b) < \varepsilon - 5\eta_1$ and an $0 < \eta \leq \eta_1$ such that for all u, v with $d(u), d(v) < \eta$, $u \wedge b \neq 0 \neq v \wedge b$, and $d(u \vee v) < \varepsilon - 5\eta$. Set

$$S = \{u \mid u \wedge b \neq 0, d(u) < \eta\}.$$

For each $u \in S$ choose an $u' \in M$ such that $d'(u) < \eta$ and $h(u') \geq u$. For $u, v \in S$ choose, further, a z such that $u \vee v \leq f(z)$ and $d'(z) < \varepsilon - 5\eta$. We can have $z \leq u' \vee v'$ (otherwise we can replace it by $z \wedge (u' \vee v')$). Now $f(u' \wedge x) \geq u \wedge (u \vee v) = u \neq 0$, hence $u' \wedge x \neq 0$ and similarly $v' \wedge z \neq 0$, and we obtain

$$d(u' \vee v') \leq d'(u' \vee z \vee v') < \varepsilon - 3\eta.$$

Set $a = \bigvee\{u' \mid u \in S\}$. Then obviously $h(a) \geq b$. Since d' is a metric diameter there are x, y such that $x \wedge a \neq 0 \neq y \wedge a$, $d'(x), d'(y) < \eta$, and

$$d'(x \vee y) > d'(a) - \eta.$$

Choose $u, v \in S$ such that $x \wedge u' \neq 0 \neq y \wedge v'$. Now we can conclude that

$$d'(a) < d'(x \vee y) + \eta \leq d'(x \vee u' \vee v' \vee y) + \eta < \varepsilon - 3\eta + 2\eta + \eta = \varepsilon. \qquad \square$$

5.2. Metric sublocales. Let (L, d) be a metric frame and let $h \colon L \to M$ be a sublocale map (frame quotient homomorphism). Define $\overline{d} \colon M \to \mathbb{R}_+$ by setting

$$\overline{d}(b) = \inf\{d(a) \mid h(a) \geq b\}.$$

Note that if $(L, d) = \Omega(X, \rho)$ and if Y is a subset of X then $\overline{d}$ is precisely the diameter on the subspace $(Y, \rho|_{Y \times Y})$.

5.2.1. Proposition. $\overline{d}$ *is an admissible diameter on* M *and* $h \colon (L, d) \to (M, \overline{d})$ *is a contractive homomorphism.*

Proof. We trivially have (D1) and (D2). Now let $a, b \in M$ and $a \wedge b \neq 0$. Choose a', b' such that $h(a') \geq a$, $h(b') \geq b$, $d(a') < \overline{d}(a) + \varepsilon$ and $d(b') < \overline{d}(b) + \varepsilon$. Then $a' \wedge b' \neq 0$ and $a \vee b \leq h(a' \vee b')$ and we have

$$\overline{d}(a \vee b) \leq d(a' \vee b') \leq d(a') + d(b') < \overline{d}(a) + \overline{d}(b) + \varepsilon$$

and as $\varepsilon > 0$ was arbitrary, we have (D3).

Since $\overline{d}(h(a)) \leq d(a)$ we have $h[U_\varepsilon^d] \subseteq U_\varepsilon^{\overline{d}}$ and hence each $U_\varepsilon^{\overline{d}}$ is a cover.

Now let $x \lhd_{\mathcal{U}(d)} y$, say $U_\varepsilon^d x \leq y$. Choose a $u \in M$ such that $\overline{d}(u) < \varepsilon$ and $u \wedge h(x) \neq 0$. We have a $v \in L$ such that $h(v) \geq u$ and $d(v) < \varepsilon$. Then $h(v \wedge x) \geq u \wedge h(x) \neq 0$ and hence $v \leq y$. Thus, $u \leq h(v) \leq h(y)$ and we see that $h(x) \lhd_{\mathcal{U}(\overline{d})} h(y)$ and finally

$$h(y) = \bigvee\{h(x) \mid x \lhd_{\mathcal{U}(d)} y\} \leq \bigvee\{z \mid z \lhd_{\mathcal{U}(\overline{d})} h(y)\},$$

and $\overline{d}$ is admissible.

The contractibility of $h: (L, d) \to (M, \overline{d})$ follows immediately from the definition of $\overline{d}$. $\qquad\square$

5.2.2. Proposition. *If d is star-additive then $\overline{d}$ is star-additive, and if d is metric then $\overline{d}$ is metric as well.*

Proof. I. Let $a \in M$ and $S \subseteq M$ be such that for all $x \in S$, $a \wedge x \neq 0$. Choose $x' \in L$ for $x \in S$ or $x = a$ such that $h(x') \geq x$ and $d(x') < \overline{d}(x) + \varepsilon$. Then for $x \in S$ obviously $a' \wedge x' \neq 0$ and $h(a' \vee \bigvee\{x' \mid x \in S\}) \geq \bigvee S$. We conclude that

$$\overline{d}(a \vee \bigvee S) \leq d(a' \vee \bigvee\{x' \mid x \in S\}) \leq d(a') + \sup\{d(x') + d(y') \mid x, y \in S\}$$
$$\leq \overline{d}(a) + \sup\{\overline{d}(x) + \overline{d}(y) \mid x, y \in S\} + 3\varepsilon.$$

II. Now let d be a metric diameter. Take $a \in M$, $a \neq 0$, and an $\varepsilon = 4\eta > 0$. Set

$$S = \{x \in L \mid d(x) < \eta \text{ and } h(x) \wedge a \neq 0\}.$$

Since U_ε^d is a cover we have $h(\bigvee S) \geq a$.

By 1.4(c) there are $x, y \in S$ such that

$$d(\bigvee S) < d(x \vee y) + \eta.$$

Set $u = h(x)$ and $v = h(y)$. Then $u, v \neq 0$ by the definition of S. Consider a $z \in L$ such that

$$h(z) \geq u \vee v \quad \text{and} \quad d(z) < \overline{d}(u \vee v) + \eta.$$

Then $h(z \wedge x) \geq u \neq 0$ and hence $z \wedge x \neq 0$, and similarly $z \wedge y \neq 0$ and we have

$$\overline{d}(a) \leq d(\bigvee S) < d(x \vee y) + \eta \leq d(x \vee z \vee y) + \eta$$
$$\leq d(z) + d(x) + d(y) + \eta < \overline{d}(u \vee v) + 4\eta = \overline{d}(u \vee v) + \varepsilon. \qquad\square$$

5.2.3. Proposition. *In any of* **DFrm**, **D$_S$Frm** *or* **D$_M$Frm**, *each morphism*

$$h\colon (L, d) \to (M, d')$$

can be decomposed as

$$h = \; ((L, d) \xrightarrow{\;\;f\;\;} (\overline{L}, \overline{d}) \xrightarrow{\;\;g\;\;} (M, d'))$$

with f surjective and g one-one.

Proof. Decompose the frame homomorphism h as

$$L \xrightarrow{\;\;f\;\;} \overline{L} = h[L] \xrightarrow{\;g=\subseteq\;} M$$

and endow $\overline{L}$ with the diameter $\overline{d}$ as above. Now let $b \in M$, $d'(b) < \varepsilon$. Take an $a \in L$ such that $h(a) = g(f(a)) \geq b$ and $d(a) < \varepsilon$. Then for $\overline{a} = h(a) = f(a)$, $g(\overline{a}) \geq b$ and $\overline{d}(\overline{a}) < \varepsilon$. Thus, also g is contractive. $\qquad\square$

5.3. Coproducts. This will be slightly more technical. We will need an auxiliary category

$$\mathbf{P_S Frm}$$

of *star-additive prediametric frames*. The objects (L, d) satisfy (D1), (D2), (DS), and the admissibility condition is replaced by

$$\forall a \in L, \quad a \leq \bigvee_{\varepsilon > 0} \vartriangleleft_\varepsilon^d a. \tag{p-adm}$$

Further, we will need the following consequences of the adjunction between $U_\varepsilon^d(-)$ and $\vartriangleleft_\varepsilon^d(-)$ (recall 1.1.1):

- $\vartriangleleft_\varepsilon^d$ preserves meets,
- if $\varepsilon \leq \eta$ then $\vartriangleleft_\varepsilon^d \geq \vartriangleleft_\eta^d$ (since obviously $U_\varepsilon^d a \leq U_\eta^d a$), and
- $\vartriangleleft_{\varepsilon+\eta}^d(a) \leq \vartriangleleft_\varepsilon^d(\vartriangleleft_\eta^d(a))$ (since obviously $U_\varepsilon^d(U_\eta^d a) \leq U_{\varepsilon+\eta}^d a$).

5.3.1. Lemma. **D$_S$Frm** *is epireflective in* **P$_S$Frm**.

Proof. For $(L, d) \in \mathbf{P_S Frm}$ consider

$$R_d = \{(\bigvee U_\varepsilon^d, 1) \mid \varepsilon > 0\},$$

set $\overline{L} = L/R_d$ (recall III.11), and define $\overline{d}$ as in 5.2. Then $\overline{d}$ is a star-additive diameter in $\overline{L}$ and the quotient homomorphism

$$\kappa = \kappa_{(L,d)}\colon (L, d) \to (\overline{L}, \overline{d})$$

is contractive (the same reasoning as in 5.2.1 and 5.2.2 shows that $\bar{d}$ satisfies (D1), (D2), (DS) and (p-adm), and that κ is contractive; because of the definition of R_d, $U_\varepsilon^{\bar{d}}$ are covers, and hence in particular also $a \geq \bigvee_{\varepsilon>0} \vartriangleleft_\varepsilon^{\bar{d}} a$).

Now let (M, d') be in $\mathbf{D_SFrm}$ and let $h\colon (L, d) \to (M, d')$ be a contractive homomorphism. Take an $\varepsilon > 0$. For each $b \in M$ with $d'(b) < \varepsilon$ there is an $a \in U_\varepsilon^d$ with $h(a) \geq b$. Thus, $h(\bigvee U_\varepsilon^d) \geq \bigvee U_\varepsilon^{d'} = 1$ so that h respects R_d and by III.11.3 there is precisely one $\bar{h}$ such that $\bar{h} \cdot \kappa = h$. $\qquad\square$

5.3.2. Let (L_i, d_i), $i \in J$, be in $\mathbf{D_SFrm}$. We will use the construction (and notation) from IV.4 with the whole of $\prod_{i \in J} L_i$ instead of $\prod_{i \in J}' L_i$, and with the relation R extended to

$$R_d = R \cup \left\{ \left(\bigcup_{\varepsilon>0} {\downarrow}(\vartriangleleft_\varepsilon^i a_i)_{i \in J}, {\downarrow}a \right) \mid a \in \prod_{i \in J} L_i \right\}$$

where we abbreviate

$$\vartriangleleft_\varepsilon^{d_i} \quad \text{to} \quad \vartriangleleft_\varepsilon^i .$$

We denote

$$\tilde{L} = \mathfrak{D}\left(\prod_{i \in J} L_i\right)/R_d, \quad \text{and} \quad \tilde{\mu}\colon \mathfrak{D}\left(\prod_{i \in J} L_i\right) \to \tilde{L} \text{ the quotient morphism,}$$

and further define (notation from IV.4.3)

$$\tilde{\iota}_j = \tilde{\mu}\alpha\kappa_j\colon L_j \to \tilde{L}$$

Because of the R-part of the relation R_d,

$$\tilde{\iota}_j \text{ are frame homomorphisms.}$$

It is easy to see that the saturated elements $U \subseteq \prod_i L_i$ are the down-sets such that (using the symbol $*_j$ from IV.4.2)

(S1) for all j and $\{x^k \mid k \in K\} \subseteq L_j$ and for all $u \in \prod_i L_i$

$$\{x^k *_j u \mid k \in K\} \subseteq U \;\Rightarrow\; \left(\bigvee_k x^k\right) *_j u \in U,$$

(S2) if $(\vartriangleleft_\varepsilon^j(a_j) \wedge x_j)_j \in U$ for all $\varepsilon > 0$ then $a \wedge x \in U$.

(Compare (S1) with the saturation condition in IV.4.3.)

5.3.3. Set

$$\tilde{\mathbf{n}} = \{u \in \prod_{i \in J} L_i \mid \forall \varepsilon > 0 \; \exists j, \; \vartriangleleft_\varepsilon^j u_j = 0\}.$$

Note that this is bigger than the $\mathbf{n}$ from IV.5.1; but combining the two saturation conditions (taking into account the empty K again) we easily infer that

each saturated U contains $\tilde{\mathbf{n}}$.

We have

Lemma. $\widetilde{\mu}(\emptyset) = \widetilde{n}$, and for each $a \in \prod_i L_i$, $\widetilde{\mu}(\downarrow a) = \downarrow a \cup \widetilde{n}$.

Proof. We need to show that $\widetilde{n}$ and $\downarrow a \cup \widetilde{n}$ are saturated.

$\widetilde{n}$: Let $\{x^k *_j u \mid k \in K\} \subseteq \widetilde{n}$ and suppose that for $x = \bigvee_k x^k$ there is an $\varepsilon > 0$ such that for each i one has $\lhd^i_\varepsilon(x *_j u) \neq 0$. Thus in particular $\bigvee_k x^k \neq 0$ and $\lhd^i_\varepsilon u_i \neq 0$ for $i \neq j$. But then for some k and $0 < \eta < \varepsilon$, $\lhd^i_\eta(x^k *_j u) \neq 0$ for all i.

Let $b \wedge x \notin \widetilde{n}$. Then there is an $\varepsilon > 0$ such that $\lhd^j_{2\varepsilon}(b_j \wedge x_j) \neq 0$ for all j and consequently

$$\lhd^j_\varepsilon(\lhd^j_\varepsilon(b_j) \wedge x_j) \geq \lhd^j_{2\varepsilon}(b_j) \wedge \lhd^j_\varepsilon(x_j) \geq \lhd^j_{2\varepsilon}(b_j) \wedge \lhd^j_{2\varepsilon}(x_j) \geq \lhd^j_{2\varepsilon}(b_j \wedge x_j) \neq 0$$

and hence $(\lhd^j_\varepsilon(b_j) \wedge x_j) \notin \widetilde{n}$.

$\downarrow a \cup \widetilde{n}$: Let for $x = \bigvee_k x^k$, $x *_j u \notin \downarrow a \cup \widetilde{n}$. Thus, there is an $\varepsilon > 0$ and an $r \in J$ such that $(x *_j u)_r \not\leq a_r$ and $\lhd^i_\varepsilon(x *_j u) \neq 0$ for all i. If $r = j$ we have a k such that $x^k \not\leq a_j$ so that also for some $0 < \eta \leq \varepsilon$, $\lhd^j_\varepsilon x^k \not\leq a_j$, and $\lhd^i_\eta(x^k *_j u) \neq 0$ for all i. If $r \neq j$ take an $\eta \leq \varepsilon$ and a k such that $\lhd^i_\eta(x^k) \neq 0$. Since $\lhd^i_\eta(x^k *_j u) = \lhd^i_\eta(u_i)$ for $i \neq j$ we have $\lhd^i_\eta(x^k) \neq 0$ for all i, and $u_r = (x^k *_j u)r \not\leq a_r$.

Let $b \wedge x \not\leq \downarrow a \cup \widetilde{n}$. Then, first, $b \wedge x \notin \widetilde{n}$ and hence for some $\varepsilon > 0$, $(\lhd^i_\varepsilon b_i \wedge x_i)_i \notin \widetilde{n}$. Further, there is an r such that $b_r \wedge x_r \not\leq a_r$, and hence for some $0 < \eta \leq \varepsilon$ we have $\lhd^r_\eta(b_r \wedge x_r) \not\leq a_r$. Thus, $(\lhd^i_\eta b_i \wedge x_i)_i \notin \downarrow a \cup \widetilde{n}$. □

5.3.4. Similarly as in IV.5.1 we will write

$$\oplus_i a_i \quad \text{or} \quad \oplus a_i \quad \text{for} \quad \downarrow (a_i)_i \cup \widetilde{n}.$$

From the saturatedness condition (S2) we immediately infer

Fact. *For each* $a \in \prod_i L_i$ *one has* $\bigvee_{\varepsilon > 0} (\oplus \lhd^i_\varepsilon (a_i)) = \oplus a_i$.

5.3.5. Define $\widetilde{d}$ on $\widetilde{L}$ by setting

$$\widetilde{d}(U) = \inf\{\sup_j d_j(a_j) \mid U \subseteq \oplus a_j\}.$$

Set $\lhd_\varepsilon = \lhd^{\widetilde{d}}_\varepsilon$. We have

Lemma. *For each* $a \in \prod_i L_i$ *and each* $\varepsilon > 0$,

$$\oplus \lhd^i_\varepsilon (a_i) \leq \lhd_\varepsilon(\oplus a_i).$$

Proof. Let $\widetilde{d}(U) < \varepsilon$ and $U \cup \oplus \lhd^i_\varepsilon(a_i) \neq \widetilde{n}$. Thus there is an x such that $U \leq \downarrow x \cup \widetilde{n}$ and $d_i(x_i) < \varepsilon$ for all i. Hence $\downarrow x \cap \downarrow(\lhd^i_\varepsilon(a_i))_i \neq \widetilde{n}$ so that $x_i \wedge \lhd^i_\varepsilon(a_i) \neq 0$ for all i, and hence $x_i \leq a_i$, and finally $U \subseteq \downarrow a \cup \widetilde{n} = \oplus a_i$. □

5.3.6. Proposition. $\tilde{d}$ *is an admissible star-additive prediameter on* $\tilde{L}$ *and the homomorphisms* $\tilde{\iota}_j \colon (L_j, d_j) \to (\tilde{L}, \tilde{d})$ *are contractive.*

Proof. (D1) and (D2) are obvious and (DS) easily follows from the fact that $\oplus a_j \wedge \oplus b_j \neq 0$ (that is, $\tilde{n}$) only if $a_i \wedge b_i \neq 0$ for all i.

By 5.3.5 and 5.3.4 we have

$$\bigvee_{\varepsilon>0} \vartriangleleft_\varepsilon(U) \geq \bigvee_{\varepsilon>0} \bigvee\{\vartriangleleft_\varepsilon(\oplus a_j) \mid a \in U\} \geq \bigvee_{\varepsilon>0} \bigvee\{\oplus \vartriangleleft_\varepsilon^j(a_j) \mid a \in U\}$$

$$= \bigvee\{\bigvee_{\varepsilon>0} \oplus \vartriangleleft_\varepsilon^j(a_j) \mid a \in U\} = \bigvee\{\oplus a_j \mid a \in U\} = U$$

proving (p-adm). Finally, if $\tilde{d}(U) < \varepsilon$ take an $a = (a_j)_j$ with $d_j(a_j) < \varepsilon$ such that $U \subseteq {\downarrow} a \cup \tilde{n}$; we have $\tilde{\iota}_k(a_k) \geq \oplus a_j$. $\square$

5.3.7. Lemma. *Let* (M, d) *be in* $\mathbf{D_SFrm}$ *and let* $f_j \colon (L_j, d_j) \to (M, d)$ *be contractive homomorphisms. Then there is precisely one contractive homomorphism* $g \colon (\tilde{L}, \tilde{d}) \to (M, d)$ *such that* $g\tilde{\iota}_j = f_j$ *for all* j.

Proof. I. First we will prove that there exists a contractive homomorphism $f \colon (\tilde{L}, \tilde{d}) \to (M, d)$ such that

(a) $f\tilde{\iota}_j = f_j$ and

(b) $f(\oplus_j a_j) = \bigvee_{\varepsilon>0} \bigwedge_j f_j(\vartriangleleft_\varepsilon^j(a_j))$.

Define $h \colon \mathfrak{D}(\prod_{j \in J} L_j) \to M$ by setting

$$h(U) = \bigvee\{\bigvee_{\varepsilon>0} \bigwedge_{j \in J} f_j \vartriangleleft_\varepsilon^j(a_j) \mid a \in U\}.$$

Obviously $h(\bigcup_k U_k) = \bigvee_k h(U_k)$, $h(1) = 1$ and $h(U_1 \cap U_2) \leq h(U_1) \wedge h(U_2)$ and since, further

$$h(U_1) \wedge h(U_2) = \bigvee\{\bigvee_{\varepsilon>0} \bigwedge_j f_j \vartriangleleft_\varepsilon^j(a_j^1) \wedge \bigvee_\eta \bigwedge_j f_j \vartriangleleft_\eta^j(a_j^2) \mid a^i \in U_i\}$$

$$= \bigvee\{\bigvee_{\varepsilon>0} \bigwedge_j f_j \vartriangleleft_\varepsilon^j(a_j^1 \wedge a_j^2) \mid a^i \in U_i\} \leq h(U_1 \cap U_2),$$

h is a frame homomorphism.

If $\bigvee_k a_j^k = a_j$ and $a_i^k = a_i$ for $i \neq j$ then

$$h(\bigcup_k {\downarrow} a^k) = \bigvee_k h({\downarrow} a^k) = \bigvee_k \bigvee_{\varepsilon>0} \bigwedge_i f_i \vartriangleleft_\varepsilon^i(a_i^k)$$

$$= \bigvee_{\varepsilon>0} \bigvee_k \bigwedge_{i \neq j} f_i \vartriangleleft_\varepsilon^i(a_i^k) \wedge f_j \vartriangleleft_\varepsilon^j(a_j^k) = \bigvee_{\varepsilon>0} (\bigwedge_{i \neq j} f_i \vartriangleleft_\varepsilon^i(a_i) \wedge f_j \vartriangleleft_\varepsilon^j(\bigvee_k a_j^k))$$

$$= \bigvee_{\varepsilon>0} \bigwedge_{i \neq j} f_i \vartriangleleft_\varepsilon^i(a_i) \wedge \bigvee_{\varepsilon>0} f_j \vartriangleleft_\varepsilon^j(\bigvee_k a_j^k) = \bigvee_{\varepsilon>0} \bigwedge_{i \neq j} f_i \vartriangleleft_\varepsilon^i(a_i^k) \wedge f_j(\bigvee_k \bigvee_{\varepsilon>0} \vartriangleleft_\varepsilon^j(a_j^k))$$

$$= \bigwedge_{i \neq j} f_i \vartriangleleft_\varepsilon^i(a_i^k) \wedge f_j(a_j) \geq h({\downarrow} a) \geq h(\bigcup_k {\downarrow} a^k).$$

Using 5.3.4 we obtain

$$h(\bigcup_{\varepsilon>0} \downarrow \sphericalangle_\varepsilon^j (a_j))_j = \bigvee_{\varepsilon>0} h(\downarrow \sphericalangle_\varepsilon^j (a_j)_j) = \bigvee_{\varepsilon>0} \bigvee_\eta \bigwedge_j f_j(\sphericalangle_\eta^j \sphericalangle_\varepsilon^j (a_j))$$

$$\geq \bigvee_{\varepsilon>0} \bigvee_\eta \bigwedge_j f_j(\sphericalangle_{\varepsilon+\eta}^j(a_j)) = \bigvee_{\varepsilon>0} \bigwedge_j f_j(a_j) = h(\downarrow a).$$

Thus, h respects the relation R_d and hence there is a frame homomorphism $f\colon \widetilde{L} \to M$ such that $f \cdot \widetilde{\mu} = h$. In particular we have $f(\oplus a_j) = h(\downarrow a) = \bigvee_{\varepsilon>0} \bigwedge_j f_j \sphericalangle_\varepsilon^j (a_j)$. Further,

$$f\widetilde{\iota}_j(a_j) = f\widetilde{\mu}(a_j *_j \overline{1}) = h(\downarrow(a_j *_j \overline{1})) = \bigvee_{\varepsilon>0} f_j \sphericalangle_\varepsilon^j (a_j) = f_j(\bigvee_{\varepsilon>0} \sphericalangle_\varepsilon^j(a_j)) = f_j(a_j).$$

Finally let $d(b) < \varepsilon$. Choose $\eta > 0$ such that $d(b) < \varepsilon - 2\eta$. There exist $x_j \in L_j$ such that $d_j(x_j) < \varepsilon - 3\eta$ and $f_j(x_j) \geq b$. Put $a_j = U_\eta^{d_j} x_j$. Since d_j are star-additive, we have $d_j(a_j) < \varepsilon$. Now $\widetilde{d}(\oplus a_j) < \varepsilon$ and $b \leq \bigwedge_j f_j(\sphericalangle_\varepsilon^j(a_j)) \leq f(\oplus a_j)$. Thus, f is contractive.

II. Now let a contractive homomorphism $g\colon (\widetilde{L}, \widetilde{d}) \to (M, d)$ be such that $g\widetilde{\iota}_j = f_j$ for all j.

Claim. *For all $\varepsilon > 0$, $\sphericalangle_\varepsilon^d(\bigwedge_j f_j \sphericalangle_\varepsilon^j (a_j)) \leq g(\oplus a_j)$.*

(Indeed, let $b \neq 0$ be such that $d(b) < \varepsilon$ and $b \leq f_j(\sphericalangle_\varepsilon^j(a_j))$ for all j. Since g is contractive, there is a c such that $\widetilde{d}(c) < \varepsilon$ — hence, $d_j(c_j) < \varepsilon$ — and $b \leq g(\oplus c_j)$. We have $c_j \wedge \sphericalangle_\varepsilon^j(a_j) \neq 0$: else $b = f_j(c_j) \wedge b \leq f_j(c_j) \wedge f_j(\sphericalangle_\varepsilon^j(a_j)) = 0$ contradicting the assumption. Thus, we have for all j, $c_j \leq a_j$ and $b \leq g(\oplus c_j) \leq g(\oplus a_j)$. Now

$$\sphericalangle_\varepsilon^d(\bigwedge_j f_j \sphericalangle_\varepsilon^j (a_j)) \leq \bigvee\{b \mid d(b) < \varepsilon \text{ and } b \leq f_j(\sphericalangle_\varepsilon^j(a_j))\}$$

and hence the inequality follows.)

Since for $\eta \leq \varepsilon$ we have

$$\sphericalangle_\eta^j(f_j(\sphericalangle_\varepsilon^j(a_j))) \leq \sphericalangle_\eta^j(f_j(\sphericalangle_\varepsilon^j(a_j))) \leq g(\oplus a_j)$$

it follows that $\bigwedge_j f_j \sphericalangle_\varepsilon^j (a_j) \leq g(\oplus a_j)$ and we conclude that

$$f(\oplus a_j) \leq \bigvee_{\varepsilon>0} \bigwedge_j f_j \sphericalangle_\varepsilon^j (a_j) \leq g(\oplus a_j).$$

On the other hand,

$$g(\oplus a_j) = g(\bigvee_{\varepsilon>0} \oplus \sphericalangle_\varepsilon^j (a_j)) = \bigvee_{\varepsilon>0} g(\oplus \sphericalangle_\varepsilon^j (a_j))$$

$$= \bigvee_{\varepsilon>0} g(\bigwedge_j \widetilde{\iota}_j \sphericalangle_\varepsilon^j (a_j)) \leq \bigvee_{\varepsilon>0} \bigwedge_j f_j \sphericalangle_\varepsilon^j (a_j) = f(\oplus a_j).$$

Since the elements $\oplus a_j$ form a basis of $\widetilde{L}$, $f = g$. $\qquad\square$

5.3.8. Theorem. *The categories* $\mathbf{D_S Frm}$ *and* $\mathbf{D_M Frm}$ *are cocomplete.*

Proof. For a system (L_i, d_i), $i \in J$, in $\mathbf{D_S Frm}$ consider the $(\widetilde{L}, \widetilde{d})$ from 5.3.2 and 5.3.5, and define ι, $\bigoplus$ and d by taking the reflections

$$\iota_j \colon (L_j, d_j) \to (\bigoplus_{i \in J} L_i, d) \quad \text{of} \quad \widetilde{\iota}_j \colon (L_j, d_j) \to (\widetilde{L}, \widetilde{d})$$

by 5.3.1. Then, by 5.3.7, the system

$$\left(\iota_j \colon (L_j, d_j) \to (\bigoplus_{i \in J} L_i, d) \right)_{j \in J}$$

is a coproduct in $\mathbf{D_S Frm}$.

Let $f, g \colon (L, d) \to (M, d')$ be morphisms in $\mathbf{D_S Frm}$. The quotient homomorphism

$$\kappa \colon (M, d') \to (M/R, \overline{d'}) \quad \text{with} \quad R = \{(f(a), g(a)) \mid a \in L\}$$

and $\overline{d'}$ as in 5.2 constitutes by 5.2.1 and 5.2.2 a coequalizer in $\mathbf{D_S Frm}$. Thus we have coproducts and coequalizers, and hence $\mathbf{D_S Frm}$ is cocomplete (see AII.5.3.2). Consequently, also $\mathbf{D_M Frm}$ is, by 5.1 and AII.8.3. $\qquad\square$

5.4. Finite coproducts. Let (L_i, d_i), $i = 1, 2, \ldots, n$, be in $\mathbf{D_S Frm}$ or in $\mathbf{D_M Frm}$.

5.4.1. Lemma. *Consider the relation R and the quotient map*

$$\mu \colon \mathfrak{D}(\prod_{i=1}^{n} L_i) \to \mathfrak{D}(\prod_{i=1}^{n} L_i)/R$$

from IV.4. Then for all $a \in \prod_i L_i$ we have

$$\mu(\bigcup_{\varepsilon > 0} \downarrow (\lhd_\varepsilon^j a_j)_j) = \mu(\downarrow a).$$

Thus μ coincides with the $\widetilde{\mu}$ from 5.3.2.

Proof. We have

$$U = \bigcup_{\varepsilon > 0} \downarrow (\lhd_\varepsilon^1 a_1, \ldots, \lhd_\varepsilon^n a_n) = \bigcup_{\varepsilon_1 > 0} \bigcup_{\varepsilon_2 > 0} \cdots \bigcup_{\varepsilon_n > 0} \downarrow (\lhd_{\varepsilon_1}^1 a_1, \lhd_{\varepsilon_2}^2 a_2, \ldots, \lhd_{\varepsilon_n}^n a_n)$$

(consider $\varepsilon = \min\{\varepsilon_1, \ldots, \varepsilon_n\}$). Thus, we obtain

$$U \, R \, \bigcup_{\varepsilon_2} \cdots \bigcup_{\varepsilon_n} \downarrow (a_1, \lhd_{\varepsilon_2}^2 a_2, \ldots, \lhd_{\varepsilon_n}^n a_n) \, R \, \bigcup_{\varepsilon_3} \cdots \bigcup_{\varepsilon_n} \downarrow (a_1, a_2, \lhd_{\varepsilon_3}^3 a_3, \ldots, \lhd_{\varepsilon_n}^n a_n) \, R$$

$$\cdots R \downarrow (a_1, \ldots, a_{n-1}, \lhd_{\varepsilon_n}^n a_n)) \, R \downarrow (a_1, \ldots, a_n)$$

so that the second part of R_d in 5.3.2 is in our case redundant. $\qquad\square$

5.4.2. Lemma. *For the system* $((L_i, d_i))_{i=1,2,\ldots,n}$ *the prediameter* $\widetilde{d}$ *is a diameter.*

Proof. We have $\widetilde{d}(\oplus a_i) = \max\{d_i(a_i) \mid i = 1, 2, \ldots, n\}$ and hence

$$U_\varepsilon^{\widetilde{d}} \supseteq \left\{ \bigoplus_{i=1}^n a_i \mid a_i \in U_\varepsilon^{d_i} \right\}$$

and the latter is obviously a cover.   □

5.4.3. Proposition. *Finite coproducts in both* $\mathbf{D_S Frm}$ *and* $\mathbf{D_M Frm}$ *are the standard coproducts of frames as in IV.4 endowed with the diameters*

$$d(U) = \inf\{\max_i d_i(a_i) \mid U \subseteq \oplus a_i\}.$$

Proof. In the finite case, $\prod'_i L_i$ coincides with $\prod_i L_i$. By 5.4.1 we have

$$\widetilde{L} = L_1 \oplus \cdots \oplus L_n$$

as in IV.4. Since $\widetilde{d}$ is by 5.4.2 a diameter, the reflection from 5.3.1 is now identical. Finally, because of the coreflection from 5.1, the coproducts in $\mathbf{D_M Frm}$ are obtained as in $\mathbf{D_S Frm}$.   □

Chapter XII

Entourages. Asymmetric Uniformity

The uniform structure as discussed in chapters VIII and X is an extension of the classical one introduced by Tukey ([259]) in 1940. This was, however, not the first approach to modeling uniform phenomena as known from metric spaces. It was preceded by the Weil uniformities ([272], 1938) based on distinguished neighbourhoods of the diagonal of the space in question, the so-called *entourages*. In fact in the first decades the topologists preferred the Weil model and it took until 1964 when the cover approach was brought to the focus of interest by the celebrated Isbell's monograph [135].

In this chapter we will discuss, first, the point-free variant of the entourage uniformity. It should be emphasized that, unlike the cover one, which is a genuine extension (covers correspond precisely to the classical concept), here we have, rather, an imitation: localic products $L \times L$ (frame coproducts $L \oplus L$) do not precisely correspond to products of spaces. Nevertheless, the two structures *are* equivalent and produce the same class of uniform homomorphisms. Due to the nature of the frame coproduct this equivalence is a deeper fact than the classical one (and it is in fact somewhat surprising: the square of the underlying locale – coproduct of the frame with itself – is, first, by far not such a simple structure as the square of spaces $X \times X$, and, second and more important, in the spatial case it does not generally correspond to the space $X \times X$) and can be used to advantage (e.g., we will see an instance in Chapter XV where the uniformity of certain maps is almost immediate when using entourages while one does not really know how to prove it in terms of covers).

We have already mentioned that the entourage principle predominated in the first decades. Surprisingly enough, one of the most important features seems to have played little or no role in the preference. At least until 1948 when Nachbin ([186]) observed that dropping the symmetry requirement (which is impossible in the inherently symmetric cover definition) leads to a rather interesting structure that, a.o., can be carried by general spaces, not only by completely regular ones. A non-symmetric modification of the cover approach turned out to be possible, after all, much later (Gantner and Steinlage [111], 1972); it did not find a general response, probably because in the classical setting there was not much demand for it (unlike in the point-free setting). The non-symmetric uniformities will be treated in the second part of this chapter.

1. Entourages

1.1. Entourages. Let L be a frame. An *entourage* in L is an element $E \in L \oplus L$ such that

$$\{u \mid u \oplus u \le E\}$$

is a cover of L.

The composite of entourages E and F is given by

$$\begin{aligned} E \circ F &= \bigvee\{a \oplus c \mid \exists b \neq 0, \ a \oplus b \le E \text{ and } b \oplus c \le F\} \\ &= \bigvee\{a \oplus c \mid \exists b \neq 0, \ (a,b) \in E \text{ and } (b,c) \in F\}. \end{aligned} \tag{1.1.1}$$

A word of caution: unions of saturated sets are not necessarily saturated and the join above is typically bigger than the corresponding union.

Evidently this definition makes sense for arbitrary $E, F \in L \oplus L$, and the operators $E \circ (-)$ and $(-) \circ E$ are clearly order-preserving.

For an entourage E set

$$E^{-1} = \{(a,b) \mid (b,a) \in E\}$$

(which is obviously an entourage again). An entourage E is *symmetric* if $E^{-1} = E$.

1.2. Proposition. *Let $a, b \in L$ and $E, F \in L \oplus L$. Then:*

(a) $(a \oplus b)^{-1} = b \oplus a$.

(b) $(E \circ F)^{-1} = F^{-1} \circ E^{-1}$.

Proof. (a) is obvious.

(b): By (a), it suffices to check that

$$\left(\bigvee_{i \in I} E_i\right)^{-1} = \bigvee_{i \in I} (E_i)^{-1}$$

for every $\{E_i \mid i \in I\} \subseteq L \oplus L$. Of course, $E_i^{-1} \le (\bigvee_{i \in I} E_i)^{-1}$ for every $i \in I$. On the other hand, if $E_i^{-1} \le F$ for every $i \in I$ then $E_i \le F^{-1}$ and thus $\bigvee_{i \in I} E_i \le F^{-1}$, that is, $(\bigvee_{i \in I} E_i)^{-1} \le F$. $\qquad\square$

1.3. Proposition. *For each entourage E of L we have:*

(a) $E \le E \circ E$; *more generally, for any entourage E and any $F \in L \oplus L$, $F \le (E \circ F) \cap (F \circ E)$.*

(b) *If E is symmetric and $0 \neq a \oplus b \le E$, then $(a \vee b) \oplus (a \vee b) \le E \circ E$.*

Proof. (a): It suffices to show the inequality $F \leq F \circ E$ (the inequality $F \leq E \circ F$ may be proved in a similar way). Let $a \oplus b \leq F$ and $b \neq 0$. We have

$$b = \bigvee\{b \wedge u \mid u \oplus u \leq E\} = \bigvee\{b \wedge u \mid u \oplus u \leq E, \ u \wedge b \neq 0\}.$$

Now we have, for $b \wedge u \neq 0$, $a \oplus (b \wedge u) \leq F$ and $(b \wedge u) \oplus (b \wedge u) \leq E$, hence $a \oplus (b \wedge u) \leq F \circ E$ and hence

$$a \oplus b = a \oplus \bigvee\{b \wedge u \mid u \oplus u \leq E\} = \bigvee\{a \oplus (b \wedge u) \mid u \oplus u \leq E\} \leq F \circ E.$$

(b): Let $a \oplus b \leq E$, so that also $b \oplus a \leq E$ and hence $a \oplus a, b \oplus b \leq E \circ E$; since $a, b \neq 0$, by (a) also $a \oplus b, b \oplus a \leq E \circ E$. Thus $a \oplus (a \vee b) \leq E$, $b \oplus (a \vee b) \leq E$ and finally $(a \vee b) \oplus (a \vee b) \leq E$. $\qquad\square$

1.4. Remarks. (1) By the formula for Δ^* in V.2.1 it follows that an $E \in L \oplus L$ is an entourage if and only if $\Delta^*(E) = 1$. Therefore, E is an entourage if and only if the diagonal $\Delta \colon L \to L \oplus L$ factorizes through the embedding $j_E \colon \mathfrak{o}(E) \to L \oplus L$:

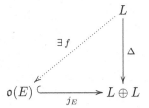

Indeed, if $j_E f = \Delta$ then $\Delta^*(E) = f^* j_E^*(E) = f^*(1) = 1$. Conversely, if $\Delta^*(E) = 1$ then, for every $x \in L$,

$$E \to \Delta(x) = \Delta(\Delta^*(E) \to x) = \Delta(1 \to x) = \Delta(x).$$

Thus $\Delta(x) \in \mathfrak{o}(E)$ and, by Proposition II.2.3, $\Delta \colon L \to \mathfrak{o}(E)$ is localic.

Using Corollary IV.5.2.1, we may then conclude that the collection $Ent(L)$ of all entourages of L is completely described by the open sublocales $\mathfrak{o}(E)$ of the product locale $L \oplus L$ for which the diagonal $\Delta \colon L \to L \oplus L$ factorizes through the embedding $j_E \colon \mathfrak{o}(E) \to L \oplus L$.

This shows that the entourage definition in 1.1 is just a mimicking (or an extension by analogy) in the category of locales of the classical definition. In fact, recall VIII.1.1: an entourage of a set X is a subset E of $X \times X$ for which the diagonal $\Delta_X \colon X \to X \times X$ factorizes through the inclusion $E \hookrightarrow X \times X$:

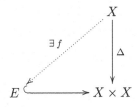

(Recall further that the entourages of a uniform space $(X, \mathcal{E})$ that are open in the product topology $X \times X$ constitute a basis for the uniformity $\mathcal{E}$ [83].)

(2) There is also a categorical justification for the formula (1.1.1). Indeed, in any category $\mathcal{C}$ with finite limits and images, a *relation*

$$r \colon X \twoheadrightarrow Y$$

is a subobject $r \colon R \to X \times Y$, or equivalently, a pair $(r_1 \colon R \to X, r_2 \colon R \to Y)$ of morphisms with $\langle r_1, r_2 \rangle \colon R \to X \times Y$ a subobject [106]. The *composite $s \circ r$* of $r \colon X \twoheadrightarrow Y$ and $s \colon Y \twoheadrightarrow Z$ is the image of

$$\langle r_1 \cdot s_1', s_2 \cdot r_2' \rangle \colon R \times_Y S \to X \times Z,$$

where $(R \times_Y S, s_1', r_2')$ is the pullback of (r_2, s_1).

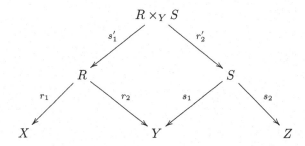

Now, when applied in **Loc** to relations defined by open sublocales this composite is given precisely by the formula (1.1.1) (see [93] for the details).

(3) Notice also that the inverse E^{-1} of an entourage E is precisely $r^*(E)$ where r is the automorphism $L \oplus L \to L \oplus L$ interchanging the two product projections $L \oplus L \to L$:

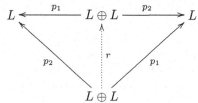

2. Uniformities via entourages

2.1. The "uniformly below" relation. If E is an entourage (resp. $\mathcal{E}$ a set of entourages) write

$$b \lhd_E a \ \text{ if } \ E \circ (b \oplus b) \leq a \oplus a, \quad \text{and} \quad b \lhd_{\mathcal{E}} a \ \text{ if } \ \exists E \in \mathcal{E}, \ b \lhd_E a.$$

This is the entourage counterpart of the "uniformly below" relation $\lhd_A$ of VIII.2.3.

2.1.1. Remark. By 1.2, in case $\mathcal{E}$ is *symmetric* (i.e., $E \in \mathcal{E} \Rightarrow E^{-1} \in \mathcal{E}$) then

$$b \lhd_{\mathcal{E}} a \ \text{ iff } \ \exists E \in \mathcal{E}, \ (b \oplus b) \circ E \leq a \oplus a.$$

2.1.2. Proposition. *Let $\mathcal{E}$ be a set of entourages of L. Then:*

(a) $a \leq b \lhd_{\mathcal{E}} c \leq d \Rightarrow a \lhd_{\mathcal{E}} d$.

(b) *If for any $E_1, E_2 \in \mathcal{E}$ there is an $F \in \mathcal{E}$ such that $F \leq E_1 \cap E_2$ then* $a \lhd_{\mathcal{E}} b_1, b_2 \Rightarrow a \lhd_{\mathcal{E}} (b_1 \wedge b_2)$ *and* $a_1, a_2 \lhd_{\mathcal{E}} b \Rightarrow (a_1 \vee a_2) \lhd_{\mathcal{E}} b$.

Proof. (a) is obvious.

(b): From $E_1 \circ (a \oplus a) \leq b_1 \oplus b_1$ and $E_2 \circ (a \oplus a) \leq (b_2 \oplus b_2)$ it follows that

$$
\begin{aligned}
(E_1 \cap E_2) \circ (a \oplus a) \ &\leq \ (E_1 \circ (a \oplus a)) \cap (E_2 \circ (a \oplus a)) \\
&\leq \ (b_1 \oplus b_1) \cap (b_2 \oplus b_2) \leq (b_1 \wedge b_2) \oplus (b_1 \wedge b_2).
\end{aligned}
$$

The other assertion may be proved similarly. $\qquad\square$

2.2. Admissible systems. A system of entourages $\mathcal{E}$ of L is said to be *admissible* if

$$\forall x \in L, \ x = \bigvee \{y \mid y \lhd_{\mathcal{E}} x\}.$$

2.3. Entourage uniformity on a frame. An *entourage uniformity* on a frame L is a non-empty admissible set of entourages $\mathcal{E}$ such that

(E1) if $E \in \mathcal{E}$ and $E \leq F$ then $F \in \mathcal{E}$,

(E2) if $E, F \in \mathcal{E}$ then $E \cap F \in \mathcal{E}$,

(E3) if $E \in \mathcal{E}$ then $E^{-1} \in \mathcal{E}$, and

(E4) for every $E \in \mathcal{E}$ there is an $F \in \mathcal{E}$ such that $F \circ F \leq E$.

An *entourage uniform frame* is a couple $(L, \mathcal{E})$ where $\mathcal{E}$ is an entourage uniformity on L.

An entourage uniformity $\mathcal{E}$ is often described by a *basis* $\mathcal{E}'$, that is, a subset $\mathcal{E}' \subseteq \mathcal{E}$ such that

$$\mathcal{E} = \{F \mid E \leq F \text{ for an } E \in \mathcal{E}'\}.$$

Note that an entourage uniformity has for instance the basis constituted by the *symmetric* entourages: indeed, for a general E consider the $E^{-1} \cap E \leq E$.

2.3.1. Example. Let (L, d) be a metric frame. For any real $\varepsilon > 0$ let

$$E_\varepsilon = \bigvee \{x \oplus x \mid d(x) < \varepsilon\}.$$

The family $(E_\varepsilon)_{\varepsilon > 0}$ is a basis for an entourage uniformity on L.

2.3.2. For an entourage E define

$$\widetilde{E} = \bigvee \{u \oplus u \mid u \oplus u \leq E\}.$$

It follows immediately from 1.3(b) that

if $\mathcal{E}$ is a uniformity then for each $E \in \mathcal{E}$, $\widetilde{E} \in \mathcal{E}$.

2.4. Uniform homomorphisms. For any pair of frame homomorphisms $f_i \colon L_i \to M_i$ $(i = 1, 2)$ let $f_1 \oplus f_2$ be the unique morphism from $L_1 \oplus L_2$ to $M_1 \oplus M_2$ that makes the following diagram commutative.

$$
\begin{array}{ccccc}
L_1 & \xrightarrow{\;u_{L_1}\;} & L_1 \oplus L_2 & \xleftarrow{\;u_{L_2}\;} & L_2 \\
\Big\downarrow{\scriptstyle f_1} & & \Big\downarrow{\scriptstyle f_1 \oplus f_2} & & \Big\downarrow{\scriptstyle f_2} \\
M_1 & \xrightarrow{\;u_{M_1}\;} & M_1 \oplus M_2 & \xleftarrow{\;u_{M_2}\;} & M_2
\end{array}
$$

Clearly,

$$(f_1 \oplus f_2)\left(\bigvee_{j \in J} (a_j \oplus b_j) \right) = \bigvee_{j \in J} \Big(f_1(a_j) \oplus f_2(b_j) \Big).$$

Let $(L, \mathcal{E}), (M, \mathcal{F})$ be entourage uniform frames. A *uniform homomorphism* $h \colon (L, \mathcal{E}) \to (M, \mathcal{F})$ is a frame homomorphism $h \colon L \to M$ such that

$$(h \oplus h)(E) \in \mathcal{F} \quad \text{for every } E \in \mathcal{E}. \tag{2.4.1}$$

The motivation is clear: recalling III.7.3 this means, in terms of open sublocales, that $(h \oplus h)_{-1}(\mathfrak{o}(E)) = \mathfrak{o}((h \oplus h)^*(E)) \in \mathcal{E}$ for any $\mathfrak{o}(E) \in \mathcal{F}$.

When the uniformities are replaced by their bases (usually very naturally described) the formula (2.4.1) is modified to

$$\forall E \in \mathcal{E} \; \exists F \in \mathcal{F} \text{ such that } (h \oplus h)(E) \geq F. \tag{2.4.2}$$

The resulting category of entourage uniform frames and uniform homomorphisms is denoted by

EUniFrm.

3. Entourages versus covers

3.1. In this section we will prove that the concepts of entourage uniformity and covering uniformity (from Chapter VIII) are equivalent in the sense that there are natural translations of one into the other, with satisfactory properties (for instance frame homomorphisms are uniform iff they are uniform when the uniformities are replaced by the associated ones).

Recall from the preamble to this chapter that while the cover uniformities are genuine generalization of the classical ones, the entourage ones are, rather, just an analogy of the classical definition (since products of locales – coproducts of frames – do not quite correspond to products of classical spaces). Hence the fact of equivalence is somewhat deeper than it sounds.

3.2. The following will play a crucial role.

3.2.1. Lemma. *Let U be a cover of L and let $x \oplus y \leq \bigvee\{u \oplus u \mid u \in U\}$. Let $y \neq 0$. Then $x \leq Uy$.*

Proof. We have

$$x \oplus y = \bigvee\{(u \wedge x) \oplus (u \wedge y) \mid u \in U\}$$
$$= \bigvee\{(u \wedge x) \oplus (u \wedge y) \mid u \in U, \ u \wedge y \neq 0\} \leq (Uy \wedge x) \oplus y.$$

Thus, if $y \neq 0$, $x \leq x \wedge Uy$, and finally $x \leq Uy$. $\qquad\square$

3.3. Translations. For a cover U define an entourage E_U, and for an entourage E define a cover U_E as follows:

$$E_U = \bigvee\{x \oplus x \mid x \in U\}, \quad U_E = \{x \mid x \oplus x \leq E\}.$$

This correspondence is of course not one-one but we have the following

3.3.1. Lemma. (a) $U \leq U_{E_U} \leq UU$.

(b) $\widetilde{E} = E_{U_E} \leq E$.

Proof. (a): If $x \oplus x \leq E_U$ then for any $u_0 \in U$ such that $u_0 \wedge x \neq 0$ we have, by 3.2.1, $x \leq U(u_0 \wedge x) \leq Uu_0$. Thus,

$$U_{E_U} = \{x \mid x \oplus x \leq E_U\} \leq UU.$$

On the other hand, trivially, $U \leq U_{E_U}$ (because $U \subseteq U_{E_U}$).

(b): We have, by the definitions, $E_{U_E} = \bigvee\{x \oplus x \mid x \oplus x \leq E\} = \widetilde{E}$. $\qquad\square$

3.3.2. Lemma. (a) $Ub \leq a \quad \Rightarrow \quad b \vartriangleleft_{E_U} a$.

(b) $b \vartriangleleft_E a \quad \Rightarrow \quad U_E b \leq a$.

Proof. (a): Let $Ub \leq a$ and let $x \oplus y \leq E_U$ and $y \oplus z \leq b \oplus b$ for some $y \neq 0$. Then $x \oplus y \leq \bigvee \{u \oplus u \mid u \in U\}$ and by 3.2.1, $x \leq Uy$. Thus, $x \oplus z \leq Uy \oplus b \leq Ub \oplus b \leq a \oplus a$.

(b): Let $u \in U_E$ and $u \wedge b \neq 0$. Then $u \oplus (u \wedge b) \leq E$ and $(u \wedge b) \oplus b \leq b \oplus b$, and hence $u \oplus b \leq E \circ (b \oplus b) \leq a \oplus a$; thus, as $b \neq 0$, $u \leq a$ and we conclude that $U_E b \leq a$. $\qquad \square$

3.3.3. Lemma. *Let U be a cover and let E be a symmetric entourage. Then:*

(a) $E_U \circ E_U \leq E_{UU}$.

(b) *If* $(F \circ F) \circ (F \circ F) \leq E$ *then* $U_F U_F \leq U_E$.

Proof. (a): Take an $x \oplus z \leq E_U \circ E_U$; hence there is a $y \neq 0$ such that $x \oplus y \leq E_U$ and $y \oplus z \leq E_U$. Choose a $u \in U$ such that $u \wedge y \neq 0$. Since $x \oplus (u \wedge y) \leq E_U$ and $(u \wedge y) \oplus z \leq E_U$ we obtain from 3.2.1 that $x \oplus z \leq Uu \oplus Uu \leq E_{UU}$.

(b): Set $F_1 = F \circ F$. Fix a $u \in U_F$ and take an arbitrary $v \in U_F$ such that $u \wedge v \neq 0$. Then because of $u \wedge v \neq 0$, $v \oplus u \leq F_1 = F \circ F$ and since F_1 is saturated,

$$U_F u \oplus u = \left(\bigvee \{v \mid v \wedge u \neq 0\} \right) \oplus u \leq F_1.$$

Since F is symmetric, F_1 is symmetric as well and hence by 1.3(b) $U_F u \oplus U_F u \leq F_1 \circ F_1 \leq E$, and $U_F u \in U_E$. $\qquad \square$

For an entourage uniformity $\mathcal{E}$ set

$$\mathcal{U}_{\mathcal{E}} = \{V \mid V \geq U_E, \ E \in \mathcal{E}\}$$

and for a cover uniformity $\mathcal{U}$ define

$$\mathcal{E}_{\mathcal{U}} = \{F \mid F \geq E_U, \ U \in \mathcal{U}\}.$$

3.3.4. Theorem. *The correspondences $\mathcal{E} \mapsto \mathcal{U}_{\mathcal{E}}$ and $\mathcal{U} \mapsto \mathcal{E}_{\mathcal{U}}$ constitute a one-one correspondence between the entourage uniformities and the cover ones. More explicitly, $\mathcal{E}_{\mathcal{U}_{\mathcal{E}}} = \mathcal{E}$ and $\mathcal{U}_{\mathcal{E}_{\mathcal{U}}} = \mathcal{U}$.*

Proof. By 3.3.2, $\triangleleft_{\mathcal{U}} = \triangleleft_{\mathcal{E}_{\mathcal{U}}}$ (and $\triangleleft_{\mathcal{E}} = \triangleleft_{\mathcal{U}_{\mathcal{E}}}$). Thus, if any of the associated uniformities is admissible then the other is as well.

By 3.3.1 we have the formulas $\mathcal{E}_{\mathcal{U}_{\mathcal{E}}} = \mathcal{E}$ and $\mathcal{U}_{\mathcal{E}_{\mathcal{U}}} = \mathcal{U}$.

Thus, it remains to be proved that the systems $\mathcal{U}_{\mathcal{E}}$ and $\mathcal{E}_{\mathcal{U}}$ are uniformities (of their types). Condition (E4) follows from 3.3.3(a) whereas (U3) follows from 3.3.3(b); the other facts are straightforward. $\qquad \square$

3.4. The concrete isomorphism UniFrm $\cong$ EUniFrm. As we now show, the translation above between entourages and covers provides a concrete isomorphism between the concrete categories **UniFrm** and **EUniFrm**.

3.4.1. The functor Φ: UniFrm $\to$ EUniFrm. The following proposition yields a functor

$$\Phi\colon \mathbf{UniFrm} \to \mathbf{EUniFrm}.$$

Proposition. (a) *Let $(L,\mathcal{U})$ be an object of* **UniFrm**. *Then $(L,\mathcal{E}_{\mathcal{U}})$ is an object of* **EUniFrm**.

(b) *Let $h\colon (L,\mathcal{U}) \to (M,\mathcal{V})$ be a morphism of* **UniFrm**. *Then h is a* **EUniFrm**-*morphism $(L,\mathcal{E}_{\mathcal{U}}) \to (M,\mathcal{E}_{\mathcal{V}})$.*

Proof. (a) was already proved in 3.3.4.

(b): Consider an $E \in \mathcal{E}_{\mathcal{U}}$. Then $U_E \in \mathcal{U}_{\mathcal{E}_{\mathcal{U}}} = \mathcal{U}$ so $h[U_E] \in \mathcal{V} = \mathcal{U}_{\mathcal{E}_{\mathcal{V}}}$, that is, $h[U_E] \geq U_F$ for some $F \in \mathcal{E}_{\mathcal{V}}$. Since $\widetilde{F} \in \mathcal{E}_{\mathcal{V}}$ (recall 2.3.2), it now suffices to show that $(h \oplus h)(E) \geq \widetilde{F}$. Let $v \oplus v \leq F$. Then $v \in U_F$ and hence there is a $u \in U_E$ such that $v \leq h(u)$. Then $v \oplus v \leq h(u) \oplus h(u) \leq (h \oplus h)(u \oplus u) \leq (h \oplus h)(E)$ and hence $(h \oplus h)(E) \geq \widetilde{F}$. $\qquad\square$

3.4.2. The functor Ψ: EUniFrm $\to$ UniFrm. In the reverse direction we have:

Proposition. (a) *Let $(L,\mathcal{E})$ be an object of* **EUniFrm**. *Then $(L,\mathcal{U}_{\mathcal{E}})$ is an object of* **UniFrm**.

(b) *Let $h\colon (L,\mathcal{E}) \to (M,\mathcal{F})$ be a morphism of* **EUniFrm**. *Then h is a* **UniFrm**-*morphism $(L,\mathcal{U}_{\mathcal{E}}) \to (M,\mathcal{U}_{\mathcal{F}})$.*

Proof. (a): See 3.3.4.

(b): Consider a $U \in \mathcal{U}_{\mathcal{E}}$ and a $V \in \mathcal{U}_{\mathcal{E}}$ such that $VV \leq U$. Then $E_V \in \mathcal{E}_{\mathcal{U}_{\mathcal{E}}} = \mathcal{E}$ so $(h \oplus h)(E_V) \in \mathcal{F} = \mathcal{E}_{\mathcal{U}_{\mathcal{F}}}$, that is, $(h \oplus h)(E_V) \geq E_W$ for some $W \in \mathcal{U}_{\mathcal{F}}$. It then suffices to show that $h[U] \geq W$. Take a $w \in W$. Then $w \oplus w \leq \bigvee\{h(v) \oplus h(v) \mid v \in V\}$ and hence, by 3.2.1, if we take a $v_0 \in V$ such that $y = w \wedge h(v_0) \neq 0$ we obtain that $w \leq h[V]y \leq h[V]h(v_0) \leq h(Vv_0) \leq h(u)$ for some $u \in U$. Thus, $W \leq h[U]$. $\qquad\square$

It follows immediately from 3.3.4, 3.4.1 and 3.4.2 that

3.4.3. Corollary. *The functors Φ and Ψ establish a concrete isomorphism between the concrete categories* **UniFrm** *and* **EUniFrm**. $\qquad\square$

4. Asymmetric uniformity: the classical case

4.1. Entourages. The study of quasi-uniformities began in 1948 with Nachbin's investigations on uniform preordered spaces (see [186] and [188]). The classical theory of quasi-uniform spaces – as opposed to the theory of uniform spaces – is usually presented in the literature in terms of entourages: a quasi-uniformity on a set is defined just by dropping the symmetry requirement from the set of axioms of Weil for a uniform space (see Chapter VIII, Section 1). This is a theory which has achieved much success (as a justification for this statement see, for example, [169] and [170]). We refer the reader to Fletcher and Lindgren ([103]) for a general reference on quasi-uniform spaces.

Let X be a set and let $\mathcal{U}$ be a family of entourages (i.e., reflexive relations) of X. The pair $(X,\mathcal{U})$ is a *quasi-uniform space* provided that:

(1) $V \supseteq U \in \mathcal{U} \;\;\Rightarrow\;\; V \in \mathcal{U}$,

(2) $U, V \in \mathcal{U} \;\;\Rightarrow\;\; U \cap V \in \mathcal{U}$, and

(3) for every $U \in \mathcal{U}$ there is a $V \in \mathcal{U}$ such that $V \circ V \subseteq U$.

One has then two associated topologies $\mathfrak{T}_1(\mathcal{U})$ and $\mathfrak{T}_2(\mathcal{U})$:

- an $M \subseteq X$ is $\mathfrak{T}_1(\mathcal{U})$-*open* iff for each $x \in M$ there is a $U \in \mathcal{U}$ such that $Ux = \{y \mid (y,x) \in U\} \subseteq M$, and

- an $M \subseteq X$ is $\mathfrak{T}_2(\mathcal{U})$-*open* iff for each $x \in M$ there is a $U \in \mathcal{U}$ such that $xU = U^{-1}x = \{y \mid (x,y) \in U\} \subseteq M$.

We have thus the *induced bitopological space*

$$(X, \mathfrak{T}_1(\mathcal{U}), \mathfrak{T}_2(\mathcal{U})). \tag{4.1.1}$$

A map f from a quasi-uniform space $(X,\mathcal{U})$ to a quasi-uniform space $(Y,\mathcal{V})$ is *uniformly continuous* if $(f \times f)^{-1}[V] \in \mathcal{U}$ for every $V \in \mathcal{V}$. Each uniformly continuous map $f \colon (X,\mathcal{U}) \to (Y,\mathcal{V})$ defines immediately a bicontinuous map $f \colon (X, \mathfrak{T}_1(\mathcal{U}), \mathfrak{T}_2(\mathcal{U})) \to (Y, \mathfrak{T}_1(\mathcal{V}), \mathfrak{T}_2(\mathcal{V}))$ (see 5.1 below).

4.2. Paircovers. The symmetry axiom is not explicit in the definition via covers of a uniformity (Chapter VIII, Section 1). In order to describe quasi-uniformities via covers, since covers are present in the form of sections of entourages and entourages are no longer symmetric, one is forced to consider pairs of covers interlocking each other. This cover-like approach for quasi-uniform spaces was introduced by Gantner and Steinlage ([111]). A *conjugate pair of covers* (briefly, *paircover*) of a set X is a subset $\mathcal{U}$ of $\mathfrak{P}(X) \times \mathfrak{P}(X)$ satisfying

$$\bigcup\{U_1 \cap U_2 \mid (U_1,U_2) \in \mathcal{U}\} = X.$$

The paircover $\mathcal{U}$ is called *strong* if, in addition,

$$U_1 \cap U_2 \neq \emptyset \text{ whenever } (U_1,U_2) \in \mathcal{U} \text{ and either } U_1 \neq \emptyset \text{ or } U_2 \neq \emptyset.$$

Let $\mathcal{U}$ and $\mathcal{V}$ be paircovers. A paircover $\mathcal{U}$ is a *refinement* of a paircover $\mathcal{V}$ (and one writes $\mathcal{U} \leq \mathcal{V}$) if for each $(U_1, U_2) \in \mathcal{U}$, there is $(V_1, V_2) \in \mathcal{V}$ with $U_1 \subseteq V_1$ and $U_2 \subseteq V_2$. Further,

$$\mathcal{U} \wedge \mathcal{V} = \{(U_1 \cap V_1, U_2 \cap V_2) \mid (U_1, U_2) \in \mathcal{U}, (V_1, V_2) \in \mathcal{V}\}.$$

Finally, we write for a paircover $\mathcal{U}$ and a subset $A \subseteq X$ (resp. an element $x \in X$)

$$\mathcal{U} * A = \bigcup\{U_1 \mid (U_1, U_2) \in \mathcal{U} \text{ and } U_2 \cap A \neq \emptyset\} \ (\text{resp. } \mathcal{U} * x = \mathcal{U} * \{x\})$$

and

$$A * \mathcal{U} = \bigcup\{U_2 \mid (U_1, U_2) \in \mathcal{U} \text{ and } U_1 \cap A \neq \emptyset\} \ (\text{resp. } x * \mathcal{U} = \{x\} * \mathcal{U}).$$

Now, a non-empty family $\mathfrak{U}$ of paircovers of a set X is a *quasi-uniformity* on X provided that:

(C1) $\mathcal{U} \in \mathfrak{U}$ and $\mathcal{U} \leq \mathcal{V}$ implies $\mathcal{V} \in \mathfrak{U}$,

(C2) if $\mathcal{U}, \mathcal{V} \in \mathfrak{U}$ then there exists a strong paircover $\mathcal{W} \in \mathfrak{U}$ such that $\mathcal{W} \leq \mathcal{U} \wedge \mathcal{V}$, and

(C3) for each $\mathcal{U} \in \mathfrak{U}$ there is a $\mathcal{V} \in \mathfrak{U}$ such that

$$\mathcal{V}^* = \{(\mathcal{V} * V_1, V_2 * \mathcal{V})) \mid (V_1, V_2) \in \mathcal{V}\} \leq \mathcal{U}.$$

For the associated topologies, a set $M \subseteq X$ is:

- $\mathfrak{T}_1(\mathfrak{U})$-*open* iff for each $x \in M$ there is a $\mathcal{U} \in \mathfrak{U}$ such that $\mathcal{U} * x \subseteq M$,
- $\mathfrak{T}_2(\mathfrak{U})$-*open* iff for each $x \in M$ there is a $\mathcal{U} \in \mathfrak{U}$ such that $x * \mathcal{U} \subseteq M$.

The correspondence between entourages and (strong) paircovers that provides the equivalence between the two approaches is the following:

If $\mathcal{U}$ is a paircover of X then $\mathcal{U}$ determines an entourage

$$R_{\mathcal{U}} = \bigcup\{U_1 \times U_2 \mid (U_1, U_2) \in \mathcal{U}\};$$

conversely, for any entourage R of X, R determines a strong paircover $\mathcal{U}_R$ given by

$$\mathcal{U}_R = \{(xR, Rx) \mid x \in X\}.$$

The two definitions of quasi-uniformity are then easily seen to be equivalent (see [111]).

A mapping $f \colon (X, \mathfrak{U}) \to (Y, \mathfrak{V})$ is *uniformly continuous* if

$$\text{for each } \mathcal{V} \in \mathfrak{V}, \quad f^{-1}[\mathcal{V}] = \{(f^{-1}[V_1], f^{-1}[V_2]) \mid (V_1, V_2) \in \mathcal{V}\} \in \mathfrak{U}.$$

We denote the category of quasi-uniform spaces and uniformly continuous maps by

QUnif.

We have

4.2.1. Fact. *A uniformly continuous mapping $f \colon (X, \mathfrak{U}) \to (Y, \mathfrak{V})$ is continuous in the associated bitopologies, that is, $f \colon (X, \mathfrak{T}_1(\mathfrak{U}), \mathfrak{T}_2(\mathfrak{U})) \to (Y, \mathfrak{T}_1(\mathfrak{V}), \mathfrak{T}_2(\mathfrak{V}))$ is a bicontinuous mapping.*

Proof. It suffices to show that $f\colon (X, \mathfrak{T}_i(\mathfrak{U})) \to (Y, \mathfrak{T}_i(\mathfrak{V}))$ is continuous for $i = 1$ (the case $i = 2$ can be treated similarly).

Let $M \subseteq Y$ be $\mathfrak{T}_1(\mathfrak{V})$-open and $x \in f^{-1}[M]$. There is a $\mathcal{V} \in \mathfrak{V}$ such that $\mathcal{V} * f(x) \subseteq M$ and, by hypothesis, $f^{-1}[\mathcal{V}] = \{(f^{-1}[V_1], f^{-1}[V_2]) \mid (V_1, V_2) \in \mathcal{V}\} \in \mathfrak{U}$. It suffices to show that $f^{-1}[\mathcal{V}] * x \subseteq f^{-1}[M]$:

$$
\begin{aligned}
f^{-1}[\mathcal{V}] * x &= \bigcup\{f^{-1}[V_1] \mid (V_1, V_2) \in \mathcal{V}, x \in f^{-1}[V_2]\} \\
&= f^{-1}(\bigcup\{V_1 \mid (V_1, V_2) \in \mathcal{V}, x \in f^{-1}[V_2]\}) \\
&= f^{-1}(\mathcal{V} * f(x)) \subseteq f^{-1}[M]. \qquad \square
\end{aligned}
$$

4.2.2. Note. In particular this means that

$$
f\colon (X, \mathfrak{T}_1(\mathfrak{U}) \vee \mathfrak{T}_2(\mathfrak{U})) \to (Y, \mathfrak{T}_1(\mathfrak{V}) \vee \mathfrak{T}_2(\mathfrak{V}))
$$

is continuous.

4.3. Take a quasi-uniformity $\mathfrak{U}$ on a set X and a bitopology (τ_1, τ_2) on X. What makes $\mathfrak{U}$ to induce (τ_1, τ_2) as the associated bitopology?

First we will see that one can restrict the quasi-uniformity $\mathfrak{U}$ to a system of *open paircovers*, that is, paircovers $\mathcal{U}$ such that $U_i \in \tau_i$ $(i = 1, 2)$ for every $(U_1, U_2) \in \mathcal{U}$. More precisely, we have

4.3.1. Proposition. *For each $\mathcal{U} \in \mathfrak{U}$ define $\mathrm{int}(\mathcal{U})$ as the system of all pairs of interiors $(\mathrm{int}_{\tau_1}(U_1), \mathrm{int}_{\tau_2}(U_2))$ of the pairs $(U_1, U_2) \in \mathcal{U}$. Set $\mathrm{int}(\mathfrak{U}) = \{\mathrm{int}(\mathcal{U}) \mid \mathcal{U} \in \mathfrak{U}\}$. Then*

(a) *each $\mathrm{int}(\mathcal{U})$ is in $\mathfrak{U}$ and (of course) $\mathrm{int}(\mathcal{U}) \leq \mathcal{U}$,*

(b) *a mapping $f\colon (X, \mathfrak{U}) \to (Y, \mathfrak{V})$ is uniformly continuous iff*

for each $\mathcal{V} \in \mathrm{int}(\mathfrak{V})$, $\{(f^{-1}[V_1], f^{-1}[V_2]) \mid (V_1, V_2) \in \mathcal{V}\} \in \mathrm{int}(\mathfrak{U})$,

(c) *$\mathrm{int}(\mathfrak{U})$ satisfies the conditions (C1)–(C3) above, with the proviso that in (C1) we consider only the open paircovers $\mathcal{V} \geq \mathcal{U}$.*

Proof. (a): For $\mathcal{U} \in \mathfrak{U}$ consider the $\mathcal{V}$ given by (C3). We show that $\mathcal{V} \leq \mathrm{int}(\mathcal{U})$, which suffices to prove the assertion. So suppose $(V_1, V_2) \in \mathcal{V}$; by assumption, there is $(U_1, U_2) \in \mathcal{U}$ such that $\mathcal{V} * V_1 \subseteq U_1$ and $V_2 * \mathcal{V} \subseteq U_2$ but this also shows that $V_i \subseteq \mathrm{int}_{\tau_i}(U_i)$ $(i = 1, 2)$ since $x \in V_1$ implies $\mathcal{V} * x \subseteq \mathcal{V} * V_1 \subseteq U_1$ and, similarly, $x \in V_2$ implies $x * \mathcal{V} \subseteq V_2 * \mathcal{V} \subseteq U_2$.

(b): Let $\mathcal{V} \in \mathrm{int}(\mathfrak{V})$ and $(V_1, V_2) \in \mathcal{V}$. By (a), $\mathcal{V} \in \mathfrak{V}$. By assumption, $f^{-1}[\mathcal{V}] \in \mathfrak{U}$. But by 4.2.1, $\mathrm{int}(f^{-1}[V_i]) = f^{-1}[V_i]$ $(i = 1, 2)$. Hence $f^{-1}[\mathcal{V}] \in \mathrm{int}(\mathfrak{U})$.

Conversely, let $\mathcal{V} \in \mathfrak{V}$. Since by (a) $\mathrm{int}(\mathcal{V}) \leq \mathcal{V}$, then

$$
f^{-1}[\mathcal{V}] \geq \{(f^{-1}[V_1], f^{-1}[V_2]) \mid (V_1, V_2) \in \mathrm{int}(\mathcal{V})\} \in \mathrm{int}(\mathfrak{U}) \subseteq \mathfrak{U}.
$$

Hence $f^{-1}[\mathcal{V}] \in \mathfrak{U}$.

(c): Let us show that $\text{int}(\mathfrak{U})$ satisfies conditions (C1)–(C3).

(C1): If $\mathcal{U} \in \text{int}(\mathfrak{U})$ and $\mathcal{U} \leq \mathcal{V}$ with $\mathcal{V}$ open then, by (a), $\mathcal{V} \in \mathfrak{U}$ and therefore $\mathcal{V} \in \text{int}(\mathfrak{U})$, since $\text{int}(\mathcal{V}) = \mathcal{V}$.

(C2): Obvious since $\mathcal{U} \wedge \mathcal{V}$ is open whenever $\mathcal{U}$ and $\mathcal{V}$ are open.

(C3): Let $\mathcal{U} \in \text{int}(\mathfrak{U})$. Consider $\mathcal{V} \in \mathfrak{U}$ such that $\mathcal{V}^* \leq \mathcal{U}$. For each

$$\left(\text{int}_{\tau_1}(V_1), \text{int}_{\tau_2}(V_2)\right) \in \text{int}(\mathcal{V})$$

with $(V_1, V_2) \in \mathcal{V}$, we have

$$\text{int}(\mathcal{V}) * (\text{int}_{\tau_1}(V_1))$$
$$= \bigcup\{\text{int}_{\tau_1}(W_1) \mid (W_1, W_2) \in \mathcal{V},$$
$$\text{int}_{\tau_2}(W_2) \cap \text{int}_{\tau_1}(V_1) \neq \emptyset\} \subseteq \mathcal{V} * V_1.$$

Similarly, $(\text{int}_{\tau_2}(V_2)) * \text{int}(\mathcal{V}) \subseteq V_2 * \mathcal{V}$. Then, from $\mathcal{V}^* \leq \mathcal{U}$ it follows immediately that $(\text{int}(\mathcal{V}))^* \leq \mathcal{U}$. $\qquad\square$

4.3.2. Note. For each $U \subseteq X$ let U° denote the interior of U with respect to the topology $\tau_1 \vee \tau_2$. Then, since $\text{int}_{\tau_i}(U) \subseteq U^\circ$, it follows immediately from 4.3.1(a) that each $\mathcal{U}^\circ = \{(U_1^\circ, U_2^\circ) \mid (U_1, U_2) \in \mathcal{U}\} \leq \mathcal{U}$ is also in $\mathfrak{U}$ and that the collection $\mathfrak{U}^\circ = \{\mathcal{U}^\circ \mid \mathcal{U} \in \mathfrak{U}\}$ satisfies conditions similar to (b) and (c) in 4.3.1.

Second, we have the following property of the open sets.

4.3.3. Proposition. (a) $U \in \tau_1$ iff $U = \bigcup\{V \in \tau_1 \mid \mathcal{U}*V \subseteq U$ for some $\mathcal{U} \in \text{int}(\mathfrak{U})\}$.

(b) $U \in \tau_2$ iff $U = \bigcup\{V \in \tau_2 \mid V * \mathcal{U} \subseteq U$ for some $\mathcal{U} \in \text{int}(\mathfrak{U})\}$.

Proof. We prove only the first equivalence, the second can be proved in a similar way. The union is trivially in τ_1. Now let $U \in \tau_1$ and let $x \in U$. Hence there is a $\mathcal{W} \in \mathfrak{U}$ such that $\mathcal{W} * x \subseteq U$. By the preceding proposition there is an open paircover $\mathcal{V} \in \mathfrak{U}$ such that $\mathcal{V}^* \leq \mathcal{W}$. Take a $(V_1, V_2) \in \mathcal{V}$ such that $x \in V_1 \cap V_2$, and a $(W_1, W_2) \in \mathcal{W}$ such that $\mathcal{V} * V_1 \subseteq W_1$ and $V_2 * \mathcal{V} \subseteq W_2$. Then $x \in W_1 \cap W_2$ and hence $W_1 \subseteq U$. Thus, for each $x \in U$ we have a $\mathcal{V} \in \mathfrak{U}$ and a $V_1 \in \tau_1$ such that $x \in V_1 \subseteq \mathcal{V} * V_1 \subseteq W_1 \subseteq U$. $\qquad\square$

5. Biframes

5.1. Bitopological spaces. Bitopological spaces (briefly, bispaces), introduced in 1963 by Kelly ([162]), and studied by many authors since then, play an important role in point-set topology. Their point-free counterpart are the biframes introduced by Banaschewski, Brümmer and Hardie in a 1983 article ([24]).

A *bispace* (or *bitopological space*)

$$(X, \mathfrak{T}_1, \mathfrak{T}_2)$$

is, as before, a triple consisting of a set X and two topologies $\mathfrak{T}_1$ and $\mathfrak{T}_2$ on X. A *bicontinuous map*

$$f \colon (X, \mathfrak{T}_1, \mathfrak{T}_2) \to (Y, \mathfrak{T}'_1, \mathfrak{T}'_2)$$

between bispaces $(X, \mathfrak{T}_1, \mathfrak{T}_2)$ and $(Y, \mathfrak{T}'_1, \mathfrak{T}'_2)$ is a function $f \colon X \to Y$ between their underlying sets for which $f \colon (X, \mathfrak{T}_i) \to (Y, \mathfrak{T}'_i)$ is continuous for $i = 1, 2$. The category of bispaces and bicontinuous maps is usually denoted as

BiTop.

5.2. The category of biframes and biframe homomorphisms. Let $(X, \mathfrak{T}_1, \mathfrak{T}_2)$ be a bispace. We have two frames, $\Omega_1(X) = \mathfrak{T}_1$ and $\Omega_2(X) = \mathfrak{T}_2$, and a third one, the topology $\Omega(X) = \mathfrak{T}_1 \vee \mathfrak{T}_2$ on X generated by the union $\mathfrak{T}_1 \cup \mathfrak{T}_2$ (the coarsest topology finer than $\mathfrak{T}_1$ and $\mathfrak{T}_2$), from which $\Omega_1(X)$ and $\Omega_2(X)$ are subframes (which together generate it). Thus, aiming for "generalized bispaces", one is led to the following definition:

A *biframe* is a triple

$$(L, L_1, L_2)$$

in which L is a frame, L_1 and L_2 are subframes of L and $L_1 \cup L_2$ generates L (by joins of finite meets). This means that any element x of L can be expressed as a join of finite meets of elements of $L_1 \cup L_2$, that is, $x = \bigvee_{i \in J}(a_i \wedge b_i)$ for some $a_i \in L_1$ and $b_i \in L_2$.

A *biframe homomorphism* $h \colon (L, L_1, L_2) \to (M, M_1, M_2)$ is a frame homomorphism from L to M such that the image of L_i ($i = 1, 2$) under h is contained in M_i. Biframes and biframe homomorphisms are the objects and morphisms of the category

BiFrm.

5.3. The functor $D \colon \mathbf{Frm} \to \mathbf{BiFrm}$ given by $D(L) = (L, L, L)$ (and the obvious action on maps) is an embedding of **Frm** into **BiFrm** which allows us to regard biframe notions as extensions of frame notions. The left adjoint to D is given by $T \colon \mathbf{BiFrm} \to \mathbf{Frm}$, $T(L, L_1, L_2) = L$ and $T(h) = h \colon L \to M$.

The basic adjunction between **Top** and **Frm** from Chapter II is immediately extendable to one between bispaces and biframes (the details may be found in [24] or [235]).

5.4. Pseudocomplements. If (L, L_1, L_2) is a biframe and $a \in L_i$ ($i = 1, 2$), we denote by $a^\bullet$ the element

$$\bigvee\{b \in L_j \mid a \wedge b = 0\} \quad (j \in \{1, 2\}, j \neq i).$$

This is the analogue on biframes ([235], [236]) of the pseudocomplement a^* of an element a of a frame L.

5.5. Regular and completely regular biframes. For $a, b \in L_i$ $(i = 1, 2)$ we write

$$a \prec_i b$$

(and say that a is i-*rather below* b) if $a^{\bullet} \vee b = 1$, or equivalently, if there exists $c \in L_j$ $(j \in \{1, 2\}, j \neq i)$ such that $a \wedge c = 0$ and $c \vee b = 1$.
   A biframe (L, L_1, L_2) is *regular* if

$$a = \bigvee \{b \in L_i \mid b \prec_i a\} \quad \text{for every } a \in L_i \quad (i = 1, 2).$$

Observe that the total part L of a regular biframe is always a regular frame.

   Note. A bitopological space $(X, \mathfrak{T}_1, \mathfrak{T}_2)$ is regular ([172]) iff the biframe $(\mathfrak{T}_1 \vee \mathfrak{T}_2, \mathfrak{T}_1, \mathfrak{T}_2)$ is regular.

   Let a and b be elements of L_i $(i \in \{1, 2\})$. We say that a is i-*completely below* b in (L, L_1, L_2) and write

$$a \lll_i b$$

if there are $a_r \in L_i$ (r rational, $0 \leq r \leq 1$) such that

$$a_0 = a, \quad a_1 = b \quad \text{and} \quad a_r \prec_i a_s \quad \text{for } r < s.$$

The biframe (L, L_1, L_2) is *completely regular* if

$$a = \bigvee \{b \in L_i \mid b \lll_i a\} \quad \text{for every } a \in L_i \quad (i = 1, 2).$$

   Note. A bitopological space $(X, \mathfrak{T}_1, \mathfrak{T}_2)$ is completely regular ([172]) iff the biframe $(\mathfrak{T}_1 \vee \mathfrak{T}_2, \mathfrak{T}_1, \mathfrak{T}_2)$ is completely regular.

6. Quasi-uniformity in the point-free context via paircovers

6.1. Paircovers for frames. Let L be a frame. In analogy with 4.2 a $U \subseteq L \times L$ is a *paircover* of L if
$$\bigvee \{u_1 \wedge u_2 \mid (u_1, u_2) \in U\} = 1.$$

A paircover U of L is *strong* if, for any $(u_1, u_2) \in U$, $u_1 \vee u_2 = 0$ whenever $u_1 \wedge u_2 = 0$. For any $U, V \subseteq L \times L$ we write

$$U \leq V \quad (\text{and say that } U \text{ refines } V)$$

if for any $(u_1, u_2) \in U$ there is $(v_1, v_2) \in V$ with $u_1 \leq v_1$ and $u_2 \leq v_2$. Further, for any paircovers U and V,

$$U \wedge V = \{(u_1 \wedge v_1, u_2 \wedge v_2) \mid (u_1, u_2) \in U, (v_1, v_2) \in V\}$$

is a paircover. The set of all paircovers of L has also non-empty arbitrary joins given just by union.

6.2. For $a \in L$ and $U, V \subseteq L \times L$, we set

$$\mathsf{st}_1(a, U) = \bigvee\{u_1 \mid (u_1, u_2) \in U \text{ and } u_2 \wedge a \neq 0\},$$

$$\mathsf{st}_2(a, U) = \bigvee\{u_2 \mid (u_1, u_2) \in U \text{ and } u_1 \wedge a \neq 0\},$$

$$U^{-1} = \{(u_2, u_1) \mid (u_1, u_2) \in U\}$$

and

$$\mathsf{st}(V, U) = \{(\mathsf{st}_1(v_1, U), \mathsf{st}_2(v_2, U)) \mid (v_1, v_2) \in V\}.$$

The particular case $\mathsf{st}(U, U)$ is denoted by U^*. We say that U *star-refines* V if $U^* \leq V$.

6.2.1. Lemma. *Let* $U, V, W \subseteq L \times L$ *and* $a, b \in L$. *Then, for* $i, j = 1, 2$, *we have:*

(a) $a \leq b \Rightarrow \mathsf{st}_i(a, U) \leq \mathsf{st}_i(b, U)$.

(b) $U \leq V \Rightarrow \mathsf{st}_i(a, U) \leq \mathsf{st}_i(a, V)$.

(c) *If* U *is a paircover then* $a \leq \mathsf{st}_i(a, U)$ *and* $U \leq U^*$.

(d) *If* U *is a paircover then* $\mathsf{st}_i(\mathsf{st}_i(a, U), U) \leq \mathsf{st}_i(a, U^*)$.

(e) $\mathsf{st}_i(a, U^{-1}) = \mathsf{st}_j(a, U)$ $(j \neq i)$.

(f) *If* U *is a paircover then* $\mathsf{st}(\mathsf{st}(V, U), U) \leq \mathsf{st}(V, U^*)$.

(g) *For any frame homomorphism* $h \colon L \to M$, $\mathsf{st}_i(h(a), h[U]) \leq h(\mathsf{st}_i(a, U))$.

(h) *For any frame homomorphism* $h \colon L \to M$, $h[U]^* \leq h[U^*]$.

Proof. (a), (b), (d) and (e) are trivial.

(c): For each $a \in L$ we have $a = a \wedge 1 = a \wedge \bigvee\{u_1 \wedge u_2 \mid (u_1, u_2) \in U\} = \bigvee\{a \wedge u_1 \wedge u_2 \mid (u_1, u_2) \in U, a \wedge u_1 \wedge u_2 \neq 0\} \leq \bigvee\{u_1 \mid (u_1, u_2) \in U, a \wedge u_2 \neq 0\}$. Thus $a \leq \mathsf{st}_1(a, U)$. The case $i = 2$ is similar. Hence $U \leq U^*$.

(f) follows from property (d), (g) is obvious (since $h(a) \neq 0$ implies $a \neq 0$ for every frame homomorphism h) and (h) is an immediate consequence of (g). $\qquad\square$

6.3. The relations "quasi-uniformly below". Given a non-empty family $\mathcal{U}$ of paircovers of L, we write

$$a \lhd_{\mathcal{U}}^i b \quad (i = 1, 2)$$

whenever $\mathsf{st}_i(a, U) \leq b$ for some $U \in \mathcal{U}$, and define

$$L_i(\mathcal{U}) = \{a \in L \mid a = \bigvee\{b \in L \mid b \lhd_{\mathcal{U}}^i a\}\} \quad (i = 1, 2).$$

6.3.1. Remark. Note that by 6.2.1(e), if the family $\mathcal{U}$ is *symmetric* (that is, $U \in \mathcal{U} \Rightarrow U^{-1} \in \mathcal{U}$) then $\lhd_{\mathcal{U}}^1 \equiv \lhd_{\mathcal{U}}^2$; in that case we denote the common $\lhd_{\mathcal{U}}^1 \equiv \lhd_{\mathcal{U}}^2$ just by $\lhd_{\mathcal{U}}$.

6.3.2. Lemma. *Let $\mathcal{U}$ be a basis for a filter of paircovers of L. Then, for $i = 1, 2$, the relations $\lhd_{\mathcal{U}}^i$ are sublattices of $L \times L$, stronger than $\leq$, satisfying the following properties for any $a, b, c, d \in L$.*

(a) *$a \leq b \lhd_{\mathcal{U}}^i c \leq d$ implies $a \lhd_{\mathcal{U}}^i d$.*

(b) *$L_i(\mathcal{U})$ is a subframe of L.*

Proof. The fact that each $\lhd_{\mathcal{U}}^i$ is stronger than $\leq$ follows from 6.2.1(c). Clearly $0 \lhd_{\mathcal{U}}^i 0$ and $1 \lhd_{\mathcal{U}}^i 1$. If $\mathsf{st}_i(a_1, U_1) \leq b_1$ and $\mathsf{st}_i(a_2, U_2) \leq b_2$ with $U_1, U_2 \in \mathcal{U}$ then, immediately, $\mathsf{st}_i(a_1 \wedge a_2, U_1 \wedge U_2) \leq \mathsf{st}_i(a_1, U_1) \wedge \mathsf{st}_i(a_2, U_2) \leq b_1 \wedge b_2$. Since $\mathcal{U}$ is a filter basis, there exists some $V \in \mathcal{U}$ such that $V \leq U_1 \wedge U_2$. Hence, using property 6.2.1(b), we may conclude that $a_1 \wedge a_2 \lhd_{\mathcal{U}}^i b_1 \wedge b_2$. On the other hand, as can be easily checked, $\mathsf{st}_i(a_1 \vee a_2, U_1 \wedge U_2) \leq \mathsf{st}_i(a_1, U_1) \vee \mathsf{st}_i(a_2, U_2) \leq b_1 \vee b_2$. Hence $a_1 \vee a_2 \lhd_{\mathcal{U}}^i b_1 \vee b_2$.

(a) is obvious from 6.2.1(a) and (b) is an immediate consequence of the fact that each $\lhd_{\mathcal{U}}^i$ is a sublattice of $L \times L$, stronger than $\leq$, and assertion (a). $\square$

6.4. Admissible systems. A set of paircovers $\mathcal{U}$ of L is *admissible* if

$$\forall a \in L, \; a = \bigvee \{b \in L \mid b \lhd_{\overline{\mathcal{U}}} a\},$$

where $\overline{\mathcal{U}}$ denotes the filter of paircovers of L generated by $\{U \wedge U^{-1} \mid U \in \mathcal{U}\}$ (which is clearly symmetric since $(U \wedge U^{-1})^{-1} = U \wedge U^{-1}$).

6.5. Quasi-uniformity on a frame. A *quasi-uniformity* on L is a non-empty admissible set $\mathcal{U}$ of paircovers of L such that

(QU1) for any $U \in \mathcal{U}$ and any paircover V with $U \leq V$, then $V \in \mathcal{U}$,

(QU2) for any $U, V \in \mathcal{U}$ there exists a strong $W \in \mathcal{U}$ such that $W \leq U \wedge V$, and

(QU3) for any $U \in \mathcal{U}$ there is a $V \in \mathcal{U}$ such that $V^* \leq U$.

> **Note.** This definition contains, of course, the particular case of uniform structures defined by covers: it is one that has a basis consisting of pairs of covers, both coordinates of which are the same (i.e., of the form (U, U) where U is a cover of L); certainly such paircovers are strong, moreover $(U, U)^* = (UU, UU)$, where UU is the usual star of the cover U (VIII.2.2) and $\lhd_{\overline{\mathcal{U}}}$ is the usual $\lhd_{\mathcal{U}}$ of VIII.2.3.

> The pair $(L, \mathcal{U})$ is called a *quasi-uniform frame* and $\mathcal{B} \subseteq \mathcal{U}$ is a *basis* for $\mathcal{U}$ if, for each $U \in \mathcal{U}$, there is a $V \in \mathcal{B}$ such that $V \leq U$.

6.5.1. Remark. If $\mathcal{U}$ is a quasi-uniformity on L, then

$$\mathcal{U}^{-1} = \{U^{-1} \mid U \in \mathcal{U}\}$$

is also a quasi-uniformity on L. Of course, $\overline{\mathcal{U}}$ is then a uniformity and it is the coarsest uniformity containing $\mathcal{U}$.

6.6. Uniform homomorphisms. Let $(L, \mathcal{U})$ and $(M, \mathcal{V})$ be quasi-uniform frames. A frame homomorphism $h \colon L \to M$ is *uniform* if $h[U] \in \mathcal{V}$ for every $U \in \mathcal{U}$. Quasi-uniform frames and uniform homomorphisms constitute a category that we denote by

QUniFrm.

6.7. The "interior" of a paircover. Let $\mathcal{U}$ be a quasi-uniformity on L. By 6.2.1(d) the relation $\overset{\mathcal{U}}{\Subset}$ on $\mathfrak{P}(L \times L)$ defined by

$$C \overset{\mathcal{U}}{\Subset} D \equiv_{\mathsf{df}} \mathsf{st}(C, U) \leq D \text{ for some } U \in \mathcal{U}$$

is stronger than $\leq$.

Further we consider the following (interior) operator on $\mathfrak{P}(L \times L)$:

$$\mathsf{int}(C) = \bigcup \{ D \subseteq L \times L \mid D \overset{\mathcal{U}}{\Subset} C \}.$$

6.7.1. Lemma. *Let $\mathcal{U}$ be a quasi-uniformity on L. For each $U \in \mathcal{U}$,*

  (a) $\mathsf{int}(U) \leq U \leq \mathsf{int}(U^*)$,

  (b) $\mathsf{int}(U) \in \mathcal{U}$, *and*

  (c) $\mathsf{st}_i(a, \mathsf{int}(U)) \in L_i(\mathcal{U})$ *for every $a \in L$.*

Proof. (a): The inequality $\mathsf{int}(U) \leq U$ is trivial. The other follows from the obvious fact that for every $U \in \mathcal{U}$, $U \overset{\mathcal{U}}{\Subset} U^*$.

(b) is immediate from (a), using condition (QU3).

(c): We only prove the case $i = 1$ (the case $i = 2$ may be proved in a similar way). We need to show that

$$\mathsf{st}_1(a, \mathsf{int}(U)) \leq \bigvee \{ y \in L \mid y \triangleleft^1_{\mathcal{U}} \mathsf{st}_1(a, \mathsf{int}(U)) \}.$$

By definition, $\mathsf{st}_1(a, \mathsf{int}(U)) = \bigvee \{ d_1 \mid (d_1, d_2) \in \mathsf{int}(U), d_2 \wedge a \neq 0 \}$. So, let $(d_1, d_2) \in \mathsf{int}(U)$ such that $d_2 \wedge a \neq 0$. Then $(d_1, d_2) \in D \subseteq L \times L$ and there exists $V \in \mathcal{U}$ such that $\mathsf{st}(D, V) \subseteq U$. We need to show that $d_1 \triangleleft^1_{\mathcal{U}} \mathsf{st}_1(a, \mathsf{int}(U))$. To see this consider $W \in \mathcal{U}$ such that $W^* \leq V$. It suffices then to prove that $\mathsf{st}_1(d_1, W) \leq \mathsf{st}_1(a, \mathsf{int}(U))$.

By property (f) of Lemma 6.2.1 we have

$$\mathsf{st}(\mathsf{st}(D, W), W) \leq \mathsf{st}(D, W^*) \leq \mathsf{st}(D, V) \leq U,$$

which shows that $\mathsf{st}(D, W) \overset{\mathcal{U}}{\Subset} U$. Thus $\mathsf{st}(D, W) \subseteq \mathsf{int}(U)$. Therefore we only need to check that $\mathsf{st}_1(d_1, W) \leq \mathsf{st}_1(a, \mathsf{st}(D, W))$, which is easy since

$$(\mathsf{st}_1(d_1, W), \mathsf{st}_2(d_2, W)) \in \mathsf{st}(D, W)$$

and $\mathsf{st}_2(d_2, W) \wedge a \geq d_2 \wedge a \neq 0$. $\qquad\qquad\qquad\square$

6.7.2. Proposition. *Let $\mathcal{U}$ be a non-empty family of paircovers of L satisfying axioms* (QU1), (QU2) *and* (QU3). *Then $\mathcal{U}$ is admissible if and only if*

$$\text{the triple } (L, L_1(\mathcal{U}), L_2(\mathcal{U})) \text{ is a biframe.} \tag{6.7.1}$$

Proof. $\Rightarrow$: By 6.3.2, each $L_i(\mathcal{U})$ $(i = 1, 2)$ is a subframe of L. It remains to show that each $a \in L$ is a join of finite meets in $L_1(\mathcal{U}) \cup L_2(\mathcal{U})$.

Let $a \in L$. Then $a = \bigvee S$ where $S = \{b \in L \mid b \triangleleft_{\overline{\mathcal{U}}} a\}$. For each $b \in S$ there exists $\overline{U}_b \in \overline{\mathcal{U}}$ and $U_b \in \mathcal{U}$ such that $\mathsf{st}_1(b, \overline{U}_b) \leq a$ and $U_b \wedge U_b^{-1} \leq \overline{U}_b$. Consider $V_b \in \mathcal{U}$ such that $V_b^* \leq \overline{U}_b$. Then $\overline{V}_b = V_b \wedge V_b^{-1} \in \overline{\mathcal{U}}$ and $\overline{V}_b^* \leq \overline{U}_b$. Therefore $\mathrm{int}(\overline{V}_b^*) \leq \mathrm{int}(\overline{U}_b)$. Thus

$$a = \bigvee S \leq \bigvee_{b \in S} (\mathsf{st}_1(b, \overline{V}_b) \wedge \mathsf{st}_2(b, \overline{V}_b)) \leq \bigvee_{b \in S} (\mathsf{st}_1(b, \mathrm{int}(\overline{V}_b^*)) \wedge \mathsf{st}_2(b, \mathrm{int}(\overline{V}_b^*)))$$

$$\leq \bigvee_{b \in S} (\mathsf{st}_1(b, \mathrm{int}(\overline{U}_b)) \wedge \mathsf{st}_2(b, \mathrm{int}(\overline{U}_b))) \leq \mathsf{st}_1(b, \overline{U}_b) \leq a.$$

Hence

$$a = \bigvee_{b \in S} (\mathsf{st}_1(b, \mathrm{int}(\overline{U}_b)) \wedge \mathsf{st}_2(b, \mathrm{int}(\overline{U}_b)))$$

and by 6.7.1 $\mathsf{st}_i(b, \mathrm{int}(\overline{U}_b)) \in L_i(\mathcal{U})$ $(i = 1, 2)$.

$\Leftarrow$: For each $a \in L$ we may write $a = \bigvee_{i \in I}(a_i^1 \wedge a_i^2)$ for some

$$\{a_i^1 \mid i \in I\} \subseteq L_1(\mathcal{U}) \quad \text{and} \quad \{a_i^2 \mid i \in I\} \subseteq L_2(\mathcal{U}).$$

Taking into account that, for any $i \in I$,

$$a_i^1 = \{b \in L \mid b \triangleleft_{\mathcal{U}}^1 a_i^1\} \quad \text{and} \quad a_i^2 = \{b \in L \mid b \triangleleft_{\mathcal{U}}^2 a_i^2\},$$

it suffices to show that $b_1 \wedge b_2 \triangleleft_{\overline{\mathcal{U}}} a_1 \wedge a_2$ whenever $b_1 \triangleleft_{\mathcal{U}}^1 a_1$ and $b_2 \triangleleft_{\mathcal{U}}^2 a_2$. But this is an immediate consequence of the fact that $\triangleleft_{\overline{\mathcal{U}}}^1 = \triangleleft_{\overline{\mathcal{U}}}^2 = \triangleleft_{\overline{\mathcal{U}}}$ (recall 6.3.1) and properties 6.2.1(a) and (b). $\qquad \square$

6.8. Aside: construction of quasi-uniformities. For each paircover U define

$$U^s = \{(u_1, u_2) \in U \mid u_1 \wedge u_2 \neq 0\}.$$

We have

6.8.1. Lemma. (a) U^s *is a paircover of* L.

(b) $(U \wedge V)^s \leq U^s \wedge V^s$.

(c) $U^* \leq V \Rightarrow (U^s)^* \leq V^s$.

(d) *If $h \colon L \to M$ is a frame homomorphism, then $(h[U])^s \leq h[U^s]$.*

Proof. (a) and (b) are trivial; (c) is also obvious because $U^s \subseteq U$ so, for each $(u_1, u_2) \in U^s$, $\mathsf{st}_i(u_i, U^s) \leq \mathsf{st}_i(u_i, U) \leq v_i$ $(i = 1, 2)$ for some $(v_1, v_2) \in V$ and $u_1 \wedge u_2 \neq 0$ implies $v_1 \wedge v_2 \neq 0$. For (d) suppose $h(u_1) \wedge h(u_2) \neq 0$, where $(u_1, u_2) \in U$. Then $h(u_1 \wedge u_2) \neq 0$, so $u_1 \wedge u_2 \neq 0$ and $(u_1, u_2) \in U^s$. Consequently $(h(u_1), h(u_2)) \in h[U^s]$. $\qquad\qquad\square$

6.8.2. Proposition. (a) *Let $\mathcal{U}$ be a filter of paircovers of L with the properties* (QU3) *and* (6.7.1) *and let $\mathcal{U}^s$ be the filter of paircovers of L that has $\{U^s \mid U \in \mathcal{U}\}$ as subbases. Then $(L, \mathcal{U}^s)$ is a quasi-uniform frame.*

(b) *If $(L, \mathcal{U})$ and $(M, \mathcal{V})$ are structures as in* (a) *and $h\colon L \to M$ is a frame homomorphism with the property that*

$$U \in \mathcal{U} \Rightarrow h[U] \in \mathcal{V},$$

then $h\colon (L, \mathcal{U}^s) \to (M, \mathcal{V}^s)$ is a uniform homomorphism.

Proof. (a): $\mathcal{U}^s$ is a filter which, by 6.8.1, clearly satisfies (QU3) and (6.7.1) (because $\mathcal{U}^s \supseteq \mathcal{U}$). Since each U^s is strong, there are enough strong paircovers in order to fulfill (QU2).

(b) is an immediate consequence of 6.8.1(d). $\qquad\qquad\square$

6.8.3. Remark. Note that if $(L, \mathcal{U})$ is a quasi-uniform frame then $\mathcal{U}^s = \mathcal{U}$.

6.9. The biframe induced by a quasi-uniformity. The biframe

$$(L, L_1(\mathcal{U}), L_2(\mathcal{U}))$$

is the *induced biframe* of $(L, \mathcal{U})$. This is the point-free counterpart of the classical fact in (4.1.1).

6.10. Quasi-uniformizability and complete regularity. Let $(L, \mathcal{U})$ be a quasi-uniform frame. Since $\overline{\mathcal{U}}$ is a uniformity on L then L is necessarily regular (recall VII.2.4.1). But we have more:

6.10.1. Proposition. *Let $\mathcal{U}$ be a quasi-uniformity on L. Then, for every $a, b \in L_i(\mathcal{U})$,*

$$a \triangleleft_{\mathcal{U}}^i b \Rightarrow a \prec_i b.$$

Proof. Let $a \triangleleft_{\mathcal{U}}^i b$. Then $a \leq \mathsf{st}_i(a, U) \leq b$ for some $U \in \mathcal{U}$. Take $V \in \mathcal{U}$ such that $V^* \leq U$. Then $\mathsf{int}(V) \in \mathcal{U}$ and

$$\mathsf{int}(V) \leq \{(\mathsf{st}_1(v_1, \mathsf{int}(V)), \mathsf{st}_2(v_2, \mathsf{int}(V))) \mid (v_1, v_2) \in V\}.$$

Therefore

$$1 = \bigvee\{\mathsf{st}_1(v_1, \mathrm{int}(V)) \wedge \mathsf{st}_2(v_2, \mathrm{int}(V)) \mid (v_1, v_2) \in V\}$$

$$= \bigvee\{\mathsf{st}_1(v_1, \mathrm{int}(V)) \wedge \mathsf{st}_2(v_2, \mathrm{int}(V)) \mid (v_1, v_2) \in V, \mathsf{st}_2(v_2, \mathrm{int}(V)) \wedge a = 0\}$$

$$\vee \bigvee\{\mathsf{st}_1(v_1, \mathrm{int}(V)) \wedge \mathsf{st}_2(v_2, \mathrm{int}(V)) \mid (v_1, v_2) \in V, \mathsf{st}_2(v_2, \mathrm{int}(V)) \wedge a \neq 0\}$$

$$\leq \bigvee\{\mathsf{st}_2(v_2, \mathrm{int}(V)) \mid (v_1, v_2) \in V, \mathsf{st}_2(v_2, \mathrm{int}(V)) \wedge a = 0\}$$

$$\vee \bigvee\{\mathsf{st}_1(v_1, \mathrm{int}(V)) \mid (v_1, v_2) \in V, \mathsf{st}_2(v_2, \mathrm{int}(V)) \wedge a \neq 0\}.$$

But $\bigvee\{\mathsf{st}_2(v_2, \mathrm{int}(V)) \mid (v_1, v_2) \in V, \mathsf{st}_2(v_2, \mathrm{int}(V)) \wedge a = 0\} \leq a^{\bullet}$. On the other hand, since $V^* \leq U$,

$$\bigvee\{\mathsf{st}_1(v_1, \mathrm{int}(V) \mid (v_1, v_2) \in V, \mathsf{st}_2(v_2, \mathrm{int}(V)) \wedge a \neq 0\} \leq \mathsf{st}_1(a, U) \leq b.$$

Hence $a^{\bullet} \vee b = 1$, that is, $a \prec_i b$. $\qquad\square$

6.10.2. Corollary. *Let $\mathcal{U}$ be a quasi-uniformity on L. Then:*

(a) *Each $\lhd_{\mathcal{U}}^i$ interpolates, that is, for every $a, b \in L$, $a\lhd_{\mathcal{U}}^i b$ implies $a\lhd_{\mathcal{U}}^i c\lhd_{\mathcal{U}}^i b$ for some $c \in L_i(\mathcal{U})$ $(i = 1, 2)$.*

(b) *For every $a, b \in L_i(\mathcal{U})$, $a\lhd_{\mathcal{U}}^i b \Rightarrow a \prec_i b$.*

Proof. (a): Let $a\lhd_{\mathcal{U}}^i b$. Then $a \leq \mathsf{st}_i(a, U) \leq b$ for some $U \in \mathcal{U}$ and if we choose a $V \in \mathcal{U}$ such that $V^* \leq U$ we have immediately $a\lhd_{\mathcal{U}}^i \mathsf{st}_i(a, \mathrm{int}(V)) \in L_i(\mathcal{U})$ and by 6.2.1(d) $\mathsf{st}_i(a, \mathrm{int}(V))\lhd_{\mathcal{U}}^i b$.

(b) follows now immediately from Proposition 6.10.1. $\qquad\square$

The following is the asymmetric counterpart to VIII.2.8.2.

6.10.3. Theorem. *A biframe (L, L_1, L_2) admits a quasi-uniformity $\mathcal{U}$ whose induced biframe is precisely (L, L_1, L_2) if and only if it is completely regular.*

Proof. $\Rightarrow$ follows immediately from 6.10.2(b).

$\Leftarrow$: Let (L, L_1, L_2) be completely regular. For each $a \prec_i b$ set

$$U(a, b) = \{(b, 1), (1, a^*)\}.$$

This is a strong paircover of L. If we interpolate $a \prec_i u_1 \prec_i v_1 \prec_i b$ and set

$$U = U(a, u_1) \wedge U(u_1, v_1) \wedge U(v_1, b)$$

$$= \{(u_1, 1), (u_1, v_1^*), (u_1, u_1^*), (v_1, a^*), (v_1, v_1^*), (b, u_1^*), (1, v_1^*)\}$$

we easily check that $U^* \leq U(a, b)$. Let $\mathcal{U}$ be the filter of paircovers generated by all paircovers of the form $U(a, b)$ where $a \prec_i b$ $(i = 1, 2)$. Since the U's above are not strong, $\mathcal{U}$ is not necessarily a quasi-uniformity but if we consider the $\mathcal{U}^s$ of 6.8.2 we get the required quasi-uniform structure. $\qquad\square$

7.   The adjunction QUnif $\rightleftarrows$ QUniFrm

7.1. The adjunction from VIII.3.7 can be easily extended to quasi-uniform frames. We will describe the expected spectrum and open functors between **QUnif** and **QUniFrm**.

7.2. The functor Ω: QUnif $\to$ QUniFrm. Let (X, μ) be a quasi-uniform space and let $\mathcal{U} \in \mu$. A $\mathcal{U}$ is an *open* paircover of (X, μ) if for each $(U_1, U_2) \in \mathcal{U}$, U_1 is $\mathfrak{T}_1(\mu)$-open and U_2 is $\mathfrak{T}_2(\mu)$-open. Set

$$\Omega(X, \mu) = (\mathfrak{T}_1(\mu) \vee \mathfrak{T}_2(\mu), \mathcal{C}_\mu)$$

where $\mathcal{C}_\mu$ is the set of all open paircovers of (X, μ). It is not hard to check that $\Omega(X, \mu)$ is a quasi-uniform space. Furthermore, if $f: (X, \mu) \to (Y, \nu)$ is uniformly continuous then $\Omega(f): \Omega(Y, \nu) \to \Omega(X, \mu)$, defined by $\Omega(f)(B) = f^{-1}(B)$ for any $B \in \mathfrak{T}_1(\nu) \vee \mathfrak{T}_2(\nu)$, is a uniform frame homomorphism. Thus, Ω is a contravariant functor from **QUnif** into **QUniFrm**.

7.3. The spectrum functor Σ: QUniFrm $\to$ QUnif. On the other hand, let $(L, \mathcal{U})$ be a quasi-uniform frame and consider the spectrum

$$\big(\mathsf{Pt}(L), \{\Sigma_a \mid a \in L\}\big)$$

of L. For each $U \in \mathcal{U}$ let

$$\Sigma_U = \{(\Sigma_{u_1}, \Sigma_{u_2}) \mid (u_1, u_2) \in U\},$$

and consider the filter $\Sigma_{\mathcal{U}}$ of paircovers of ΣL generated by the system $\{\Sigma_U \mid U \in \mathcal{U}\}$.

7.3.1. Proposition. *Let $(L, \mathcal{U})$ be a quasi-uniform frame. Then $\Sigma(L, \mathcal{U}) = (\Sigma L, \Sigma_{\mathcal{U}})$ is a quasi-uniform space.*

Proof. (1) Take a $U \in \mathcal{U}$. Then

$$\bigcup \{\Sigma_{u_1} \cap \Sigma_{u_2} \mid (u_1, u_2) \in U\} = \bigcup \{\Sigma_{u_1 \wedge u_2} \mid (u_1, u_2) \in U\}$$

$$= \Sigma_{\bigvee \{u_1 \wedge u_2 \mid (u_1, u_2) \in U\}} = \Sigma_1 = \Sigma L.$$

(2) Let $U, V \in \mathcal{U}$. Trivially $U \leq V$ implies $\Sigma_U \leq \Sigma_V$ and $\Sigma_{U \wedge V} = \Sigma_U \wedge \Sigma_V$.

(3) Let $U^* \leq V$. Then $(\Sigma_U)^* \leq \Sigma_V$. Indeed: for each $(u_1, u_2) \in U$ there exists $(v_1, v_2) \in V$ satisfying $\mathsf{st}_1(u_1, U) \leq v_1$ and $\mathsf{st}_2(u_2, U) \leq v_2$; then

$$\mathsf{st}_1(\Sigma_{u_1}, \Sigma_U) = \bigcup \{\Sigma_{u_1'} \mid \Sigma_{u_2'} \cap \Sigma_{u_1} \neq \emptyset, (u_1', u_2') \in U\}$$

$$= \bigcup \{\Sigma_{u_1'} \mid \Sigma_{u_2' \wedge u_1} \neq \emptyset, (u_1', u_2') \in U\}$$

$$\subseteq \bigcup \{\Sigma_{u_1'} \mid u_2' \wedge u_1 \neq 0\} = \Sigma_{\mathsf{st}_1(u_1, U)} \subseteq \Sigma_{v_1}.$$

Similarly, $\mathsf{st}_2(\Sigma_{u_2}, \Sigma_U) \subseteq \Sigma_{v_2}$. $\square$

For a uniform homomorphism $h\colon (L,\mathcal{U}) \to (M,\nu)$ define $\Sigma h\colon \Sigma(M,\nu) \to \Sigma(L,\mathcal{U})$ by $\Sigma h(p) = p \cdot h$. It is easy to check that $\Sigma h \in$ **QUnif**. Thus we obtain a contravariant functor

$$\Sigma\colon \textbf{QUniFrm} \to \textbf{QUnif}.$$

7.3.2. Remark. It is also easy to check that the topologies $\mathfrak{T}_1(\Sigma_{\mathcal{U}})$ and $\mathfrak{T}_2(\Sigma_{\mathcal{U}})$ induced by the quasi-uniformity $\Sigma_{\mathcal{U}}$ on ΣL coincide with the spectral topologies of the spectra of $L_1(\mathcal{U})$ and $L_2(\mathcal{U})$ respectively.

7.4. Theorem. *The two above contravariant functors Ω and Σ constitute a dual adjunction, with units*

$$\eta_{(X,\mu)}\colon (X,\mu) \to \Sigma\Omega(X,\mu) \quad and \quad \xi_{(L,\mathcal{U})}\colon (L,\mathcal{U}) \to \Omega\Sigma(L,\mathcal{U})$$

given by

$$\eta_{(X,\mu)}(a)(U) = 1 \text{ iff } a \in U \quad and \quad \xi_{(L,\mathcal{U})}(a) = \Sigma_a.$$

Proof. The checking that each $\eta_{(X,\mu)}$ is uniformly continuous and that each $\xi_{(L,\mathcal{U})}$ is a uniform frame homomorphism is left to the reader. They define natural transformations

$$\eta\colon \mathrm{Id}_{\textbf{QUnif}} \dot{\to} \Sigma\Omega \quad and \quad \xi\colon \mathrm{Id}_{\textbf{QUniFrm}} \dot{\to} \Omega\Sigma,$$

that is, the following diagrams commute, for every $f\colon (X,\mu) \to (Y,\nu)$ and every $h\colon (L,\mathcal{U}) \to (M,\mathcal{V})$.

$$
\begin{array}{ccc}
(X,\mu) & \xrightarrow{\;\eta_{(X,\mu)}\;} & \Sigma\Omega(X,\mu) \\
{\scriptstyle f}\downarrow & (1) & \downarrow{\scriptstyle \Sigma\Omega(f)} \\
(Y,\nu) & \xrightarrow[\;\eta_{(Y,\nu)}\;]{} & \Sigma\Omega(Y,\nu)
\end{array}
\qquad
\begin{array}{ccc}
(L,\mathcal{U}) & \xrightarrow{\;\xi_{(L,\mathcal{U})}\;} & \Omega\Sigma(L,\mathcal{U}) \\
{\scriptstyle h}\downarrow & (2) & \downarrow{\scriptstyle \Omega\Sigma(h)} \\
(M,\mathcal{V}) & \xrightarrow[\;\xi_{(M,\mathcal{V})}\;]{} & \Omega\Sigma(M,\mathcal{V})
\end{array}
$$

Indeed, for each $f\colon (X,\mu) \to (Y,\nu)$ in **QUnif** and every $x \in X$,

$$\Sigma\Omega(f)(\eta_{(X,\mu)}(x)) = \eta_{(X,\mu)}(x)\Omega(f)$$

is the map $F\colon \Omega(Y,\nu) \to \{0,1\}$ given by $F(B) = 1$ iff $x \in f^{-1}(B)$. Since $x \in f^{-1}(B)$ iff $f(x) \in B$, this is precisely the map $\eta_{(Y,\nu)}(f(x))$ and diagram (1) commutes.

On the other hand, for each $h\colon (L,\mathcal{U}) \to (M,V)$ in **QUniFrm** and every $a \in L$,

$$\Omega\Sigma(h)(\xi_{(L,\mathcal{U})}(a)) = \Omega\Sigma(h)(\Sigma_a) = (\Sigma(h))^{-1}(\Sigma_a) = \{p \in \mathsf{Pt}(M) \mid \Sigma(h)(p) \in \Sigma_a\}.$$

Since

$$\Sigma(h)(p) \in \Sigma_a \text{ iff } ph \in \Sigma_a \text{ iff } p(h(a)) = 1 \text{ iff } p \in \Sigma_{h(a)},$$

then diagram (2) also commutes.

Finally,

$$\eta\colon \mathrm{Id}_{\mathbf{QUnif}} \overset{\cdot}{\to} \Sigma\Omega \quad \text{and} \quad \xi\colon \mathrm{Id}_{\mathbf{QUniFrm}} \overset{\cdot}{\to} \Omega\Sigma$$

satisfy the triangular identities

$$\text{(a)} \ \ \Omega(\eta_{(X,\mu)}) \cdot \xi_{\Omega(X,\mu)} = 1 \quad \text{and} \quad \text{(b)} \ \ \Sigma(\xi_{(L,\mathcal{U})}) \cdot \eta_{\Sigma(L,\mathcal{U})} = 1$$

for every $(X,\mu) \in \mathbf{QUnif}$ and every $(L,\mathcal{U}) \in \mathbf{QUniFrm}$:

(a): For each $A \in \mathfrak{T}_1(\mu) \vee \mathfrak{T}_2(\mu)$,

$$(\Omega(\eta_{(X,\mu)}) \cdot \xi_{\Omega(X,\mu)})(A) = \Omega(\eta_{(X,\mu)})(\Sigma_A) = \eta^{-1}_{(X,\mu)}(\Sigma_A)$$

and

$$x \in \eta^{-1}_{(X,\mu)}(\Sigma_A) \text{ iff } \eta_{(X,\mu)}(x) \in \Sigma_A \text{ iff } \eta_{(X,\mu)}(x)(A) = 1 \text{ iff } x \in A.$$

(b): For each $p\colon L \to \{0,1\}$ in $\mathsf{Pt}(L)$,

$$(\Sigma(\xi_{(L,\mathcal{U})}) \cdot \eta_{\Sigma(L,\mathcal{U})})(p) = q \cdot \xi_{(L,\mathcal{U})}$$

where $q\colon \Omega\Sigma(L,\mathcal{U}) \to \{0,1\}$ maps A into 1 iff $p \in A$. But $(q \cdot \xi_{(L,\mathcal{U})})(a) = q(\Sigma_a)$ is equal to 1 iff $p \in \Sigma_a$, that is, iff $p(a) = 1$. Hence $q \cdot \xi_{(L,\mathcal{U})} = p$ as required. □

8. Quasi-uniformity in the point-free context via entourages

8.1. Entourage quasi-uniformities on a frame. The approach via entourages in classical topology has the nice feature of allowing the passage from symmetric uniform structures to asymmetric ones just by dropping the symmetry axiom. The availability of entourages in the point-free setting allows us to do a similar thing here.

8.1.1. Definition. An *entourage quasi-uniformity* on a frame L is just an admissible system of entourages $\mathcal{E}$ satisfying axioms (E1), (E2) and (E4) of 2.3. An *entourage quasi-uniform frame* is a couple $(L, \mathcal{E})$ where $\mathcal{E}$ is a quasi-uniformity on L.

The resulting category of entourage quasi-uniform frames and uniform homomorphisms is denoted by

EQUniFrm.

For examples of entourage quasi-uniformities on a frame (namely the Pervin quasi-uniformity, the locally finite quasi-uniformity, the well-monotone quasi-uniformity and the fine-transitive quasi-uniformity) see [94].

8.2. The relations $\lhd^1_{\mathcal{E}}$ and $\lhd^2_{\mathcal{E}}$. After dropping the symmetry axiom (E3) in the definition of an entourage quasi-uniform frame the equivalence in Remark 2.1.1 is no longer valid and we have, instead of $\lhd_{\mathcal{E}}$, two order relations $\lhd^1_{\mathcal{E}}$ and $\lhd^2_{\mathcal{E}}$ defined by setting

$$b \lhd^1_{\mathcal{E}} a \equiv_{df} \exists E \in \mathcal{E}, E \circ (b \oplus b) \leq a \oplus a$$

and

$$b \lhd^2_{\mathcal{E}} a \equiv_{df} \exists E \in \mathcal{E}, (b \oplus b) \circ E \leq a \oplus a.$$

As in 2.1.2 the following is easy to check.

8.2.1. Proposition. *Let $\mathcal{E}$ be a set of entourages of L. Then:*

(1) $a \leq b \lhd^i_{\mathcal{E}} c \leq d \Rightarrow a \lhd^i_{\mathcal{E}} d$.

(2) *If for any $E_1, E_2 \in \mathcal{E}$ there is an $F \in \mathcal{E}$ such that $F \subseteq E_1 \cap E_2$, then $a \lhd^i_{\mathcal{E}} b_1, b_2 \Rightarrow a \lhd^i_{\mathcal{E}} (b_1 \wedge b_2)$ and $a_1, a_2 \lhd^i_{\mathcal{E}} b \Rightarrow (a_1 \vee a_2) \lhd^i_{\mathcal{E}} b$.* □

8.3. For each $E \in L \oplus L$ and each $a \in L$ let

$$\mathsf{st}_1(a, E) = \bigvee\{x \in L \mid x \oplus y \leq E, y \wedge a \neq 0\}$$

and

$$\mathsf{st}_2(a, E) = \bigvee\{y \in L \mid x \oplus y \leq E, x \wedge a \neq 0\}.$$

8.3.1. Proposition. *For each $a, b \in L$ and $E, F \in L \oplus L$ we have:*

(1) *If $a \leq b$ then $\mathsf{st}_i(a, E) \leq \mathsf{st}_i(b, E)$.*

(2) *If $E \leq F$ then $\mathsf{st}_i(a, E) \leq \mathsf{st}_i(a, F)$.*

(3) $\mathsf{st}_i(a, E^{-1}) = \mathsf{st}_j(a, E)$ $(j \neq i)$.

(4) $\mathsf{st}_1(a, E) \wedge b = 0$ *iff* $a \wedge \mathsf{st}_2(b, E) = 0$.

(5) $\mathsf{st}_1(\mathsf{st}_1(a, E), F) \leq \mathsf{st}_1(a, F \circ E)$ *and* $\mathsf{st}_2(\mathsf{st}_2(a, E), F) \leq \mathsf{st}_2(a, E \circ F)$.

(6) *If E is an entourage then $a \leq \mathsf{st}_1(a, E) \wedge \mathsf{st}_2(a, E)$.*

Proof. Facts (1)–(5) are straightforward. Let E be an entourage and $a \in L$. Then $a = a \wedge 1 = a \wedge \bigvee\{x \mid x \oplus x \leq E\} = \bigvee\{a \wedge x \mid x \oplus x \leq E, a \wedge x \neq 0\} \leq \mathsf{st}_1(a, E) \wedge \mathsf{st}_2(a, E)$. □

The elements $\mathsf{st}_i(a, E)$ provide a useful characterization of the relation $\lhd^i_{\mathcal{E}}$:

8.3.2. Corollary. *Let $\mathcal{E}$ be a filter of entourages. Then $a \lhd^i_{\mathcal{E}} b$ iff $\mathsf{st}_i(a, E) \leq y$ for some $E \in \mathcal{E}$.*

Proof. Assume $E \circ (a \oplus a) \leq b \oplus b$ for some $E \in \mathcal{E}$. If $x \oplus y \leq E$ with $y \wedge a \neq 0$ then $x \oplus (y \wedge a) \leq E$ and $(y \wedge a) \oplus a \leq a \oplus a$ thus $x \oplus a \leq E \circ (a \oplus a) \leq b \oplus b$. Therefore $x \leq b$, which shows that $\mathsf{st}_1(a, E) \leq b$.

Conversely, suppose that $\mathsf{st}_1(a, E) \leq b$. If $x \oplus y \leq E$ and $y \oplus z \leq a \oplus a$ with $y \neq 0$, then $y \wedge a = y \neq 0$ which implies that $x \leq \mathsf{st}_1(a, E) \leq b$. On the other hand, by fact (5) above, $z \leq a \leq \mathsf{st}_1(a, E) \leq b$. Thus $x \oplus z \leq b \oplus b$, as required. □

8.4. Entourages versus paircovers: translation. For a paircover U define the entourage

$$E_U = \bigvee\{u_1 \oplus u_2 \mid (u_1, u_2) \in U\},$$

and for an entourage E define the paircover

$$U_E = \{(\mathsf{st}_1(x_1, E), \mathsf{st}_2(x_2, E)) \mid x_1 \oplus x_2 \leq E\}.$$

Further, for a quasi-uniformity $\mathcal{U}$ set

$$\mathcal{E}_{\mathcal{U}} = \{F \mid F \geq E_U, U \in \mathcal{U}\}$$

and for an entourage quasi-uniformity $\mathcal{E}$ set

$$\mathcal{U}_{\mathcal{E}} = \{V \mid V \geq U_E, E \in \mathcal{E}\}.$$

Then, similarly to 3.3.4 and 3.4.3, we have (see [207] for details):

8.4.1. Theorem. (a) *The correspondences $\mathcal{E} \mapsto \mathcal{U}_{\mathcal{E}}$ and $\mathcal{U} \mapsto \mathcal{E}_{\mathcal{U}}$ constitute a one-one correspondence between the entourage quasi-uniformities and the cover ones. More explicitly, $\mathcal{E}_{\mathcal{U}_{\mathcal{E}}} = \mathcal{E}$ and $\mathcal{U}_{\mathcal{E}_{\mathcal{U}}} = \mathcal{U}$.*

(b) *For the associated entourage and cover uniformities, the concepts of a uniform homomorphism coincide.* □

8.4.2. Corollary. *The concrete categories* **QUniFrm** *and* **EQUniFrm** *are concretely isomorphic.* □

Chapter XIII

Connectedness

Connectedness seems to promise no surprises; after all it is an inherently point-free question, the question of non-trivial complemented elements. And the basics are indeed just as in spaces (Section 2). But when we go further, the ways part. We have facts that hold in locales similarly as in spaces but for different reasons (Section 3), and facts that are quite different; of those we present an example in Section 4; so far we have avoided lengthy tedious counterexamples, but in this section, just for once, we present a weird (co)product in detail: besides showing that connectedness is not behaving as one would expect, it also indicates why the fact from Section 3, although parallel to the classical one, cannot be proved by imitating the classical procedure.

1. A few observations about sublocales

Recall the co-frame $\mathcal{S\!\ell}(L)$ of sublocales of L and the open and closed sublocales from Chapter III.

1.1. Proposition. *The following statements are equivalent for a frame L.*

(1) *The elements a and b are mutual complements in L.*

(2) *The closed sublocales $\mathfrak{c}(a) = {\uparrow}a$ and $\mathfrak{c}(b) = {\uparrow}b$ are mutual complements in $\mathcal{S\!\ell}(L)$.*

(3) *The open sublocales $\mathfrak{o}(a)$ and $\mathfrak{o}(b)$ are mutual complements in $\mathcal{S\!\ell}(L)$.*

Proof. The equivalence (2)⇔(3) is obvious.

(1)⇒(2): If $a \vee b = 1$ and $a \wedge b = 0$ then (recall III.5.2.2) ${\uparrow}a \cap {\uparrow}b = {\uparrow}(a \vee b) = \{1\} = \mathsf{O}$ and ${\uparrow}a \vee {\uparrow}b = {\uparrow}(a \wedge b) = L$.

(2)⇒(1): If ${\uparrow}a \cap {\uparrow}b = {\uparrow}(a \vee b) = \mathsf{O} = \{1\}$ then $a \vee b = 1$ and if ${\uparrow}a \vee {\uparrow}b = {\uparrow}(a \wedge b) = L$ then $a \wedge b = 0$. $\qquad\square$

1.2. Let S be a sublocale of L. From III.6.2 recall that

1.2.1. *The closed sublocales c_S in S are precisely the intersections of the sublocales closed in L with S and we have $c(a) \cap S = c_S(\nu_S(a))$ (ν_S is the nucleus associated with S); similarly the open sublocales $o_S(a)$ in S are precisely the intersections of the sublocales open in L with S and we have $o(a) \cap S = o_S(\nu_S(a))$.*

Recall further (from III.8.4) the inclusion

1.2.2. $f[\overline{S}] \subseteq \overline{f[S]}$ *for localic maps $f \colon L \to M$ and sublocales $S \subseteq L$.*

1.2.3. Finally, we will use the following

Observation. *If $S \subseteq L$ is a dense sublocale, $\overline{o(a) \cap S} = c(a^*)$.*

   (Indeed, $\bigwedge(o(a) \cap S) = \bigwedge\{a \to s \mid s \in S\} = a \to 0 = a^*$, since $0 \in S$. Notice that this fact extends III.8.1.2.)

1.3. Convention. To assimilate the terminology with the one used while working with subspaces in classical spaces, we will speak of the trivial sublocale $O = \{1\}$ as *void*, or *empty*, and say that sublocales S and T are *disjoint* if $S \cap T = O$.

2. Connected and disconnected locales

2.1. A locale (frame) is *connected* if there are precisely two complemented elements, namely 0 and 1 (thus, a connected locale is supposed to be non-trivial). Otherwise we speak of a *disconnected* locale.

2.2. Proposition. *The following statements on a non-trivial locale L are equivalent.*

   (1) *L is disconnected.*
   (2) *There are non-void disjoint open sublocales $S, T \subseteq L$ with $S \vee T = L$.*
   (3) *There are non-void disjoint closed sublocales $S, T \subseteq L$ with $S \vee T = L$.*

Proof. It follows immediately from 1.1.   □

2.2.1. From 1.2.1 and 2.2 we immediately obtain

Corollary. *The following statements on a non-void sublocale $S \subseteq L$ are equivalent.*

   (1) *S is connected.*
   (2) *For any disjoint open sublocales S_1, S_2 of L such that $S \subseteq S_1 \vee S_2$ either $S \subseteq S_1$ or $S \subseteq S_2$.*
   (3) *For any disjoint closed sublocales S_1, S_2 of L such that $S \subseteq S_1 \vee S_2$ either $S \subseteq S_1$ or $S \subseteq S_2$.*
   (4) *If for $a, b \in L$ such that $a \vee b = 1$, $S \subseteq \uparrow(a \wedge b)$ then either $S \subseteq \uparrow a$ or $S \subseteq \uparrow b$.*   □

   (Note that (4) is just (3) formulated more explicitly.)

2.2.2. We will speak of an *element* $a \in L$ as *connected* (resp. *disconnected*) if the associated open sublocale $\mathfrak{o}(a)$ is connected (resp. disconnected). From 2.2.1(2) we obtain

Corollary. *A non-zero element* $a \in L$ *is connected iff whenever* $a \leq b \vee c$ *and* $b \wedge c = 0$ *then* $a \leq b$ *or* $a \leq c$. $\qquad\square$

2.3. From 2.2.1(4) we easily infer

Proposition. *Let* $S \subseteq L$ *be a connected sublocale. Then the closure* $\overline{S}$ *is connected as well.*

Proof. Suppose $\overline{S} \subseteq {\uparrow}(a \vee b)$ with $a \vee b = 1$ Then $S \subseteq {\uparrow}(a \vee b)$ and hence, say, $S \subseteq {\uparrow}a$; consequently $\overline{S} \subseteq {\uparrow}a$. $\qquad\square$

2.3.1. In fact one has more.

Proposition. *Let* L *be connected and let* $f: L \to M$ *be a dense localic map. Then* M *is connected.*

Proof. For the associated frame homomorphism $h: M \to L$ we have the implication $h(x) = 0 \Rightarrow x = 0$. Now let $a \vee b = 1$ and $a \wedge b = 0$ in L. Then $h(a) \vee h(b) = 1$ and $h(a) \wedge h(b) = 0$ and hence, say $h(a) = 0$. By density, then, $a = 0$. $\qquad\square$

2.4. Quite analogously like in spaces we have

Proposition. *Let* S_i, $i \in J$, *be a system of connected sublocales of* L. *Let for any two* $i, j \in J$ *there exist* $i_1, \ldots, i_n \in J$ *such that*

$$i_1 = i, i_n = j, \text{ and for } k = 1, \ldots, n-1, \; S_{i_k} \cap S_{i_{k+1}} \neq 0.$$

Then $S = \bigvee_{i \in J} S_i$ *is connected.*

Proof. Let $S \subseteq \mathfrak{c}(a) \vee \mathfrak{c}(b)$ and $\mathfrak{c}(a) \cap \mathfrak{c}(b) = 0$. Then for each $i \in J$, $S_i \subseteq \mathfrak{c}(a) \vee \mathfrak{c}(b)$ and hence either $S_i \subseteq \mathfrak{c}(a)$ or $S_i \subseteq \mathfrak{c}(b)$. Denote by J_a (resp. J_b) the set of the indices i for which $S_i \subseteq \mathfrak{c}(a)$ (resp. $S_i \subseteq \mathfrak{c}(b)$). At least one of the sets J_a, J_b is non-empty, say $J_a \neq \emptyset$. Choose an $i \in J_a$ so that $S_i \subseteq \mathfrak{c}(a)$. Consider an arbitrary $j \in J$ and take the $i_1, \ldots, i_n$ from the assumption. If $i_k \in J_a$ then $i_{k+1} \in J_a$ as well since otherwise $0 \neq S_{i_k} \cap S_{i_{k+1}} \subseteq \mathfrak{c}(a) \cap \mathfrak{c}(b) = 0$. Thus, ultimately, for every $j \in J$, $S_j \subseteq \mathfrak{c}(a)$ and hence $S \subseteq \mathfrak{c}(a)$. $\qquad\square$

2.5. Consequently,

2.5.1. *for every connected sublocale* $S \subseteq L$ *there is the largest connected sublocale containing* S, *namely the join of all connected sublocales intersecting* S.

These maximal connected sublocales are called *connected components* or simply *components* of L. By 2.4 again,

2.5.2. *two distinct components are disjoint,*

and by 2.3,

2.5.3. *every component is a closed sublocale.*

2.6. Note. Unlike in classical spaces the join (in $\mathscr{S}\ell(L)$) of all the components of L is not necessarily the whole of L. In fact, L may contain no connected sublocale whatsoever (for instance, in a Boolean algebra without atoms every non-zero element is disconnected). One has only the trivial general observation that L is a disjoint sum of its components iff there is a cover by connected sublocales.

2.7. Disconnectedness and localic maps. In classical topology, a non-void space X is disconnected iff there is a surjective continuous map $f\colon X \to \{0,1\}$ (the latter with discrete topology). We have the same here, with

$$B_2 = \Omega(\{0,1\}) = \{0 < a, b < 1\} \quad (a \neq b).$$

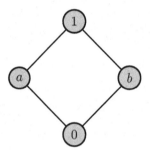

Figure 3: The frame B_2.

2.7.1. Proposition. *A locale L is disconnected iff there is a surjective localic map $f\colon L \to B_2$.*

Proof. If f is onto then the associated frame homomorphism $h\colon B_2 \to L$ is one-one and hence there are non-trivial $h(a), h(b) \in L$, complements of each other.   □

For applications a more detailed statement will be expedient.

2.7.2. Proposition. *Let $S, T \subseteq L$ be sublocales such that $S \vee T = L$. Let f be a surjective localic map such that $f[S] \subseteq \uparrow a$ and $f[T] \subseteq \uparrow b$. Then S, T are non-trivial closed and complements of each other.*
   If we denote by h the associated frame homomorphism we have $S = f^{-1}[\uparrow a] = \uparrow h(a)$ and $T = f^{-1}[\uparrow b] = \uparrow h(b)$.

Proof. Recall III.7.2 and III.7.3, in particular also the fact that for closed sublocales the sublocalic preimage coincides with the set-theoretical one.

From the adjunction between $f[-]$ and $f_{-1}[-]$ (III.7.2.2) and from III.7.3 we obtain

$$S \subseteq f^{-1}[{\uparrow}a] = {\uparrow}h(a) \quad \text{and} \quad T \subseteq f^{-1}[{\uparrow}b] = {\uparrow}h(b).$$

Since f is onto, h is one-one, $h(a), h(b) \neq 0, 1$ and ${\uparrow}h(a), {\uparrow}h(b)$ are non-trivial and complements of each other. Now, since T is a sublocale of the complement of ${\uparrow}h(a)$, we obtain

$${\uparrow}h(a) = (S \vee T) \cap {\uparrow}h(a) = (S \cap {\uparrow}h(a)) \vee \mathsf{O} = S,$$

and similarly ${\uparrow}h(b) = T$. $\square$

2.8. Extremal disconnectedness. Similarly to spaces, a locale is said to be *extremally disconnected* if the closure of each open sublocale is open.

Proposition. *For any locale L, the following statements are equivalent.*

(1) *L is extremally disconnected.*

(2) *For all $a \in L$, $\overline{\mathsf{o}(a)} = \mathsf{o}(a^{**})$.*

(3) *For all $a \in L$, $a^* \vee a^{**} = 1$ (thus, L is a De Morgan lattice).*

Proof. (2)$\Rightarrow$(1) is trivial.

(1)$\Rightarrow$(3): If $\overline{\mathsf{o}(a)} = \mathsf{o}(b)$ then by 1.2.3, $\mathsf{c}(a^*)^\mathsf{c} = \mathsf{o}(b)^\mathsf{c} = \mathsf{c}(b)$ so that ${\uparrow}a^* \cap {\uparrow}b = \{1\}$, and hence $a^* \vee b = 1$, and ${\uparrow}a^* \vee {\uparrow}b = L$, and hence in particular $0 = x \wedge y$ for some $x \geq a^*$ and $y \geq b$, and so $a^* \wedge b = 0$. Thus, $b \leq a^{**}$ and $a^* \vee a^{**} = 1$.

(3)$\Rightarrow$(2): By 1.2.3, we have to prove that $x \geq a^*$ iff $a^{**} \to x = x$. Since $a^* \wedge a^{**} = 0 \leq x$ and hence $a^* \leq a^{**} \to x$, the implication $\Leftarrow$ is trivial. Now let $a^* \leq x$ (so that $a^* \to x = 1$) and $a^* \vee a^{**} = 1$. Then

$$x = (a^* \vee a^{**}) \to x = (a^* \to x) \wedge (a^{**} \to x) = a^{**} \to x.$$ $\square$

2.9. Disconnectedness of Stone-Čech compactification. Recall VII.4. To simplify the notation we will think, in this subsection, of L as a sublocale of $\mathfrak{R}L$, a dense one. The embedding will be denoted by $j \colon L \subseteq \mathfrak{R}L$, and translating the frame homomorphism fact from VII.4.4 into the localic language we infer that for each localic map $f \colon L \to M$ with M compact there is a localic map $\widehat{f} \colon \mathfrak{R}L \to M$ such that $\widehat{f} \cdot j = f$.

2.9.1. Proposition. *Let S, T be non-trivial closed sublocales that are mutual complements. Then $\overline{S}, \overline{T}$ are non-trivial closed sublocales of $\mathfrak{R}L$ that are mutual complements.*

Proof. Let $f \colon L \to B_2$ be the localic map from 2.7.1. Take the localic map $\widetilde{f} \colon \mathfrak{R}L \to B_2$ such that $\widetilde{f}j = f$. Then, by 1.2.2, $\widetilde{f}[\overline{S}] \subseteq \overline{\widetilde{f}[S]} = \overline{f[S]} \subseteq {\uparrow}a$ and similarly $\widetilde{f}[\overline{T}] \subseteq \overline{\widetilde{f}[T]} = \overline{f[T]} \subseteq {\uparrow}b$. Use 2.7.2. $\square$

2.9.2. Proposition. *Let L be extremally disconnected. Then the Stone-Čech compactification $\Re L$ is also extremally disconnected.*

Proof. Let U be open in $\Re L$. Consider the $V = U \cap L$ open in L and the closure $\overline{V}^L$ in L, that is, $\overline{(U \cap L)} \cap L$. By 1.2.3 we have

$$\overline{V}^L = \overline{U} \cap L$$

and this sublocale has a closed complement T (in L), as L is extremally disconnected. By 2.9.1, $\overline{V}^L$ and $\overline{S}$ are closed and mutual complements, and we have

$$U = \overline{U \cap L} \subseteq \overline{V}^L = \overline{\overline{U} \cap L} \subseteq \overline{U}. \qquad \square$$

3. Locally connected locales

3.1. A locale (frame) is said to be *locally connected* if it has a basis consisting of connected elements.

3.2. Proposition. *A locally connected locale is a join of its components and all of its components are (both closed and) open.*

Proof. Let B be a basis of L such that each $b \in B$ is connected. Since obviously $L = \bigvee B$, L is the join of all of its components.

On the set B define an equivalence relation

$$a \sim b \quad \equiv_{\mathrm{df}} \quad \exists a_1 = a, a_2, \ldots, a_n = b, \quad a_j \wedge a_{j+1} \neq 0 \text{ for } k = 1, \ldots, n-1.$$

Denote by $[a]$ the equivalence class containing a. By 2.4 (since $\mathfrak{o}(a) \cap \mathfrak{o}(b) \neq \mathsf{O}$ iff $a \wedge b \neq 0$), $\bigvee[a]$ is connected.

If $\bigvee[a] \cap \bigvee[b] \neq 0$ we have $a' \sim a$ and $b' \sim b$ such that $a' \wedge b' \neq 0$. Then $a' \sim b'$ and hence $a \sim b$. Thus,

$$\text{for any two } a, b \in B, \text{ either } \bigvee[a] = \bigvee[b] \text{ or } \bigvee[a] \wedge \bigvee[b] = 0.$$

Thus we have a decomposition $1 = \bigvee_{i \in J} b_j$ (b_i are the distinct $\bigvee[a]$) with b_i connected and $b_i \wedge b_j = 0$ for $i \neq j$; in terms of sublocales,

$$L = \bigvee_{i \in J} \mathfrak{o}(b_i)$$

with disjoint open connected sublocales $\mathfrak{o}(b_i)$.

Now consider a component S of L. Denote by J_0 the set of the indices i such that $\mathfrak{o}(b_i) \cap S = \mathsf{O}$ and by J_1 the set of the indices i such that $\mathfrak{o}(b_i) \cap S \neq \mathsf{O}$. For $i \in J_1$, $\mathfrak{o}(b_i) \vee S$ is connected and hence $\mathfrak{o}(b_i) \subseteq S$. Thus,

$$\bigvee_{i \in J_1} \mathfrak{o}(b_i) \subseteq S \quad \text{and} \quad \left(\bigvee_{i \in J_0} \mathfrak{o}(b_i) \right) \cap S = \mathsf{O}$$

(the distributivity needed for the latter follows from VI.5.4.1: S is closed, hence complemented and hence linear; $\mathscr{S}\ell(L)$ is not a frame!). Consequently,

$$\bigvee_{i \in J_1} \mathfrak{o}(b_i) = S. \qquad \qquad \square$$

3.2.1. Note. In fact the components S are precisely the $\mathfrak{o}([a])$: if the J_1 in the last formula had more than one element we had $S = \mathfrak{o}(b_k) \vee \bigvee_{i \in J_1 \smallsetminus \{k\}} \mathfrak{o}(b_i)$ disconnected.

3.3. Coproducts of locally connected frames. We will use the notation from Chapter IV, Sections 4 and 5, and also speak of the down-sets $U \in \bigoplus_{i \in J} L_i$ as saturated down-sets.

3.3.1. Lemma. *Let* $c = \bigvee A$ *with all* $a \in A$ *connected. Let* c *be connected. Then* A *is chained (in the sense that* $a \sim b$ *for any two* $a, b \in A$, *see 3.2).*

Proof. Take an $a_1 \in A$ and set $A_1 = \{a \in A \mid a_1 \sim a\}$, $A_2 = A \smallsetminus A_1$. Then $c = \bigvee A_1 \vee \bigvee A_2$ and since $0 \neq a_1 \leq A_1$ and c is connected, $A_2 = \emptyset$. $\qquad \square$

3.3.2. Let L_i be connected locally connected frames, let $L = \bigoplus_{i \in J} L_i$ and let $1_L = u \vee v$ with $u \wedge v = 0_L$ (that is $u \wedge v = \mathbf{n}$). These u, v will be fixed during the following two paragraphs.

An $(a_i)_{i \in J} \in \prod'_{i \in J} L_i$ will be called *exact* if

$$\text{either } (a_i)_{i \in J} \in u \ \text{ or } \ (a_i)_{i \in J} \in v.$$

Lemma. *Let* $d *_j a$ *be exact for all* $d \in D$, *let* $a \notin \mathbf{n}$, *and let* D *be chained. Then* $(\bigvee D) *_j a$ *is exact.* $(x *_j a$ *from IV.4.1.)*

Proof. Denote by D_u (resp. D_v) the set of the $d \in D$ for which

$$d *_j a \in u \quad (\text{resp.} \quad d *_j a \in v).$$

If $d_1 *_j a \in u$ and $d_2 *_j a \in v$ we have $(d_1 \wedge d_2) *_j a \in \mathbf{n}$ and since $a_i \neq 0$ for $i \neq j$, $d_1 \wedge d_2 = 0$. Thus, as D is chained, we have, say, all $d *_j a \in u$ and hence $(\bigvee D) *_j a \in u$. $\qquad \square$

3.3.3. Theorem. *Let* L_i, $i \in J$, *be connected and locally connected frames. Then* $L = \bigoplus_{i \in J} L_i$ *is connected and locally connected.*

Proof. Take u, v as above. An $(a_i)_{i \in J} \in \prod'_{i \in J} L_i$ will be called *c-exact* if each $(c_i)_{i \in J} \leq (a_i)_{i \in J}$ with connected c_i is exact. Then

$$U = \{(a_i)_{i \in J} \mid (a_i)_{i \in J} \text{ is c-exact}\}$$

is saturated. Indeed, first, it is obviously a down-set. Now let $a_j^k *_j b$ be in U for $k \in K$; set $a_j = \bigvee_{k \in K} a_j^k$. Let $c_i \leq b_i$, $i \neq j$, and $c_j \leq a_j$ be connected. Using local connectedness we obtain

$$c_j \wedge a_j^k = \bigvee\{d_m^k \mid m \in M(k)\} \quad \text{with } d_m^k \text{ connected.}$$

Thus,

$$c_j = \bigvee D \quad \text{where} \quad D = \{d_m^k \mid k \in K, \ m \in M(k)\}$$

and by 3.3.1, D is chained, so that by 3.3.2, $(c_i)_{i \in J} = c_j *_j c$ is exact.

Now we have

$$u = \bigvee\{(a_i)_{i \in J} \mid (a_i)_{i \in J} \leq u\}, \quad v = \bigvee\{(b_i)_{i \in J} \mid (b_i)_{i \in J} \leq v\}$$

and all the $(a_i)_{i \in J} \leq u$ and $(b_i)_{i \in J} \leq v$ are, trivially, c-exact so that $U \geq u \vee v = L$ and hence in particular $\overline{1} \in U$ so that $\overline{1}$ is c-exact, and since the 1_{L_i} themselves are connected, it is exact. Thus, it is either in u or in v and the other is zero; we see that L is connected.

Now let B_i be bases of L_i consisting of connected elements. Then each $\oplus_{i \in J} b_i$, $b_i \in B_i$ is connected (since $\downarrow b_i$ are locally connected we can apply the first part of this proof for $\oplus_{i \in J} \downarrow b_i$), and $B = \{\oplus_{i \in J} b_i \mid b_i \in B_i\}$ is obviously a basis of L. $\square$

3.3.4. Corollary. *Let L_i, $i \in J$, be locally connected frames. Then $L = \bigoplus_{i \in J} L_i$ is locally connected.* $\square$

(If B_i are bases of L_i consisting of connected elements, apply 3.3.3 for the individual members of the basis $B = \{\oplus_{i \in J} b_i \mid b_i \in B_i\}$ of L.)

4. A weird example

4.1. In classical topology, a closed open set U in $X \times Y$ where X is connected is always of the form $X \times V$ where V is open and closed in Y. Indeed, if $(x, y) \in U$ then $U \cap (X \times \{y\})$ is a non-empty open and closed subset of $X \times \{y\}$, which is homeomorphic with X, hence connected, and consequently $X \times \{y\} \subseteq U$, and we conclude that $U = X \times V$ for some $V \subseteq Y$; now for any $x \in X$, $\{x\} \times Y$ is homeomorphic with Y and contemplating $\{x\} \times V = U \cap (\{x\} \times Y)$ we conclude that V is closed and open in Y.

We will say that a frame L is *p-connected* (short for *product-connected*) if for any frame M each complemented element a in $L \oplus M$ is of the form $1 \oplus b$ with b complemented in M. From the observation above we might expect that each connected frame is p-connected. But this is not the case as we will see in this section.

4.2. The frame $L_{\mathbb{B}}$. Let $\mathbb{B}$ be a complete Boolean algebra. Denote by

$$\mathrm{dc}(\mathbb{B})$$

the set of all disjoint covers of $\mathbb{B}$ by non-zero elements. Further set

$$X_{\mathbb{B}} = \{(A, u, \varepsilon) \mid A \in \mathrm{dc}(\mathbb{B}), \ u \in A, \ \varepsilon = 1, 2\}$$

(this set will be unstructured: we disregard the preorder of covers and the order of the $u \in \mathbb{B}$; to avoid too many commas we will write simply $(Au\varepsilon)$ instead of (A, u, ε)), and denote by

$$S_{\mathbb{B}}$$

the free meet-semilattice generated by $X_{\mathbb{B}}$. This may be defined as the set of all finite subsets of $X_{\mathbb{B}}$ ordered by the inverse inclusion, but it will be more transparent if we think of its elements as of the formal meets

$$x = \bigwedge_{i=1}^{n} (A_i u_i \varepsilon_i); \quad \text{we will write} \quad \gamma(x) = \{A_1, \ldots, A_n\}.$$

Further consider the down-set lattice

$$\mathfrak{D}(S_{\mathbb{B}})$$

(thus, we have here the free frame generated by $X_{\mathbb{B}}$ – recall IV.2.4).

Now consider the binary relation R on $\mathfrak{D}(S_{\mathbb{B}})$ consisting of all the pairs

$$(\downarrow(Au1) \cap \downarrow(Bv2), \emptyset) \quad \text{such that} \quad u \wedge v \neq 0,$$

and all the

$$(\downarrow(Au1) \cup \downarrow(Au2), \downarrow(Av1) \cup \downarrow(Av2))$$

and, finally, set

$$L_{\mathbb{B}} = \mathfrak{D}(S_{\mathbb{B}})/R.$$

4.2.1. From the description of the relation R we easily infer (recall III.11) the following

Observation. *The elements of $L_{\mathbb{B}}$ (the R-saturated elements) are the down-sets $U \subseteq S_{\mathbb{B}}$ such that*

(S1) *for all $A, B \in \mathrm{dc}(\mathbb{B})$ and $u \in A$, $v \in B$ such that $u \wedge v \neq 0$, $(Au1) \wedge (Bv2) \in U$,*

(S2) *and for all $x \in S_{\mathbb{B}}$ and $u, v \in A \in \mathrm{dc}(\mathbb{B})$,*

$$\{(Au1) \wedge x, (Au2) \wedge x\} \subseteq U \quad \Rightarrow \quad \{(Av1) \wedge x, (Av2) \wedge x\} \subseteq U.$$

4.3. Further notation. The chain relation from 3.2 will be modified for specific covers A by

$$a \sim_A b \quad \equiv_{df} \quad \exists a_1 = a, a_2, \dots, a_n = b, \ a_i \in A, \quad a_j \wedge a_{j+1} \neq 0 \ \text{ for } 1 \leq k \leq n-1.$$

We set

$$N = \left\{ x = \bigwedge_{i=1}^{n} (A_i u_i \varepsilon_i) \mid u_i \sim_{\bigcup_{k=1}^{n} A_k} u_j \ \Rightarrow \ \varepsilon_i = \varepsilon_j \right\}$$

and for $A \in \mathrm{dc}(\mathbb{B})$ and $\emptyset \neq J \subseteq A$ define

$$\zeta(A, J) = \{ \bigwedge_{u \in J} (Au\varepsilon_u) \mid \varepsilon_u \text{ arbitrary}\}.$$

The equivalence class of $x \in S_{\mathbb{B}}$ in $L_{\mathbb{B}}$ will be denoted by

$$\langle x \rangle, \quad \text{and } \langle (Au\varepsilon) \rangle \text{ will be simplified to } \langle Au\varepsilon \rangle.$$

From (S1) and (S2) we immediately obtain

4.3.1. Observation. (1) *If $u \wedge v \neq 0$ then $\langle Au1 \rangle \wedge \langle Bv2 \rangle = 0$.*

(2) $a(A) = \langle Au1 \rangle \vee \langle Au2 \rangle$ *does not depend on the choice of $u \in A$.*

(3) $\bigvee \{a(A) \mid A \in \mathrm{dc}(\mathbb{B})\} = 1$.

4.3.2. Lemma. *For each J we have $\bigvee\{\langle x \rangle \mid x \in \zeta(A, J)\} = a(A)$, and consequently for $x = \bigwedge_{i=1}^{n}(A_i u_i \varepsilon_i) \in S_{\mathbb{B}}$ and any choice of $J_i \subseteq A_i$ we have*

$$\langle x \rangle \leq \bigwedge_{i=1}^{n} \bigvee\{\langle y \rangle \mid y \in \zeta(A_i, J_i)\}.$$

Proof. We have $\bigvee\{\langle x \rangle \mid x \in \zeta(A, \{u\})\} = \langle Au1 \rangle \vee \langle Au2 \rangle = a(A)$ and since obviously

$$\bigvee\{\langle x \rangle \mid x \in \zeta(A, J \cup \{v\})\} = \bigvee\{\langle x \rangle \mid x \in \zeta(A, J)\} \wedge (\langle Av1 \rangle \vee \langle Av2 \rangle)$$

the first statement follows by induction. For the second statement, $\langle x \rangle \leq \langle A_i u_i \varepsilon_i \rangle \leq \langle A_i u_i 1 \rangle \vee \langle A_i u_i 2 \rangle = a(A_i) = \bigvee\{\langle y \rangle \mid y \in \zeta(A_i, J_i)\}$ for all i. $\qquad\square$

4.3.3. Lemma. *The set $U = S_{\mathbb{B}} \setminus N$ is saturated. Consequently, for each $x \in N$, $\langle x \rangle \neq 0$.*

Proof. Since N is obviously an up-set, U is a down-set. If $u \wedge v \neq 0$ then $u \sim v$ and hence $(Au1) \wedge (Bv2) \notin N$, and (S1) holds.

Now let

$$y = (Av\varepsilon) \wedge \bigwedge_{i=1}^{n} (A_i u_i \varepsilon_i) \notin U,$$

that is, $y \in N$. Take a $u \in A$. If

$$u \sim_{A \cup \bigcup_{k=1}^n A_k} u_i$$

put $\eta = \varepsilon_i$ (uniquely determined since $y \in N$), otherwise choose η arbitrarily. Now $(Av\eta) \wedge \bigwedge_{i=1}^n (A_i u_i \varepsilon_i) \in N$, hence not in U and we see that also (S2) holds.

For $x \in N$, $\langle x \rangle \nleq U$ and hence it cannot be 0. $\square$

4.3.4. Proposition. *In* $L_{\mathbb{B}} \oplus \mathbb{B}$ *set*

$$a_i = \bigvee \{ \langle Aui \rangle \oplus u \mid A \in \mathsf{dc}(\mathbb{B}), \ u \in A \} \quad (i = 1, 2).$$

Then a_2 *is the complement of* a_1 *and for both* $i = 1, 2$, *if* $a_i \leq 1 \oplus u$ *then* $u = 1$.

Proof. All the summands in

$$a_1 \wedge a_2 = \bigvee \{ (\langle Au1 \rangle \wedge \langle Bv2 \rangle) \oplus (u \wedge v) \mid u \in A, v \in B \}$$

are zero because if $u \wedge v \neq 0$ then $\langle Au1 \rangle \wedge \langle Bv2 \rangle = 0$, by 4.3.1(1). Thus $a_1 \wedge a_2 = 0$. Further we have, by 4.3.1(3),

$$a_1 \vee a_2 = \bigvee \{ (\langle Au1 \rangle \vee \langle Au2 \rangle) \oplus u \mid A \in \mathsf{dc}(\mathbb{B}), \ u \in A \}$$

$$= \bigvee_{A \in \mathsf{dc}(\mathbb{B})} (a(A) \oplus \bigvee_{u \in A} u) = \bigvee_{A \in \mathsf{dc}(\mathbb{B})} (a(A) \oplus 1)$$

$$= \left(\bigvee_{A \in \mathsf{dc}(\mathbb{B})} a(A) \right) \oplus 1 = 1.$$

Finally let $a_i \leq 1 \oplus u$. Then $\langle Avi \rangle \oplus v \leq 1 \oplus u$ for all $v \in A \in \mathsf{dc}(\mathbb{B})$ and since $\langle Av_i \rangle \neq 0$ by 4.3.1, for all the v we have $v \leq u$ (recall IV.5.2(4)) and hence $u = 1$. $\square$

4.4. Complete Boolean algebras with property (P). Denote by $\mathfrak{P}_{\mathsf{fin}}(X)$ the set of all finite subsets of X. A complete Boolean algebra is said to *satisfy* (P) if

(P) for every $\mathcal{C} \subseteq \mathsf{dc}(\mathbb{B})$ containing refinements of all the $A \in \mathsf{dc}(\mathbb{B})$, and for each mapping $\phi \colon \mathcal{C} \to \mathfrak{P}_{\mathsf{fin}}(\mathbb{B})$ such that $\phi(A) \subseteq A$ for all $A \in \mathcal{C}$ there exist $A, B \in \mathcal{C}$ for which there are no $u \in \phi(A)$ and $v \in \phi(A) \cup \phi(B)$ satisfying $u \sim_{A \cup B} v \neq u$.

Denote by ν the nucleus associated with the relation R (of 4.2).

4.4.1. Lemma. *Let* $\nu(\emptyset) \subseteq M \in \mathfrak{D}(S_{\mathbb{B}})$. *Set*

$$U = \{ x \in S_{\mathbb{B}} \mid \exists \phi \colon \gamma(x) \to \mathfrak{P}_{\mathsf{fin}}(\mathbb{B}) \quad s. \ t. \ \forall A \in \gamma(x), \phi(A) \subseteq A$$

$$\textit{and s.t. for any choice of } y_A \in \zeta(A, \phi(A)), \quad \bigwedge_{A \in \gamma(x)} y_A \wedge x \in M \}.$$

Then U *is* R-*saturated.*

Proof. Let $x \in U$ and $z \leq x$. Then $\gamma(x) \subseteq \gamma(z)$. For $A \in \gamma(z) \smallsetminus \gamma(x)$ choose $\psi(A) \subseteq A$ arbitrarily and for $A \in \gamma(x)$ set $\psi(A) = \phi(A)$. Now if we choose $y_A \in \zeta(A, \psi(A))$ we have $\bigwedge_{A \in \gamma(z)} y_A \wedge z \leq \bigwedge_{A \in \gamma(x)} y_A \wedge x \in M$; thus, U is a down-set.

(S1) is obvious since $\nu(\emptyset) \subseteq M$.

(S2): Let $(Au\varepsilon) \wedge x \in U$ for $\varepsilon = 1, 2$. For $B \in \{A\} \cup \gamma(x)$ $(=\gamma((Au\varepsilon) \wedge x) = \gamma((Av\varepsilon) \wedge x))$ set

$$J_B((Av\varepsilon) \wedge x) = J_B((Au1) \wedge x) \cup J_B((Au2) \wedge x) \text{ for } B \neq A, \text{ and}$$

$$J_A((Av\varepsilon) \wedge x) = J_A((Au1) \wedge x) \cup J_A((Au2) \wedge x) \cup \{u\}.$$

Choose $y_B \in \zeta(B, J_B((Av\varepsilon) \wedge x))$. It can be written as

$$y_B = y_B^1 \wedge y_B^2$$

with $y_B^\varepsilon \in \zeta(B, J_B((Av\varepsilon) \wedge x))$ for $B \neq A$ and

$$y_A = y_A^1 \wedge y_A^2 \wedge (Au\eta)$$

with $y_A^\varepsilon \in \zeta(B, J_B((Av\varepsilon) \wedge x))$.

We have $\bigwedge_B y_B^\varepsilon \wedge (Au\varepsilon) \wedge x \in M$ and hence

$$\bigwedge_B y_B \wedge x \leq \bigwedge_B y_B^1 \wedge \bigwedge_B y_B^2 \wedge (Au\eta) \wedge x \in M. \qquad \square$$

4.4.2. Lemma. *Let $\nu(\emptyset) \subseteq M$ and $\nu(M) = S_{\mathbb{B}}$. Then there is a mapping $\phi \colon \mathsf{dc}(\mathbb{B}) \to \mathfrak{P}_{\mathsf{fin}}(\mathbb{B})$ such that for all A, $\phi(A) \subseteq A$ and $\zeta(A, \phi(A)) \subseteq M$.*

Proof. The set U from 4.4.1 obviously contains M; thus, it is the whole of $S_{\mathbb{B}}$ and each $x \in S_{\mathbb{B}}$ has the property indicated. For $A \in \mathsf{dc}(\mathbb{B})$ choose a $u \in A$ and set

$$\phi(A) = J_A((Au1)) \cup J_A((Au2)) \cup \{u\}.$$

Now a $y \in \zeta(A, \phi(A))$ is of the form $y = y_1 \wedge y_2 \wedge (Au\varepsilon)$ with $y_\varepsilon \in \zeta(A, J_A((Au\varepsilon)))$, and $y \leq y_\varepsilon \wedge (Au\varepsilon) \in M$. $\qquad \square$

4.4.3. Lemma. *Let a Boolean algebra $\mathbb{B}$ satisfy (P). Then the frame $L_{\mathbb{B}}$ is connected.*

Proof. Let for some $M_1, M_2 \in L_{\mathbb{B}}$ there be $M_1 \wedge M_2 = 0$ and $M_1 \vee M_2 = 1$. Then $\nu(\emptyset) \subseteq M_i \subseteq M_1 \cup M_2$ and $\nu(M_1 \cup M_2) = M_1 \vee M_2 = S_{\mathbb{B}}$. By 4.3.2 we have a $\phi \colon \mathsf{dc}(\mathbb{B}) \to \mathfrak{P}_{\mathsf{fin}}(\mathbb{B})$ such that for all A, $\phi(A) \subseteq A$ and $\zeta(A, \phi(A)) \subseteq M_1 \cup M_2$. Set

$$\mathcal{C}_0 = \{A \in \mathsf{dc}(\mathbb{B}) \mid \zeta(A, \phi(A)) \subseteq M_i \text{ for some } i\} \quad \text{and} \quad \mathcal{C} = \mathsf{dc}(\mathbb{B}) \smallsetminus \mathcal{C}_0.$$

I. Let $\mathcal{C}$ contain a subdivision of any $A \in \mathsf{dc}(\mathbb{B})$. By (P) we have $A, B \in \mathcal{C}$ such that one never has $u \sim_{A \cup B} v$ with $u \in \phi(A)$ and $v \in \phi(A) \cup \phi(B)$. Since $\zeta(A, \phi(A)) \subseteq M_1 \cup M_2$ we have, say, an $x = \bigwedge_{u \in \phi(A)}(Au1) \in M_1$. Since $B \notin \mathcal{C}_0$ there is a $y = \bigwedge_{v \in \phi(B)}(Av\varepsilon_v) \in M_2$. Now $x \wedge y \in N$ and hence by 4.3.3, $\langle x \wedge y \rangle \neq 0$ while $\langle x \wedge y \rangle = \langle x \rangle \wedge \langle y \rangle \leq M_1 \wedge M_2 = 0$.

II. Consequently there has to be an $A \in \mathsf{dc}(\mathbb{B})$ such that each of its subdivisions is in $\mathcal{C}_0$. We will show that then $\mathcal{C} = \emptyset$ and hence $\mathcal{C}_0 = \mathsf{dc}(\mathbb{B})$.

Let $B \in \mathcal{C}$. The subdivision $A \wedge B = \{u \wedge v \mid u \in A, v \in \mathcal{B}\}$ of A is, however, in $\mathcal{C}_0$. Let, say, $\zeta(A \wedge B, \phi(A \wedge B)) \subseteq M_1$. Since B is not in $\mathcal{C}_0$ there exists ε_v such that $y = \bigwedge_{v \in \phi(B)}(Av\varepsilon_v) \in M_2$.

Now for $u \in \phi(A \wedge B)$ take $\varepsilon_u = \varepsilon_v$ whenever $u \leq v \in \phi(B)$ and set

$$x = \bigwedge_{u \in \phi(A \wedge B)} ((A \wedge B)u\varepsilon_u).$$

We have (again) $x \wedge y \in N$, hence $\langle x \wedge y \rangle \neq 0$ contradicting $\langle x \wedge y \rangle \leq M_1 \wedge M_2 = 0$.

III. Thus, $\mathsf{dc}(\mathbb{B}) = \mathcal{C}_0$. If we had A, B such that $\zeta(A, \phi(A)) \subseteq M_1$ and $\zeta(B, \phi(B)) \subseteq M_2$ we would obtain a contradiction

$$0 \neq \langle \bigwedge_{u \in \phi(A)}(Au1) \wedge \bigwedge_{v \in \phi(B)}(Bv1) \rangle \leq M_1 \wedge M_2.$$

Thus, say, $\zeta(A, \phi(A)) \subseteq M_1$ for all A and we have by 4.3.1 and 4.3.2

$$M_1 \geq \bigvee_{A \in \mathsf{dc}(\mathbb{B})} \bigvee\{\langle x \rangle \mid y \in \zeta(A, \phi(A))\} = \bigvee_{A \in \mathsf{dc}(\mathbb{B})} a(A) = 1. \qquad \square$$

4.5. A connected frame which is not p-connected.

Proposition. *There exists a frame $L_\mathbb{B}$ that is connected but not p-connected.*

Proof. In view of 4.3.4 and the previous lemmas it suffices to find a complete Boolean algebra $\mathbb{B}$ satisfying (P).

Denote by ω the set of all natural numbers. Consider a set Y of cardinality at least 2^ω; make it to a discrete metric space (say, with distance 1 for any two distinct points). Consider the metric space

$$(X, \rho) = Y^\omega \quad \text{and} \quad \mathbb{B} \text{ the Booleanization of } \Omega(X, \rho)$$

(that is, $\mathbb{B}$ is the Boolean algebra of all the regular open subsets of (X, ρ)). Then:

(1) the cardinality of each $0 \neq a \in \mathbb{B}$ (in fact, of any non-empty open set in (X, ρ)) is $\geq 2^{2^\omega}$.

Let $\mathcal{C}$ contain subdivisions of all the $A \in \mathsf{dc}(\mathbb{B})$; we easily find covers $S_n \in \mathcal{C}$ such that

(2) if $m \geq n$ then S_m is a refinement of S_n, and

(3) the diameter of each $u \in S_n$ is $\leq \frac{1}{n}$.

Choose finite subsets $Z \subseteq X$ meeting all the elements of $\phi(S_n)$ and set $Z = \bigcup Z_n$.
     If $\rho(x, Z) > \frac{1}{k}$ then $x \notin \bigcup_{n \geq k} \bigcup \phi(S_n)$ and hence we have

$$\overline{Z} \supseteq \bigcap_{k=1}^{\infty} \overline{\bigcup_{n \geq k} \bigcup \phi(S_n)} \geq \bigcap_{k \in \omega} (\bigvee_{n \geq k} \bigvee \phi(S_n))$$

where $\bigvee$ is the join in $\mathbb{B}$. Since Z is countable, the cardinality of $\overline{Z}$ is $\leq 2^\omega$, the meet of the $a_n = \bigvee_{n \geq k} \bigvee \phi(S_n)$ in $\mathbb{B}$ cannot be non-empty. Thus we have

$$\bigwedge_{k=1}^{\infty} a_k = 0 \quad \text{and hence} \quad \bigvee_{k=1}^{\infty} a_k^{\mathsf{c}} = 1. \tag{4.5.1}$$

Consider the system

$$C = \{a_1^{\mathsf{c}}\} \cup \bigcup_{k=2}^{m} \{a_{k-1} \wedge a_k^{\mathsf{c}} \wedge x \mid x \in S_{k-1}\}.$$

For $k \geq m$ we have $a_k \leq a_m$, hence $a_k^{\mathsf{c}} \geq a_m^{\mathsf{c}}$ so that

$$\bigvee C \wedge a_m^{\mathsf{c}} = a_1^{\mathsf{c}} \vee \bigvee_{k=2}^{m} (\{a_{k-1} \wedge a_k^{\mathsf{c}} \wedge x \mid x \in S_{k-1}\}) = a_1^{\mathsf{c}} \vee \bigvee_{k=2}^{m} (a_{k-1} \wedge a_k^{\mathsf{c}}) = a_m^{\mathsf{c}}.$$

Thus, C is a cover, by (4.5.1), and it is obviously disjoint. Moreover,

$$\forall n, \quad C \text{ refines } \phi(S_n) \cup \{(\bigvee \phi(S_n))^{\mathsf{c}}\} \tag{4.5.2}$$

(indeed, take $u = a_{k-1} \wedge a_k^{\mathsf{c}} \wedge x$ with $x \in S_{k-1}$; if $k \leq n$ then $a_k \geq \bigvee \phi(S_n)$ and $u \leq (\phi(S_n))^{\mathsf{c}}$, and if $k - 1 \geq n$ there is a $y \in S_n$ such that $x \leq y$ and we have either $y \in \phi(S_n)$ or $y \leq (\phi(S_n))^{\mathsf{c}}$ again).
     Choose a refinement $A \in \mathcal{C}$ of C. For $u \in A$ choose a number $k(u)$ such that $u \leq a_{k(u)}^{\mathsf{c}}$. Set

$$B = S_n \quad \text{where} \quad n = \max\{k(u) \mid u \in \phi(A)\}.$$

By (4.5.2), A refines $\phi(B) \cap \{(\phi(B))^{\mathsf{c}}\}$ and we have

$$\bigvee \phi(A) \leq a_n^{\mathsf{c}} = (\bigvee_{j \geq n} \bigvee \phi(S_n))^{\mathsf{c}} \leq (\bigvee \phi(B))^{\mathsf{c}}$$

so that $\bigvee \phi(A) \wedge \bigvee \phi(B) = 0$ and hence we cannot have $u \sim_{A \cup B} v \neq u$ with $u \in \phi(A)$ and $v \in \phi(A) \cup \phi(B)$. $\square$

5.   A few notes

5.1. The example in Section 4 was constructed by I. Kříž. This author (by similar, but much finer methods) also produced two connected locales the product of which is disconnected. This example was never published, though.

It should be noted that the locale $L_\mathbb{B}$ is in fact locally compact and hence spatial (see [167]).

5.2. Using the techniques of Section 3 one can prove that each connected locally connected locale is p-connected – see [165]. Also, there are classes of locales (e.g., the hereditarily Lindelöf ones) in which each (plain) connected locale is p-connected ([167]).

5.3. An extremely interesting result was published by Moerdijk and Wraith in [184]. Using categorical reasoning one can present a quite legitimate definition of path connectedness independent on the existence of points; this property can be then, in turn, shown to be equivalent with connected local connectedness.

5.4. The fact that locally connected connected locales behave better than the (plain) connected ones should not be quite surprising. Remembering the standard examples of connected spaces that are not locally connected the reader may realize that they do not really look "quite holding together". Point-free topology is in a sense more geometrical than the classical one, and the discrepancy between just connected and connected locally connected locales is more pronounced.

Chapter XIV

The Frame of Reals and Real Functions

The description of frames "by generators and defining relations" from IV.2.4.1 can be used to introduce the reals in a point-free way. André Joyal ([156]) was the first to emphasize the benefits of such a point-free approach to $\mathbb{R}$ in the constructive context of topos theory; it was first explicitly carried out in [104]. The idea is to take the standard totally ordered set $\mathbb{Q}$ of rationals and to take the set of open intervals with rational endpoints for the basic generators. One can then deal with continuous real functions and even more general real functions as we will see in this chapter.

1. The frame $\mathfrak{L}(\mathbb{R})$ of reals

1.1. The *frame of reals* is the frame $\mathfrak{L}(\mathbb{R})$ generated by all ordered pairs (p, q), with $p, q \in \mathbb{Q}$, subject to the following relations:

(R$_1$) $(p, q) \wedge (r, s) = (p \vee r, q \wedge s)$.

(R$_2$) $(p, q) \vee (r, s) = (p, s)$ whenever $p \leq r < q \leq s$.

(R$_3$) $(p, q) = \bigvee \{(r, s) \mid p < r < s < q\}$.

(R$_4$) $1 = \bigvee \{(p, q) \mid p, q \in \mathbb{Q}\}$.

1.1.1. Remarks. (1) This definition follows Banaschewski [16] (see also [33]) but it is equivalent to the one used by Johnstone [145].

(2) Note that, by (R$_3$),

$$(p, q) = \bigvee \emptyset = 0 \text{ whenever } p \geq q.$$

Hence, by (R$_1$), it follows that $(p, q) \wedge (r, s) = 0$ whenever $p \vee r \geq q \wedge s$.

(3) A map h from the generating set of $\mathfrak{L}(\mathbb{R})$ into a frame L defines a frame homomorphism $\mathfrak{L}(\mathbb{R}) \to L$ if and only if it sends the relations (R_1)–(R_4) above into identities in L. In particular, this holds for the frame $L = \Omega(\mathbb{Q})$ of the usual open subsets of $\mathbb{Q}$ and the h sending (p, q) to

$$\langle p, q \rangle = \{ t \in \mathbb{Q} \mid p < t < q \}$$

("the natural interpretation of (p, q)" in $\Omega(\mathbb{Q})$ [16]).

By this canonical homomorphism $h \colon \mathfrak{L}(\mathbb{R}) \to \Omega(\mathbb{Q})$, if $(p, q) = 0$ in $\mathfrak{L}(\mathbb{R})$ then $\langle p, q \rangle = h((p, q)) = h(0) = \emptyset$ which implies $p \geq q$.

(4) Assuming $p < q$, then $(p, q) \leq (r, s)$ iff $r \leq p < q \leq s$. Indeed, if $(p, q) \leq (r, s)$ then, by (R_1), $(p, q \wedge r) = (p, q) \wedge (p, r) \leq (r, s) \wedge (p, r) = (r \vee p, s \wedge r) = 0$ (by (2)). Hence, now by (3), $p \geq q \wedge r$ and thus $r \leq p$. Similarly it follows that $q \leq s$. Conversely, if $r \leq p < q \leq s$ then, by (R_1), $(p, q) \wedge (r, s) = (p, q)$ and thus $(p, q) \leq (r, s)$.

It follows immediately that for any $p < q$,

$$(p, q) = (r, s) \text{ iff } p = r \text{ and } q = s.$$

In summary, the only generators of $\mathfrak{L}(\mathbb{R})$ that are equal to others are the (p, q) with $p \geq q$, which are all equal to the bottom element 0 of the frame.

So the $(p, q) \in \mathfrak{L}(\mathbb{R})$ behave like the corresponding rational open intervals $\langle p, q \rangle \in \Omega(\mathbb{Q})$ (in fact, h induces a partial order isomorphism between them).

1.1.2. Consider the following elements of $\mathfrak{L}(\mathbb{R})$:

$$(p, -) = \bigvee \{ (p, q) \mid q \in \mathbb{Q} \} = \bigvee \{ (p, q) \mid p < q \in \mathbb{Q} \},$$
$$(-, q) = \bigvee \{ (p, q) \mid p \in \mathbb{Q} \} = \bigvee \{ (p, q) \mid q > p \in \mathbb{Q} \}.$$

Note that $(p, -) \wedge (-, q) = (p, q)$ for every $p, q \in \mathbb{Q}$. In particular, $(p, -) \wedge (-, q) = 0$ whenever $p \geq q$. Again, the following looks like facts in $\mathbb{Q}$.

Proposition. *For any $p, q \in \mathbb{Q}$, we have:*

(1) $(p, -) \wedge (-, q) = 0$ *whenever $p \geq q$.*

(2) $(p, -) \vee (-, q) = 1$ *whenever $p < q$.*

(3) $(p, -) = \bigvee_{r > p} (r, -)$ *and* $(-, q) = \bigvee_{s < q} (-, s)$.

(4) $\bigvee_{p \in \mathbb{Q}} (p, -) = 1$ *and* $\bigvee_{q \in \mathbb{Q}} (-, q) = 1$.

(5) $(p, q)^* = (-, p) \vee (q, -)$ *whenever $p < q$.*

Proof. (1) follows immediately from the second remark in 1.1.1.

(2): By (R_4), it is enough to show that $(r, s) \leq (p, -) \vee (-, q)$ for any $r < s$. Now, if $p \leq r$ then $(r, s) \leq (p, -)$ and if $s \leq q$ then $(r, s) \leq (-, q)$, trivially. On the other hand, if $r < p$ and $q < s$ then $(r, s) = (r, q) \vee (p, s) \leq (-, q) \vee (p, -)$ by (R_2).

(3): $\bigvee_{r > p} (r, -) = \bigvee_{r > p} \bigvee_{q \in \mathbb{Q}} (r, q) = \bigvee_{q \in \mathbb{Q}} \bigvee_{r > p} (r, q)$ and then, by (R_3), $\bigvee_{r > p} (r, -) = \bigvee_{q \in \mathbb{Q}} (p, q) = (p, -)$. The second assertion follows similarly.

(4) is an immediate consequence of condition (R_4).

(5): Since $(p,q) \wedge ((-,p) \vee (q,-)) = ((-,p) \wedge (p,q)) \vee ((p,q) \wedge (q,-)) = 0$, we have $(-,p) \vee (q,-) \leq (p,q)^*$. Further, for any (r,s) such that $r < s$, if $0 = (r,s) \wedge (p,q) = (r \vee p, s \wedge q)$ then $s \wedge q \leq r \vee p$. By going through all possible relations between p, q, r and s one sees that $s \leq p$ or $q \leq r$ and hence $(r,s) \leq (-,p) \vee (q,-)$, proving that $(p,q)^* \leq (-,p) \vee (q,-)$. $\qquad\square$

1.2. A useful (equivalent) description of $\mathfrak{L}(\mathbb{R})$

1.2.1. Proposition. $\mathfrak{L}(\mathbb{R})$ *may be equivalently defined as the frame generated by the* $(p,-)$ *and* $(-,q)$, $p,q \in \mathbb{Q}$, *subject to the relations*

(R_1') $(p,-) \wedge (-,q) = 0$ *whenever* $p \geq q$,

(R_2') $(p,-) \vee (-,q) = 1$ *whenever* $p < q$,

(R_3') $(p,-) = \bigvee_{r>p}(r,-)$,

(R_4') $(-,q) = \bigvee_{s<q}(-,s)$,

(R_5') $\bigvee_{p\in\mathbb{Q}}(p,-) = 1$ *and*

(R_6') $\bigvee_{q\in\mathbb{Q}}(-,q) = 1$.

Proof. Let $\mathfrak{L}'(\mathbb{R})$ be the frame specified by generators $(p,-)$ and $(-,q)$, $p,q \in \mathbb{Q}$, and relations (R_1')–(R_6') and define $f\colon \mathfrak{L}'(\mathbb{R}) \to \mathfrak{L}(\mathbb{R})$ on generators by $f(p,-) = \bigvee_{q\in\mathbb{Q}}(p,q)$ and $f(-,q) = \bigvee_{p\in\mathbb{Q}}(p,q)$. By 1.1.2, it turns (R_1')–(R_6') into identities in $\mathfrak{L}(\mathbb{R})$, so it is a frame homomorphism.

On the other hand, define $g\colon \mathfrak{L}(\mathbb{R}) \to \mathfrak{L}'(\mathbb{R})$ by $g(p,q) = (p,-) \wedge (-,q)$. Then g turns defining relations (R_1)–(R_4) into identities in $\mathfrak{L}'(\mathbb{R})$:

(R_1): It suffices to check that $(p,-) \wedge (r,-) = (p \vee r, -)$ (and similarly $(-,q) \wedge (-,s) = (-, q \wedge s)$). If, e.g., $p \leq r$ then $(p,-) \geq (r,-)$ by (R_3') and so $(p,-) \wedge (r,-) = (r,-) = (p \vee r, -)$.

(R_2): Let $p \leq r < q \leq s$. Then $g((p,q) \vee (r,s)) = ((p,-) \wedge (-,q)) \vee ((r,-) \wedge (-,s)) = ((p,-) \vee (r,-)) \wedge ((p,-) \vee (-,s)) \wedge ((-,q) \vee (r,-)) \wedge ((-,q) \vee (-,s))$. Then, applying (R_2'), (R_3') and (R_4'), we get $g((p,q) \vee (r,s)) = (p,-) \wedge (-,s) = g(p,s)$.

(R_3): This case follows from the fact that, by (R_3'), (R_4') and (R_1'), we have successively $(p,-) \wedge (-,q) = \bigvee_{r>p}(r,-) \wedge \bigvee_{s<q}(-,s) = \bigvee_{p<r,s<q}((r,-) \wedge (-,s)) = \bigvee_{p<r<s<q}((r,-) \wedge (-,s))$.

(R_4): $1 = \bigvee_{p,q\in\mathbb{Q}}((p,-) \wedge (-,q))$ holds by (R_5') and (R_6').

Therefore g is a frame homomorphism. Finally, fg and gf are the identity maps. Indeed, $gf(p,-) = \bigvee_{q\in\mathbb{Q}}(p,-) \wedge (-,q) = (p,-) \wedge \bigvee_{q\in\mathbb{Q}}(-,q) = (p,-)$ by (R_6'). Similarly, using (R_5'), we obtain $gf(-,q) = (-,q)$. On the other hand,

$$fg(p,q) = f(p,-) \wedge f(-,q) = \bigvee_{s>p}(p,s) \wedge \bigvee_{r<q}(r,q) = \bigvee_{p<s, r<q}(p,s) \wedge (r,q)$$

$$= \bigvee_{p<s, r<q}(p \vee r, q \wedge s) = \bigvee_{p\leq r<s\leq q}(r,s) = (p,q). \qquad\square$$

2. Properties of $\mathfrak{L}(\mathbb{R})$

2.1. Complete regularity.

Lemma. *If $p < r < s < q$ then $(r, s) \twoheadprec (p, q)$.*

Proof. First we have $(r, s) \prec (p, q)$: by 1.1.2, $(p, q) \vee (r, s)^* = (p, q) \vee ((-, r) \vee (s, -)) = (-, q) \vee (p, -) = 1$. From this it then follows immediately that $\prec$ interpolates:

$$(r, s) \prec (\tfrac{1}{2}(p + r), \tfrac{1}{2}(s + q)) \prec (p, q).$$

Since $\twoheadprec$ is the largest interpolative relation contained in $\prec$ (see V.5.7), then $\prec = \twoheadprec$. $\qquad\qquad\qquad\square$

By (R$_3$) we have immediately the following

Proposition. $\mathfrak{L}(\mathbb{R})$ *is completely regular.* $\qquad\qquad\qquad\square$

2.2. Local compactness.
For the purposes of this subsection we will use another representation of the frame $\mathfrak{L}(\mathbb{R})$, the L below in 2.2.2.

Let G be the sublattice of $\mathfrak{L}(\mathbb{R})$ consisting of all finite joins of basic generators (p, q) and $\mathfrak{J}G$ the frame of all ideals of G. For each $I \in \mathfrak{J}G$,

$$\nu(I) = \{(p_1, q_1) \vee \cdots \vee (p_n, q_n) \mid p_i, q_i \in \mathbb{Q} \text{ and } p_i < r_i < s_i < q_i \Rightarrow (r_i, s_i) \in I\}$$

is clearly closed under binary joins and easily seen to be a down-set so that it is an ideal of G. This defines an operator $\nu \colon \mathfrak{J}G \to \mathfrak{J}G$.

2.2.1. Lemma. *$(p, q) \in \nu(I)$ if and only if every (r, s) with $p < r < s < q$ belongs to I.*

Proof. Let $(p, q) = (p_1, q_1) \vee \cdots \vee (p_n, q_n)$ be in $\nu(I)$. Then every (r_i, s_i) satisfying $p_i < r_i < s_i < q_i$ is in I and it is a straightforward exercise to show, by induction, after removing redundancies, that the (p_i, q_i) can be arranged so that the corresponding $\langle p_i, q_i \rangle = \{t \in \mathbb{Q} \mid p_i < t < q_i\}$ are successively overlapping intervals. Hence any (r, s), with $p < r < s < q$, may be expressed as some $(r_1, s_1) \vee \cdots \vee (r_n, s_n) \in I$. The converse is obvious by the definition of ν. $\qquad\square$

2.2.2. Lemma. *ν is a nucleus.*

Proof. Clearly, $I \subseteq \nu(I)$ for any I, ν preserves inclusion and satisfies the weak form of preserving meets, $\nu(I) \cap J = \nu(I \cap J)$ for any I and J, by an obvious argument. Further, ν is idempotent: given $(p_1, q_1) \vee \cdots \vee (p_n, q_n) \in \nu(\nu(I))$, we have all (r_i, s_i), with $p_i < r_i < s_i < q_i$, in $\nu(I)$; then by 2.2.1 all $(r_i', s_i') \in I$ (for $r_i < r_i' < s_i' < s_i$). Since $\mathbb{Q}$ is dense in itself this implies that all $(r_i, s_i) \in I$, and consequently $(p_1, q_1) \vee \cdots \vee (p_n, q_n)$ belongs to $\nu(I)$, as desired. It follows that ν is an idempotent prenucleus, and therefore a nucleus (recall III.11.5.1). $\qquad\square$

In particular, this implies that

$$L = \nu[\mathfrak{J}G] = \{I \in \mathfrak{J}G \mid \nu(I) = I\}$$

is a frame, where binary meets are given by intersection and arbitrary joins $\bigsqcup_{\alpha \in \Lambda} I_\alpha$ are given by $\nu(\bigvee_{\alpha \in \Lambda} I_\alpha)$.

2.2.3. Lemma. *There is a frame isomorphism $h \colon \mathfrak{L}(\mathbb{R}) \to L$ given by*

$$h(a) = \bigvee \{\downarrow(p, q) \mid (p, q) \leq a\}$$

(in particular, $h(p, q) = \downarrow(p, q)$).

Proof. Clearly, any principal ideal belongs to L and the correspondence from the basic generators of $\mathfrak{L}(\mathbb{R})$ into L mapping (p, q) to $\downarrow(p, q)$ turns the relations (R$_1$)–(R$_4$) into identities in L. This determines a frame homomorphism $h \colon \mathfrak{L}(\mathbb{R}) \to L$ such that $h(a) = \bigvee \{\downarrow(p, q) \mid (p, q) \leq a\}$, which is obviously onto. Moreover, $h(a)$ is the ideal of all elements of G below a, and since $\mathfrak{L}(\mathbb{R})$ is generated by G, this makes h one-one. Hence h is an isomorphism. $\square$

Recall VII.6. Next result is the point-free extension of the classical fact that the space $\mathbb{R}$ is locally compact.

2.2.4. Proposition. $\mathfrak{L}(\mathbb{R})$ *is a continuous lattice.*

Proof. In view of (R$_3$) it will be enough to prove that $(r, s) \ll (p, q)$ whenever $p < r < s < q$.

For any updirected subset D of $\mathfrak{L}(\mathbb{R})$, if $(p, q) \leq \bigvee D$ in $\mathfrak{L}(\mathbb{R})$ then, applying the isomorphism $h \colon \mathfrak{L}(\mathbb{R}) \to L$, we get

$$h(p, q) \leq \bigsqcup_{d \in D} h(d) = \nu(\bigvee_{d \in D} h(d)),$$

where $\{h(d) \mid d \in D\}$ is an up-directed subset of $\mathfrak{J}G$. Therefore $\bigvee_{d \in D} h(d) = \bigcup_{d \in D} h(d)$ and thus we have $\downarrow(p, q) \leq \nu(\bigcup_{d \in D} h(d))$. By 2.2.1 this implies that $(r, s) \in \bigcup_{d \in D} h(d)$, that is, $(r, s) \in h(d)$ for some $d \in D$ and thus $h(r, s) = \downarrow(r, s) \leq h(d)$, from which it follows that $(r, s) \leq d$, as desired. $\square$

2.3. The closed interval frame. For any $p < q$ in $\mathbb{Q}$, the *closed interval frame*

$$\mathfrak{L}[p, q]$$

is the frame $\uparrow((-, p) \vee (q, -))$. Since $\mathbb{Q}$ has order automorphisms that assign p to 0 and q to 1, all these frames are isomorphic to $\mathfrak{L}[0, 1]$.

Proposition. *For any $p < q$, $\mathfrak{L}[p, q]$ is compact.*

Proof. By 2.1 and 2.2.4 we have

$$(p, q) \prec (p - 1, q + 1) \ll 1 \quad \text{for any } p < q.$$

Let $1 = \bigvee_{i \in J} x_i$ in $\mathfrak{L}[p, q] = \uparrow((p, q)^*)$; then $(p - 1, q + 1) \leq x_{i_1} \vee \cdots \vee x_{i_n}$ for some $i_1, \ldots, i_n \in J$, and therefore, joining with $(p, q)^*$ produces $1 = i_1 \vee \cdots \vee i_n$. $\square$

3. $\mathfrak{L}(\mathbb{R})$ versus the usual space of reals

3.1. Background: Dedekind cuts. Dedekind originally defined a *cut* (A, B) as a *partition* of the rationals into two non-empty classes where every member of A is smaller than every member in B ([66]). It is important to notice, as Dedekind observed, that each rational number p produces two cuts which one should not look at as essentially different:

$$(\downarrow p, \mathbb{Q} \smallsetminus \downarrow p) \quad \text{and} \quad (\mathbb{Q} \smallsetminus \uparrow p, \uparrow p)).$$

Note. In Appendix I.4.5 the Dedekind cuts as considered above correspond to the $(\varphi(M), \mathbb{Q} \smallsetminus \varphi(M))$. However, in this section it will be convenient in the rational cuts to replace $(\varphi(M), \mathbb{Q} \smallsetminus \varphi(M)) = (\downarrow p, \mathbb{Q} \smallsetminus \downarrow p)$ by

$$(\downarrow p \smallsetminus \{p\}, \uparrow p \smallsetminus \{p\})$$

as a representative for p, i.e., to work with the *open Dedekind cuts*. Specifically, these are the pairs (A, B) of non-empty subsets of $\mathbb{Q}$ such that

(D$_1$) $A \cap B = \emptyset$,

(D$_2$) A is a down-set and B is an up-set,

(D$_3$) A contains no greatest element and B has no least element,

(D$_4$) if $p < q$ then either $p \in A$ or $q \in B$.

$\mathbb{R}$ is then the set of Dedekind cuts with the total order $(A, B) \leq (C, D)$ iff $A \subseteq C$. The real number r represented by the Dedekind cut (A, B) is characterized by the condition

$$p < r \text{ iff } p \in A, \quad r < q \text{ iff } q \in B. \tag{3.1.1}$$

The irrational numbers are the cuts (A, B) for which (A, B) is a partition of $\mathbb{Q}$.

In this section the symbol $\mathbb{R}$ (alone) will designate the usual space of real numbers defined by the Dedekind cuts of $\mathbb{Q}$.

3.2. Proposition. *The spectrum* $\Sigma\mathfrak{L}(\mathbb{R})$ *is homeomorphic to* $\mathbb{R}$. *More specifically, there is a homeomorphism* $\varphi \colon \Sigma\mathfrak{L}(\mathbb{R}) \to \mathbb{R}$ *satisfying*

$$p < \varphi(h) < q \quad \text{iff} \quad h(p, -) = 1 \text{ and } h(-, q) = 1,$$

for all $h \colon \mathfrak{L}(\mathbb{R}) \to \mathbf{2}$ *and all* $p, q \in \mathbb{Q}$.

Proof. Given any $h \colon \mathfrak{L}(\mathbb{R}) \to \mathbf{2}$, define subsets A_h and B_h of $\mathbb{Q}$ by

$$A_h = \{p \in \mathbb{Q} \mid h(p, -) = 1\} \quad \text{and} \quad B_h = \{q \in \mathbb{Q} \mid h(-, q) = 1\}.$$

By 1.2.1, the pair (A_h, B_h) is a Dedekind cut. Indeed, by (R_5'), (R_5') and the compactness of $\mathbf{2}$, A_h and B_h are non-empty; (D_1) follows from (R_1'), (D_2) and (D_3) follow from (R_3') and (R_4'), and (D_4) follows from (R_2').

Now, if r_h is the real number represented by (A_h, B_h) then (3.1.1) implies that

$$p < r_h < q \text{ iff } p \in A_h \text{ and } q \in B_h$$

$$\text{iff } h(p, -) = 1 = h(-, q),$$

as desired.

Conversely, for any $r \in \mathbb{R}$ represented by the Dedekind cut (A, B), the correspondence $h_r \colon \mathfrak{L}(\mathbb{R}) \to \mathbf{2}$ defined on generators $(p, -)$ and $(-, q)$ by $h_r(p, -) = 1$ iff $p \in A$ and $h_r(-, q) = 1$ iff $q \in B$, turns the relations (R_1')–(R_6') into identities in $\mathbf{2}$ (this is straightforward to check being (A, B) a Dedekind cut). Hence we have a frame homomorphism $h_r \colon \mathfrak{L}(\mathbb{R}) \to \mathbf{2}$ such that $h_r(p, -) = 1$ and $h_r(-, q) = 1$ iff $p < r < q$.

The correspondences $h \mapsto r_h$ and $r \mapsto h_r$ thus established are inverse to each other because any $r \in \mathbb{R}$ is determined by the $p, q \in \mathbb{Q}$ such that $p < r < q$. Hence we have a one-one onto map $\varphi \colon \Sigma\mathfrak{L}(\mathbb{R}) \to \mathbb{R}$ satisfying $p < \varphi(h) < q$ iff $h(p, -) = 1$ and $h(-, q) = 1$, for all $p, q \in \mathbb{Q}$. It remains to show that φ is a homeomorphism: the topology of $\Sigma\mathfrak{L}(\mathbb{R})$ is generated by the sets

$$\Sigma_{(p,q)} = \{h \in \Sigma\mathfrak{L}(\mathbb{R}) \mid h(p, -) = 1 = h(-, q)\},$$

and φ maps these to

$$\varphi[\Sigma_{(p,q)}] = \{\varphi(h) \mid h(p, -) = 1 = h(-, q)\} = \{\varphi(h) \mid p < \varphi(h) < q\} = \,]p, q[,$$

the real open intervals with endpoints p and q, which generate the topology of $\mathbb{R}$.
$\square$

3.2.1. Note. The preceding result suggests that the space $\mathbb{R}$ may be constructed as $\Sigma\mathfrak{L}(\mathbb{R})$ since the latter requires no prior knowledge of $\mathbb{R}$. This provides a new method of introducing the reals (see [16] for more details). We will see that the usual order and algebraic structure of the reals are obtainable via $\mathfrak{L}(\mathbb{R})$.

3.2.2. Recall from Chapter II that the spatial reflection of a frame L is the frame homomorphism $\eta_L \colon L \to \Omega\Sigma L$ given by $a \mapsto \Sigma_a$. In the present situation, the homeomorphism $\varphi^{-1} \colon \mathbb{R} \to \Sigma\mathfrak{L}(\mathbb{R})$ of Proposition 3.2 induces an isomorphism $\Omega\Sigma\mathfrak{L}(\mathbb{R}) \to \Omega(\mathbb{R})$ taking $\Sigma_{(p,q)}$ to $]p, q[$, as seen in the proof of the proposition. Combining this we obtain the following

Corollary. *The frame homomorphism $h \colon \mathfrak{L}(\mathbb{R}) \to \Omega(\mathbb{R})$ defined by $(p, q) \mapsto]p, q[$ is the spatial reflection map. It is an isomorphism.*

Proof. It remains to show that h is an isomorphism. It is clearly onto; in order to show that it is one-one it is enough to show that it is co-dense (recall V.5.6). Consider then any

$$a = \bigvee\{(p, q) \mid (p, q) \le a\} \in \mathfrak{L}(\mathbb{R})$$

mapped to the top element of $\Omega(\mathbb{R})$, that is,

$$h(a) = \bigcup\{]p, q[\mid (p, q) \leq a\} = \mathbb{R}.$$

But the closed intervals in $\mathbb{R}$ are compact, so, for any $r, s \in \mathbb{Q}$,

$$[r, s] \subseteq]p_1, q_1[\cup \cdots \cup]p_n, q_n[$$

for some $(p_i, q_i) \leq a$. After removing redundancies, joining overlapping intervals and using (R$_2$), we may assume the $]p_i, q_i[$ are disjoint; hence $[r, s] \subseteq]p_i, q_i[$ for some i, showing that $(r, s) \leq (p_i, q_i)$ in $\mathfrak{L}(\mathbb{R})$ and therefore $(r, s) \leq a$. By (R$_4$), this implies $a = 1$.   □

4.   The metric uniformity of $\mathfrak{L}(\mathbb{R})$

4.1. For each $n \in \mathbb{N}$ let

$$U_n = \left\{(p, q) \in \mathfrak{L}(\mathbb{R}) \mid 0 < q - p < \tfrac{1}{n}\right\}.$$

Each U_n is a cover of $\mathfrak{L}(\mathbb{R})$. Indeed, by successive application of (R$_2$) any (r, s) may be written as the join of some $(p_1, q_1), \ldots, (p_t, q_t)$ where

$$p_1 = r < p_2 < q_1 < p_3 < \cdots < p_t < q_{t-1} < q_t = s, \ 0 < q_i - p_i < \tfrac{1}{n}.$$

Proposition. $\{U_n \mid n \in \mathbb{N}\}$ *is a basis for a uniformity* $\mathcal{U}$ *in* $\mathfrak{L}(\mathbb{R})$.

Proof. U_{n+1} is obviously contained in U_n, hence $U_{n+1} \leq U_n$ and we have a filter basis of covers. Further, for any $(p, q) \in U_{3n}$,

$$U_{3n}(p, q) = \bigvee\{(r, s) \mid 0 < s - r < \tfrac{1}{3n}, (r, s) \wedge (p, q) \neq 0\}$$
$$\leq \left(p - \tfrac{1}{3n}, q + \tfrac{1}{3n}\right) \in U_n,$$

showing that $U_{3n}^* \leq U_n$. Finally, if $p < r < s < q$ take a n sufficiently large such that $p \leq r - \tfrac{1}{n} < s + \tfrac{1}{n} \leq q$; then $U_n(r, s) \leq (r - \tfrac{1}{n}, s + \tfrac{1}{n}) \leq (p, q)$, and hence $(r, s) \vartriangleleft_\mathcal{U} (p, q)$. Using (R$_3$), this proves the admissibility condition.   □

4.2. Proposition. *The uniform frame* $(\mathfrak{L}(\mathbb{R}), \mathcal{U})$ *is complete.*

Proof. Recall VIII.6.2. We need to show that any dense ue-surjection

$$h \colon (M, \mathcal{V}) \to (\mathfrak{L}(\mathbb{R}), \mathcal{U})$$

of uniform frames is an isomorphism. Since h is dense, it suffices to exhibit a right inverse f for it. Let h_* be the right (Galois) adjoint of h. Since h being onto implies $hh_* = \mathrm{id}$, it will be enough to check that the $f \colon \mathfrak{L}(\mathbb{R}) \to M$ defined on generators by

$f(p, q) = h_*(p, q)$ is in fact a frame homomorphism, that is, it turns the conditions (R$_1$)–(R$_4$) into identities in M.

(R$_1$): It is obvious because h_* preserves arbitrary meets.

(R$_2$): Given $p \leq r < q \leq s$, take $\frac{1}{n} < q - r$ and any $(u, v) \in U_n$. Then

$$(p, s) \wedge (u, v) = (p \vee u, s \wedge v) \leq \begin{cases} (p, q) & \text{if } s \wedge v \leq q \\ (r, s) & \text{if } r \leq p \vee u. \end{cases}$$

On the other hand, if $q < s \wedge v$ and $p \vee u < r$ then $q < v$ and $u < r$ so that $q - r < v - u < \frac{1}{n}$, a contradiction. This shows that $(p, s) \wedge (u, v)$ is below (p, q) or (r, s), and hence $h_*(p, s) \wedge h_*(u, v) \leq h_*(p, q) \vee h_*(r, s)$. On the other hand, since h is a dense ue-surjection, $h_*[U_n]$ is a cover; consequently $h_*(p, s) \leq h_*(p, q) \vee h_*(r, s)$. Since the reverse inequality holds by the (R$_1$) condition, this proves the desired identity.

(R$_3$): Note that by the properties of dense ue-surjections,

$$h_*(p, q) = \bigvee \{h_*(a) \mid a \triangleleft_{\mathcal{U}} (p, q)\}.$$

On the other hand, using that a is a join of basic generators, one sees that $U_n a \leq (p, q)$ implies $a \leq (p + \frac{1}{n}, q - \frac{1}{n})$ for any n so that $a \triangleleft_{\mathcal{U}} (p, q)$ implies $a \leq (r, s)$ where $p < r < s < q$. Hence we also have

$$h_*(p, q) = \bigvee \{h_*(r, s) \mid p < r < s < q\},$$

as required.

(R$_4$): For any n, $h_*[U_n]$ is a cover of M and therefore

$$1 = \bigvee \{h_*(p, q) \mid p, q \in \mathbb{Q}\}. \qquad \square$$

4.2.1. Note. The classical completeness of $\mathbb{R}$ may now be seen as a consequence of Propositions 3.2 and 4.2, using the fact that the spectrum of a complete uniform frame is a complete uniform space in the induced uniformity ([36]; see Chapter X).

Recall from 1.1.1(4) that there is a partial order isomorphism between the $(p, q) \in \mathfrak{L}(\mathbb{R})$ and the rational open intervals $\langle p, q \rangle \in \Omega(\mathbb{Q})$ that defines a canonical frame homomorphism $h \colon \mathfrak{L}(\mathbb{R}) \to \Omega(\mathbb{Q})$. It maps each of the uniform covers U_n to the analogous covers of $\Omega(\mathbb{Q})$, which means that the uniformity on $\Omega(\mathbb{Q})$ induced by the metric uniformity $\mathcal{U}$ of $\mathfrak{L}(\mathbb{R})$ via h is the usual (metric) uniformity $\mathcal{Q}$ on the rationals and $h \colon (\mathfrak{L}(\mathbb{R}), \mathcal{U}) \to (\Omega(\mathbb{Q}), \mathcal{Q})$ is a uniform homomorphism.

4.2.2. Corollary. $h \colon (\mathfrak{L}(\mathbb{R}), \mathcal{U}) \to (\Omega(\mathbb{Q}), \mathcal{Q})$ *is the completion of* $(\Omega(\mathbb{Q}), \mathcal{Q})$.

Proof. It suffices to check that the uniform homomorphism h is a dense ue-surjection (recall VIII.3.3), which is straightforward. $\qquad \square$

This is the point-free counterpart of the classical fact that $\mathbb{R}$ is the uniform completion of $\mathbb{Q}$.

5. Continuous real functions

5.1. From the adjunction between locales and topological spaces (II.4.6)

$$\mathbf{Top} \underset{\mathsf{Sp}}{\overset{\mathsf{Lc}}{\rightleftarrows}} \mathbf{Loc}$$

we have a natural equivalence (see AII.6.1)

$$\mathbf{Top}(X, \mathsf{Sp}(L)) \cong \mathbf{Loc}(\mathsf{Lc}(X), L) = \mathbf{Frm}(L, \Omega(X)).$$

Applying this for $L = \mathfrak{L}(\mathbb{R})$ and combining it with the homeomorphism of Proposition 3.2 we get a bijection

$$\mathbf{Top}(X, \mathbb{R}) \cong \mathbf{Frm}(\mathfrak{L}(\mathbb{R}), \Omega(X))$$

under which each $h\colon \mathfrak{L}(\mathbb{R}) \to \Omega(X)$ corresponds to the $\widetilde{h}\colon X \to \mathbb{R}$ given by

$$p < \widetilde{h}(x) < q \text{ iff } x \in h(p,q). \tag{5.1.1}$$

This shows that continuous real functions on a space X are represented by frame homomorphisms $h\colon \mathfrak{L}(\mathbb{R}) \to \Omega(X)$, and hence regarding the frame homomorphisms $\mathfrak{L}(\mathbb{R}) \to L$ for a general frame L as the continuous real functions on L provides a natural extension of the classical notion: a *continuous real function* on a frame L is a frame homomorphism $\mathfrak{L}(\mathbb{R}) \to L$. We will speak of

$$\mathcal{C}(L)$$

as the set of all continuous real functions on L, partially ordered by

$$f \le g \equiv_{\mathsf{df}} f(p,-) \le g(p,-) \text{ for every } p \in \mathbb{Q}$$
$$\Leftrightarrow g(-,q) \le f(-,q) \text{ for every } q \in \mathbb{Q}.$$

5.2. Scales. A *scale* in a frame L is a map $\sigma\colon \mathbb{Q} \to L$ such that $\sigma(s) \prec \sigma(r)$ whenever $r < s$ and $\bigvee\{\sigma(r) \mid r \in \mathbb{Q}\} = 1 = \bigvee\{\sigma(r)^* \mid r \in \mathbb{Q}\}$.

5.2.1. Note. The name *scale* used here differs from its use in [145] where it refers to maps from the unit interval of $\mathbb{Q}$ into L. In [16] the term *descending trail* is used.

Scales provide us with a general method for defining continuous real functions on a frame, as follows.

5.2.2. Lemma. *For each scale σ in L the formulas*

$$f(r,-) = \bigvee\{\sigma(s) \mid s > r\} \quad and \quad f(-,r) = \bigvee\{\sigma(s)^* \mid s < r\} \quad (r \in \mathbb{Q})$$

determine a frame homomorphism $f\colon \mathfrak{L}(\mathbb{R}) \to L$ such that $f(r,-) \le \sigma(r) \le f(-,r)^*$ for every $r \in \mathbb{Q}$.

Proof. The inequality $f(r,-) \leq \sigma(r)$ is obvious; the other is also easy since $f(-,r)^* = \bigwedge_{s<r} \sigma(s)^{**}$ and for each $s < r$, $\sigma(r) \leq \sigma(s) \leq \sigma(s)^{**}$. Now we check that f turns relations (R_1')–(R_6') into identities in L.

(R_1'): For each $r \geq s$ we have $f(r,-) \wedge f(-,s) = \bigvee_{s'<s\leq r<r'} \sigma(r') \wedge \sigma(s')^* = 0$.

(R_2'): Let $r < s$. Then

$$f(r,-) \vee f(-,s) = \bigvee_{r'>r} \sigma(r') \vee \bigvee_{s'<s} \sigma(s')^* \geq \bigvee_{r<r'<s'<s} \sigma(r') \vee \sigma(s')^* = 1$$

since $\sigma(s') \prec \sigma(r')$.

(R_3') and (R_4') follow because $\mathbb{Q}$ is dense in itself.

(R_5') and (R_6') follow immediately from the property $\bigvee\{\sigma(r) \mid r \in \mathbb{Q}\} = 1 = \bigvee\{\sigma(r)^* \mid r \in \mathbb{Q}\}$ of a scale. □

5.2.3. Examples: (1) Constant functions. For each $p \in \mathbb{Q}$ let σ_p be defined by $\sigma_p(r) = 1$ if $r < p$ and $\sigma_p(r) = 0$ otherwise. This is clearly a scale. The corresponding function in $\mathcal{C}(L)$ provided by 5.2.2 is given for each $r \in \mathbb{Q}$ by

$$\mathbf{p}(r,-) = \begin{cases} 1 & \text{if } r < p \\ 0 & \text{if } r \geq p \end{cases} \quad \text{and} \quad \mathbf{p}(-,r) = \begin{cases} 0 & \text{if } r \leq p \\ 1 & \text{if } r > p. \end{cases}$$

(2) Characteristic functions. Let a be a complemented element of L. Then σ_a defined by $\sigma_a(r) = 1$ if $r < 0$, $\sigma_a(r) = a^c$ if $0 \leq r < 1$ and $\sigma_a(r) = 0$ if $r \geq 1$, is a scale. It generates the *characteristic function* $\chi_a \in \mathcal{C}(L)$ defined for each $r \in \mathbb{Q}$ by

$$\chi_a(r,-) = \begin{cases} 1 & \text{if } r < 0 \\ a^c & \text{if } 0 \leq r < 1 \\ 0 & \text{if } r \geq 1 \end{cases} \quad \text{and} \quad \chi_a(-,r) = \begin{cases} 0 & \text{if } r \leq 0 \\ a & \text{if } 0 < r \leq 1 \\ 1 & \text{if } r > 1. \end{cases}$$

5.2.4. Lemma. *Let $f, g \in \mathcal{C}(L)$ be determined by scales σ_1 and σ_2 respectively. Then $f \leq g$ if and only if $\sigma_1(p) \leq \sigma_2(q)$ for every $p > q$ in $\mathbb{Q}$.*

Proof. $f \leq g$ iff $\bigvee\{\sigma_1(s) \mid s > r\} \leq \bigvee\{\sigma_2(s) \mid s > r\}$ for every $r \in \mathbb{Q}$. Let $p > q$ in $\mathbb{Q}$. If $f \leq g$ then $\sigma_1(p) \leq \bigvee_{s>q} \sigma_1(s) \leq \bigvee_{s>q} \sigma_2(s) \leq \sigma_2(q)$. Conversely, for each $r \in \mathbb{Q}$ and $s > r$ take q such that $r < q < s$; then $\sigma_1(s) \leq \sigma_2(q)$ with $q > r$ thus $f \leq g$. □

5.3. Calculating in $\mathcal{C}(L)$. We now describe some of the algebraic aspects of the reals that give rise to function algebras.

5.3.1. Additive inverse. For $f \in \mathcal{C}(L)$ define

$$\sigma_{-f}(r) = f(-,-r) \quad \text{for every } r \in \mathbb{Q}.$$

Since $\sigma_{-f}(r)^* = f(-,-r)^* \geq f(-r,-)$, it is clear that σ_{-f} is a scale. The real function generated by it, denoted $-f$, is defined for each $r \in \mathbb{Q}$ by

$$-f(r,-) = f(-,-r) \quad \text{and} \quad -f(-,r) = f(-r,-).$$

5.3.2. Product with a scalar. For $0 < \lambda \in \mathbb{Q}$ and $f \in \mathcal{C}(L)$ define

$$\sigma(p) = f(\tfrac{p}{\lambda},-) \quad \text{for each } p \in \mathbb{Q}.$$

For every $p < q$, $\sigma(p) \vee \sigma(q)^* = f(\tfrac{p}{\lambda},-) \vee f(\tfrac{q}{\lambda},-)^* \geq f(\tfrac{p}{\lambda},-) \vee f(-,\tfrac{q}{\lambda}) = 1$, $\bigvee_{p \in \mathbb{Q}} \sigma(p) = \bigvee_{p \in \mathbb{Q}} f(\tfrac{p}{\lambda},-) = 1$ and $\bigvee_{p \in \mathbb{Q}} \sigma(p)^* \geq \bigvee_{p \in \mathbb{Q}} f(-,\tfrac{p}{\lambda}) = 1$. Hence σ is a scale. The real function generated by it, denoted $\lambda \cdot f$, is defined for each $r \in \mathbb{Q}$ by

$$(\lambda \cdot f)(r,-) = f(\tfrac{r}{\lambda},-) \quad \text{and} \quad (\lambda \cdot f)(-,r) = f(-,\tfrac{r}{\lambda}).$$

5.3.3. (Binary) join and meet. Given $f, g \in \mathcal{C}(L)$, let

$$\sigma(p) = f(p,-) \vee g(p,-) \quad \text{for each } p \in \mathbb{Q}.$$

For any $p < q$,

$$\sigma(p) \vee \sigma(q)^* = f(p,-) \vee g(p,-) \vee \left(f(q,-)^* \wedge g(q,-)^* \right)$$
$$= \left(f(p,-) \vee g(p,-) \vee f(q,-)^* \right) \wedge \left(f(p,-) \vee g(p,-) \vee g(q,-)^* \right) = 1.$$

Moreover $\bigvee_{p \in \mathbb{Q}} \sigma(p) = 1$ and

$$\bigvee_{p \in \mathbb{Q}} \sigma(p)^* = \bigvee_{p \in \mathbb{Q}} \left(f(p,-)^* \wedge g(p,-)^* \right)$$
$$\geq \bigvee_{p \in \mathbb{Q}} \left(f(-,p) \wedge g(-,p) \right) \geq \bigvee_{r,s \in \mathbb{Q}} \left(f(-,r) \wedge g(-,s) \right)$$

(since for any $r, s \in \mathbb{Q}$, $p = r \vee s \in \mathbb{Q}$ and $f(-,r) \wedge g(-,s) \leq f(-,p) \wedge g(-,p)$), from which it follows that $\bigvee_{p \in \mathbb{Q}} \sigma(p)^* = \left(\bigvee_{p \in \mathbb{Q}} f(-,p) \right) \wedge \left(\bigvee_{p \in \mathbb{Q}} g(-,p) \right) = 1$. Hence σ is a scale. It is straightforward to check that the frame homomorphism determined by σ is precisely the *supremum* $f \vee g$ in $\mathcal{C}(L)$. It is given by the formulas

$$(f \vee g)(r,-) = \bigvee_{s > r} \left(f(s,-) \vee g(s,-) \right) = f(r,-) \vee g(r,-)$$

and

$$(f \vee g)(-,r) = \bigvee_{s < r} \left(f(s,-) \vee g(s,-) \right)^* = f(-,r) \wedge g(-,r).$$

(For the latter identity, if $s < r$ then $(f(s,-) \vee g(s,-))^* = f(s,-)^* \wedge g(s,-)^* \leq f(-,r) \wedge g(-,r)$; conversely, $f(-,r) \wedge g(-,r) = \bigvee_{s_1,s_2 < r}(f(-,s_1) \wedge g(-,s_2)) \leq \bigvee_{s < r}(f(s,-)^* \wedge g(s-)^*)$.)

Concerning meets, since $f \leq g$ iff $-g \leq -f$ for every $f, g \in \mathcal{C}(L)$, the *infimum* $f \wedge g$ exists and is given by $f \wedge g = -(-f \vee -g)$. Hence

$$(f \wedge g)(r, -) = -f(-, -r) \wedge -g(-, -r) = f(r, -) \wedge g(r, -)$$

and

$$(f \wedge g)(-, r) = -f(-r, -) \vee -g(-r, -) = f(-, r) \vee g(-, r).$$

5.3.4. Sum. Let $f, g \in \mathcal{C}(L)$. For each $p \in \mathbb{Q}$ define

$$\sigma(p) = \bigvee_{r \in \mathbb{Q}} \big(f(r, -) \wedge g(p - r, -)\big).$$

Proposition. σ *is a scale.*

Proof. For each $p \in \mathbb{Q}$ let $\delta(p) = \bigvee_{s \in \mathbb{Q}} \big(f(-, s) \wedge g(-, p - s)\big)$.

Claim 1. *If* $p \geq q \in \mathbb{Q}$ *then* $\sigma(p) \wedge \delta(q) = 0$.

(Indeed, for any $p, q, r, s \in \mathbb{Q}$ with $p \geq q$, then either $s \leq r$ or $q - s < p - r$ and thus either $f(r, -) \wedge f(-, s) = 0$ or $g(p - r, -) \wedge g(-, q - s) = 0$.)

Claim 2. *If* $p < q \in \mathbb{Q}$ *then* $\sigma(p) \vee \delta(q) = 1$.

(Indeed: Let $p < q \in \mathbb{Q}$ and $t = \frac{q-p}{2} > 0$. Then $\bigvee_{r \in \mathbb{Q}} f(r, r + t) = \bigvee_{s \in \mathbb{Q}} g(s, s + t) = 1$. Let $r, s \in \mathbb{Q}$. If $r + s > p$ then $f(r, r + t) \wedge g(s, s + t) \leq f(r, -) \wedge g(p - r, -) \leq \sigma(p)$. Otherwise, if $r + s \leq p$ then $s + t \leq q - r - t$ and so $f(r, r + t) \wedge g(s, s + t) \leq f(-, r + t) \wedge g(-, q - (r + t)) \leq \delta(q)$. Hence $1 = \bigvee_{r, s \in \mathbb{Q}} \big(f(r, r + t) \wedge g(s, s + t)\big) \leq \sigma(p) \vee \delta(q)$.)

Finally, let $p < q \in \mathbb{Q}$ and consider $r \in \mathbb{Q}$ such that $p < r < q$. It follows immediately from claims 1 and 2 that $\sigma(p) \vee \sigma(q)^* \geq \sigma(p) \vee \delta(r) = 1$. On the other hand

$$\bigvee_{p \in \mathbb{Q}} \sigma(p) = \bigvee_{p, r \in \mathbb{Q}} \big(f(r, -) \wedge g(p - r, -)\big)$$

$$= \bigvee_{r \in \mathbb{Q}} \big(f(r, -) \wedge \bigvee_{p \in \mathbb{Q}} g(p - r, -)\big) = \bigvee_{r \in \mathbb{Q}} f(r, -) = 1$$

and

$$\bigvee_{p \in \mathbb{Q}} \sigma(p)^* \geq \bigvee_{p \in \mathbb{Q}} \delta(p) = \bigvee_{p, s \in \mathbb{Q}} \big(f(-, s) \wedge g(-, p - s)\big)$$

$$= \bigvee_{s \in \mathbb{Q}} \big(f(-, s) \wedge \bigvee_{p \in \mathbb{Q}} g(-, p - s)\big) = \bigvee_{s \in \mathbb{Q}} f(-, s) = 1. \qquad \square$$

The real function generated by the scale σ, denoted $f + g$ (the *sum* of f and g), is given by the formulas

$$(f + g)(r, -) = \bigvee_{t \in \mathbb{Q}} \big(f(t, -) \wedge g(r - t, -)\big)$$

and

$$(f + g)(-, r) = \bigvee_{t \in \mathbb{Q}} \left(f(-, t) \wedge g(-, r - t) \right).$$

(The former is obvious; for the latter, $(f+g)(-, r) = \bigvee_{s<r} \sigma(s)^*$ and therefore $(f + g)(-, r) \leq \delta(r)$ (by Claim 2); on the other hand, $\delta(r) = \bigvee_{t \in \mathbb{Q}} \bigvee_{s<r} (f(-, t) \wedge g(-, s - t)) = \bigvee_{s<r} \delta(s) \leq \bigvee_{s<r} \sigma(s)^*$.)

Remark. For any $f, g \in \mathcal{C}(L)$, $f - g$ is defined as $f + (-g)$ and is given by the formulas

$$(f - g)(r, -) = \bigvee_{t \in \mathbb{Q}} f(t, -) \wedge g(-, t - r)$$

and

$$(f - g)(-, r) = \bigvee_{t \in \mathbb{Q}} f(-, t) \wedge g(t - r, -).$$

Notice that, immediately, $f - f = \mathbf{0}$.

5.3.5. Product. We start with the case $f, g \geq \mathbf{0}$. For each $p \in \mathbb{Q}$ define

$$\sigma(p) = \begin{cases} \bigvee_{r>0} \left(f(r, -) \wedge g(\frac{p}{r}, -) \right) & \text{if } p \geq 0 \\ 1 & \text{if } p < 0. \end{cases}$$

Proposition. σ *is a scale.*

Proof. Let

$$\delta(q) = \begin{cases} \bigvee_{s>0} \left(f(-, s) \wedge g(-, \frac{q}{s}) \right) & \text{if } q > 0 \\ 0 & \text{if } q \leq 0. \end{cases}$$

Claim 1. *If* $p \geq q \in \mathbb{Q}$ *then* $\sigma(p) \wedge \delta(q) = 0$.

(Indeed, for any $p, q, r, s \in \mathbb{Q}$ with $p \geq q > 0$ (the case $q \leq 0$ is trivial) and $r, s > 0$, either $s \leq r$ or $\frac{q}{s} \leq \frac{p}{r}$ and thus either $f(r, -) \wedge f(-, s) = 0$ or $g(\frac{p}{r}, -) \wedge g(-, \frac{q}{s}) = 0$.)

Claim 2. *If* $p < q \in \mathbb{Q}$ *then* $\sigma(p) \vee \delta(q) = 1$.

(Indeed: Let $0 \leq p < q \in \mathbb{Q}$ (the case $p < 0$ is trivial) and $t \in \mathbb{Q}$ such that $1 < t^2 \leq \frac{q}{p}$. We have that $\bigvee_{r>0} f(r, rt) = f(0, -)$ and $\bigvee_{s>0} g(s, st) = g(0, -)$. Let $0 < r, s \in \mathbb{Q}$. If $rs > p$ then $f(r, rt) \wedge g(s, st) \leq f(r, -) \wedge g(\frac{p}{r}, -) \leq \sigma(p)$. Otherwise, if $rs \leq p$ then $st \leq \frac{q}{rt}$ and so $f(r, rt) \wedge g(s, st) \leq f(-, rt) \wedge g(-, \frac{q}{rt}) \leq \delta(q)$. Hence $f(0, -) \wedge g(0, -) = \bigvee_{r,s>0} (f(r, rt) \wedge g(s, st)) \leq \sigma(p) \vee \delta(q)$. On the other hand, $\delta(q) \vee f(0, -) = \bigvee_{s>0} ((f(-, s) \vee f(0, -)) \wedge (g(-, \frac{q}{s}) \vee f(0, -))) = \bigvee_{s>0} (g(-, \frac{q}{s}) \vee f(0, -)) = 1$ and, similarly, $\delta(q) \vee g(0, -) = 1$, thus $1 = (\delta(q) \vee f(0, -)) \wedge (\delta(q) \vee g(0, -)) = \delta(q) \vee (f(0, -) \wedge g(0, -)) \leq \sigma(p) \vee \delta(q)$.)

Finally, let $p < q \in \mathbb{Q}$. Take $r \in \mathbb{Q}$ such that $p < r < q$. It follows immediately from claims 1 and 2 that $\sigma(p) \vee \sigma(q)^* \geq \sigma(p) \vee \delta(r) = 1$. On the other hand,

$\bigvee_{p\in\mathbb{Q}}\sigma(p) = 1$ and

$$\bigvee_{p\in\mathbb{Q}}\sigma(p)^* \geq \bigvee_{p\in\mathbb{Q}}\delta(p) = \bigvee_{p,s>0}\left(f(-,s)\wedge g(-,\tfrac{p}{s})\right)$$

$$= \bigvee_{s>0}\left(f(-,s)\wedge \bigvee_{p>0} g(-,\tfrac{p}{s})\right) = \bigvee_{s>0} f(-,s) = 1. \qquad \square$$

The real function $f \cdot g$ (the *product* of f and g) is the frame map generated by σ, which is given by the formulas

$$(f\cdot g)(r,-) = \begin{cases} \bigvee_{s>0}\left(f(s,-)\wedge g(\tfrac{r}{s},-)\right) & \text{if } r \geq 0 \\ 1 & \text{if } r < 0 \end{cases}$$

and

$$(f\cdot g)(-,r) = \begin{cases} \bigvee_{s>0}\left(f(-,s)\wedge g(-,\tfrac{r}{s})\right) & \text{if } r > 0 \\ 0 & \text{if } r \leq 0. \end{cases}$$

(The former is obvious; for the latter, notice that by Claim 2, $(f \cdot g)(-,r) = \bigvee_{s<r}\sigma(s)^* \leq \delta(r)$; on the other hand, applying Claim 1, $\bigvee_{s>0}\left(f(-,s)\wedge g(-,\tfrac{r}{s})\right) = \bigvee_{s>0}\bigvee_{0<p<r}\left(f(-,s)\wedge g(-,\tfrac{p}{s})\right) = \bigvee_{0<p<r}\delta(p) \leq \bigvee_{0<p<r}\sigma(p)^*$.)

In order to extend this to the product of two arbitrary f and g define

$$f^+ = f\vee\mathbf{0} \quad\text{and}\quad f^- = (-f)\vee\mathbf{0}$$

for any $f \in \mathcal{C}(L)$. Notice that $f = f^+ - f^-$. Then $f\cdot g = (f^+ - f^-)\cdot(g^+ - g^-)$ is defined as

$$(f^+\cdot g^+) - (f^+\cdot g^-) - (f^-\cdot g^+) + (f^-\cdot g^-).$$

In particular, if $f, g \leq \mathbf{0}$, then $f\cdot g = f^-\cdot g^- = (-f)\cdot(-g)$.

5.4. The function algebra $\mathcal{C}(L)$. Recall that a *lattice-ordered ring* is a ring A with a lattice structure such that, for all $f, g, h \in A$, $(f\wedge g) + h = (f + h)\wedge(g + h)$ and $fg \geq 0$ whenever $f \geq 0$ and $g \geq 0$ ([113]).

5.4.1. Theorem. $\mathcal{C}(L)$ *is an archimedean commutative lattice-ordered ring with unit.*

Proof. All properties can be easily checked by application of the formulas in 5.3. We illustrate this with archimedeaness. Let $f \geq \mathbf{0}$ and $nf \leq g$ for all $n \geq 1$. Then, for all $p > 0$, $f(\tfrac{p}{n},-) \leq g(p,-)$ and consequently $f(0,-) \leq g(p,-)$ by (R'_3). It follows that $f(0,-)\wedge g(-,p) = 0$ for all $p > 0$, and since $\bigvee\{g(-,p) \mid p \in \mathbb{Q}\} = 1$ by (R'_6) this shows $f(0,-) = 0$. Hence $f \leq \mathbf{0}$ and consequently $f = \mathbf{0}$, as required. $\qquad\square$

5.4.2. Remark. In fact one has more: $\mathcal{C}(L)$ is an *f-ring* (i.e., $(f\wedge g)h = (fh)\wedge(gh)$ for any f, g, h with $h \geq \mathbf{0}$ ([248])), *strong* (i.e., every $f \geq \mathbf{1}$ is invertible) and *bounded* (i.e., each f satisfies $f\vee(-f) \leq \mathbf{n}$ for some natural n); see [16] for details.

6.   Cozero elements

6.1. A *cozero element* of L is an element of the form

$$f(-,0) \vee f(0,-)$$

for some $f \in \mathcal{C}(L)$. For any such f we refer to $f(-,0) \vee f(0,-)$ as $\mathrm{coz}(f)$.

6.1.1. Note. For any topological space X, the cozero elements of the frame $\Omega(X)$ are exactly the usual cozero sets of X; the *cozero map* $\mathrm{coz} \colon \mathcal{C}(L) \to L$ is the extension to arbitrary $\mathcal{C}(L)$ of the correspondence between functions f on spaces and their cozero sets $f^{-1}(\mathbb{R} \setminus \{0\})$. Cozero elements play a particular role in the theory of σ-frames ([26],[27]), the point-free counterpart to Alexandroff spaces (recall IV.6.7.2; a *σ-frame* is a κ-frame for $\kappa = \aleph_1$). In fact, the collection of all cozeros of L, usually denoted by $\mathrm{Coz}\, L$, forms a sub-σ-frame of L (see 6.2.4 below).

The following summarizes some of the properties of coz.

6.1.2. Proposition. *Let* $f, g \in \mathcal{C}(L)$.

(1) *If* $\mathbf{0} \leq f \leq g$ *then* $\mathrm{coz}(f) \leq \mathrm{coz}(g)$.

(2) $\mathrm{coz}(f + g) \leq \mathrm{coz}(f) \vee \mathrm{coz}(g)$ *and* $\mathrm{coz}(\mathbf{0}) = 0$. *If* $f, g \geq \mathbf{0}$ *then* $\mathrm{coz}(f + g) = \mathrm{coz}(f) \vee \mathrm{coz}(g)$.

(3) $\mathrm{coz}(f \cdot g) = \mathrm{coz}(f) \wedge \mathrm{coz}(g)$ *and* $\mathrm{coz}(\mathbf{1}) = 1$.

(4) *If* $f, g \geq \mathbf{0}$ *then* $\mathrm{coz}(f \wedge g) = \mathrm{coz}(f) \wedge \mathrm{coz}(g)$.

(5) $\mathrm{coz}(f - \mathbf{p}) = f(-,p) \vee f(p,-)$ *for every* $p \in \mathbb{Q}$.

(6) $\mathrm{coz}((f - \mathbf{p})^+ \wedge (\mathbf{q} - f)^+) = f(p,q)$ *for every* $p, q \in \mathbb{Q}$.

Proof. (1): $\mathrm{coz}(f) = f(0,-) \leq g(0,-) = \mathrm{coz}(g)$.

(2): Obviously $\mathrm{coz}(\mathbf{0}) = 0$. For $\mathrm{coz}(f + g)$, $(f + g)(-,0) \leq f(-,0) \vee g(-,0)$ and $(f + g)(0,-) \leq f(0,-) \vee g(0,-)$. Moreover, if $f, g \geq \mathbf{0}$ then $f + g \geq f, g$ and hence $\mathrm{coz}(f) \vee \mathrm{coz}(g) \leq \mathrm{coz}(f + g)$ by (1).

(3): Trivially, $\mathrm{coz}(\mathbf{1}) = 1$. Further, calculating $fg(-,0)$ and $fg(0,-)$ one gets

$$fg(-,0) = (f(0,-) \wedge g(-,0)) \vee (f(-,0) \wedge g(0,-))$$

and

$$fg(0,-) = (f(0,-) \wedge g(0,-)) \vee (f(-,0) \wedge g(-,0)),$$

which by distributivity gives the desired identity since

$$\mathrm{coz}(f) \wedge \mathrm{coz}(g) = (f(-,0) \vee f(0,-)) \wedge (g(-,0) \vee g(0,-)).$$

(4): By the rules of f-rings (in 5.4.2), $fg \leq fh$ whenever $f \geq \mathbf{0}$ and $g \leq h$. Therefore $fg \leq (f(f \vee g)) \wedge (g(f \vee g)) = (f \wedge g)(f \vee g)$ and hence

$$\mathrm{coz}(f) \wedge \mathrm{coz}(g) = \mathrm{coz}(fg) \leq \mathrm{coz}(f \wedge g) \wedge \mathrm{coz}(f \vee g) \leq \mathrm{coz}(f \wedge g).$$

(5): $\mathrm{coz}(f-\mathbf{p})=(f-\mathbf{p})(-,0)\vee(f-\mathbf{p})(0,-)=\bigvee_{s\in\mathbb{Q}}f(-,s)\wedge\mathbf{p}(s,-)\vee\bigvee_{r\in\mathbb{Q}}f(r,-)\wedge$
$\mathbf{p}(-,r)=\bigvee_{s<p}f(-,s)\vee\bigvee_{r>p}f(r,-)=f(-,p)\vee f(p,-)$.

(6): For any f, f^+ is given by

$$f^+(r,-)=1\ (r<0),\quad f^+(r,-)=f(r,-)\ (r\geq 0),$$

and

$$f^+(-,r)=0\ (r\leq 0),\quad f^+(-,r)=f(-,r)\ (r>0).$$

Therefore $\mathrm{coz}((f-\mathbf{p})^+)=f(p,-)$ and $\mathrm{coz}(\mathbf{q}-f)^+)=f(-,q)$. The conclusion follows by application of (4). □

6.2. Complete regularity via cozero elements. There is a neat connection between the relation $\prec\!\!\prec$ and continuous real functions:

6.2.1. Proposition. *For $a,b\in L$, $a\prec\!\!\prec b$ if and only if there is an $f\in\mathcal{C}(L)$ such that $a\leq f(\frac{1}{2},-)\leq f(0,-)\leq b$.*

Proof. Let a_r $(r\in[0,1]\cap\mathbb{Q})$ be a sequence witnessing that $a\prec\!\!\prec b$ and define for each $p\in\mathbb{Q}$, $\sigma(p)=1$ if $p<0$, $\sigma(p)=a_{1-p}$ if $p\in[0,1]$ and $\sigma(p)=0$ if $p>1$. This is clearly a scale in L that determines the frame homomorphism $f\colon\mathfrak{L}(\mathbb{R})\to L$ given by formulas

$$f(r,-)=\begin{cases}1 & \text{if } r<0\\ \bigvee_{r<s\leq 1}a_{1-s} & \text{if } 0\leq r<1,\\ 0 & \text{if } r\geq 1\end{cases}\quad f(-,r)=\begin{cases}0 & \text{if } r\leq 0\\ \bigvee_{0\leq s<r}a^*_{1-s} & \text{if } 0<r\leq 1\\ 1 & \text{if } r>1.\end{cases}$$

This is the desired f. For the converse notice that $(\frac{1}{2},-)\prec\!\!\prec(0,-)$ (by the corresponding variant of 2.1) and that any frame homomorphism preserves $\prec\!\!\prec$ (by V.5.8). □

6.2.2. Remark. Applied to spaces this means that for any space X, $U\prec\!\!\prec V$ in $\Omega(X)$ iff there exists a continuous real function $\tilde{f}$ on X such that $\tilde{f}(x)>\frac{1}{2}$ for all $x\in U$ and $\tilde{f}(x)\leq 0$ for all $x\notin V$ (just take $f\colon\mathfrak{L}(\mathbb{R})\to\Omega(X)$ for $U\prec\!\!\prec V$ as in the proposition and then the corresponding $\tilde{f}$ of (5.1.1)). This readily implies that

A space X is completely regular iff the frame $\Omega(X)$ is completely regular.

(Note that this characterizes the complete regularity of a space X by an internal lattice condition on $\Omega(X)$ irrespective of the space $\mathbb{R}$, external to X; it appeared firstly in the 1953 doctoral dissertation of B. Banaschewski ([8]).)

Now, the cozero elements of any frame may be described without reference to $\mathfrak{L}(\mathbb{R})$ as follows.

6.2.3. Proposition. *The following are equivalent for $a \in L$.*

(1) $a \in \mathrm{Coz}\,L$.

(2) $a = \bigvee_{n \in \mathbb{N}} x_n$ *where* $x_n \prec\!\!\prec a$ *for all* $n \in \mathbb{N}$.

(3) $a = \bigvee_{n \in \mathbb{N}} a_n$ *where* $a_n \prec\!\!\prec a_{n+1}$ *for all* $n \in \mathbb{N}$.

Proof. (1)$\Rightarrow$(2): If $a = \mathrm{coz}(f)$ for some $f \in \mathcal{C}(L)$ then $a = f(-,0) \vee f(0,-) = \bigvee_{n \in \mathbb{N}} f((-,\frac{1}{n}) \vee (\frac{1}{n},-))$. Since $(-,\frac{1}{n}) \vee (\frac{1}{n},-) \prec\!\!\prec (-,0) \vee (0,-)$ and any homomorphism preserves $\prec\!\!\prec$ we conclude that $x_n = f((-,\frac{1}{n}) \vee (\frac{1}{n},-)) \prec\!\!\prec a$.

(2)$\Rightarrow$(3): Given $a = \bigvee_{n \in \mathbb{N}} x_n$ in the stated conditions, define a_n inductively by

$$a_1 = x_1, \ a_1 \vee x_2 \prec\!\!\prec a_2 \prec\!\!\prec a, \ \ldots, \ a_n \vee x_{n+1} \prec\!\!\prec a_{n+1} \prec\!\!\prec a,$$

using the fact that $\prec\!\!\prec$ interpolates and is stable under binary joins.

(3)$\Rightarrow$(1): For each n, let c_r^n $(r \in [0,1] \cap \mathbb{Q})$ be an interpolating sequence witnessing $a_n \prec\!\!\prec a_{n+1}$. For each $r \in [0,1] \cap \mathbb{Q}$ define

$$c_r = c_{\tau_n(r)}^n \quad \text{if } \tfrac{n-1}{n} \leq r < \tfrac{n}{n+1}$$

(where τ_n is an increasing bijection between $[0,1] \cap \mathbb{Q}$ and $[\frac{n-1}{n}, \frac{n}{n+1}] \cap \mathbb{Q}$); this defines an interpolating sequence c_r $(r \in [0,1] \cap \mathbb{Q})$ between a_1 and a such that $a = \bigvee_{r<1} c_r$. Proposition 6.2.1 then produces an f such that $\mathrm{coz}(f) = f(0,-) = \bigvee_{0<s\leq 1} c_{1-s} = \bigvee_{0\leq r<1} c_r = a$. $\qquad\square$

6.2.4. Remarks. (1) As an obvious consequence of this characterization, $\mathrm{Coz}\,L$ is a sub-σ-frame of L, meaning it is a bounded sub-meet-semilattice closed under joins in L of countable subsets.

(2) Another consequence of this characterization is that $a \prec\!\!\prec b$ iff $a \prec\!\!\prec c \prec\!\!\prec b$ for some $c \in \mathrm{Coz}\,L$: for any interpolating sequence c_r $(r \in [0,1] \cap \mathbb{Q})$ witnessing that $a \prec\!\!\prec b$, $c = \bigvee\{c_r \mid r < \frac{1}{2}\}$ is a cozero element by the proposition and clearly $a \prec\!\!\prec c \prec\!\!\prec b$.

6.2.5. Corollary. *A frame is completely regular if and only if it is generated by its cozero elements.*

Proof. Assume L is completely regular, that is, $a = \bigvee\{x \in L \mid x \prec\!\!\prec a\}$ for every $a \in L$. By 6.2.1, for each such x there is an $f \in \mathcal{C}(L)$ such that $x \leq f(\frac{1}{2},-) \leq f(0,-) \leq a$. Then a is the join of these elements $f(\frac{1}{2},-)$ and from (6) in 6.1.2,

$$f(\tfrac{1}{2},-) = \bigvee\{f(\tfrac{1}{2},p) \mid p > \tfrac{1}{2}\} = \bigvee\{\mathrm{coz}((f-\tfrac{1}{2})^+) \wedge \mathrm{coz}((\mathbf{p}-f)^+) \mid p > \tfrac{1}{2}\}.$$

Conversely, if L is generated by its cozero elements then every element of L is a join of cozero elements by 6.1.2(3). The fact that L is then completely regular follows immediately from 6.2.3. $\qquad\square$

7. More general real functions

7.1. Given a topological space X, it is sometimes desirable to consider real functions on X which are not necessarily continuous. For instance, *Bluemberg spaces* are spaces X in which for *every function* $f \colon X \to \mathbb{R}$ there is a dense subspace D of X such that the restriction $f|_D \colon D \to \mathbb{R}$ is continuous.

There are also the important semicontinuous real-valued functions (introduced by Baire in 1899 for real-valued functions with domain $\mathbb{R}$):

A real function f defined in a space X is *upper* (resp. *lower*) *semicontinuous* if, for any $x \in X$ and $r \in \mathbb{R}$ satisfying $f(x) < r$ (resp. $f(x) > r$), there is a neighbourhood $U \subseteq X$ of x such that $f(y) < r$ (resp. $f(y) > r$) for every $y \in U$.

(Obviously, for each closed $F \subseteq X$, the characteristic map χ_F is upper semicontinuous, and for each open $A \subseteq X$, χ_A is lower semicontinuous.) $\qquad$ (7.1.1)

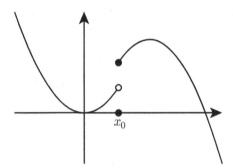

x_0

Figure 4: An upper semicontinuous function.

Denoting by $\mathfrak{T}_l$ the *lower topology* of $\mathbb{R}$, generated by intervals $]-\infty, q[$, with $q \in \mathbb{Q}$, $f \colon X \to \mathbb{R}$ is upper semicontinuous iff $f \colon (X, \Omega(X)) \to (\mathbb{R}, \mathfrak{T}_l)$ is continuous. Analogously, considering the *upper topology* $\mathfrak{T}_u$ of $\mathbb{R}$, generated by intervals $]p, +\infty[$, $p \in \mathbb{Q}$, $f \colon X \to \mathbb{R}$ is lower semicontinuous iff $f \colon (X, \Omega(X)) \to (\mathbb{R}, \mathfrak{T}_u)$ is continuous.

7.2. The frame $\mathfrak{S}(L)$. Up to now we have only considered continuous real functions. With the goal of dealing with more general functions it is convenient to introduce the co-frame $\mathcal{Sl}(L)$ of sublocales of a frame L (III.3.2.1) turned upside down. We shall denote this frame by

$$\mathfrak{S}(L).$$

From III.6.1.5 and IV.6.2 we know that the set $\mathfrak{c}L$ of all closed sublocales of L is a subframe of $\mathfrak{S}(L)$, isomorphic to the given L.

7.3. Real functions. In the point-sensitive case, the set

$$\mathcal{F}(X, \mathbb{R})$$

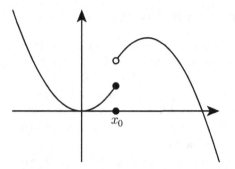

Figure 5: A lower semicontinuous function.

of all functions $X \to \mathbb{R}$ (not necessarily continuous) is clearly in bijection with $\textbf{Top}\big((X, \mathfrak{P}(X)), (\mathbb{R}, \mathfrak{T})\big)$, whatever topology $\mathfrak{T}$ we put on the reals. In particular, for the usual (Euclidean) topology $\mathfrak{T}_e$ of $\mathbb{R}$, we have

$$\mathcal{F}(X, \mathbb{R}) \cong \textbf{Top}\big((X, \mathfrak{P}(X)), (\mathbb{R}, \mathfrak{T}_e)\big).$$

Therefore, by the adjunction between $\textbf{Top}$ and $\textbf{Loc}$,

$$\mathcal{F}(X, \mathbb{R}) \cong \textbf{Loc}(\mathfrak{P}(X), \mathfrak{L}(\mathbb{R})) = \textbf{Frm}(\mathfrak{L}(\mathbb{R}), \mathfrak{P}(X)).$$

Note that $\mathfrak{P}(X)$ is the subobject lattice of X in $\textbf{Top}$. Now, when we move to $\textbf{Loc}$ replacing X by an arbitrary locale L, we should replace $\mathfrak{P}(X)$ by the sublocale lattice $\mathfrak{S}(L)$ of L. This conceptually justifies to adopt the elements of the lattice-ordered ring

$$\mathcal{C}(\mathfrak{S}(L)) = \textbf{Frm}(\mathfrak{L}(\mathbb{R}), \mathfrak{S}(L))$$

as the definition of *real functions* in an arbitrary frame L.

7.3.1. Semicontinuous real functions. The important feature of this approach is that the structure of $\mathfrak{S}(L)$ is rich enough to allow to distinguish the different types of continuities. An *upper* (resp. *lower*) *semicontinuous function* is an $f \colon \mathfrak{L}(\mathbb{R}) \to \mathfrak{S}(L)$ for which $f(-, r) \in \mathfrak{c}L$ (resp. $f(r, -) \in \mathfrak{c}L$) for every $r \in \mathbb{Q}$. The class of upper (resp. lower) semicontinuous functions in L will be denoted by

$$\mathcal{U}(L) \ \text{(resp. } \mathcal{L}(L)\text{)}.$$

Then a continuous function is just a function in $\mathcal{U}(L) \cap \mathcal{L}(L)$, i.e., an $f \colon \mathfrak{L}(\mathbb{R}) \to \mathfrak{S}(L)$ for which $f(\mathfrak{L}(\mathbb{R})) \subseteq \mathfrak{c}L$. Evidently, by the isomorphism $\mathfrak{c}L \cong L$, this definition is equivalent to the previous one in 5.1 (more precisely the two classes of functions are in a bijective correspondence via the isomorphism $\mathfrak{c}L \cong L$). We will keep the notation $\mathcal{C}(L)$ for both classes of functions in the bijection.

Examples: Characteristic functions. Let S be a complemented sublocale of L. Then σ_S defined by $\sigma_S(r) = 1$ if $r < 0$, $\sigma_S(r) = S^c$ if $0 \le r < 1$ and $\sigma_S(r) = 0$ otherwise, is a scale in $\mathfrak{S}(L)$. By application of 5.2.2 it determines the *characteristic function* χ_S defined for each $r \in \mathbb{Q}$ by

$$\chi_S(r, -) = \begin{cases} 1 & \text{if } r < 0 \\ S^c & \text{if } 0 \le r < 1 \\ 0 & \text{if } r \ge 1 \end{cases} \quad \text{and} \quad \chi_S(-, r) = \begin{cases} 0 & \text{if } r \le 0 \\ S & \text{if } 0 < r \le 1 \\ 1 & \text{if } r > 1. \end{cases}$$

Note that $\chi_S \in \mathfrak{U}(L)$ iff S is closed, $\chi_S \in \mathcal{L}(L)$ iff S is open, and $\chi_S \in \mathcal{C}(L)$ iff S is clopen (compare this with (7.1.1)).

7.4. Insertion theorem. Similarly as in spaces, there is a plethora of insertion-type results characterizing several classes of frames in terms of the insertion of a continuous real function in between certain pairs of (semicontinuous) real functions. Of those we present the one characterizing normality. We will need two auxiliary results for normal frames.

7.4.1. Lemma. *A frame L is normal if and only if for any countable subsets $\{a_i\}_{i \in \mathbb{N}}$ and $\{b_i\}_{i \in \mathbb{N}}$ of L satisfying $a_i \vee (\bigwedge_{j \in \mathbb{N}} b_j) = 1$ and $b_i \vee (\bigwedge_{j \in \mathbb{N}} a_j) = 1$ for every $i \in \mathbb{N}$, there exists $u \in L$ such that $a_i \vee u = 1$ and $b_i \vee u^* = 1$ for every $i \in \mathbb{N}$.*

Proof. Let L be a normal frame. Then, for each $i \in \mathbb{N}$, $a_i \vee (\bigwedge_{j \in \mathbb{N}} b_j) = 1$ implies, by normality, the existence of $u_i \in L$ satisfying $a_i \vee u_i = 1$ and $(\bigwedge_{j \in \mathbb{N}} b_j) \vee u_i^* = 1$. Similarly, from $b_i \vee (\bigwedge_{j \in \mathbb{N}} a_j) = 1$ it follows that there is $v_i \in L$ such that $b_i \vee v_i = 1$ and $(\bigwedge_{j \in \mathbb{N}} a_j) \vee v_i^* = 1$. Then, for each $i \in \mathbb{N}$, we have

$$a_i \vee \left(\bigwedge_{k=1}^{i} v_k^* \right) \ge \left(\bigwedge_{j \in \mathbb{N}} a_j \right) \vee \left(\bigwedge_{k=1}^{i} v_k^* \right) = \bigwedge_{k=1}^{i} \left(\left(\bigwedge_{j \in \mathbb{N}} a_j \right) \vee v_k^* \right) = 1$$

and, similarly,

$$b_i \vee \left(\bigwedge_{k=1}^{i} u_k^* \right) \ge \left(\bigwedge_{j \in \mathbb{N}} b_j \right) \vee \left(\bigwedge_{k=1}^{i} u_k^* \right) = \bigwedge_{k=1}^{i} \left(\left(\bigwedge_{j \in \mathbb{N}} b_j \right) \vee u_k^* \right) = 1.$$

Now define, for each $i \in \mathbb{N}$,

$$u_i' = u_i \wedge \bigwedge_{k=1}^{i} v_k^* \quad \text{and} \quad v_i' = v_i \wedge \bigwedge_{k=1}^{i} u_k^*$$

and consider $u = \bigvee_{i \in \mathbb{N}} u_i'$ and $v = \bigvee_{i \in \mathbb{N}} v_i'$. Clearly,

$$a_i \vee u \ge a_i \vee u_i' = (a_i \vee u_i) \wedge (a_i \vee (\bigwedge_{k=1}^{i} v_k^*)) = 1,$$

$$b_i \vee v \ge b_i \vee v_i' = (b_i \vee v_i) \wedge (b_i \vee (\bigwedge_{k=1}^{i} u_k^*)) = 1$$

and

$$u \wedge v = \bigvee_{i \in \mathbb{N}} \bigvee_{j \in \mathbb{N}} (u'_i \wedge v'_j) = \bigvee_{i \in \mathbb{N}} \bigvee_{j \in \mathbb{N}} (u_i \wedge v_j \wedge \bigwedge_{k=1}^{i} v^*_k \wedge \bigwedge_{l=1}^{j} u^*_l) = 0.$$

Thus $v \leq u^*$ and we conclude that $b_i \vee u^* \geq b_i \vee v = 1$.

The converse is trivial.   □

In the following, given $f \in \mathfrak{U}(L)$, $g \in \mathcal{L}(L)$ and $r \in \mathbb{Q}$, let f_r, g_r be the elements in L such that $f(-, r) = \mathfrak{c}(f_r)$ and $g(r, -) = \mathfrak{c}(g_r)$.

7.4.2. Lemma. *Let L be a normal frame and let $\{\alpha_i \mid i \in \mathbb{N}\}$ be an enumeration of $\mathbb{Q}$. If $f \in \mathfrak{U}(L)$ and $g \in \mathcal{L}(L)$ satisfy $f \leq g$, then there exists $\{u_{\alpha_i}\}_{i \in \mathbb{N}} \subseteq L$ such that*

$$(q > \alpha_i) \quad \Rightarrow \quad (f_q \vee u_{\alpha_i} = 1), \tag{7.4.1}$$

$$(p < \alpha_i) \quad \Rightarrow \quad (g_p \vee u^*_{\alpha_i} = 1), \tag{7.4.2}$$

$$(\alpha_{j_1} < \alpha_{j_2}) \quad \Rightarrow \quad (u_{\alpha_{j_1}} \vee u^*_{\alpha_{j_2}} = 1). \tag{7.4.3}$$

Proof. We will prove this by showing by induction over $\mathbb{N}$ that, for each $i \in \mathbb{N}$, there exists $u_{\alpha_i} \in L$ such that

$$(q > \alpha_i) \Rightarrow (f_q \vee u_{\alpha_i} = 1), \quad (p < \alpha_i) \Rightarrow (g_p \vee u^*_{\alpha_i} = 1) \text{ and}$$

$$(\alpha_{j_1} < \alpha_{j_2}) \Rightarrow (u_{\alpha_{j_1}} \vee u^*_{\alpha_{j_2}} = 1), \text{ for all } j_1, j_2 \leq i.$$

Since $f \leq g$, we have $f_q \vee \bigwedge_{r < \alpha_1} g_r \geq f_q \vee g_{\alpha_1} = 1$ for every $q > \alpha_1$. Similarly, for every $p < \alpha_1$, $(\bigwedge_{r > \alpha_1} f_r) \vee g_p = 1$. Then 7.4.1 provides $u_{\alpha_1} \in L$ satisfying $f_q \vee u_{\alpha_1} = 1$, for every $q > \alpha_1$, and $g_p \vee u^*_{\alpha_1} = 1$, for every $p < \alpha_1$, which shows the first step of the induction.

Now, consider $k \in \mathbb{N}$, and assume, by inductive hypothesis, that for any $i < k$ there is $u_{\alpha_i} \in L$ satisfying $f_q \vee u_{\alpha_i} = 1$, for every $q > \alpha_i$, $g_p \vee u^*_{\alpha_i} = 1$, for every $p < \alpha_i$, and $(\alpha_{j_1} < \alpha_{j_2}) \Rightarrow (u_{\alpha_{j_1}} \vee u^*_{\alpha_{j_2}} = 1)$, for all $j_1, j_2 \leq k - 1$. Let

$$\{a_n\}_{n \in \mathbb{N}} = \{f_q \mid q > \alpha_k\} \cup \{u^*_{\alpha_i} \mid i < k, \alpha_k < \alpha_i\}$$

and

$$\{b_n\}_{n \in \mathbb{N}} = \{g_p \mid p < \alpha_k\} \cup \{u_{\alpha_i} \mid i < k, \alpha_i < \alpha_k\};$$

Then $\{a_n\}_{n \in \mathbb{N}}$ and $\{b_n\}_{n \in \mathbb{N}}$ satisfy the conditions in 7.4.1:

I. For each $q > \alpha_k$, $f_q \vee (\bigwedge_{p < \alpha_k} g_p \wedge \bigwedge_{i < k, \, \alpha_i < \alpha_k} u_{\alpha_i})$ is equal to

$$\left(f_q \vee (\bigwedge_{p < \alpha_k} g_p)\right) \wedge \left(f_q \vee (\bigwedge_{i < k, \, \alpha_i < \alpha_k} u_{\alpha_i})\right) = 1$$

since $f_q \vee (\bigwedge_{p < \alpha_k} g_p) = 1$, and, by inductive hypothesis,

$$f_q \vee \left(\bigwedge_{i < k, \, \alpha_i < \alpha_k} u_{\alpha_i}\right) = \bigwedge_{i < k, \, \alpha_i < \alpha_k} (f_q \vee u_{\alpha_i}) = 1.$$

II. For each $i < k$ such that $\alpha_k < \alpha_i$, $u_{\alpha_i}^* \vee (\bigwedge_{p<\alpha_k} g_p \wedge \bigwedge_{j<k,\,\alpha_j<\alpha_k} u_{\alpha_j})$ is equal to

$$\left(u_{\alpha_i}^* \vee \left(\bigwedge_{p<\alpha_k} g_p\right)\right) \wedge \left(u_{\alpha_i}^* \vee \left(\bigwedge_{j<k,\,\alpha_j<\alpha_k} u_{\alpha_j}\right)\right) = 1,$$

since $u_{\alpha_i}^* \vee (\bigwedge_{p<\alpha_k} g_p) \geq u_{\alpha_i}^* \vee g_{\alpha_k} = 1$ and, by inductive hypothesis,

$$u_{\alpha_i}^* \vee \left(\bigwedge_{j<k,\,\alpha_j<\alpha_k} u_{\alpha_j}\right) = \bigwedge_{j<k,\,\alpha_j<\alpha_k} (u_{\alpha_i}^* \vee u_{\alpha_j}) = 1.$$

III. Similarly to I and II, respectively, one can prove that, for each $p < \alpha_k$, $(\bigwedge_{q>\alpha_k} f_q \wedge \bigwedge_{i<k,\alpha_k<\alpha_i} u_{\alpha_i}^*) \vee g_p = 1$, and that, for each $i < k$ satisfying $\alpha_i < \alpha_k$, $(\bigwedge_{q>\alpha_k} f_q \wedge \bigwedge_{i<k,\alpha_k<\alpha_i} u_{\alpha_i}^*) \vee u_{\alpha_i} = 1$. Thus, it follows from 7.4.1 that there exists $u_{\alpha_k} \in L$ such that

$$f_q \vee u_{\alpha_k} = 1 \text{ for all } q > \alpha_k, \qquad (\alpha_k < \alpha_i) \Rightarrow (u_{\alpha_i}^* \vee u_{\alpha_k} = 1) \text{ for all } i < k,$$
$$g_p \vee u_{\alpha_k}^* = 1 \text{ for all } p < \alpha_k \text{ and } (\alpha_i < \alpha_k) \Rightarrow (u_{\alpha_i} \vee u_{\alpha_k}^* = 1) \text{ for all } i < k.$$

This, together with the inductive hypothesis, provides us with the desired $u_{\alpha_i} \in L$.
□

7.4.3. Theorem. *A frame L is normal if and only if for each $f \in \mathcal{U}(L)$ and $g \in \mathcal{L}(L)$ satisfying $f \leq g$, there is an $h \in \mathcal{C}(L)$ such that $f \leq h \leq g$.*

Proof. Let $f \in \mathcal{U}(L)$ and $g \in \mathcal{L}(L)$ such that $f \leq g$. By normality, there exists $\{u_{\alpha_i}\}_{i\in\mathbb{N}} \subseteq L$ satisfying conditions (7.4.1), (7.4.2) and (7.4.3) of 7.4.2. (7.4.1) implies that $\mathfrak{c}(u_{\alpha_i}) \geq f(-,q)^* \geq f(q,-)$ whenever $\alpha_i < q$. Similarly, from (7.4.2) it follows that $\mathfrak{c}(u_{\alpha_i}^*) \geq g(p,-)^* \geq g(-,p)$ whenever $p < \alpha_i$. Thus $\bigvee_{i\in\mathbb{N}} \mathfrak{c}(u_{\alpha_i}) = 1 = \bigvee_{i\in\mathbb{N}} \mathfrak{c}(u_{\alpha_i}^*)$, that is, $\bigvee_{i\in\mathbb{N}} u_{\alpha_i} = 1 = \bigvee_{i\in\mathbb{N}} u_{\alpha_i}^*$. Together with (7.4.3) this means that $\{u_{\alpha_i}\}_{i\in\mathbb{N}}$ is a scale in L. Consequently, the formulas

$$\varphi(r,-) = \bigvee_{\alpha_i>r} u_{\alpha_i} \quad \text{and} \quad \varphi(-,r) = \bigvee_{\alpha_i<r} u_{\alpha_i}^*$$

determine a frame homomorphism $\varphi \colon \mathcal{L}(\mathbb{R}) \to L$. Consider the corresponding $h \colon \mathcal{L}(\mathbb{R}) \to \mathcal{S}(L)$ given by

$$h(r,-) = \mathfrak{c}(\varphi(r,-)) = \bigvee_{\alpha_i>r} \mathfrak{c}(u_{\alpha_i}) \quad \text{and} \quad h(-,r) = \mathfrak{c}(\varphi(-,r)) = \bigvee_{\alpha_i<r} \mathfrak{c}(u_{\alpha_i}^*).$$

It satisfies $f \leq h \leq g$. Indeed, condition (7.4.1) means that, for each $r \in \mathbb{Q}$ and each $\alpha_i < r$, $f(-,r) \geq \mathfrak{c}(u_{\alpha_i}^*)$. Then, of course, $f(-,r) \geq \bigvee_{\alpha_i<r} \mathfrak{c}(u_{\alpha_i}^*) = h(-,r)$, which means that $f \leq h$. Similarly, by condition (7.4.2) we have $g(r,-) \geq \mathfrak{c}(u_{\alpha_i}^{**}) \geq \mathfrak{c}(u_{\alpha_i})$ whenever $\alpha_i > r$. Thus, for each $r \in \mathbb{Q}$, $g(r,-) \geq \bigvee_{\alpha_i>r} \mathfrak{c}(u_{\alpha_i}) = h(r,-)$.

Conversely, suppose $a \vee b = 1$ in L. Then $\mathfrak{o}(b) \leq \mathfrak{c}(a)$, that is,

$$\chi_{\mathfrak{c}(a)} \leq \chi_{\mathfrak{o}(b)}.$$

By the hypothesis, there exists $h \in \mathcal{C}(L)$ such that $\chi_{\mathfrak{c}(a)} \leq h \leq \chi_{\mathfrak{o}(b)}$. Let $u, v \in L$ such that $\mathfrak{c}(u) = h(\frac{1}{2}, -)$ and $\mathfrak{c}(v) = h(-, \frac{1}{2})$. Then

$$1 = \chi_{\mathfrak{c}(a)}(-, \tfrac{3}{4}) \vee \chi_{\mathfrak{c}(a)}(\tfrac{1}{2}, -) \leq \chi_{\mathfrak{c}(a)}(-, \tfrac{3}{4}) \vee h(\tfrac{1}{2}, -) = \mathfrak{c}(a) \vee \mathfrak{c}(u) = \mathfrak{c}(a \vee u)$$

and

$$1 = \chi_{\mathfrak{o}(b)}(\tfrac{1}{4}, -) \vee \chi_{\mathfrak{o}(b)}(-, \tfrac{1}{2}) \leq h(\tfrac{1}{4}, -) \vee \chi_{\mathfrak{o}(b)}(-, \tfrac{1}{2}) = \mathfrak{c}(b) \vee \mathfrak{c}(v) = \mathfrak{c}(b \vee v).$$

Therefore $a \vee u = 1 = b \vee v$. On the other hand, $0 = \mathfrak{c}(u) \wedge \mathfrak{c}(v) = \mathfrak{c}(u \wedge v)$, hence $u \wedge v = 0$. This proves the normality of L. $\square$

7.4.4. Note. Theorem 7.4.3 applied to $L = \Omega(X)$ for a normal space X yields the celebrated Katětov-Tong Insertion Theorem ([159], [254]). Several other point-free insertion theorems have been obtained recently, such as the monotone version of 7.4.3 characterizing monotone normality, the strict insertion theorem characterizing perfect normality, the strict insertion theorem characterizing countable paracompactness and the insertion theorem for general functions characterizing complete normality ([97], [116], [117], [119] [123]). Most interestingly, the last one characterizes completely normal frames in terms of an insertion condition for general (not necessarily semicontinuous) real functions $f_1, f_2 \colon \mathfrak{L}(\mathbb{R}) \to \mathfrak{S}(L)$.

7.5. Urysohn's Separation Lemma. Let L be a normal frame and consider $a, b \in L$ satisfying $a \vee b = 1$. Then $\chi_{\mathfrak{c}(a)} \leq \chi_{\mathfrak{o}(b)}$. Consequently, the application of 7.4.3 yields an $h \in \mathcal{C}(L)$ such that

$$\chi_{\mathfrak{c}(a)} \leq h \leq \chi_{\mathfrak{o}(b)}.$$

Rephrasing this internally in L we get immediately the non-trivial implication of the following characterization of normality due to Dowker and Papert ([68]), which is the point-free extension of the separation lemma of Urysohn.

7.5.1. Corollary. *A frame L is normal if and only if, for every $a, b \in L$ satisfying $a \vee b = 1$, there exists $h \colon \mathfrak{L}(\mathbb{R}) \to L$ such that $h((-, 0) \vee (1, -)) = 0$, $h(-, 1) \leq a$ and $h(0, -) \leq b$.* $\square$

7.6. Tietze's Extension Theorem. For any sublocale S of L, let $\nu_S \colon L \to S$ denote the corresponding frame quotient, given by $\nu_S(a) = \bigwedge \{ s \in S \mid a \leq s \}$ (recall III.5). A map $\widetilde{f}$ in $\mathcal{C}(L)$ is a *continuous extension* of $f \in \mathcal{C}(S)$ whenever the following diagram commutes

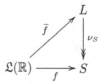

Theorem 7.4.3 also implies the point-free version of Tietze's Extension Theorem. We only state it for bounded functions (the proof for the general case is rather technical so we omit it).

An $f \in \mathcal{C}(L)$ is said to be *bounded* if

$$f(p, q) = 1 \text{ for some } p < q$$

(or equivalently, if $f(-, p) \vee f(q, -) = 0$ for some $p < q$). Let

$$\mathcal{C}^*(L)$$

denote the set of bounded members f of $\mathcal{C}(L)$ satisfying $f(-, 0) \vee f(1, -) = 0$.

7.6.1. Corollary. *A frame L is normal iff for each closed sublocale S of L and each $h \in \mathcal{C}^*(S)$, there exists a continuous extension $\widetilde{h} \in \mathcal{C}^*(L)$ of h.*

Proof. The implication "$\Leftarrow$" being straightforward, we only prove "$\Rightarrow$". Let $S = \mathfrak{c}(a)$ and consider a bounded continuous $h \colon \mathcal{L}(\mathbb{R}) \to S$. Define $f, g \colon \mathcal{L}(\mathbb{R}) \to \mathfrak{S}(L)$ by

$$f(-, r) = \begin{cases} 0 & \text{if } r \leq 0 \\ \mathfrak{c}(h(-, r)) & \text{if } 0 < r \leq 1, \\ 1 & \text{if } r > 1 \end{cases} \quad f(r, -) = \begin{cases} 1 & \text{if } r < 0 \\ \bigvee\limits_{s > r} \mathfrak{o}(h(-, s)) & \text{if } 0 \leq r < 1 \\ 0 & \text{if } r \geq 1 \end{cases}$$

and

$$g(r, -) = \begin{cases} 1 & \text{if } r < 0 \\ \mathfrak{c}(h(r, -)) & \text{if } 0 \leq r < 1, \\ 0 & \text{if } r \geq 1. \end{cases} \quad g(r, -) = \begin{cases} 0 & \text{if } r \leq 0 \\ \bigvee\limits_{s < r} \mathfrak{o}(h(s, -)) & \text{if } 0 < r \leq 1 \\ 1 & \text{if } r > 1. \end{cases}$$

Checking that $f \in \mathcal{U}(L)$, $g \in \mathcal{L}(L)$ and $f \leq g$ is straightforward. Therefore, by 7.4.3, there is a $\kappa \in \mathcal{C}(L)$ such that $f \leq \kappa \leq g$. Since $\kappa((-, 0) \vee (1, -)) \leq f(-, 0) \vee g(1, -) = 0$, κ is bounded. Then the desired extension of h to L is the

$$\widetilde{h} \colon \mathcal{L}(\mathbb{R}) \to L \quad \text{where } \mathfrak{c}(\widetilde{h}(p, q)) = \kappa(p, q).$$

Indeed, $\widetilde{h}(p, q) \vee a = h(p, q)$, that is, $\kappa(p, q) \vee \mathfrak{c}(a) = \mathfrak{c}(h(p, q))$, for every $p, q \in \mathbb{Q}$:

First $\kappa(p, q) \vee \mathfrak{c}(a) \leq (g(p, -) \wedge f(-, q)) \vee \mathfrak{c}(a) \leq \mathfrak{c}(h(p, q))$. On the other hand, for each $r < q$,

$$h(-, r) = h(-, r) \wedge (\widetilde{h}(-, q) \vee \widetilde{h}(r, -))$$
$$= (h(-, r) \wedge \widetilde{h}(-, q)) \vee (h(-, r) \wedge \widetilde{h}(r, -))$$
$$\leq \widetilde{h}(-, q) \vee (h(-, r) \wedge h(r, -)) = \widetilde{h}(-, q) \vee h(0) = \widetilde{h}(-, q) \vee a.$$

Similarly, $h(s, -) \leq \widetilde{h}(p, -) \vee a$ for every $s > p$. Since

$$h(p, q) = \left(\bigvee\limits_{r < q} h(-, r) \right) \wedge \left(\bigvee\limits_{s > p} h(s, -) \right),$$

we conclude that $h(p, q) \leq (\widetilde{h}(-, q) \vee a) \wedge (\widetilde{h}(p, -) \vee a) = \widetilde{h}(p, q) \vee a$, as required. $\square$

8. Notes

8.1. Extended real functions. The discussion of continuous real functions with possibly infinite values can be brought to point-free topology by replacing the frame $\mathfrak{L}(\mathbb{R})$ by the *frame of extended reals* $\mathfrak{L}(\overline{\mathbb{R}})$, the point-free counterpart of the extended real line $\overline{\mathbb{R}} = \mathbb{R} \cup \{\pm\infty\}$, see the recent [28].

8.2. The real uniformity of a frame. The functions in $\mathcal{C}(L)$ and the metric uniformity of $\mathfrak{L}(\mathbb{R})$ can be used to define a uniformity on some frames L. For that consider the *real-uniform covers* of L, that is, the covers refined by some $f_1[U_n] \wedge \cdots \wedge f_k[U_n]$, for homomorphisms $f_i \colon \mathfrak{L}(\mathbb{R}) \to L$ and basic uniform covers U_n of the metric uniformity of $\mathfrak{L}(\mathbb{R})$. These covers form a filter relative to the refinement relation, and any cover of this kind is star refined by such a cover. What about admissibility? Using 6.2.1 it is not hard to see that $a \twoheadleftarrow b$ iff $Ca \leq b$ for some real-uniform cover C of L ([17], Lemma 1). Hence

> *The real-uniform covers of a frame L form a uniformity of L if and only if L is completely regular.*

This is the *real uniformity* of a completely regular frame L; L is called *real-complete* if it is complete in this uniformity.

8.2.1. Remarks. (1) The real uniformity of a completely regular frame is the point-free counterpart of the initial uniformity on a Tychonoff space X with respect to all continuous maps $X \to \mathbb{R}$ and the usual metric uniformity of $\mathbb{R}$, employed by Nachbin in his definition of realcompact spaces ([187]).

(2) $\mathfrak{L}(\mathbb{R})$ is real-complete since, by Proposition 4.2, it is complete in the coarser metric uniformity.

(3) Any compact completely regular frame L is real-complete. Indeed, its real uniformity is just its unique uniformity consisting of all covers and L is complete in this (VIII.6.2).

(4) Real-completeness is coreflective in the category of completely regular frames, with coreflection maps $\rho_L \colon C_{\mathbb{R}} L \to L$ determined by the completion of L with respect to its real uniformity ([16], Proposition 8).

The familiar result in classical topology that a Tychonoff space X is complete in the initial uniformity given by the continuous maps $X \to \mathbb{R}$ and the usual metric uniformity of $\mathbb{R}$ iff X is homeomorphic to a closed subspace of some power of $\mathbb{R}$ ([83], 3.11) has the following point-free counterpart:

8.2.2. Proposition. *A completely regular frame is real-complete iff it is isomorphic to a closed quotient of some copower of $\mathfrak{L}(\mathbb{R})$.*

See [16] or [17] for more details.

8.3. Realcompactness. The study of realcompactness goes back to the independent works of Hewitt ([131]) and Nachbin ([187]) in 1948–50 (this is why they are sometimes referred as "Hewitt-Nachbin spaces" [273]). Realcompact spaces play a role in the study of rings $\mathcal{C}(X)$ analogous to that played by compact spaces in the study of $\mathcal{C}^*(X)$. In Hewitt's original formulation, a Tychonoff space X is *realcompact* provided

- every unitary $\mathbb{R}$-algebra homomorphism $\mathcal{C}(X) \to \mathbb{R}$ is of the form e_x for some $x \in X$, where $e_x(f) = f(x)$ (Hewitt [131], 1948).

In Nachbin's equivalent formulation in terms of uniformities, X is realcompact if

- it is complete in the real uniformity of X defined by $\mathcal{C}(X)$ (i.e., *realcompact* $\equiv$ *real-complete*) (Nachbin [187], 1950).

To a point-set topologist, this means that

- the space is homeomorphic to a closed subspace of some power of $\mathbb{R}$ (Shirota [242], 1952).

Realcompactness for frames was first studied by Reynolds ([232]) (actually Reynolds studied realcompactness in the setting of a Grothendieck topos but this provides in particular a concept for locales and frames); his definition is equivalent to the point-free counterpart of Shirota's characterization (and then, by 8.2.2, to real-completeness). However, quite surprisingly, this concept does not faithfully mirror the classical notion since it only captures Lindelöf spaces, as was shown some years later by Madden and Vermeer ([179]) with the following

8.3.1. Theorem. *A completely regular frame is isomorphic to a closed quotient of some copower of $\mathfrak{L}(\mathbb{R})$ iff it is Lindelöf.*

(In particular, the frame $\Omega(X)$ of a Tychonoff space X is realcompact in the sense of Reynolds iff X is Lindelöf, and it is well known that there are many examples of realcompact spaces which fail to be Lindelöf, such as all powers of $\mathbb{R}$ with uncountable exponent.)

Therefore this approach does not provide a conservative extension to frames of the classical notion (nor does the point-free counterpart of Hewitt's original formulation as it is also equivalent to real-completeness [17]). This was mended by Schlitt ([237], [239]) with the weaker notion of H-realcompactness. Afterwards, Banaschewski and Gilmour ([27]) provided an alternative solution, with a particularly natural variant of Schlitt's notion in terms of the cozero parts of frames: a completely regular frame L is *realcompact* if every σ-proper maximal ideal in $\mathrm{Coz}\, L$ is completely proper (for any sublattice A of L, an ideal $I \subseteq L$ is σ-*proper* if $\bigvee S \neq 1$ for any countable $S \subseteq I$ and is said to be *completely proper* if $\bigvee I \neq 1$). Then space X is realcompact iff the frame $\Omega(X)$ is realcompact and the realcompact frames are coreflective in the category of all completely regular frames.

For more about this subject see [17] and [16].

8.4. The semicontinuous quasi-uniformity of a frame. For each $n \in \mathbb{N}$,

$$Q_n = \left\{ ((-,q),(p,-)) \mid p,q \in \mathbb{Q},\, 0 < q - p < \frac{1}{n} \right\}$$

is a strong paircover of $\mathfrak{L}(\mathbb{R})$. These paircovers generate a quasi-uniformity $\mathfrak{Q}$ on $\mathfrak{L}(\mathbb{R})$, the *metric quasi-uniformity* of the reals ([96]).

Now, for any frame L, any $f \in \mathcal{U}(L)$ and $n \in \mathbb{N}$ let

$$C_{f,n} = \left\{ (f(-,q),f(p,-)) \mid ((-,q),(p,-)) \in Q_n,\, f(p,q) \neq 0 \right\}.$$

The $C_{f,n}$ are paircovers of $\mathfrak{S}(L)$ that generate a quasi-uniformity (the *semicontinuous quasi-uniformity*). This is an important example of a transitive quasi-uniformity; it is the coarsest quasi-uniformity $\mathcal{U}$ on $\mathfrak{S}(L)$ for which each real function $h \colon \mathfrak{L}(\mathbb{R}) \to \mathfrak{S}(L)$ is a uniform homomorphism $h \colon (\mathfrak{L}(\mathbb{R}), \mathfrak{Q}) \to (\mathfrak{S}(L), \mathcal{U})$.

This can be extended to any collection $\mathcal{F}$ of real functions on L containing all characteristic functions χ_S for a closed sublocale S:

8.4.1. Proposition. $\{C_{f,n} \mid f \in \mathcal{F}, n \in \mathbb{N}\}$ *is a subbasis for a quasi-uniformity* $\mathcal{U}_{\mathcal{F}}$ *on the frame* $\mathfrak{S}(L)$.

For more information consult [96].

Chapter XV

Localic Groups

In this chapter we will discuss group structures in the point-free context. Although completely regular locales (which will turn out to carry such structures) are genuine generalizations of completely regular spaces (carriers of topological groups), a localic group is not exactly a generalization of the topological one (although the most important topological groups can be viewed as localic ones). This has in fact rather pleasant consequences: in particular, each subgroup of a localic group is closed. On the other hand there are strong analogies, in particular in the natural uniform structures obtained from the operations.

1. Basics

1.1. A *group* in a category $\mathcal{C}$ with products (recall AII.10) is a collection of data (A, m, i, e) with

$$m\colon A \times A \to A, \quad i\colon A \to A, \quad e\colon T \to A$$

satisfying

$$m(m \times \mathrm{id}) = m(\mathrm{id} \times m),$$
$$m(e \times \mathrm{id}) = m(\mathrm{id} \times e) = \mathrm{id}, \quad \text{and}$$
$$m(i \times \mathrm{id})\Delta = m(\mathrm{id} \times i)\Delta = et_A$$

($T = A^0$ is the empty product, that is, the terminal object of $\mathcal{C}$, the object such that from every $C \in \mathrm{obj}\mathcal{C}$ there is precisely one morphism $t_C\colon C \to T$).

Note. If $\mathcal{C}$ is the category of sets (where products are the Cartesian ones and $T = \{\emptyset\}$ is the one-point set this is the classical group $(A, \cdot, (-)^{-1}, e)$ and the identities above are the standard $x(yz) = (xy)z$, $xe = ex = x$ and $xx^{-1} = x^{-1}x = e$; or, e.g., in **Top** we have operations such that the maps $(x, y) \mapsto xy$ and $x \mapsto x^{-1}$ are continuous (the classical topological group).

In particular, a *localic group* is a group in **Loc**. For convenience,

we will view it as a co-group in the category of frames,

that is, an $(L, \mu, \gamma, \varepsilon)$ with

$\mu \colon L \to L \oplus L$ (the *multiplication*), $\gamma \colon L \to L$ (the *inverse*), and

$$\varepsilon \colon L \to \mathbf{2} \text{ (the \textit{unit})}$$

such that

$$(\mu \oplus \mathrm{id})\mu = (\mathrm{id} \oplus \mu)\mu,$$
$$(\varepsilon \oplus \mathrm{id})\mu = (\mathrm{id} \oplus \varepsilon)\mu = \mathrm{id}, \text{ and}$$
$$\nabla(\gamma \oplus \mathrm{id})\mu = \nabla(\mathrm{id} \oplus \gamma)\mu = \sigma\varepsilon,$$

where $\sigma \colon \mathbf{2} \to L$ sends 0 to 0 and 1 to 1, and ∇ is the codiagonal $L \oplus L \to L$ defined by $\nabla \iota_i = \mathrm{id}$, $i = 1, 2$.

1.2. Conventions. The reader has certainly observed, already at the first of the definitions, that we work with the product $A \times B$ as with an associative operation although, e.g., $m(m \times \mathrm{id})$ starts in $A \times (A \times A)$ while $m(\mathrm{id} \times m)$ starts in $(A \times A) \times A$. The products $A \times (B \times C)$ and $(A \times B) \times C$ are in general isomorphic but not necessarily identical. To avoid confusion we will make the (obvious) convention that the cartesian products in the construction of coproducts in **Frm** are associative which in turn amounts to identifying

$$(a \oplus b) \oplus c, \quad a \oplus (b \oplus c) \quad \text{and} \quad a \oplus b \oplus c$$

(in $(L \oplus L) \oplus L$, $L \oplus (L \oplus L)$ resp. $L \oplus L \oplus L$, etc.). Furthermore, we will work with the factor $\mathbf{2}$ as a void one, that is,

$$L \oplus \mathbf{2} = \mathbf{2} \oplus L = L$$

with coproduct injections

$$L \xrightarrow{\mathrm{id}} L \xleftarrow{\sigma} \mathbf{2} \quad \text{and} \quad \mathbf{2} \xrightarrow{\sigma} L \xleftarrow{\mathrm{id}} L.$$

1.3. Remarks. (1) Note that a localic group is a group in point-free topological context, but not a generalization of a topological group; that is, not every topological group can be viewed as a localic one. The reason is that we do not in general have

$$\Omega(X) \oplus \Omega(X) \cong \Omega(X \times X)$$

so that the Ω-image of the group multiplication is not necessarily a localic binary operation. In many important cases the products are preserved, though (e.g., in the locally compact, or in the complete metric case) and we can view such topological groups as localic ones.

(2) Note that a localic group always has at least one (spectrum) point, namely the unit $\varepsilon \colon L \to \mathbf{2}$. This may be the only point, even in large localic groups (see 6.2 below).

1.4. Various identities concerning the operations follow in fact from a general categorical result (see AII.10). Since we have not developed those techniques, we will indicate proofs of special identities we will need in some detail. Unless otherwise stated, $\iota_i \colon L \to L \oplus L$, $i = 1, 2$, are the coproduct injections.

1.4.1. Lemma. (1) $\gamma\gamma = \mathrm{id}$.

(2) $\varepsilon\gamma = \varepsilon$.

(3) $(\gamma \oplus \gamma)\mu = \tau\mu\gamma$, where $\tau \colon L \oplus L \to L \oplus L$ is defined by $\tau\iota_i = \iota_{3-i}$ $(i = 1, 2)$.

Proof. We obtain the identities by manipulating suitable formulas using the identities from 1.1. In the first case we will painstakingly do all the computation. Later, we will only provide the initial formula(s). The reader may do the detail as an easy (although somewhat boring) exercise.

(1): The initial formula is

$$\nabla(\nabla \oplus \mathrm{id})(\gamma\gamma \oplus \gamma\mathrm{id})(\mu \oplus \mathrm{id})\mu.$$

First, we transform it to

$$\cdots = \nabla((\nabla(\gamma\gamma \oplus \gamma)\mu) \oplus \mathrm{id})\mu = \nabla((\nabla(\gamma \oplus \gamma)(\gamma \oplus \mathrm{id})\mu) \oplus \mathrm{id})\mu$$

$$= \nabla((\gamma\nabla(\gamma \oplus \mathrm{id})\mu) \oplus \mathrm{id})\mu = \nabla(\gamma\sigma\varepsilon \oplus \mathrm{id})\mu$$

$$= \nabla(\sigma\varepsilon \oplus \mathrm{id})\mu = \nabla(\sigma \oplus \mathrm{id})(\varepsilon \oplus \mathrm{id})\mu = \mathrm{id}_{\mathbf{2}} \oplus \mathrm{id} = \mathrm{id}.$$

Secondly we obtain, using $(\mu \oplus \mathrm{id})\mu = (\mathrm{id} \oplus \mu)\mu$ and $\nabla(\nabla \oplus \mathrm{id}) = \nabla(\mathrm{id} \oplus \nabla)$,

$$\cdots = \nabla(\mathrm{id} \oplus \nabla)(\gamma\gamma \oplus \gamma\mathrm{id})(\mathrm{id} \oplus \mu)\mu$$

$$= \nabla(\gamma\gamma \oplus (\nabla(\gamma \oplus \mathrm{id})\mu))\mu = \nabla(\gamma\gamma \oplus \sigma\varepsilon)$$

$$= \nabla(\gamma\gamma \oplus \sigma)(\mathrm{id} \oplus \varepsilon)\mu = \nabla(\gamma\gamma \oplus \sigma) = \gamma\gamma.$$

(2): Here we will start with

$$\nabla^3(\varepsilon\gamma \oplus \varepsilon \oplus \varepsilon)(\mu \oplus \mathrm{id})\mu$$

where the codiagonal $\nabla^3 \colon \mathbf{2} = \mathbf{2} \oplus \mathbf{2} \oplus \mathbf{2} \to \mathbf{2}$ is the identity $\mathrm{id}_{\mathbf{2}}$ and is here only to make the computation transparent. If we think of ∇^3 as $\nabla^2(\nabla^2 \oplus \mathrm{id})$ $(\nabla^2 = \mathrm{id}^2 \colon \mathbf{2} \oplus \mathbf{2} \to \mathbf{2})$ we ultimately obtain

$$\cdots = \nabla(\varepsilon \oplus \mathrm{id})(\mathrm{id} \oplus \varepsilon)\mu = \nabla(\varepsilon \oplus \mathrm{id}) = \varepsilon.$$

If we view ∇^3 as $\nabla^2(\mathrm{id} \oplus \nabla^2)$ and replace $(\mu \oplus \mathrm{id})\mu$ by $(\mathrm{id} \oplus \mu)\mu$ we proceed to

$$\cdots = \nabla(\varepsilon\gamma \oplus \varepsilon)\mu = \nabla(\varepsilon\gamma \oplus \mathrm{id})(\mathrm{id} \oplus \varepsilon)\mu = \varepsilon\gamma.$$

(3): Here we obtain the result by a somewhat lengthier computation of

$$\widetilde{\nabla}(\widetilde{\nabla} \oplus \mathrm{id} \oplus \mathrm{id})(\mu\gamma \oplus \mu \oplus \tau(\gamma \oplus \gamma)\mu)(\mathrm{id} \oplus \mu)\mu$$

$(\widetilde{\nabla} \colon L \oplus L \oplus L \oplus L \to L \oplus L$ is the codiagonal) and the equivalent

$$\widetilde{\nabla}(\mathrm{id} \oplus \mathrm{id} \oplus \widetilde{\nabla})(\mu\gamma \oplus \mu \oplus \tau(\gamma \oplus \gamma)\mu)(\mu \oplus \mathrm{id})\mu. \qquad \square$$

1.4.2. Remark. Identity (1) corresponds to the classical identity $(x^{-1})^{-1} = x$ while identity (2) corresponds to $e^{-1} = e$ and (3) to $(xy)^{-1} = y^{-1}x^{-1}$.

1.5. The left and right system of covers. Set

$$N_L = \{a \in L \mid \varepsilon(a) = 1\}$$

(briefly, N). This is in fact the neighbourhood filter of the unit ε. One has

(1) $a, b \in N \Rightarrow a \wedge b \in N$, and

(2) $a \in N \Rightarrow \gamma(a) \in N$ (by the equation $\gamma\varepsilon = \varepsilon$).

For each $a \in N$ set

$$U(a) = \{x \in L \mid x \oplus \gamma(x) \le \mu(a)\}, \quad \mathcal{U} = \{U(a) \mid \varepsilon(a) = 1\}, \quad \text{and}$$
$$V(a) = \{x \in L \mid \gamma(x) \oplus x \le \mu(a)\}, \quad \mathcal{V} = \{V(a) \mid \varepsilon(a) = 1\}.$$

1.5.1. Lemma. *We have*

$$U(a) = \{x \wedge y \mid x \oplus y \le (\mathrm{id} \oplus \gamma)\mu(a)\}$$

and similarly

$$V(a) = \{x \wedge y \mid x \oplus y \le (\gamma \oplus \mathrm{id})\mu(a)\}.$$

Proof. If $x \oplus \gamma(x) \le \mu(a)$ then $x \oplus x \le (\mathrm{id} \oplus \gamma)(x \oplus \gamma(x)) \le (\mathrm{id} \oplus \gamma)\mu(a)$. On the other hand, if $x \oplus y \le (\mathrm{id} \oplus \gamma)\mu(a)$ then $(x \wedge y) \oplus \gamma(x \wedge y) \le (\mathrm{id} \oplus \gamma)(x \oplus y) \le (\mathrm{id} \oplus \gamma)(\mathrm{id} \oplus \gamma)\mu(a) = \mu(a)$. $\square$

1.5.2. Lemma. *If $u \oplus v \le \mu(x)$ then $U(u)v \le x$. More generally, if M is an arbitrary frame and $w \in M$ then for any $x \in M \oplus L$ (resp. $x \in L \oplus M$), if*

$$w \oplus u \oplus v \le (\mathrm{id}_M \oplus \mu)(x) \quad (\text{resp.} \quad u \oplus v \oplus w \le (\mu \oplus \mathrm{id}_M)(x))$$

then

$$w \oplus U(u)v \le x \quad (\text{resp.} \quad V(v)u \oplus w \le x).$$

Proof. It suffices to prove the more general statement. For a $t \in U(u)$ we have $t \oplus \gamma(t) \le \mu(u)$ and hence

$$w \oplus t \oplus \gamma(t) \oplus v \le w \oplus \mu(u) \oplus v = (\mathrm{id}_M \oplus \mu \oplus \mathrm{id}_L)(w \oplus u \oplus v)$$

$$\le (\mathrm{id}_M \oplus \mu \oplus \mathrm{id}_L)(\mathrm{id}_M \oplus \mu)(x) = (\mathrm{id}_M \oplus (\mu \oplus \mathrm{id}_L)\mu)(x).$$

Applying $\mathrm{id}_M \oplus \mathrm{id}_L \oplus \nabla(\gamma \oplus \mathrm{id}_L)$ to the leftmost and to the rightmost expression we obtain in the first case

$$(\mathrm{id}_M \oplus \mathrm{id}_L \oplus \nabla(\gamma \oplus \mathrm{id}_L))(w \oplus t \oplus \gamma(t) \oplus v) = w \oplus t \oplus (t \wedge v)$$

and in the latter one

$$
\begin{aligned}
(\mathrm{id}_M \oplus \mathrm{id}_L &\oplus \nabla(\gamma \oplus \mathrm{id}_L))(\mathrm{id}_M \oplus (\mu \oplus \mathrm{id}_L)\mu)(x) \\
&= (\mathrm{id}_M \oplus ((\mathrm{id}_L \oplus \nabla(\gamma \oplus \mathrm{id}_L))(\mu \oplus \mathrm{id}_L)\mu))(x) \\
&= (\mathrm{id}_M \oplus ((\mathrm{id}_L \oplus \nabla(\gamma \oplus \mathrm{id}_L))(\mathrm{id}_L \oplus \mu)\mu))(x) \\
&= (\mathrm{id}_M \oplus ((\mathrm{id}_L \oplus \nabla(\gamma \oplus \mathrm{id}_L)\mu)\mu))(x) = (\mathrm{id}_M \oplus ((\mathrm{id}_L \oplus \sigma_L \varepsilon)\mu))(x) \\
&= (\mathrm{id}_M \oplus ((\mathrm{id}_L \oplus \sigma_L)(\mathrm{id}_L \oplus \varepsilon)\mu))(x) = (\mathrm{id}_M \oplus \mathrm{id}_L \oplus \sigma_L)(x) = x \oplus 1.
\end{aligned}
$$

Hence, if $t \wedge v \neq 0$ we obtain $w \oplus t \leq x$, and finally $w \oplus U(u)v \leq x$.   □

1.5.3. For an $x \in L$ set

$$B(x) = \{v \mid \exists u, \; \varepsilon(u) = 1, u \oplus v \leq \mu(x)\}.$$

Lemma. $x = \bigvee B(x)$.

Proof. We have $x = (\varepsilon \oplus \mathrm{id})\mu(x)$ and $\mu(x) = \bigvee\{u \oplus v \mid u \oplus v \leq \mu(x)\}$ and hence

$$x = \bigvee\{\varepsilon(u) \oplus v \mid u \oplus v \leq \mu(x)\} = \bigvee B(x)$$

(recall that in the convention of 1.2, $\varepsilon(u) \oplus v = v$ if $\varepsilon(u) = 1$, and 0 otherwise).   □

1.5.4. Proposition. $\mathcal{U}$ *and* $\mathcal{V}$ *are admissible systems of covers.*

Proof. First, they are covers. We have

$$
\begin{aligned}
\bigvee U(a) &= \bigvee\{x \wedge y \mid x \oplus y \leq (\mathrm{id} \oplus \gamma)\mu(a)\} \\
&= \bigvee\{\nabla(x \oplus y) \mid x \oplus y \leq (\mathrm{id} \oplus \gamma)\mu(a)\} \\
&= \nabla\bigvee\{x \oplus y \mid x \oplus y \leq (\mathrm{id} \oplus \gamma)\mu(a)\} = \nabla(\mathrm{id} \oplus \gamma)\mu(a) = \sigma\varepsilon(a) = 1.
\end{aligned}
$$

Now by 1.5.2, if $v \in B(x)$, that is, if there is a u such that $\varepsilon(u) = 1$ and $u \oplus v \leq \mu(x)$ then $v \lhd_\mathcal{U} x$. Use 1.5.3.   □

1.5.5. From 1.5.4 we obtain

Corollary. *If* $(L, \mu, \gamma, \varepsilon)$ *is a localic group then* L *is regular.*   □

1.5.6. Remark. The regularity is what we need at the moment. In fact one has much more: the systems of covers $\mathcal{U}$ and $\mathcal{V}$ generate uniformities (and hence the frame is completely regular). This will be discussed later.

2. The category of localic groups

2.1. Let $(L,\mu,\gamma,\varepsilon),(M,\mu',\gamma',\varepsilon')$ be localic groups. A frame homomorphism $h\colon L \to M$ is said to be a (*localic*) *group homomorphism* if

$$\mu' \cdot h = (h \oplus h) \cdot \mu, \quad \gamma' \cdot h = h \cdot \gamma, \quad \text{and} \quad \varepsilon' \cdot h = \varepsilon,$$

in other words, if the diagrams

$$
\begin{array}{ccc}
L & \xrightarrow{\ h\ } & M \\
{\scriptstyle\mu}\downarrow & & \downarrow{\scriptstyle\mu'} \\
L \oplus L & \xrightarrow[h\oplus h]{} & M \oplus M
\end{array}
\quad,\quad
\begin{array}{ccc}
L & \xrightarrow{\ h\ } & M \\
{\scriptstyle\gamma}\downarrow & & \downarrow{\scriptstyle\gamma'} \\
L & \xrightarrow[\ h\ ]{} & M
\end{array}
\quad \text{and} \quad
\begin{array}{ccc}
L & \xrightarrow{\ h\ } & M \\
{\scriptstyle\varepsilon}\downarrow & & \downarrow{\scriptstyle\varepsilon'} \\
2 & = \!\!= & 2
\end{array}
$$

commute.

2.2. For a frame homomorphism $h\colon L \to M$ set

$$c(h) = \bigvee \{x \mid h(x) = 0\}.$$

We have $h(c(h)) = 0$ and hence there is a frame homomorphism

$$\overline{h}\colon L \to \uparrow c(h)$$

defined by $\overline{h}(x) = x \vee c(h)$, obviously an onto one.

Furthermore, if we define $h'\colon \uparrow c(h) \to M$ by setting $h'(x) = x$ we obtain a frame homomorphism (note that $0_{\uparrow c(h)} = c(h)$, and $h(c(h)) = 0$). If $h'(x) = 0$ then $h(x) = 0$ and $x \le c(h)$, that is, $x = c(h) = 0_{\uparrow c(h)}$. Thus,

$h'\colon \uparrow c(h) \to M$ *is a dense homomorphism, and* $h = h'\overline{h}$.

Note. Compare this decomposition with the (extremal epi, one-one) factorization in IV.1.6.1. Here we have something else, namely a

(*closed extremal epi, dense*) *factorization.*

2.2.1. Let $(L,\mu,\gamma,\varepsilon)$ and $(M,\mu',\gamma',\varepsilon')$ be localic groups, and h a group homomorphism. Consider the diagram

$$
\begin{array}{ccccc}
L & \xrightarrow{\ \overline{h}\ } & \uparrow c & \xrightarrow{\ h'\ } & M \\
{\scriptstyle\mu}\downarrow & & & & \downarrow{\scriptstyle\mu'} \\
L \oplus L & \xrightarrow[\overline{h}\oplus\overline{h}]{} & \uparrow c \oplus \uparrow c = \uparrow(c \oplus c) & \xrightarrow[h'\oplus h']{} & M \oplus M
\end{array}
$$

We can define $\overline{\mu} \colon \uparrow c \to \uparrow c \oplus \uparrow c$ by setting

$$\overline{\mu}(x) = \mu(x) \quad (\text{for } x \geq c(h)).$$

Thus $\overline{\mu}$ is obviously a homomorphism and we have, in fact $\overline{\mu} \overline{h} = (\overline{h} \oplus \overline{h})\mu$. But we also have

$$\mu' h' = (h' \oplus h')\overline{\mu}$$

since $\mu' h' \overline{h} = \mu' h = (h' \oplus h')(\overline{h} \oplus \overline{h})\mu = (h' \oplus h')\overline{\mu}\overline{h}$, and $\overline{h}$ is onto.

   Similarly we can define $\overline{\gamma} \colon \uparrow c \to \uparrow c$ by $\overline{\gamma}(x) = \gamma(x)$ (that is, $\overline{\gamma}\overline{h} = \overline{h}\gamma$) and $\overline{\varepsilon} \colon \uparrow c \to \mathbf{2}$ by $\overline{\varepsilon}\overline{h} = \varepsilon$. Thus, we have obtained a system $(\uparrow c, \overline{\mu}, \overline{\gamma}, \overline{\varepsilon})$ which is easily shown to be a localic group (recall that by V.5.6, dense homomorphisms are monomorphisms in the category of regular frames, and hence we can check the equalities $\varphi = \psi$ as $h'\varphi = h'\varphi$ resp. $(h' \oplus h')\varphi = (h' \oplus h')\varphi$). Summarizing we obtain

2.2.2. Proposition. *Each group homomorphism $h \colon (L, \mu, \gamma, \varepsilon) \to (M, \mu', \gamma', \varepsilon')$ can be decomposed as*

$$(L, \mu, \gamma, \varepsilon) \xrightarrow{\ \overline{h}\ } (\uparrow \overline{c}, \overline{\mu}, \overline{\gamma}, \overline{\varepsilon}) \xrightarrow{\ h'\ } (M, \mu', \gamma', \varepsilon')$$

with a closed onto group homomorphism $\overline{h}$ and a dense group homomorphism h'. $\square$

2.3. Note that the following facts hold generally for varieties of algebras in categories; we will discuss them only in our special case.

2.3.1. Lemma. *Let L_i, $i \in J$, be frames, let $L = \bigoplus_{i \in J} L_i$ with coproduct injections $\iota_i \colon L_i \to L$. Then $(\iota_i \oplus \iota_i \colon L_i \oplus L_i \to L \oplus L)_{i \in J}$, is a coproduct.*

Proof. Denote by $\iota_{ij} \colon L_i \to L_i \oplus L_i$, $\tau_j \colon L \to L \oplus L$, $i \in J$, $j = 1, 2$, the coproduct injections, so that the diagrams

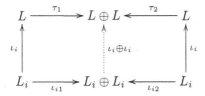

commute (i.e., $(\iota_i \oplus \iota_i) \cdot \iota_{ij} = \tau_j \cdot \iota_i$). Now let $\varphi_i \colon L_i \oplus L_i \to M$, $i \in J$, be frame homomorphisms. Define $\psi_j \colon L \to M$, $j = 1, 2$, and $\varphi \colon L \oplus L \to M$ by

$$\psi_j \cdot \iota_i = \varphi_i \cdot \iota_{ij}, \quad \varphi \cdot \tau_j = \psi_j \ (i \in J, j = 1, 2).$$

Then $\varphi \cdot (\iota_i \oplus \iota_i) \cdot \iota_{ij} = \varphi \cdot \tau_j \cdot \iota_i = \psi_j \cdot \iota_i = \varphi_i \cdot \iota_{ij}$ and hence $\varphi \cdot (\iota_i \oplus \iota_i) = \varphi_i$. The unicity is obvious. $\square$

2.3.2. Coproducts. Consider a system $(L_i, \mu_i, \gamma_i, \varepsilon_i)$, $i \in J$, of localic groups. In the coproduct $L = \bigoplus_{i \in J} L_i$ with coproduct injections ι_i, define operations

$$\mu \colon L \to L \oplus L, \quad \gamma \colon L \to L, \quad \varepsilon \colon L \to 2$$

by setting

$$\mu \cdot \iota_i = (\iota_i \oplus \iota_i) \cdot \mu_i, \quad \gamma \cdot \iota_i = \iota_i \cdot \gamma_i \text{ and } \varepsilon \cdot \iota_i = \varepsilon_i. \tag{2.3.1}$$

Checking that μ, γ, ε satisfy the formulas from 1.1 is straightforward.

Proposition. $\iota_i \colon (L_i, \mu_i, \gamma_i, \varepsilon_i) \to (L, \mu, \gamma, \varepsilon)$ are group homomorphisms and the system $(\iota_i)_{i \in J}$ is a coproduct in the category of localic groups and group homomorphisms.

Proof. The injections ι_i are group homomorphisms in the sense of 2.1 by the definition (2.3.1). Now let $h_i \colon (L_i, \mu_i, \gamma_i, \varepsilon_i) \to (M, \mu', \gamma', \varepsilon')$ be group homomorphisms. Define $h \colon L \to M$ by $h\iota_i = h_i$. We have

$$\mu' h \iota_i = \mu' h_i = (h_i \oplus h_i)\mu_i = (h \oplus h)(\iota_i \oplus \iota_i)\mu_i = (h \oplus h)\mu\iota_i,$$

hence $\mu' h = (h \oplus h)\mu$; $\gamma' h \iota_i = \gamma' h_i = h_i \gamma_i = h\iota_i \gamma_i = h\gamma\iota_i$, hence $\gamma' h = h\gamma$, and finally, $\varepsilon' h \iota_i = \varepsilon' h_i = \varepsilon_i = \varepsilon\iota_i$ and $\varepsilon' h = \varepsilon$. Thus, h is a group homomorphism. $\quad\square$

2.3.3. Coequalizers. Let $f, g \colon (M, \mu', \gamma', \varepsilon') \to (L, \mu, \gamma, \varepsilon)$ be group homomorphisms. Consider the coequalizer $h \colon L \to \overline{L}$ of f, g. We have

$$(h \oplus h)\mu f = (h \oplus h)(f \oplus f)\mu' = (h \oplus h)(g \oplus g)\mu' = (h \oplus h)\mu g$$

and hence there is a $\overline{\mu} \colon \overline{L} \to \overline{L} \oplus \overline{L}$ such that $\overline{\mu} h = (h \oplus h)\mu$.

Similarly, $h\gamma f = hf\gamma' = hg\gamma' = h\gamma g$ and hence there is a $\overline{\gamma} \colon \overline{L} \to \overline{L}$ such that $\overline{\gamma} h = h\gamma$, and since $\varepsilon f = \varepsilon' = \varepsilon g$ we also have $\overline{\varepsilon} \colon L \to 2$ such that $\overline{\varepsilon} h = \varepsilon$. By these definitions, h is a homomorphism $(L, \mu, \gamma, \varepsilon) \to (\overline{L}, \overline{\mu}, \overline{\gamma}, \overline{\varepsilon})$; it is easy to see that $(\overline{L}, \overline{\mu}, \overline{\gamma}, \overline{\varepsilon})$ is a localic group.

Proposition. *Just defined $h \colon (L, \mu, \gamma, \varepsilon) \to (\overline{L}, \overline{\mu}, \overline{\gamma}, \overline{\varepsilon})$ is a coequalizer in the category of localic groups and group homomorphisms.*

Proof. Let $k \colon (L, \mu, \gamma, \varepsilon) \to (L', \mu', \gamma', \varepsilon')$ be a group homomorphism such that $kf = kg$. Thus we have a frame homomorphism $\overline{k} \colon \overline{L} \to L'$ such that $\overline{k} h = k$. Now we have

$$\mu' \overline{k} h = \mu' k = (k \oplus k)\mu = (\overline{k} \oplus \overline{k})(h \oplus h)\mu = (\overline{k} \oplus \overline{k})\overline{\mu} h$$

and since h is onto, $\mu' \overline{k} = (\overline{k} \oplus \overline{k})\overline{\mu}$. Similarly, $\gamma' \overline{k} = \overline{k} \overline{\gamma}$ and $\varepsilon' \overline{k} = \varepsilon$. Thus, $\overline{k}$ is a group homomorphism. $\quad\square$

2.3.4. Equalizers. Let $f, g\colon (L, \mu, \gamma, \varepsilon) \to (M, \mu', \gamma', \varepsilon')$ be group homomorphisms. Since our frames are regular and the category of regular frames is well powered (V.6.6.1) there is a set $(h_i\colon (L_i, \mu_i, \gamma_i, \varepsilon_i) \to (L, \mu, \gamma, \varepsilon))$ such that each dense group homomorphism $h\colon (\overline{L}, \overline{\mu}, \overline{\gamma}, \overline{\varepsilon}) \to (L, \mu, \gamma, \varepsilon)$ is represented, up to isomorphism, by some of the h_i.

Now consider the coproduct $\iota_i\colon (L_i, \mu_i, \gamma_i, \varepsilon_i) \to (\widetilde{L}, \widetilde{\mu}, \widetilde{\gamma}, \widetilde{\varepsilon})$ in the category of localic groups (2.3.2) of all the L_i such that $fh_i = gh_i$. Then we have the homomorphism $h\colon \widetilde{L} \to L$ such that $h \cdot \iota_i = h_i$. Decomposing it as in 2.2.2 we obtain the equalizer as the h'.

2.3.5. Summarizing the facts above and using the dual of AII.5.3.2 we obtain

Theorem. *The category of localic groups and group homomorphisms is cocomplete and has equalizers.* □

3. Closed Subgroup Theorem

3.1. Lemma. *Let $h\colon L \to M$ be a dense frame homomorphism. If $h(x) = h(y)$ then $x^{**} = y^{**}$.*

Proof. Since $h(x \wedge y^*) = h(x) \wedge h(y^*) = h(y) \wedge h(y^*) = h(y \wedge y^*) = 0$, we have $x \wedge y^* = 0$. Thus, $y^* \leq x^*$, hence $x^{**} \leq y^{**}$ and by symmetry also $y^{**} \leq x^{**}$. □

3.2. Lemma. *In a localic group $(L, \mu, \gamma, \varepsilon)$ holds the implication*

$$x \oplus y \leq \mu(a) \quad \Rightarrow \quad x^{**} \oplus y^{**} \leq \mu(a).$$

Proof. For any cover A of L we have $a^{**} \leq Aa$ (since $Aa = Aa^{**}$, recall VIII.2.3.2).

Now let $x \oplus y \leq \mu(a)$. For $v \in B(y)$ (recall 1.5.3) there is a u with $\varepsilon(u) = 1$ and $u \oplus v \leq \mu(y)$. Thus we have

$$x \oplus u \oplus v \leq (\mathrm{id} \oplus \mu)(x \oplus y) \leq (\mathrm{id} \oplus \mu)\mu(a) = (\mu \oplus \mathrm{id})\mu(a)$$

and hence by 1.5.2 $V(u)x \oplus v \leq \mu(a)$ and consequently $x^{**} \oplus v \leq \mu(a)$ and finally

$$x^{**} \oplus y = x^{**} \oplus \bigvee B(y) = \bigvee \{x^{**} \oplus v \mid v \in B(y)\} \leq \mu(a).$$

Repeating the procedure we obtain $x^{**} \oplus y^{**} \leq \mu(a)$. □

3.3. Lemma. *Let $(L, \mu, \gamma, \varepsilon)$ be a localic group, let M be a frame, and let $h\colon L \to M$ be a dense onto frame homomorphism. Let $0 \neq y_1 \oplus y_2 \leq (h \oplus h)\mu(a)$. Then there are $x_i \in L$ such that $y_i = h(x_i)$ and $x_1 \oplus x_2 \leq \mu(a)$.*

Proof. Consider the subset

$$X = \{(h(u_1), h(u_2)) \mid u_1 \oplus u_2 \le \mu(a)\} \subseteq M \times M.$$

We will prove it is saturated, and hence an element of $M \oplus M$ (recall IV.4.3).

If $(w_1, w_2) \le (h(u_1), h(u_2))$ choose t_i such that $w_i = h(t_i)$; then $t_i \wedge u_i \le u_i$ and $h(t_i \wedge u_i) = w_i$; hence X is a down-set. Now let $(w_i, w) \in X$, $i \in J$. There are $t_{i,1}, t_{i,2} \in L$ such that $w_i = h(t_{i,1})$, $w = h(t_{i,2})$ and $t_{i,1} \oplus t_{i,2} \le \mu(a)$. We can assume that $J \ne \emptyset$ as obviously $(0, y)$ is in X for any y. Choose a fixed $k \in J$ and set $t = t_{k,2}$. Then by 3.1 $t^{**} = t_{i,2}^{**}$ for each $i \in J$ and we have, by 3.2, $t_{i,1} \oplus t \le t_{i,1}^{**} \oplus t_{i,2}^{**} \le \mu(a)$. Then $(\bigvee_{i \in J} t_{i,1}) \oplus t \le \mu(a)$, and thus $(\bigvee_{i \in J} w_i, w) \in X$. Now $y_1 \oplus y_2 \le X$ and hence $(y_1, y_2) \in X$. $\square$

3.4. Proposition. *Let a localic group homomorphism*

$$h: (L, \mu_L, \gamma_L, \varepsilon_L) \to (M, \mu_M, \gamma_M, \varepsilon_M)$$

be onto and dense. Then it is an isomorphism.

Proof. Suppose not. Then there are $u_2 \not\le u_1$ such that $h(u_1) = h(u_2)$. By regularity there is a $v \prec u_2$ such that $v \not\le u_1$ so that $a = v^* \vee u_1 \ne 1$ and $v^* \vee u_2 = 1$. We have $h(a) = h(v^*) \vee h(u_1) = h(v^*) \vee h(u_2) = 1$ and hence

$$(h \oplus h)\mu_L(a) = \mu_M h(a) = 1 = 1 \oplus 1$$

and hence, by 3.3, there are $x_i \in L$ such that $h(x_i) = 1$ and $x_1 \oplus x_2 \le \mu(a)$. Now $\varepsilon_L(x_1) = \varepsilon_M h(x_1) = 1$ and we can apply 1.5.2 to obtain that $x_2 \lhd_{\mathcal{U}} a$, and hence $x_2^* \vee a = 1$. As $a \ne 1$, $x_2^* \ne 0$. This yields a contradiction with the density: $h(x_2^*) = h(x_2^*) \wedge h(x_2) = h(x_2^* \wedge x_2) = 0$. $\square$

3.5. As a corollary we now obtain from the decomposition formula in 2.2.2

Theorem. *Each subgroup of a localic group is closed.* $\square$

4. The multiplication μ is open.
The semigroup of open parts

4.1. Lemma. *There is an isomorphism* $\alpha: L \oplus L \to L \oplus L$ *such that* $\alpha \iota_1 = \mu$ *and* $\alpha\alpha = \mathrm{id}$.

Proof. ($\iota_i: L \to L \oplus L$, $i = 1, 2$ are, again, the coproduct injections). First note that if $L \oplus L \oplus L$ is viewed as the product $(L \oplus L) \oplus L$ then the first coproduct injection $\iota_1^{L \oplus L, L}$ is $\mathrm{id} \oplus \iota_1$ (because it sends $x \oplus y \in L \oplus L$ to $x \oplus y \oplus 1$) and the second, $\iota_2^{L\oplus L,L}$, is $(\iota_2 \oplus \mathrm{id})\iota_2$ (sending z to $1 \oplus 1 \oplus z$). Similarly if we view $L \oplus L \oplus L$ as $L \oplus (L \oplus L)$ we have

$$\iota_1^{L,L\oplus L} = (\mathrm{id} \oplus \iota_1)\iota_1 \quad \text{and} \quad \iota_2^{L,L\oplus L} = \iota_2 \oplus \mathrm{id}.$$

Set

$$\alpha = (L \oplus L \xrightarrow{\mu \oplus \gamma} L \oplus L \oplus L \xrightarrow{\mathrm{id}\oplus\nabla} L \oplus L).$$

We have

$$\alpha\iota_1 = (\mathrm{id} \oplus \nabla)(\mu \oplus \gamma)\iota_1 = (\mathrm{id} \oplus \nabla)(\mathrm{id} \oplus \iota_1)\mu = \mu, \text{ and}$$

$$\alpha\iota_2 = (\mathrm{id} \oplus \nabla)(\iota_2 \oplus \mathrm{id})\iota_2\gamma = \iota_2\nabla\iota_2\gamma = \iota_2\gamma.$$

Consequently further

$$\alpha\alpha\iota_1 = (\mathrm{id} \oplus \nabla)(\mu \oplus \iota)\mu = (\mathrm{id} \oplus \nabla)(\mathrm{id} \oplus \mathrm{id} \oplus \gamma)(\mu \oplus \mathrm{id})\mu$$
$$= (\mathrm{id} \oplus \nabla)(\mathrm{id} \oplus \mathrm{id} \oplus \gamma)(\mathrm{id} \oplus \mu)\mu = (\mathrm{id} \oplus (\nabla(\mathrm{id} \oplus \gamma)\mu))\mu$$
$$= (\mathrm{id} \oplus \sigma)(\mathrm{id} \oplus \varepsilon)\mu = \mathrm{id} \oplus \sigma = \iota_1,$$

$$\alpha\alpha\iota_2 = \alpha\iota_2\gamma = \iota_2\gamma\gamma = \iota_2. \qquad \square$$

4.1.1. Remark. The α above corresponds to the classical mapping

$$(x, y) \mapsto (xy, y^{-1}).$$

4.1.2. Proposition. [Johnstone] *The multiplication* $\mu\colon L \to L \oplus L$ *is an open homomorphism.*

Proof. By 1.5.5, L is regular and hence, by V.5.6(3), it suffices to show that it has a left Galois adjoint. We have, in the notation of 4.1, $\mu = \alpha\iota_1$; now α, as an isomorphism, is its own adjoint, and ι_1 has, as it is easy to check, the left adjoint

$$(\iota_1)_\#(u) = \bigvee\{x \mid \exists y \neq 0, \ x \oplus y \leq u\}.$$

Thus, we have $\mu_\# = (\iota_1)_\#\alpha$. $\qquad \square$

Note. Because of other uses of the asterisk in this chapter we will use here the symbol $\phi_\#$ for left Galois adjoints of monotone maps ϕ.

4.1.3. Since ι_1 is a monomorphism,

$$\mu \text{ is a monomorphism (in fact, one-one)}$$

and hence we have

$$\mu\mu_\# \geq \mathrm{id} \quad \text{and} \quad \mu_\#\mu = \mathrm{id}.$$

In particular

$$\mu_\#(0) = 0.$$

4.2. The semigroup operation

4.2.1. On L define a (classical) binary operation $*$ and a unary operation $(-)^{-1}$ by setting

$$x * y = \mu_\#(x \oplus y), \qquad x^{-1} = \gamma(x).$$

The algebra $(L, *, (-)^{-1})$ is the counterpart of the semigroup (with involution) of open subsets of a topological group, with the operations

$$UV = \{uv \mid u \in U, v \in V\}, \quad U^{-1} = \{u^{-1} \mid u \in U\}.$$

We have the obvious

Observation. (1) If $x' \leq x$ and $y' \leq y$ then $x' * y' \leq x * y$.

(2) If $x = 0$ or $y = 0$ then $x * y = 0$.

4.2.2. Lemma. $(\mu \oplus \mathrm{id})$ *has a left adjoint and we have*

$$(\mu \oplus \mathrm{id})_\#(a \oplus b) = \mu_\#(a) \oplus b.$$

Similarly, $(\mathrm{id} \oplus \mu)$ *has a left adjoint, and*

$$(\mathrm{id} \oplus \mu)_\#(a \oplus b) = a \oplus \mu_\#(b).$$

Proof. It is easy to check that if $E \in L \oplus L$ is saturated then the *union*

$$\bigcup \{\downarrow(\mu(x), y) \mid x \oplus y \leq E\}$$

is saturated so that $(\mu \oplus \mathrm{id})(E) = \bigcup \{\downarrow(\mu(x), y) \mid x \oplus y \leq E\}$. Hence

$$F \leq (\mu \oplus \mathrm{id})(E)$$

$$\text{iff} \quad \forall a \oplus b \leq F, \ (a, b) \in \bigcup \{\downarrow(\mu(x), y) \mid x \oplus y \leq E\}$$

$$\text{iff} \quad \forall a \oplus b \leq F \ \exists x \oplus y \leq E, \ a \leq \mu(x) \text{ and } b \leq y$$

$$\text{iff} \quad \forall a \oplus b \leq F \ \exists x \oplus y \leq E, \ \mu_\#(a) \leq x \text{ and } b \leq y$$

$$\text{iff} \quad \forall a \oplus b \leq F \ \mu_\#(a) \oplus b \leq E$$

$$\text{iff} \quad \varphi(F) = \bigvee \{a \oplus b \mid \mu_\#(a) \oplus b \leq E\} \leq E.$$

In particular $(\mu \oplus \mathrm{id})_\#(a \oplus b) = \varphi(a \oplus b) = \mu_\#(a) \oplus b$. $\qquad\square$

4.2.3. Proposition. (1) *The operation* $*$ *is associative.*

(2) $x * y \geq x \wedge \varepsilon(y), \varepsilon(x) \wedge y$. *Hence, if* $y \in N$ *then* $x * y \geq x$ *and* $y * x \geq x$.

(3) *If* $x \wedge y \neq 0$ *then* $x * y^{-1} \in N$.

(4) *We have the formula* $(x * y)^{-1} = y^{-1} * x^{-1}$.

(5) *If* $x \in N$ *then* $x^{-1} \in N$.

Proof. (1): Use 4.2.2. We have

$$a * (b * c) = \mu_\#(a \oplus \mu_\#(b \oplus c)) = \mu_\#(\mathrm{id} \oplus \mu)_\#(a \oplus b \oplus c)$$
$$= ((\mathrm{id} \oplus \mu)\mu)_\#(a \oplus b \oplus c) = ((\mu \oplus \mathrm{id})\mu)(a \oplus b \oplus c)$$
$$= \mu_\#(\mu \oplus \mathrm{id})_\#(a \oplus b \oplus c) = \mu_\#(\mu_\#(a \oplus b) \oplus c)$$
$$= (a * b) * c.$$

(2): We have $\mu\mu_\#(x \oplus y) \geq x \oplus y$. Applying $\mathrm{id} \oplus \varepsilon$ on both sides we obtain

$$x * y \geq (\mathrm{id} \oplus \varepsilon)(x \oplus y) = x \oplus \varepsilon(y) = x \wedge \varepsilon(y).$$

(3): By 4.1.3 we have

$$\sigma\varepsilon(x * y^{-1}) = \nabla(\mathrm{id} \oplus \gamma)\mu\mu_\#(\mathrm{id} \oplus \gamma)(x \oplus y)$$
$$\geq \nabla(\mathrm{id} \oplus \gamma)(\mathrm{id} \oplus \gamma)(x \oplus y) = x \wedge y \neq 0,$$

so $\varepsilon(x * y^{-1})$ cannot be 0.

(4): Since $\tau_\# = \tau$ and $\gamma_\# = \gamma$, we obtain from 1.4.1(3) that $\mu_\#(\gamma \oplus \gamma) = \gamma\mu_\#\tau$, and as $\tau(x \oplus y) = y \oplus x$ we conclude

$$y^{-1} * x^{-1} = \mu_\#(\gamma(y) \oplus \gamma(x)) = \gamma\mu_\#(x \oplus y) = (x * y)^{-1}.$$

(5) follows from 1.4.1(2). □

Here is one more fact about the multiplication.

4.2.4. Lemma. *For each $a \in N$ there are $b, c \in N$ such that $b * b \leq a$ and $c * c^{-1} \leq a$.*

Proof. Any L can be viewed as a coproduct

$$\mathbf{2} \xrightarrow{\ \sigma_L\ } L \xleftarrow{\ \mathrm{id}_L\ } L$$

and since $\sigma_\mathbf{2} = \mathrm{id}_\mathbf{2}$ we have

$$\varepsilon = \mathrm{id}_\mathbf{2} \oplus \varepsilon \colon L = \mathbf{2} \oplus L \to \mathbf{2} \oplus \mathbf{2} = \mathbf{2}.$$

Hence, $\varepsilon = (\mathrm{id}_\mathbf{2} \oplus \varepsilon)(\varepsilon \oplus \mathrm{id}_L)\mu = (\varepsilon \oplus \varepsilon)\mu$ and we obtain, for $a \in N$,

$$1 = \varepsilon(a) = \bigvee\{\varepsilon(x) \oplus \varepsilon(y) \mid x \oplus y \leq \mu(a)\} = \bigvee\{\varepsilon(x) \oplus \varepsilon(y) \mid x * y \leq a\}$$

so that there are x, y such that $x * y \leq a$ and $\varepsilon(x) = \varepsilon(y) = 1$. Set $b = x \wedge y$ and $c = x \wedge \gamma(y)$. □

5. Uniformities

5.1. The covers from 1.5 can be now rewritten using the operation $*$ as

$$U(a) = \{x \in L \mid x * x^{-1} \leq a\}, \quad \text{and}$$

$$V(a) = \{x \in L \mid x^{-1} * x \leq a\}.$$

This will be used in this subsection to show that the systems $\mathcal{U}$ and $\mathcal{V}$ are subbases of uniformities.

5.1.1. Proposition. *The systems $\mathcal{U}$ and $\mathcal{V}$ generate uniformities on L.*

Proof. We know already from 1.5.4 that $\mathcal{U}$ is an admissible system of covers. Trivially $U(a \wedge b) \leq U(a) \wedge U(b)$.

For $a \in N$ choose, by 4.2.4, a $b \in N$ such that $b * b * b^{-1} * b^{-1} \leq a$. We will show that $U(b)U(b) \leq U(a)$.

Fix an $x \in U(b)$ and consider any $u \in U(b)$ such that $u \wedge x \neq 0$. Thus, $x * x^{-1} \leq b$ and $u * u^{-1} \leq b$ and, by 4.2.3, $(u \wedge x)^{-1} * (u \wedge x) \in N$. Thus,

$$u \leq u * (u \wedge x)^{-1} * (u \wedge x) \leq u * u^{-1} * x \leq b * x$$

and hence $U(b)x \leq b * x$ and finally, since also $b^{-1} \in N$, again by 4.2.3,

$$U(b)x * (U(b)x)^{-1} \leq b * x * x^{-1} * b^{-1} \leq b * b * b^{-1} \leq b * b * b^{-1} * b^{-1} \leq a$$

and $U(b)x \in U(a)$.   □

5.1.2. The uniformity $\mathcal{U}$ (resp. $\mathcal{V}$) is called the *left uniformity* (resp. *right uniformity*) on the localic group.

5.1.3. Corollary. *If $(L, \mu, \gamma, \varepsilon)$ is a localic group then L is completely regular.*   □

5.2. Alternative description of the uniformities via entourages. For an $a \in N$ set

$$E(a) = (\mathrm{id} \oplus \gamma)\mu(a) \quad \text{and} \quad F(a) = (\gamma \oplus \mathrm{id})\mu(a);$$

hence

$$E(a) = \bigvee\{x \oplus y \mid x \oplus y \leq (\mathrm{id} \oplus \gamma)\mu(a)\} = \bigvee\{x \oplus y \mid x * y^{-1} \leq a\},$$

and similarly for $F(a)$. Note that by 4.2.3(4), $E(a)^{-1} = \bigvee\{x \oplus y \mid y * x^{-1} \leq a\} = \bigvee\{x \oplus y \mid x * y^{-1} \leq a^{-1}\} = E(a^{-1})$. Recall XII.1.

5.2.1. Lemma. $E(a)$ *and* $F(a)$ *are entourages of* L.

Proof. For $x \in U(a)$ we have $x \oplus \gamma(x) \leq \mu(a)$. Hence

$$\bigvee\{x \mid x \oplus x \leq E(a)\} \geq \bigvee U(a) = 1$$

since $U(a)$ is a cover, as we already know.   □

5.2.2. Denote by $\mathcal{E}'$ (resp. $\mathcal{F}'$) the system of entourages $\{E(a) \mid a \in N\}$ (resp. $\{F(a) \mid a \in N\}$), and set

$$\mathcal{E} = \{E \mid E \text{ entourage}, E \geq E(a) \in \mathcal{E}'\},$$

$$\mathcal{F} = \{E \mid E \text{ entourage}, E \geq F(a) \in \mathcal{F}'\}.$$

Recall the equivalence of the descriptions of uniformities by entourages and by covers proved in Chapter XII. For the $\mathcal{U}$ and $\mathcal{V}$ of 1.5 and in the notation of XII.3.3 we have

Proposition. $\mathcal{E} = \mathcal{E}_{\mathcal{U}}$ and $\mathcal{F} = \mathcal{E}_{\mathcal{V}}$; therefore, the systems $\mathcal{E}$ and $\mathcal{F}$ are entourage uniformities.

Proof. We will show that

$$\mathcal{E} = \{E \mid E \text{ entourage}, E \geq E(a) \in \mathcal{E}'\}$$

$$= \mathcal{E}_{\mathcal{U}} = \{E \mid E \text{ entourage}, E \geq E_{U(a)}, a \in N\}.$$

We have $E_{U(a)} (= \bigvee\{x \oplus x \mid x \oplus \gamma(x) \leq \mu(a)\}) \leq E(a)$.

To obtain an estimate from the other side, choose by 4.2.4 $b, c \in N$ such that $b * b^{-1} \leq c$ and $c * c^{-1} \leq a$. Let $x \oplus y \leq E(b)$. We can assume $x \oplus y \neq 0$, hence $x \neq 0 \neq y$. First, as $y \neq 0$, we have by 4.2.3 ((2), (3) and (5)),

$$x * x^{-1} \leq x * y^{-1} * y * x^{-1} \leq b * b^{-1} \leq c \quad \text{and} \quad x * y^{-1} \leq b * b^{-1} \leq c$$

and hence $(x, x), (x, y) \in E(c)$ and since $E(c)$ is saturated we have, for $z = x \vee y$, $(x, z) \in E(c)$, that is, $x * z^{-1} \leq c$. Now $(x * z^{-1}) * (x * z^{-1})^{-1} \leq c * c^{-1} \leq a$, hence

$$(x * z^{-1}) \oplus (x * z^{-1}) \leq E_{U(a)} \quad \text{and} \quad (x * z^{-1}) * (x * z^{-1})^{-1} \leq \mu(E_{U(a)}).$$

Since $x \wedge z \neq 0$ we have $(x * z^{-1})^{-1} \in N$ by 4.2.3(3), and by 4.2.3(2) we obtain

$$x * z^{-1} \leq x * z^{-1} * (x * z^{-1})^{-1} \leq \mu(E_{U(a)})$$

so that $x \oplus y \leq x \oplus z \leq E_{U(a)}$. Thus, $E(b) \leq E_{U(a)}$. $\square$

5.2.3. The two-sided uniformity. The common refinement $\mathcal{E} \wedge \mathcal{F}$ of $\mathcal{E}$ and $\mathcal{F}$ is referred to as the

two-sided uniformity on L.

In general, the three uniformities are different; however, by 1.4.1(1)

$$(\gamma \oplus \gamma)[E(a)] = F(a)$$

and hence γ is a uniform isomorphism from L with its left uniformity to L with its right uniformity.

Remark. In the *commutative case* (i.e., $\mu\tau = \mu$), the three uniformities coincide because $x \oplus y \leq \mu(a)$ iff $y \oplus x \leq \mu(a)$. Indeed,

$$x \oplus y \leq \mu(a) \Rightarrow \tau(x \oplus y) \leq \tau\mu(a) = \mu(a),$$

that is $y \oplus x \leq \mu(a)$; therefore

$$E(a)^{-1} = \bigvee\{(x \oplus \gamma(y))^{-1} \mid x \oplus y \leq \mu(a)\}$$

$$= \bigvee\{\gamma(y) \oplus x \mid y \oplus x \leq \mu(a)\} = F(a).$$

5.2.4. Proposition. *Any group homomorphism $h\colon L \to M$ is uniform in any of the uniformities.*

Proof. The proof will be done for the left uniformity $\mathcal{E}$. We need to show that for every $a \in N_L$, the image $(h \oplus h)(E(a))$ of a uniform entourage $E(a) \in \mathcal{E}_L$ belongs to the uniformity $\mathcal{E}_M$ of M. But by the definition of a group homomorphism, we have

$$(h \oplus h)(E(a)) = (h \oplus h)(\mathrm{id}_L \oplus \gamma_L)\mu_L(a)$$

$$= (\mathrm{id}_M \oplus \gamma_M)(h \oplus h)\mu_L(a)$$

$$= (\mathrm{id}_M \oplus \gamma_M)\mu_M(h(a)) = E(h(a))$$

(since $\varepsilon_M h = \varepsilon_L$, $h(a) \in N_M$, and $E(h(a))$ makes sense). $\qquad\square$

5.2.5. Remarks. (1) Note that we did not have to prove that $\mathcal{E}$ is a uniformity. It followed from the fact that $\mathcal{E}_\mathcal{U}$ is one.

(2) Proving that h is uniform in terms of the covering uniformities would be considerably harder. Here we have an instance of a considerable advantage of the entourage approach. Note that the equivalence of the two is a somewhat stronger fact than the corresponding classical equivalence: the coproduct involved does not quite correspond to the classical product.

6. Notes

6.1. LT-groups. As we have already mentioned in 1.3, a topological group is not necessarily a localic one, that is, it may not always be translated into a localic group by the functor Ω. Let (X, m, i, e) be a topological group. Denote by p_i, $i = 1, 2$, the product projections $p_i\colon X \times X \to X$. One has the homomorphism

$$\pi\colon \Omega(X) \oplus \Omega(X) \to \Omega(X)$$

defined by $\pi\iota_i = \Omega(p_i)$. It may be an isomorphism and it may be not (although it is always a very special type of homomorphism – see IV.5.4). If it is, the group X immediately transforms into a localic one. Recall that by IV.5.4.2 this happens whenever $\Omega(X) \oplus \Omega(X)$ is spatial.

This concerns, e.g.,

- the locally compact groups (see [142]),
- the complete metric groups (see [142]), or
- the Čech complete groups (see [142]; a space is *Čech complete* if it is G_δ – that is, an intersection of countably many open subsets – in its Stone-Čech compactification – see also VII.7).

In fact one does not really need the homomorphism π above to be an isomorphism. It suffices that the original group operation m lifts to the commutative diagram

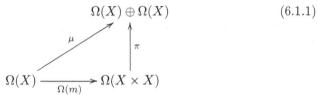

$$\tag{6.1.1}$$

Whenever a topological group can be translated onto a localic one (by an isomorphism π, or more generally by the lifting (6.1.1)) we speak of an

LT-group.

For more about this subject see [142].

One may think of the fact that not every topological group is an LT one as a drawback. In fact it is rather an advantage: it singles out the better behaved ones. In particular it is responsible for the Closed Subgroup Theorem (one does not have such anomalies like factorizing $\mathbb{R}$ by the much smaller group of rationals and obtaining 0).

6.1.1. Problem. In fact, no example of an LT-group with a non-trivial lifting is known. The question whether there is one or not seems to be a very hard problem.

6.2. Each localic group has at least one (spectrum) point, namely the unit. There exist arbitrarily large localic groups with no other point (the known construction is rather technical and cannot be presented here; see [142]).

This has an important feature. In classical topological groups we can think of the uniform covers as created from the neighbourhoods of the unit by the homeomorphic shift $x \mapsto ax$ resp. $x \mapsto xa$. This is certainly not wrong, but the absence of points other than the unit that does not impede forming of the uniformities indicates that the uniform nature of localic groups is deeper than we may think.

6.3. The Closed Subgroup Theorem leads to the following natural conjecture. A uniform frame is dense in its completion, and a group structure looks as if it could be extended over the completion. Thus, localic groups should be complete. It is almost true, but not quite so simple:

- by a result of J. Isbell there exists a localic group that is not complete in the one-sided uniformity,
- but by a result of Banaschewski and Vermeulen indeed

    each localic group is complete in its two-sided uniformity.

For details see [51].

6.4. Theorem 2.3.5 shows that the category of localic groups duly behaves as an algebraic variety, as is to be expected. One can ask whether this behaviour is completed by the existence of free algebras. It does. For the existence of the free functor see [142].

Appendix I

Posets

1. Basics

1.1. Posets. A (*partial*) *order* on a set X is a binary relation $R \subseteq X \times X$ satisfying

(1) $\forall a, aRa$ (reflexivity),
(2) $\forall a, b, c, aRb \,\&\, bRc \Rightarrow aRc$ (transitivity), and
(3) $\forall a, b, aRb \,\&\, bRa \Rightarrow a = b$ (antisymmetry).

If, moreover

(4) $\forall a, b$ either aRb or bRa

we speak of a *linear* or *total* order.

A set equipped with a partial order is called a *partially ordered set*, briefly a *poset*. In the special case above we speak of a *linearly ordered set*, or a *totally ordered set*.

A linearly ordered set, in particular if it is a subposet of a general bigger poset, is often referred to as a *chain*.

1.1.1. In mathematical praxis we often encounter relations satisfying just (1) and (2). Then we speak of a *preorder* (and of a *preordered set*). For many purposes one can replace a preordered set (X, R) by the ordered one, $(X/\sim, R)$, where $\sim$ is the equivalence relation defined by

$$a \sim b \quad \text{iff} \quad aRb \,\&\, bRa;$$

often, however, such an identification can be confusing.

1.1.2. Notation. If there is no danger of confusion we typically use for an order or preorder the symbol $\leq$, even for distinct relations on distinct sets. Sometimes one uses other symbols, like

$$\leq, \preceq, \subseteq, \sqsubseteq, \prec, \vartriangleleft$$

(typically for orders given by special definitions), or to distinguish distinct orders (in particular if defined on the same set) by indices, or dashes, etc.,

$$\leq_1, \leq_2, \leq', \leq'' .$$

Further, we use the symbol

$$a < b \quad \text{for} \quad a \leq b \text{ and } a \neq b.$$

If there is no danger of confusion we often write just X for $(X, \leq)$.

It should be noted that the partial order is typically the primary relation, and the relation $<$ is the secondary one. Thus, it is rather

$$a < b \quad \text{meaning} \quad a \text{ less-or-equal but not equal } b$$

than

$$a \leq b \quad \text{meaning} \quad a \text{ less or equal } b.$$

1.1.3. The opposite (dual) order. If $X = (X, \leq)$ is an ordered set then $(X, \leq')$ where

$$x \leq' y \quad \text{iff} \quad y \leq x$$

is also an ordered set. It will be denoted by

$$(X, \leq)^{\mathrm{op}} \quad \text{or simply} \quad X^{\mathrm{op}}$$

and referred to as the *opposite* or *dual* of $(X, \leq)$.

1.2. Examples. (a) The sets of integers, rationals or reals with the usual orders are linearly ordered.

(b) Integers with the relation $a|b$ (a divides b) form a preordered set. Here the reduction to order ($x \sim y$ means the same absolute value) is transparent, but if we consider the divisibility on the set of all polynomials over a field it is less so, although not too bad.

(c) The refinement relation on the set of covers of a space or frame (Chapter VIII) is a preorder where the identification by $\sim$ may obscure the nature of the facts.

(d) Real functions on a set X with the usual order

$$f \leq g \quad \equiv \quad \forall x \in X, \ f(x) \leq g(x).$$

1.2.1. A very important example. The set $\mathfrak{P}(X)$ of all subsets of a set X with the relation of inclusion.

This is a sort of universal partial order. That is,

each poset can be represented as a set of subsets of a set X (typically, of course, not all of them) ordered by inclusion.

(Indeed, to represent $(X, \leq)$ take the set X itself and represent $x \in X$ by $\{z \mid z \leq x\} \subseteq X$. We trivially have $x \leq y$ iff $\{z \mid z \leq x\} \subseteq \{z \mid z \leq y\}$.)

1.3. Monotone maps. Let $(X, \leq)$ and $(Y, \leq)$ be posets. A map $f\colon X \to Y$ is a *monotone* map if

$$a \leq b \Rightarrow f(a) \leq f(b).$$

It is an *isomorphism* if it has an inverse that is monotone as well.

The category of posets and monotone maps is denoted by

Pos.

1.4. Some more notation. For an element $x \in (X, \leq)$ set

$$\downarrow x = \{y \mid y \leq x\} \quad \text{and} \quad \uparrow x = \{y \mid y \geq x\}$$

and, for a subset $M \subseteq X$,

$$\downarrow M = \bigcup \{\downarrow x \mid x \in M\} \quad \text{and} \quad \uparrow M = \bigcup \{\uparrow x \mid x \in M\}.$$

A subset $M \subseteq X$ is a

down-set resp. *up-set if* $\downarrow M = M$ resp. $\uparrow M = M$.

An element $x \in (X, \leq)$ is *maximal* (resp. *minimal*) if $\uparrow x = \{x\}$ (resp. $\downarrow x = \{x\}$). Thus, x is maximal iff $\{x\}$ is an up-set and x is minimal iff $\{x\}$ is a down-set.

We say that an $x \in X$ is a *lower bound* (resp. *upper bound*) of a subset $M \subseteq X$ if

$$\forall m \in M, \ x \leq m \quad (\text{resp.} \quad \forall m \in M, \ x \geq m)$$

and denote by

$$\mathsf{lb}M \quad \text{resp.} \quad \mathsf{ub}M$$

the set of all lower resp. upper bounds of M. Note that

$$\mathsf{lb}M = \bigcap \{\downarrow x \mid x \in M\} \quad \text{and} \quad \mathsf{ub}M = \bigcap \{\uparrow x \mid x \in M\}$$

and compare it with the formulas for $\downarrow M$ and $\uparrow M$ above.

1.4.1. The following facts are very easy to check.

Observations. (a) $\downarrow \bigcup_{i \in J} M_i = \bigcup_{i \in J} \downarrow M_i$ and $\uparrow \bigcup_{i \in J} M_i = \bigcup_{i \in J} \uparrow M_i$ *while for intersections the corresponding formulas do not generally hold.*

(b) $\mathsf{lb} \bigcup_{i \in J} M_i = \bigcap_{i \in J} \mathsf{lb}M_i$ *and* $\mathsf{ub} \bigcup_{i \in J} M_i = \bigcap_{i \in J} \mathsf{ub}M_i$.

(c) *If M_i are down-sets (resp. up-sets) then both $\bigcup_{i \in J} M_i$ and $\bigcap_{i \in J} M_i$ are down-sets (resp. up-sets).*

(d) *If M is a down-set (resp. up-set) then $X \smallsetminus M$ is an up-set (resp. down-set).*

1.5. Directed sets. A subset $D \subseteq (X, \leq)$ is said to be *directed* (more precisely, *up-directed*) if every finite subset of D has an upper bound in D.

In other words, D is directed if it is non-empty and for every $a, b \in D$ there is a $c \in D$ such that $a, b \leq c$.

2. Zorn's Lemma

2.1. This appendix is not intended to substitute a course of set theory. Nevertheless, we find it useful to discuss briefly Zorn's Lemma. Since it is a fact that is
known to be equivalent with the Axiom of Choice (AC), it is often simply used as
an expedient technical trick instead of the AC. But it is, first of all, a theorem on
partially ordered sets. AC is simple and intuitively acceptable as a proof principle
(although one is, of course, often interested in whether it has been used or not);
accepting it, one basically assumes that if one has infinitely many non-void sets
X_α one can choose representatives $x_\alpha \in X_\alpha$ *all at once*. Zorn's Lemma, however,
is a statement concerning a structure (partial order) on a set; it may be intuitively acceptable, but it is, rather, in the class of statements one easily believes,
can imagine why, but feels it is something to be proved (however simple a proof
may be).

2.2. We will assume that the reader has a basic knowledge of ordinal numbers
and of cardinalities (sizes) of sets (used in the text anyway). In particular we will
use the fact that if we have sets X_α, α ordinals, and if $X_\alpha \subset X_\beta$ for $\alpha < \beta$ then
the size of X_α increases beyond any bond. The standard facts about ordinals and
cardinals, of course, are also choice dependent.

2.2.1. Zorn's Lemma. *Let $(X, \leq)$ be a poset and let every chain $C \subseteq X$ have an
upper bound. Then for each $x \in X$ there is a $y \geq x$ maximal in X.*

Proof. For each chain $C \subseteq X$ choose an $f(C) \in \text{ub}C$. Suppose the statement
above does not hold. Then there is an $x_0 \in X$ such that for each $y \geq x_0$ there is
a $g(y) > y$. For a chain $C \ni x_0$ set $\phi(C) = gf(C)$ and for ordinals α define C_α by
transfinite induction putting

$$C_0 = \{x_0\}, \quad C_{\alpha+1} = C_\alpha \cup \{\phi(C_\alpha)\}, \text{ and}$$

$$C_\alpha = \bigcup_{\beta < \alpha} C_\beta \text{ for limit ordinals } \alpha$$

(obviously the C_α are chains). For $\alpha < \beta$ we have $C_\alpha \subset C_\beta$ contradicting the fact
that the C_α, subsets of X, are bound in size.   □

> **Note.** We have actually used AC three times: in the definition of f, in
> the definition of g, and contemplating the ordinals and the cardinality
> of C_α.

2.3. How the Axiom of Choice follows from Zorn's Lemma. On the other hand we
have

Proposition. *Let Zorn's Lemma hold. Then for every surjective mapping $f \colon X \to Y$
there is a mapping $g \colon Y \to X$ such that $f \cdot g = \text{id}_Y$.*

Proof. Consider the set F of mappings $g: D(g) \to X$ such that $D(g) \subseteq Y$ and for all $y \in D(g)$, $f(g(y)) = y$; on F consider the order (of extension)

$$g_1 \leq g_2 \quad \text{iff} \quad D(g_1) \subseteq D(g_2) \text{ and } \forall y \in D(g_1), \ g_1(y) = g_2(y).$$

Let G be a chain in F. Set $D = \bigcup_{g \in G} D(g)$ and define $h: D \to X$ by setting

$$h(y) = g(y) \quad \text{for} \quad y \in D(g), \ g \in G.$$

(This is correct, if $y \in D(g_1)$ and $y \in D(g_2)$, $g_i \in G$, then, say $D(g_1) \subseteq D(g_2)$, and $g_1(y) = g_2(y)$.)

Since for each $g \in G$ and $y \in D(g)$ we have $h(y) = g(y)$, $h(f(y)) = y$, and $h \geq f$. Thus, $h \in \mathrm{ub}G$. By Zorn's Lemma there is a g maximal in F. We have $D(g) = Y$: else we can take any $y_0 \in Y \smallsetminus D(g)$ and extend the mapping g to $g': D(g) \cup \{y_0\} \to X$ by taking $g'(y_0)$ an arbitrary value in $f^{-1}[\{y_0\}]$. $\qquad\square$

3.   Suprema and infima

3.1. The *supremum* of a subset $M \subseteq (X, \leq)$ is the least upper bound of M. Similarly, the *infimum* of M is the greatest lower bound of M.

Of course, an $M \subseteq X$ does not have to possess a supremum or an infimum (there does not even have to exist any upper or lower bound). But if it does, the infimum (or supremum) is unique, for trivial reasons: we speak of a largest, not of a maximal lower bound.

3.1.1. Explicitly, s is the supremum of M if

(1) $\forall m \in M$, $m \leq s$, and

(2) if $m \leq x$ for every $m \in M$, then $s \leq x$

and similarly for the infimum.

In linearly ordered sets the condition (2) may be replaced by

(2′) if $x < s$ then there is an $m \in M$ such that $x < m$.

This is a very expedient reformulation as the reader certainly remembers from working with suprema of reals in analysis. One has to have in mind, though, that this replacement is not correct in general posets.

One often speaks of the supremum as of the *join* of M and of the infimum as of the *meet* of M.

3.2. Notation. We write $\sup M$ or $\bigvee M$ for the supremum of M (for finite sets $\{a, b\}$ or $\{a_1, \ldots, a_n\}$ we write $a \vee b$ and $a_1 \vee \ldots \vee a_n$), and $\inf M$ or $\bigwedge M$ for the infimum of M (for finite sets, $a \wedge b$ and $a_1 \wedge \ldots \wedge a_n$).

3.3. Since each x is both a lower and an upper bound of the empty set,

$$\sup \emptyset \text{ is the least element of } X$$

and

$$\inf \emptyset \text{ is the greatest element of } X.$$

For the former one often uses the symbol

$$0 \quad \text{or} \quad \bot$$

and speaks of the *bottom* of X, and for the latter one writes

$$1 \quad \text{or} \quad \top$$

and speaks of the *top* of X.

3.4. Proposition (Associativity of suprema and infima). *Let M_i, $i \in J$, be subsets of $(X, \leq)$, let the suprema $s_i = \sup M_i$ exist, and let there exist $s = \sup_{i \in J} s_i$. Then s is the supremum of $\bigcup_{i \in J} M_i$, that is,*

$$\sup \bigcup_{i \in J} M_i = \sup_{i \in J} \sup M_i.$$

Similarly for infima.

Proof. $s \geq s_i \geq m$ for all $m \in M_i$, $i \in J$; thus, s is an upper bound of $\bigcup_{i \in J} M_i$. Now let x be an upper bound of $\bigcup_{i \in J} M_i$. Then in particular it is an upper bound of M_i and hence $s_i \leq x$. Since this holds for all the i, $s \leq x$. $\qquad\qquad\square$

4. Semilattices, lattices and complete lattices. Completion

4.1. A poset X is called a *meet-semilattice* (resp. *join-semilattice*) if there is an infimum $a \wedge b$ (resp. supremum $a \vee b$) for any two $a, b \in X$.

One speaks of a *bounded meet-* resp. *join-semilattice* if there is, furthermore, top resp. bottom. Thus, bounded meet-semilattices (resp. join-semilattices) are the posets in which all finite subsets have infima (resp. suprema).

4.2. A poset X is a *lattice* if there is an infimum $a \wedge b$ and a supremum $a \vee b$ for any two $a, b \in X$.

A *bounded lattice* (one often speaks of a *lattice with 0 and 1*) is a poset in which all finite subsets have infima and suprema (thus, it is a lattice that has bottom and top).

> **Note.** The boundedness of a lattice or semilattice is often implicitly assumed. Thus, it is advisable to make sure what that or other author has in mind.

4.3. A poset is a *complete lattice* if every subset has a supremum and an infimum. It suffices to assume one of the two. We have

4.3.1. Proposition. *Let each subset of a poset X have a supremum. Then X is a complete lattice.*

Proof. For a subset M consider $N = \mathsf{lb}M$ and set $a = \sup N$. Obviously, for every $m \in M$, $m \in \mathsf{ub}N$, hence $a \leq m$, and consequently $a \in N = \mathsf{lb}M$. Thus, a is the greatest element in $\mathsf{lb}M$, that is, it is the infimum of M. $\qquad\square$

4.4. Here is a very useful theorem about complete lattices. In spite of its simplicity it has a lot of striking consequences.

Theorem (Knaster-Tarski Fixed Point Theorem). *Let L be a complete lattice. Then each monotone mapping $f\colon L \to L$ has a fixed point, that is, there is an $x \in L$ such that $f(x) = x$.*

Proof. Set $M = \{x \mid x \leq f(x)\}$ and $s = \sup M$. If $x \in M$ then $x \leq s$ and hence $x \leq f(x) \leq f(s)$. Thus, $f(s)$ is an upper bound of M and hence

$$s \leq f(s) \quad \text{(in particular, } s \in M\text{)}.$$

On the other hand, if $x \in M$ then $x \leq f(x)$ and hence $f(x) \leq f(f(x))$ so that $f(x) \in M$. In particular $f(s) \in M$ and we conclude $f(s) \leq s \leq f(s)$. $\qquad\square$

4.4.1. Notes. (1) In fact the s from our construction is the least fixed point; using $\inf\{x \mid x \geq f(x)\}$ we similarly get the greatest fixed point.

(2) To show how this very simply proved fact can yield by a very simple reasoning a fact that does not seem to be quite so easy, let us prove the Cantor-Bernstein Theorem stating that

> *if X, Y are sets and if there exist one-one maps $f\colon X \to Y$, $g\colon Y \to X$ then there exists a one-one onto map $h\colon X \to Y$.*

Take the monotone mapping $\phi\colon \mathfrak{P}(X) \to \mathfrak{P}(X)$ defined by

$$\phi(M) = X \smallsetminus g[Y \smallsetminus f[X \smallsetminus M]].$$

By 4.4 we have an $A \subseteq X$ such that

$$A = X \smallsetminus g[Y \smallsetminus f[X \smallsetminus A]], \quad \text{that is,} \quad X \smallsetminus A = g[Y \smallsetminus f[X \smallsetminus A]].$$

Define

$$h(x) = \begin{cases} f(x) & \text{if } x \in A, \\ g^{-1}(x) & \text{if } x \in X \smallsetminus A. \end{cases}$$

To see that h is a correctly defined one-one onto mapping is a matter of simple checking.

4.5. Completion. A general poset need not have suprema or infima. Can they be added? Just extending $(X, \leq)$ to a complete lattice (which is easy – representing $x \in X$ as $\downarrow x \in \mathfrak{P}(X)$ as in 1.2.1 does the job) may not be satisfactory; rather, we wish for an extension in which both the existing infima and the existing suprema are preserved (note that the embedding $(x \mapsto \downarrow x)$ does at least preserve infima).

4.5.1. Recall 1.4, define $\phi(M) = \mathsf{lb}(\mathsf{ub}(M))$, and set

$$\mathsf{DMN}(X, \leq) = \Big(\{M \subseteq X \mid \phi(M) = M\}, \subseteq\Big).$$

Lemma. (1) *If* $M \subseteq N$ *then* $\mathsf{ub}(M) \supseteq \mathsf{ub}(N)$ *and* $\mathsf{lb}(M) \supseteq \mathsf{lb}(N)$. *Consequently,* ϕ *is monotone.*

(2) $M \subseteq \phi(M) = \mathsf{lb}(\mathsf{ub}(M))$ *and* $M \subseteq \mathsf{ub}(\mathsf{lb}(M))$.

(3) $\mathsf{ub}(\downarrow a) = \uparrow a$ *and* $\mathsf{lb}(\uparrow a) = \downarrow a$.

(4) $\phi\phi(M) = \phi(M)$.

Proof. (1) through (3) are immediate observations. Now by (1) and (2), $\mathsf{lb}(M) \subseteq \mathsf{lb}(\mathsf{ub}(\mathsf{lb}(M))) \subseteq \mathsf{lb}(M)$ and $\mathsf{ub}(M) \subseteq \mathsf{ub}(\mathsf{lb}(\mathsf{ub}(M))) \subseteq \mathsf{ub}(M)$ so that $\mathsf{lb}(M) = \mathsf{lb}(\mathsf{ub}(\mathsf{lb}(M)))$ and $\mathsf{ub}(M) = \mathsf{ub}(\mathsf{lb}(\mathsf{ub}(M)))$ and finally

$$\mathsf{lb}(\mathsf{ub}(M)) = \mathsf{lb}(\mathsf{ub}(\mathsf{lb}(\mathsf{ub}(M)))). \qquad \square$$

4.5.2. Theorem (Dedekind–MacNeille completion). (1) $L = \mathsf{DMN}(X)$ *is a complete lattice. The suprema in* L *are given by the formula*

$$\bigvee_{i \in J} M_i = \phi(\bigcup_{i \in J} M_i).$$

(2) *The embedding* $(a \mapsto \downarrow a): (X, \leq) \to \mathsf{DMN}(X, \leq)$ *preserves all the existing suprema and infima.*

Proof. (1): If $\phi(M) = M$ and if $M \supseteq M_i$ for all $i \in J$ we have $M \supseteq \bigcup_{i \in J} M_i$ and hence

$$M = \phi(M) \supseteq \phi(\bigcup_{i \in J} M_i).$$

(2): By the lemma, $\phi(\downarrow a) = \mathsf{lb}(\mathsf{ub}(\downarrow a)) = \mathsf{lb}(\uparrow a) = \downarrow a$ so that indeed $\downarrow a \in \mathsf{DMN}(X)$. For $a = \inf_{i \in J} a_i$ we have $\downarrow a = \bigcap_{i \in J} \downarrow a_i$ which is the infimum already in $\mathfrak{P}(X)$ and consequently also in $\mathsf{DMN}(X)$. Finally we have for $a = \sup_{i \in J} a_i$

$$\bigvee_{i \in J} \downarrow a_i = \phi(\bigcup_{i \in J} \downarrow a_i) = \mathsf{lb}(\mathsf{ub}(\bigcup_{i \in J} \downarrow a_i)) = \mathsf{lb}\Big(\bigcap_{i \in J} \uparrow a_i\Big) = \mathsf{lb}(\uparrow a) = \downarrow a. \qquad \square$$

5. Galois connections (adjunctions)

5.1. Monotone maps $f\colon X \to Y$, $g\colon Y \to X$ are *Galois adjoint* (or are in a *Galois connection*) – f is a *left adjoint* of g, and g is a *right adjoint* of f – if

$$\forall\, x \in X, \forall\, y \in Y, \quad f(x) \le y \;\Leftrightarrow\; x \le g(y).$$

Note that a left (resp. right) Galois adjoint of a given map does not have to exist (we will learn a necessary, and in some extent also sufficient, condition shortly). If it exists, however,

it is uniquely determined

(if $f_i(x) \le y \Leftrightarrow x \le g(y)$, $i = 1, 2$, then $f_1(x) \le y \Leftrightarrow f_2(x) \le y$).

5.2. Examples. (a) Mutually inverse isomorphisms are adjoint, both to the left and to the right.

(b) "*Almost inverse functions* $\mathbb{N} \to \mathbb{N}$": Suppose a mapping $f\colon \mathbb{N} \to \mathbb{N}$ ($\mathbb{N}$ is the set of natural numbers) can be extended to an increasing real function $\widetilde{f}$ on the interval $[1, +\infty)$ and let ϕ be its inverse. Then, if we denote by $\lfloor x \rfloor$ resp. $\lceil x \rceil$ the lower resp. upper integral part of a real number x we obtain that

$$\lceil \phi(-) \rceil \text{ is a left adjoint of } f, \quad \text{and} \quad \lfloor \phi(-) \rfloor \text{ is a right adjoint of } f$$

(which is seen from the obvious fact that $\lceil \phi(m) \rceil \le n$ iff $\phi(m) \le n$, and $\lfloor \phi(m) \rfloor \ge n$ iff $\phi(m) \ge n$).

Thus for instance $\lceil \log_2 \rceil$ and $\lfloor \log_2 \rfloor$ are the left and the right adjoint of the exponentiation 2^n.

(c) Let X, Y be arbitrary sets and $f\colon X \to Y$ an arbitrary map. We have

$$f[A] \subseteq B \quad \text{if and only if} \quad A \subseteq f^{-1}[B].$$

Thus the maps

$$f[-]\colon \mathfrak{P}(X) \to \mathfrak{P}(Y), \quad f^{-1}[-]\colon \mathfrak{P}(Y) \to \mathfrak{P}(X)$$

are adjoint, $f[-]$ to the left and $f^{-1}[-]$ to the right.

(d) $f^{-1}[-]$ also has a right adjoint. Namely, we have

$$f^{-1}[B] \subseteq A \quad \text{if and only if} \quad B \subseteq Y \smallsetminus f[X \smallsetminus A].$$

The image function $f[-]$ has no left adjoint, though.

(e) Let X, Y be topological spaces and let $f\colon X \to Y$ be a continuous map. Then the adjunction from (d) can be modified to

$$f^{-1}[B] \subseteq A \quad \text{if and only if} \quad B \subseteq \mathrm{int}(Y \smallsetminus f[X \smallsetminus A])$$

while the adjunction from (c) has no counterpart.

5.3. The following proposition yields an expedient description of the adjunctions.

Theorem. *Monotone maps $f\colon X \to Y$ and $g\colon Y \to X$ are adjoint (f on the left, g on the right) if and only if there holds*

$$f(g(y)) \le y \quad \text{and} \quad x \le g(f(x)).$$

Proof. If f, g are adjoint, the inequalities follow from the fact that $g(y) \le g(y)$ and $f(x) \le f(x)$. On the other hand, let our new inequalities hold; then if $f(x) \le y$ we obtain $x \le g(f(x)) \le g(y)$, and if $x \le g(y)$ we conclude $f(x) \le f(g(y)) \le y$. $\square$

5.3.1. Corollary. *If monotone maps f, g are adjoint then*

$$fgf = f \quad \text{and} \quad gfg = g.$$ $\square$

(We have $f(gf(x)) \ge f(x)$ as $gf(x) \ge x$, and $fg(f(x)) \le f(x)$.)

5.4. Theorem. *The left Galois adjoints preserve suprema, and the right ones preserve infima.*

Proof. The proof will be done for left adjoints. Let $s = \sup M$ exist. First, $f(s)$ is an upper bound of the set $f[M]$. Now if y is a general upper bound of $f[M]$ we have for all $m \in M$, $f(m) \le y$ and hence $m \le g(y)$. Thus, $g(y)$ is an upper bound of the set M, hence $s \le g(y)$, and finally $f(s) \le y$. $\square$

5.5. Obviously we cannot expect 5.4 reversed if there are few suprema resp. infima. But we have

Theorem. *If X, Y are complete lattices then a monotone map $f\colon X \to Y$ is a left (resp. right) adjoint if and only if it preserves all suprema (resp. infima).*

Proof. Let f preserve suprema. Define a mapping $g\colon Y \to X$ by setting

$$g(y) = \sup\{x \mid f(x) \le y\}.$$

Trivially, $f(x) \le y$ implies $x \le g(y)$. But also if $x \le g(y) = \sup\{z \mid f(z) \le y\}$ we obtain, since f preserves the supremum,

$$f(x) \le \sup\{f(z) \mid f(z) \le y\} \le y.$$ $\square$

5.6. Notes. (1) Recall the Example 5.2(c), 5.3, and the standard set-theoretical formulas

$$f[f^{-1}[B]] \subseteq B \quad \text{and} \quad A \subseteq f^{-1}[f[A]],$$

$$f[f^{-1}[f[A]]] = f[A] \quad \text{and} \quad f^{-1}[f[f^{-1}[B]]] = f^{-1}[B].$$

(2) Recall the examples 5.2(c),(d),(e). Confront 5.4 with the fact that in sets the preimage function preserves both unions and intersections, while the image function preserves unions only; for continuous f and open sets, the preimage preserves unions, but generally not the meets (which do not coincide with the intersections).

6. (Semi)lattices as algebras. Distributive lattices

6.1. From 3.4 we see that in a (meet-)semilattice S,

$$a \wedge (b \wedge c) = (a \wedge b) \wedge c,$$
$$a \wedge b = b \wedge a, \qquad (\wedge\text{-eq})$$
$$a \wedge a = a.$$

If S has a largest element 1, we have, moreover

$$1 \wedge a = a. \qquad (1\text{-eq})$$

The equations $(\wedge\text{-eq})$ resp. $(\wedge\text{-eq})+(1\text{-eq})$ determine the structure of a semi-lattice resp. bounded semilattice. Namely, we have

Proposition. *Let there be an operation $\wedge$ on a set X satisfying $(\wedge\text{-eq})$. Then there is precisely one partial order on X such that $a \wedge b = \inf\{a, b\}$ and X is a bounded semilattice iff it contains an element 1 satisfying (1-eq).*

Proof. There is at most one such order since we have to have $x \leq y$ iff $x = \inf\{x, y\}$. Thus, define

$$x \leq y \equiv_{\text{df}} x \wedge y = x.$$

This relation is an order: we have $x \leq x$ since $x \wedge x = x$; if $x \leq y \leq z$ then $x \wedge y = x$ and $y \wedge z = y$ and hence $x \wedge z = (x \wedge y) \wedge z = x \wedge (y \wedge z) = x \wedge y = x$ and consequently $x \leq z$; finally if $x \leq y \leq x$ then $x = x \wedge y = y \wedge x = y$.

In this order, $x \wedge y = \inf\{x, y\}$: indeed, first of all $x \wedge y$ is a lower bound of the set $\{x, y\}$ since $(x \wedge y) \wedge x = (x \wedge x) \wedge y = x \wedge y$ and even more trivially $(x \wedge y) \wedge y = x \wedge y$; it is the largest lower bound since if $z \wedge x = z = z \wedge y$ then $z \wedge (x \wedge y) = (z \wedge x) \wedge y = z \wedge y = z$ and hence $z \leq x \wedge y$.

If $1 \wedge a = a$ then $a \leq 1$ in our definition. $\square$

6.2. If X is a lattice we have two operations $\wedge$ and $\vee$ satisfying $(\wedge\text{-eq})$, and similar equations for the join,

$$a \vee (b \vee c) = (a \vee b) \vee c,$$
$$a \vee b = b \vee a, \qquad (\vee\text{-eq})$$
$$a \vee a = a$$

and the operations $\wedge$ a $\vee$ are, furthermore, interconnected by the equalities

$$a \wedge (a \vee b) = a, \quad a \vee (a \wedge b) = a. \qquad (\wedge\vee\text{-eq})$$

We have

Proposition. *Let there be operations $\wedge$ and $\vee$ on a set X satisfying $(\wedge\text{-eq})$, $(\vee\text{-eq})$ and $(\wedge\vee\text{-eq})$. Then there is precisely one partial order on X such that $a \wedge b = \inf\{a, b\}$ and $a \vee b = \sup\{a, b\}$. Further, X is bounded iff there are elements 0 and 1 such that $0 \vee a = 1 \wedge a = a$ for all a.*

Proof. Again, there is obviously at most one such order. The operations induce two orders,

$$x \leq_1 y \quad \text{iff} \quad x \wedge y = x,$$
$$x \leq_2 y \quad \text{iff} \quad x \vee y = y$$

and we only have to prove they coincide. But if we have $x \wedge y = x$ then $y = y \vee (y \wedge x) = y \vee x$, and if $y = y \vee x$ then $x = x \wedge (x \vee y) = x \wedge y$. $\square$

6.2.1. Note. In the sequel, when speaking of (semi)lattices we will have in mind the sets equipped with $\wedge$ and $\vee$ (in the bounded case also 0 and 1) as operations. Thus, a sublattice will be a subalgebra (closed under the operations in question), not just a subposet that happens to be a (semi)lattice. Similarly, one considers lattice homomorphisms preserving the operations, more special maps than the monotone ones.

6.3. Distributive lattices. A lattice L is said to be *modular*, if there holds the implication

$$a \leq c \quad \Rightarrow \quad a \vee (b \wedge c) = (a \vee b) \wedge c.$$

It is *distributive* if there holds the equality

$$a \vee (b \wedge c) = (a \vee b) \wedge (a \vee c). \tag{distr}$$

Obviously,

every distributive lattice is modular.

6.3.1. Proposition. *A lattice L is distributive if and only if there holds the equality*

$$a \wedge (b \vee c) = (a \wedge b) \vee (a \wedge c). \tag{distr$'$}$$

In other words, L is distributive if and only if L^{op} is distributive.

Proof. If we have (distr$'$), we compute $(a \vee b) \wedge (a \vee c) = (a \wedge (a \vee c)) \vee (b \wedge (a \vee c)) = a \vee (b \wedge a) \vee (b \wedge c) = a \vee (b \wedge c)$. The other implication is deduced the same way. $\square$

6.3.2. Modular lattices play an important role in algebra, and also in other parts of mathematics. Here, however, their role is more or less auxiliary. Out of many interesting properties we will prove the following characterization.

Proposition. *A lattice L is modular if and only if it does not contain a sublattice isomorphic with the lattice C_5 in Figure 6.*

Proof. Let L contain C_5. Then, in the notation from the picture, we have

$$x \vee (a \wedge y) = x \vee b = x < y = c \wedge y = (x \vee a) \wedge y$$

although $x \leq y$. Thus, L is not modular.

Conversely, let L not be modular. Then there exist u, v, w such that $u \leq w$ and $u \vee (v \wedge w) < (u \vee v) \wedge w$. Hence, v cannot be comparable with any of $u \vee (v \wedge w)$ and $(u \vee v) \wedge w$.

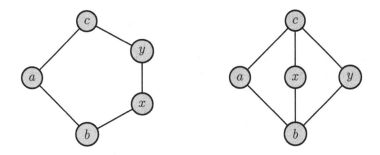

Figure 6: Prohibited configurations C_5 and D_3.

(If we had $v \geq u \vee (v \wedge w)$ there would be $v \geq u$ and hence $u \vee (v \wedge w) = v \wedge w = (u \vee v) \wedge w$; if we had $v \leq (u \vee v) \wedge w$ there would be $v \leq w$ and $u \vee (v \wedge w) = u \vee v = (u \vee v) \wedge w$.)

We have $v \vee u \vee (v \wedge w) = v \vee u$, and as also $(u \vee v) \wedge w \leq v \vee u$ we see that $v \vee u = v \vee ((u \vee v) \wedge w)$. Similarly

$$v \wedge (u \vee (v \wedge w)) = v \wedge (u \vee v) \wedge w = v \wedge w.$$

Thus, we obtain a copy of C_5 in L setting $a = v$, $b = v \wedge w$, $c = v \vee u$, $x = u \vee (v \wedge w)$ and $y = (u \vee v) \wedge w$. □

6.3.3. Lemma. *A modular lattice is distributive if and only if there holds the equality*

$$(a \wedge b) \vee (a \wedge c) \vee (b \wedge c) = (a \vee b) \wedge (a \vee c) \wedge (b \vee c).$$

Proof. Obviously, the inequality

$$(a \wedge b) \vee (a \wedge c) \vee (b \wedge c) \leq (a \vee b) \wedge (a \vee c) \wedge (b \vee c)$$

holds in any lattice. Thus, the point is in the opposite one.

I. If a lattice is distributive then

$$(a \vee b) \wedge (a \vee c) \wedge (b \vee c) = (a \vee (b \wedge (a \vee c))) \wedge (b \vee c) =$$
$$= (a \wedge (b \vee c)) \vee (b \wedge (a \vee c)) = (a \wedge b) \vee (a \wedge c) \vee (b \wedge a) \vee (b \wedge c).$$

II. Let the equality hold. Then we have

$$(a \vee b) \wedge c = (a \vee b) \wedge (a \vee c) \wedge (b \vee c) \wedge c = ((a \wedge b) \vee (a \wedge c) \vee (b \wedge c)) \wedge c.$$

If, moreover, the lattice is modular we have, as $(a \wedge c) \vee (b \wedge c) \leq c$, further,

$$\cdots = (a \wedge c) \vee (b \wedge c) \vee (a \wedge b \wedge c) = (a \wedge c) \vee (b \wedge c). \qquad \square$$

6.3.4. Theorem. *A lattice L is distributive if and only if it contains no sublattices isomorphic either with the C_5 or with the D_3 in Figure 6.*

Proof. I. If L contains the configuration C_5 it is not even modular. If it contains D_3, the distributivity is violated by the inequality $(a \wedge x) \vee (a \wedge y) = b < a = a \wedge c = a \wedge (x \vee y)$.

II. Let the lattice not be distributive. If it is not modular it contains C_5. Thus, let L be modular but not distributive. By 6.3.3 there exist $a, b, c \in L$ such that

$$x = (a \wedge b) \vee (a \wedge c) \vee (b \wedge c) < (a \vee b) \wedge (a \vee c) \wedge (b \vee c) = y.$$

Set

$$u = (a \vee (b \wedge c)) \wedge (b \vee c),$$
$$v = (b \vee (a \wedge c)) \wedge (a \vee c),$$
$$w = (c \vee (a \wedge b)) \wedge (a \vee b).$$

We will prove that

$$u \wedge v = u \wedge w = v \wedge w = x \quad \text{and} \quad u \vee v = u \vee w = v \vee w = y. \qquad (*)$$

For this it suffices to show that $u \wedge v = x$ (the rest will follow by permuting the elements a, b, c and interchanging $\wedge$ and $\vee$ which is correct since from the definition of the elements u, v, w we obtain, using modularity,

$$u = (a \wedge (b \vee c)) \vee (b \wedge c),$$
$$v = (b \wedge (a \vee c)) \vee (a \wedge c),$$
$$w = (c \wedge (a \vee b)) \vee (a \wedge b). \quad)$$

Indeed we have (employing modularity again)

$$u \wedge v = (a \vee (b \wedge c)) \wedge (b \vee c) \wedge (b \vee (a \wedge c)) \wedge (a \vee c)$$
$$= (a \vee (b \wedge c)) \wedge (b \vee (a \wedge c)) = (a \wedge (b \vee (a \wedge c))) \vee (b \wedge c)$$
$$= (a \wedge b) \vee (a \wedge c) \vee (b \wedge c).$$

By $(*)$, as $x \neq y$, we now have D_3 represented by the incomparable elements u, v, w, the common infimum x of the pairs $\{u, v\}$, $\{u, w\}$ and $\{v, w\}$, and the common supremum y of these pairs. $\qquad \square$

6.3.5. Proposition. *A lattice L is distributive if and only if each pair of equations of the form*

$$a \wedge x = b$$
$$a \vee x = c$$

has at most one solution x.

Proof. I. Let L be distributive and let

$$a \wedge x = b, \ a \vee x = c, \ a \wedge y = b \text{ and } a \vee y = c.$$

Then $x = x \wedge (a \vee x) = x \wedge (a \vee y) = (x \wedge a) \vee (x \wedge y) = (y \wedge a) \vee (x \wedge y) = y \wedge (a \vee x) = y \wedge (a \vee y) = y$.

II. Let L not be distributive. Then in any of the configurations C_5 or D_3 in Fig. 6, both the x and y solve the equations above. $\square$

6.4. Ideals and filters in distributive lattices. An *ideal* in a bounded distributive lattice L is a subset $J \subseteq L$ such that

$$\begin{aligned}
&0 \in J, \\
&a, b \in J \ \Rightarrow \ a \vee b \in J, \qquad\qquad\qquad\qquad\qquad \text{(idl)} \\
&b \le a \ \& \ a \in J \ \Rightarrow \ b \in J.
\end{aligned}$$

A *filter* in L is a subset $F \subseteq L$ such that

$$\begin{aligned}
&1 \in F, \\
&a, b \in F \ \Rightarrow \ a \wedge b \in F, \qquad\qquad\qquad\qquad\qquad \text{(fltr)} \\
&b \ge a \ \& \ a \in F \ \Rightarrow \ b \in F.
\end{aligned}$$

An ideal resp. filter is *proper* if it is not the whole of the lattice L, hence in the case of an ideal if $1 \notin J$ and in the case of a filter if $0 \notin F$. Often, ideals resp. filters are automatically assumed to be proper.

6.4.1. A *prime ideal* resp. *prime filter* is a proper ideal J resp. filter F such that

whenever $a \wedge b \in J$, then either $a \in J$ or $b \in J$

resp. whenever $a \vee b \in F$, then either $a \in F$ or $b \in F$.

A *maximal* ideal resp. filter is a proper ideal resp. filter that is not contained in a bigger proper ideal resp. filter. Often one considers maximality with respect to some more special condition – see, e.g., Birkhoff's Theorem below.

6.4.2. Lemma. *Let J be an ideal and let F be a filter in L, and let $J \cap F = \emptyset$. Then there exists a filter $\overline{F} \supseteq F$ maximal with respect to the condition $\overline{F} \cap J = \emptyset$, and this $\overline{F}$ is prime.*

Proof. Let $\mathcal{F}$ be a system of filters containing F and disjoint with J, such that for any two $F_1, F_2 \in \mathcal{F}$ either $F_1 \subseteq F_2$ or $F_2 \subseteq F_1$. Then obviously $\bigcup \mathcal{F}$ is a filter, and it is still disjoint with J. By Zorn's Lemma there exists a maximal filter $\overline{F}$ from the first part of the statement. We will prove it is a prime filter.

Let $a \notin \overline{F}$, $b \notin \overline{F}$ and $a \vee b \in \overline{F}$. Define $G = \{x \mid x \vee b \in \overline{F}\}$. If $x, y \in G$ then $(x \wedge y) \vee b = (x \vee b) \wedge (y \vee b) \in \overline{F}$ and hence $x \wedge y \in G$; if $z \ge x$ then

$z \vee b \geq x \vee b \in \overline{F}$ and hence $z \in G$. Thus, G is a filter and since obviously $G \supseteq \overline{F}$ and $G \ni a \notin \overline{F}$, it is larger than $\overline{F}$ and hence cannot be disjoint with J. Choose $c_1 \in G \cap J$ (hence in particular $c_1 \vee b \in \overline{F}$) and set $H = \{x \mid x \vee c_1 \in \overline{F}\}$. Again, by the same reasoning as before, we see that H is a filter and that it is larger than $\overline{F}$ (since it contains b). Hence it has to contain some $c_2 \in J$. This is a contradiction since then $c_1 \vee c_2 \in J \cap \overline{F}$. $\qquad\square$

6.4.3. Theorem (Birkhoff's Theorem on prime filters and ideals). *Let J be an ideal and let F be a filter in a bounded distributive lattice and let $J \cap F = \emptyset$. Then there exists a prime filter $\overline{F} \supseteq F$ and a prime ideal $\overline{J} \supseteq J$ such that $\overline{J} \cap \overline{F} = \emptyset$.*

Proof. By 6.4.2 take, first, the $\overline{F}$ and J and then apply this lemma again dually (that is, interchanging ideals and filters and the operations $\vee$ and $\wedge$) for these disjoint sets. $\qquad\square$

6.4.4. Corollary. *Let $a \nleq b$ in a bounded distributive lattice. Then there exists a prime filter F such that $b \notin F \ni a$.* $\qquad\square$

7. Pseudocomplements and complements. Heyting and Boolean algebras

7.1. Pseudocomplements. A *pseudocomplement* of an element a in a meet-semilattice L with 0 is the largest element b such that $b \wedge a = 0$, if it exists. It is usually denoted by

$$a^*.$$

Thus, a^* is characterized by the formula

$$b \wedge a = 0 \quad \text{iff} \quad b \leq a^* \qquad\qquad \text{(PSC)}$$

(note that a pseudocomplement, if it exists, is necessarily unique).

A meet-semilattice in which any element has a pseudocomplement is said to be *pseudocomplemented*.

> **Note.** (Semi)lattice homomorphisms do not necessarily preserve pseudo-complements. One has obviously $f(a^*) \leq f(a)^*$, but the other inequality generally need not hold.

7.1.1. Facts. (a) *The correspondence $a \mapsto a^*$ is antimonotone.*

(b) *The correspondence $a \mapsto a^{**}$ is monotone, and $a \leq a^{**}$.*

(c) $a^{***} = a^*$.

Proof. (a): If $a \leq b$ then $a \wedge b^* = 0$ and hence $b^* \leq a^*$.

(b): $a \wedge a^* = 0$ and therefore $a \leq a^{**}$.

(c): By (b), $a^* \leq (a^*)^{**}$; by (b) and (a), $a^* \geq (a^{**})^*$. $\qquad\square$

7.1.2. Proposition. $(a \wedge b)^{**} = a^{**} \wedge b^{**}$.

Proof. Trivially $(a \wedge b)^{**} \leq a^{**} \wedge b^{**}$ by the monotony of $(-)^{**}$. Now by 7.1.1(c) we have

$$a \wedge b = 0 \text{ iff } a^{**} \wedge b = 0. \tag{$*$}$$

Since $a \wedge b \leq (a \wedge b)^{**}$, we have $a \wedge b \wedge (a \wedge b)^* = 0$ and hence, using twice $(*)$ we obtain $a^{**} \wedge b^{**} \wedge (a \wedge b)^* = 0$ from which it follows that $a^{**} \wedge b^{**} \leq (a \wedge b)^{**}$.   □

7.1.3. Proposition (first De Morgan law). *In a distributive lattice, if a and b have pseudocomplements then $a \vee b$ has, and we have*

$$(a \vee b)^* = a^* \wedge b^*.$$

Proof. By distributivity, $(a \vee b) \wedge a^* \wedge b^* = 0$. Moreover, for any c satisfying $(a \vee b) \wedge c = 0$ we have $(a \wedge c) \vee (b \wedge c) = 0$ and hence $c \leq a^* \wedge b^*$.   □

7.2. Complements. A *complement* of an element a in a lattice L is an element $b \in L$ such that

$$b \wedge a = 0 \quad \text{and} \quad b \vee a = 1.$$

In a general lattice, a complement is not uniquely determined. However,

7.2.1. Proposition. *In a distributive lattice, each complement is a pseudocomplement. Hence, in particular, it is uniquely determined.*

Proof. Let $b \wedge a = 0$ and $b \vee a = 1$. If $x \wedge a = 0$ then by distributivity

$$x = x \wedge 1 = x \wedge (b \vee a) = (x \wedge b) \vee (x \wedge a) = x \wedge b$$

and hence $x \leq b$.   □

Note. The unicity, of course, also follows immediately from 6.3.5.

7.2.2. A pseudocomplement is not necessarily a complement since $a \vee a^*$ may be smaller than 1. Nevertheless, this join is always *dense* in the following sense:

Proposition. $(a \vee a^*)^* = 0$.

Proof. If $b \wedge (a \vee a^*) = 0$ then in particular $b \wedge a = 0$ and $b \wedge a^* = 0$ (by monotony). Thus, $b \leq a^*$ and $b \leq a^{**}$ and hence $b \leq a^* \wedge a^{**} = 0$.   □

7.3. Heyting algebras. A *Heyting semilatice* is a meet-semilattice with 0 equipped with a binary operation $\to$ satisfying

$$c \leq a \to b \quad \text{iff} \quad c \wedge a \leq b. \tag{H}$$

If L is a bounded lattice we speak of a *Heyting algebra*. In particular, setting $b = 0$ we see that

a Heyting semilattice resp. algebra is a pseudocomplemented semilattice resp. lattice, with $a^* = a \to 0$.

7.3.1. The formula (H) states that each $a \to (-)$ is a right Galois adjoint to $a \wedge (-)$ (it is a monotone map: if $b_1 \leq b_2$ use $a \to b_1 = a \to b_1$ to obtain $(a \to b_1) \wedge a \leq b_1 \leq b_2$ and then $a \to b_1 \leq a \to b_2$). We have

Proposition. (1) *A meet-semilattice admits at most one Heyting operation.*

(2) *In a Heyting semilattice the meet distributes over all existing joins; in particular, every Heyting algebra is a distributive lattice.*

(3) *In a meet-semilattice we have* $(a \wedge b) \to c = a \to (b \to c)$.

Proof. (1) and (2) follow from 5.1 and 5.4.

(3): We have $x \leq (a \wedge b) \to c$ iff $x \wedge a \wedge b \leq c$ iff $x \wedge a \leq b \to c$ iff $x \leq a \to (b \to c)$. $\square$

7.3.2. From 5.5 we obtain, furthermore,

Proposition. *A complete lattice admits a Heyting operation iff there holds the distributive law*

$$\left(\bigvee_{i \in J} a_i \right) \wedge b = \bigvee_{i \in J} (a_i \wedge b)$$

for any b and any system a_i, $i \in J$.

Thus, complete Heyting algebras are in fact the same as frames. The *Heyting homomorphisms* (preserving also the operation $\to$) differ from the frame ones (see III.7.2), though.

7.3.3. Proposition (first De Morgan law). *In a Heyting algebra,*

$$\left(\bigvee_{i \in J} a_i \right)^* = \bigwedge_{i \in J} a_i^*$$

whenever the supremum $\bigvee_{i \in J} a_i$ exists.

Proof. $c \leq \left(\bigvee_{i \in J} a_i \right)^*$ iff $c \wedge \left(\bigvee_{i \in J} a_i \right) = 0$ iff $\bigvee_{i \in J} (c \wedge a_i) = 0$ iff $c \wedge a_i = 0$ for all i iff $c \leq a_i^*$ for all i. $\square$

7.3.4. Lemma. *Let L be a lattice. A binary operation $\to$ on L makes L into a Heyting algebra iff the following equations hold for all $a, b, c \in L$:*

(1) $a \to a = 1$,

(2) $a \wedge (a \to b) = a \wedge b$,

(3) $b \wedge (a \to b) = b$,

(4) $a \to (b \wedge c) = (a \to b) \wedge (a \to c)$.

Proof. Suppose L is a Heyting algebra. Equation (1) is clear since $c \wedge a \leq a$ is always true. By definition $a \wedge (a \to b) \leq b$, so $a \wedge (a \to b) \leq a \wedge b$. But $(a \wedge b) \wedge a \leq b$, hence $a \wedge b \leq a \to b$, and hence $a \wedge b = a \wedge (a \to b)$. Since $b \wedge a \leq b$, we have

$b \leq a \to b$ and therefore $b \wedge (a \to b) = b$. Finally, $a \to (b \wedge c) \leq (a \to b) \wedge (a \to c)$ (because $a \to (-)$ is monotone). On the other hand,

$$(a \to b) \wedge (a \to c) \wedge a = (a \wedge (a \to b)) \wedge (a \wedge (a \to c)) \leq b \wedge c,$$

so that $(a \to b) \wedge (a \to c) \leq a \to (b \wedge c)$.

Now suppose the equations hold. Then if $c \leq a \to b$, we have

$$c \wedge a \leq a \wedge (a \to b) = a \wedge b \leq b.$$

Conversely, if $c \wedge a \leq b$, we have

$$c = c \wedge (a \to c) \leq (a \to a) \wedge (a \to c) = a \to (a \wedge c) \leq a \to b$$

(by equations (3), (1) and (4), respectively, and the fact that by (4) $a \to (-)$ is monotone). $\qquad\square$

7.4. Boolean algebras. A *Boolean algebra* is a distributive lattice with complements.

7.4.1. Proposition. *Each Boolean algebra is Heyting.*

Proof. Since distributivity is assumed, the complement of x is the pseudocomplement x^* (see 7.2.1). Put

$$a \to b = a^* \vee b.$$

If $c \leq a \to b$, then $c \wedge a \leq (a^* \vee b) \wedge a = b \wedge a \leq b$; conversely, if $c \wedge a \leq b$, then $c = c \wedge (a \vee a^*) = (c \wedge a) \vee (c \wedge a^*) \leq b \vee a^*$. $\qquad\square$

7.4.2. Proposition. *In a Boolean algebra B we have $a^{**} = a$ for all a. Consequently, the mapping $(a \mapsto a^*) \colon B \to B^{\mathrm{op}}$ is an isomorphism.*

Proof. Since $a \vee a^* = 1$ and $a \wedge a^* = 0$ and B is distributive, a is by 6.3.5 the complement (and by 7.2.1 the pseudocomplement) of a^*. $\qquad\square$

7.4.3. Proposition. *A Heyting algebra L is a Boolean algebra iff $a^{**} = a$ for all $a \in L$.*

Proof. It suffices to verify that if $a^{**} = a$ then $a \vee a^* = 1$. The equation makes $(-)^* \colon L \to L^{\mathrm{op}}$ an isomorphism that, hence, carries joins into meets and vice versa. Thus, from $a \wedge a^* = 0$ we obtain $1 = a^* \vee a^{**} = a^* \vee a$. $\qquad\square$

7.4.4. De Morgan formulas. Since for a Boolean algebra the mapping $a \mapsto a^*$ is an isomorphism $B \to B^{\mathrm{op}}$ we have immediately

Proposition. *In a Boolean algebra,*

$$\left(\bigvee_{i \in J} a_i\right)^* = \bigwedge_{i \in J} a_i^*, \quad \left(\bigwedge_{i \in J} a_i\right)^* = \bigvee_{i \in J} a_i^*$$

whenever the supremum $\bigvee_{i \in J} a_i$ resp. the infimum $\bigwedge_{i \in J} a_i$ exists. $\qquad\square$

7.4.5. Proposition. *A lattice homomorphism from a Boolean algebra into any lattice sends complements to complements.*

Proof. It follows immediately from 6.3.5. □

7.4.6. Recalling the formula for the Heyting operation in 7.4.1 we obtain

Corollary. *A lattice homomorphism between Boolean algebras is always a Heyting homomorphism.* □

7.5. Ultrafilters. In Boolean algebras all the prime filters and prime ideals are maximal. We have

Proposition. *Let F be a proper filter in a Boolean algebra B. Then the following statements are equivalent.*

(1) *F is maximal.*

(2) *F is prime.*

(3) *For every $a \in B$ either $a \in F$ or $a^* \in F$.*

Proof. $(1) \Rightarrow (2)$ is implicitly contained in 6.4.2 but let us present a direct proof. If F is a maximal proper filter, $a \vee b \in F$ and $a \notin F$, set $G = \{x \mid x \vee b \in F\}$. Then G is a filter and it is obviously larger then F (as $a \in G \smallsetminus F$); hence it is not proper so that $0 \in G$ and consequently $b = 0 \vee b \in G$.

$(2) \Rightarrow (3)$ follows from the fact that $a \vee a^* = 1 \in F$.

$(3) \Rightarrow (1)$: Let $F \subset G$ for a filter G. Choose an $a \in G \smallsetminus F$. Since $a \notin F$ there has to be $a^* \in F \subseteq G$, hence $a, a^* \in G$ and finally $0 = a \wedge a^* \in G$. Thus, G is not proper. □

7.5.1. The Boolean Ultrafilter Theorem. A prime ($\equiv$ maximal) filter in a Boolean algebra is called *ultrafilter*. By 6.4.2,

> *every proper filter in a Boolean algebra is contained in a proper ultrafilter.*

This is known as the "Boolean Ultrafilter Theorem".

7.6. Booleanization. Let L be a Heyting semilattice. Set

$$\mathfrak{B}L = \{a \mid a^{**} = a\} \subseteq L.$$

7.6.1. Proposition. $\mathfrak{B}L$ *is a Boolean algebra, with the join given by*

$$a \sqcup b = (a^* \wedge b^*)^*.$$

Proof. For $x = a, b$, since $x \wedge a^* \wedge b^* = 0$ we have $x \leq (a^* \wedge b^*)^*$. Now if $x = x^{**}$ and $a, b \leq x$ then $a^*, b^* \geq x^*$, hence $a^* \wedge b^* \geq x^*$, and $(a^* \wedge b^*)^* \leq x^{**} = x$. By 7.3.1(3), if $c = c^{**}$ then $b \to c = b \to ((c \to 0) \to 0) = (b \wedge (c \to 0)) \to 0 = (b \wedge c^*)^* \in \mathfrak{B}L$. Thus, $\mathfrak{B}L$ admits a Heyting operation, namely the original one from L, and hence it is distributive.

Finally, if $a = a^{**}$ we have $a \sqcup a^* = (a^* \wedge a)^* = 0^* = 1$. □

Note. If L is a Heyting algebra then $a \sqcup b$ is obtained from the original join by the formula $a \sqcup b = (a \vee b)^{**}$: indeed, by the first De Morgan law, $a^* \wedge b^* = (a \vee b)^*$.

7.6.2. Proposition. *The following conditions on a Heyting algebra L are equivalent.*

(1) *The second De Morgan law $(a \wedge b)^* = a^* \vee b^*$ holds.*

(2) *The identity $a^* \vee a^{**} = 1$ holds.*

(3) *Every element of $\mathcal{B}L$ has a complement in L.*

(4) *The identity $(a \vee b)^{**} = a^{**} \vee b^{**}$ holds.*

(5) *$\mathcal{B}L$ is a sublattice of L.*

Proof. (1)$\Rightarrow$(2): $a^* \vee a^{**} = (a \wedge a^*)^* = 0^* = 1$.

(2)$\Rightarrow$(3): Let $a \in \mathcal{B}L$. Then $a \vee a^* = a^{**} \vee a^* = 1$.

(3)$\Rightarrow$(4): Since $(a \vee b)^{**} = (a^* \wedge b^*)^*$, it suffices to check that

$$(a^* \wedge b^*)^* = a^{**} \vee b^{**}.$$

Clearly, $(a^* \wedge b^*) \wedge (a^{**} \vee b^{**}) = 0$. Moreover, since $a^*, b^* \in \mathcal{B}L$, they are complemented (that is, $a^* \vee a^{**} = b^* \vee b^{**} = 1$); so if $x \wedge (a^* \wedge b^*) = 0$ then $a^{**} \vee b^{**} = (a^{**} \vee b^{**}) \vee (x \wedge a^* \wedge b^*) = a^{**} \vee b^{**} \vee x$, and hence $x \leq a^{**} \vee b^{**}$.

(4)$\Rightarrow$(5): For each $a, b \in \mathcal{B}L$, $a \vee b = a^{**} \vee b^{**} = (a \vee b)^{**}$ thus $a \vee b \in \mathcal{B}L$.

(5)$\Rightarrow$(1): $a^* \vee b^* = (a^* \vee b^*)^{**} = (a^{**} \wedge b^{**})^* = (a \wedge b)^*$. $\square$

Appendix II

Categories

We include in this appendix a brief account of the basic notions and facts of category theory used throughout the text, although no doubt it will already be familiar to many of the readers. Our purpose here is only to outline the territory in category theory that we assume familiar. For more information the reader can consult the books [175] or [1].

1. Categories

The basic idea of category theory is to shift attention from the study of objects to the study of the structure of special maps and relations between objects.

1.1. Categories. A *category* $\mathcal{C}$ consists of:

(1) A class obj$\mathcal{C}$ of *objects* (notation: $A, B, C, \dots$).

(2) A class mph$\mathcal{C}$ of *morphisms* (notation: $f, g, h, \dots$). Each morphism f has a *domain* or *source* A (notation: $\mathrm{dom}(f)$) and a *codomain* or *target* B (notation: $\mathrm{codom}(f)$) which are objects of $\mathcal{C}$; this is indicated by writing $f\colon A \to B$.

(3) A *composition law* that assigns to each pair (f, g) of morphisms satisfying $\mathrm{dom}(g) = \mathrm{codom}(f)$ a morphism $g \cdot f\colon \mathrm{dom}(f) \to \mathrm{codom}(g)$, satisfying

    (a) $(h \cdot g) \cdot f = h \cdot (g \cdot f)$ whenever the compositions are defined;

    (b) For each object A of $\mathcal{C}$ there is an *identity morphism* $\mathrm{id}_A\colon A \to A$ such that $f \cdot \mathrm{id}_A = f$ and $\mathrm{id}_A \cdot g = g$ whenever the compositions are defined.

If there is no danger of confusion one writes simply gf for $g \cdot f$; the identity id_A is sometimes denoted by 1_A.

1.1.1. Examples. (a) The category **Set** of sets as objects and functions (mappings) between them. Keep in mind that a function $f\colon A \to B$ in our context is not only a relation $f \subseteq A \times B$ with the standard properties, but such a relation *together with the information* on B (the A can be deduced).

(b) Numerous so-called *concrete categories*: Groups or abelian groups with homomorphisms, other types of algebras with homomorphisms, spaces with continuous maps, etc. Again, a morphism $f\colon A \to B$ carries the information on A and B (it can very well happen that distinct morphisms are defined by the same formula as mappings).

Usually, one denotes such categories by abbreviations indicating the structure in question. Thus, **Top**, **Frm**, **SLat$_1$** etc., but the name can refer to the structure of morphisms like, e.g., in **Loc** vs. **Frm**.

(c) A poset $(X, \leq)$ can be viewed as a category with $\mathsf{obj}(X, \leq) = X$ and morphisms $x \to y$ expressing the facts that $x \leq y$ (thus, a morphism is not a mapping but a statement). It is surprising how many phenomena concerning general categories can be observed in these very simple categories.

(d) The category of relations in which the objects are sets, and morphisms $R\colon A \to B$ are binary relations $R \subseteq A \times B$, with the standard composition.

1.1.2. Dual category. If $\mathcal{C}$ is a category we can take the same classes of objects and morphisms, and interchange the domains and codomains (which leads to an inverted composition). Thus,

$$f\colon A \to B \text{ is now } f\colon B \to A \quad \text{and we have a composition} \quad f * g = g \cdot f.$$

Thus obtained category is called the *dual* or *opposite* of $\mathcal{C}$ and denoted by

$$\mathcal{C}^{\mathrm{op}}.$$

This extremely simple construction is very useful: whenever something is proved for general categories, then it holds in an individual category (in the obvious modification) both for a concept and for the dually defined one (monomorphisms and epimorphisms, limits and colimits, etc.; see below).

1.1.3. Product of categories. If $\mathcal{C}, \mathcal{D}$ are categories we have the category $\mathcal{C} \times \mathcal{D}$ with objects pairs (A, B), $A \in \mathsf{obj}\mathcal{C}, B \in \mathsf{obj}\mathcal{D}$, and morphisms $(A_1, A_2) \to (B_1, B_2)$ pairs (f_1, f_2) with $f_i\colon A_i \to B_i$ in $\mathcal{C}$ resp. $\mathcal{D}$, with obvious composition.

1.1.4. Hom-sets. For objects A, B in a category $\mathcal{C}$ set

$$\mathcal{C}(A, B) = \{f \mid f\colon A \to B \in \mathsf{mph}\mathcal{C}\}.$$

This may, in a quite general case, not be a set in the universe one works in. But in the typical cases the reader may encounter (for instance, in all the examples above),

$$\mathcal{C}(A, B) \quad \text{is a set.}$$

This is usually referred to as stating that the category $\mathcal{C}$ is *locally small* (a category is *small* if, furthermore, also obj$\mathcal{C}$ is a set – this, in contrast, is not the case in the examples above, with the exception of 1.1.1(c)).

Local smallness is often assumed and not explicitly mentioned.

1.2. Subcategories. A category $\mathcal{D}$ is a *subcategory* of a category $\mathcal{C}$ if obj$\mathcal{D} \subseteq$ obj$\mathcal{C}$, $\mathcal{D}(A, B) \subseteq \mathcal{C}(A, B)$ for all $A, B \in$ obj$\mathcal{D}$ and if the multiplication and the units are as in $\mathcal{C}$. If, moreover,

$$\forall A, B \in \text{obj}\mathcal{D}, \quad \mathcal{D}(A, B) = \mathcal{C}(A, B)$$

we speak of a *full subcategory*.

Thus, e.g., the category of abelian groups is a full subcategory of that of all groups, and the category of metric spaces and uniformly continuous maps is a subcategory of the category of metric spaces and continuous maps but not a full one.

1.3. The most basic special morphisms. A morphism $f: A \to B$ is a *monomorphism* (resp. *epimorphism*) if

$$\forall g, h, \quad f \cdot g = f \cdot h \implies g = h$$

$$(\text{resp. } \forall g, h, \quad g \cdot f = h \cdot f \implies g = h).$$

Thus, monomorphisms in $\mathcal{C}$ are epimorphisms in $\mathcal{C}^{\text{op}}$, and vice versa.

Note that in **Set**, monomorphisms coincide with one-one maps, and epimorphisms are precisely the onto maps. In standard categories of algebras, monomorphisms coincide with the embeddings of subalgebras up to isomorphism. In more general concrete categories monomorphisms are often the general one-one maps not necessarily imprinting the structure of the bigger objects; see Section 4 below.

A morphism $f: A \to B$ is an *isomorphism* if there is a morphism $g: B \to A$ such that $fg = \text{id}_B$ and $gf = \text{id}_A$. The morphism g is uniquely determined, usually denoted by f^{-1}, and called the *inverse* of f.

1.3.1. The following are straightforward observations.

(a) *Each isomorphism is both a monomorphism and an epimorphism while in general categories morphisms that are both mono- and epimorphisms (sometimes one speaks of bimorphisms) need not be isomorphisms.*

(b) *If fg is a monomorphism then g is a monomorphism; if fg is an epimorphism then f is an epimorphism.*

(c) *If $fg = \text{id}$ and if either f is a monomorphism or g is an epimorphism then f and g are (mutually inverse) isomorphisms.*

1.3.2. Exercise. Check all the statements in 1.3 and 1.3.1.

2. Functors and natural transformations

2.1. A *functor* $F: \mathcal{C} \to \mathcal{D}$ consists of maps $\text{obj}\mathcal{C} \to \text{obj}\mathcal{D}$ and $\text{mph}\mathcal{C} \to \text{mph}\mathcal{D}$ (both are usually denoted by the symbol of the whole of the functor) that respect the domains, codomains, unit and composition. Thus,

$$\text{if } f: A \to B \text{ then } F(f): F(A) \to F(B),$$

$$F(\text{id}_A) = \text{id}_{F(A)} \text{ and } F(gf) = F(g)F(f).$$

2.1.1. Examples. (a) The forgetful functors $U: \mathcal{C} \to \mathbf{Set}$ from a concrete category $\mathcal{C}$ associating with an object the underlying set, and with a morphism the underlying map.

(b) Free functors $F: \mathbf{Set} \to \mathcal{A}$ where $\mathcal{A}$ is a (nice) category of algebras and $F(M)$ is the free algebra generated by M.

(c) Functors of algebraic topology.

(d) The Stone-Čech compactification from the category of completely regular spaces into that of compact ones.

(e) Monotone maps $f: (X, \leq) \to (Y, \leq)$ viewed in the sense of 1.1.1(c).

2.1.2. A very important example – the hom-functor. Let $\mathcal{C}$ be a category. Define

$$\mathcal{C}: \mathcal{C}^{\text{op}} \times \mathcal{C} \to \mathbf{Set}$$

with $\mathcal{C}(A, B)$ as in 1.1.4 and by setting for $f: A' \to A$ and $g: B \to B'$ (morphisms in $\mathcal{C}$),

$$\mathcal{C}(f, g)(\phi) = g\phi f.$$

2.1.3. Contravariant functors. Often we encounter a construction F with the properties similar like in 2.1, only modified by "reversing the arrows", that is, such that

$$\text{if } f: A \to B \text{ then } F(f): F(B) \to F(A),$$

$$F(\text{id}_A) = \text{id}_{F(A)} \text{ and } F(gf) = F(f)F(g).$$

Then one speaks of a *contravariant functor*.

The reader may wonder why such concept should be introduced at all: a contravariant functor may be surely viewed as a standard functor

$$F: \mathcal{C}^{\text{op}} \to \mathcal{D} \quad \text{or} \quad F: \mathcal{C} \to \mathcal{D}^{\text{op}} \text{ ?}$$

The hitch is in the "or". Sometimes it is obvious which of the categories should be inverted – thus for instance the contravariant functor $\Omega: \mathbf{Top} \to \mathbf{Frm}$ is naturally a functor $\mathbf{Top} \to \mathbf{Loc}$ rather than $\mathbf{Top}^{\text{op}} \to \mathbf{Frm}$, but more often than not we have constructions that do not call for inverting preferably any of the categories involved (for instance the typical dualities, but let us take a much simple case, the preimage functor $\mathfrak{P}\colon \mathbf{Set} \to \mathbf{Set}$ with $\mathfrak{P}(X) = \{A \mid A \subseteq X\}$, $\mathfrak{P}(f)(B) = f^{-1}[B]$).

2.2. Natural transformations. A *natural transformation*

$$\alpha\colon F \xrightarrow{\;\cdot\;} G$$

between two functors $F, G\colon \mathcal{C} \to \mathcal{D}$ is a collection of morphisms in $\mathcal{D}$

$$\alpha = (\alpha_A)_{A\in\mathrm{obj}\mathcal{C}}, \quad \alpha_A\colon F(A) \to G(A)$$

such that for every morphism $f\colon A \to B$ in $\mathcal{C}$, the following diagram commutes in $\mathcal{D}$:

$$
\begin{array}{ccc}
F(A) & \xrightarrow{\;\alpha_A\;} & G(A) \\
{\scriptstyle F(f)}\big\downarrow & & \big\downarrow{\scriptstyle G(f)} \\
F(B) & \xrightarrow[\;\alpha_B\;]{} & G(B)
\end{array}
$$

If each α_A is an isomorphism, one speaks of a *natural equivalence* and writes

$$F \cong G.$$

Convention. If there is no danger of confusion one often omits the dot in $\xrightarrow{\;\cdot\;}$, or speaks just of a *transformation*.

2.2.1. Examples. (a) Define $Q\colon \mathbf{Set} \to \mathbf{Set}$ by setting

$$Q(X) = X \times X, \quad Q(f) = ((x,y) \mapsto (f(x), f(y))).$$

The diagonal maps $\Delta_X = (x \mapsto (x,x))$ constitute a natural transformation $\Delta\colon \mathrm{Id} \xrightarrow{\;\cdot\;} Q$; similarly one has natural transformations "first projection" and "second projection" $Q \xrightarrow{\;\cdot\;} \mathrm{Id}$.

(b) For a vector space V consider the dual V^*, the vector space of linear mappings $V \to \mathbb{R}$, and for linear mapping $f\colon V \to W$ the $f^*\colon W^* \to V^*$ defined by $f^*(\alpha) = \alpha \cdot f$. The maps

$$\kappa_X = (x \mapsto \widetilde{x})\colon V \to V^{**}$$

constitute a natural transformation κ from the identical functor into that of double dual.

In the category of *finite-dimensional* vector spaces, κ is a natural equivalence. Note that in this category already the V^* is isomorphic with V but there is no natural equivalence of the identity functor and the first dual.

(c) (A contravariant example: the characteristic function.) Consider the functor $\mathfrak{P}$ from 2.1.3 and $\mathcal{C}(-, \{0,1\})$ viewed as a contravariant functor $\mathbf{Set} \to \mathbf{Set}$. The

$$\chi = (\chi_X\colon \mathfrak{P}(X) \to \mathcal{C}(X, \{0,1\}))_X \text{ defined by } \chi_X(A)(x) = 1 \text{ iff } x \in A$$

constitutes a natural equivalence.

2.2.2. Proposition. *Let* $\mathcal{C}(A, -) \cong \mathcal{C}(B, -)$. *Then* A *and* B *are isomorphic. More generally, if* $\mathcal{C}(F(-), -) \cong \mathcal{C}(G(-), -)$ *for functors* $F, G\colon \mathcal{D} \to \mathcal{C}$ *then* $F \cong G$. *Similarly for* $\mathcal{C}(-, A) \cong \mathcal{C}(-, B)$ *and* $\mathcal{C}(-, F(-)) \cong \mathcal{C}(-, G(-))$.

Proof. I. Take a natural equivalence $\varepsilon\colon \mathcal{C}(A, -) \cong \mathcal{C}(B, -)$ with inverse $\bar{\varepsilon}$, and set

$$\beta = \varepsilon_A(\mathrm{id}_A)\colon B \to A \quad \text{and} \quad \alpha = \bar{\varepsilon}_B(\mathrm{id}_B)\colon A \to B.$$

Now

$$\alpha \cdot \beta = \mathcal{C}(\mathrm{id}, \alpha)(\varepsilon_A(\mathrm{id}_A)) = \varepsilon_B(\mathcal{C}(\mathrm{id}, \alpha)(\mathrm{id}_A)) = \varepsilon_B(\alpha) = \varepsilon_B(\bar{\varepsilon}_B(\mathrm{id}_B)) = \mathrm{id}_B$$

and similarly $\beta \cdot \alpha = \mathrm{id}_A$.

II. Let $\varepsilon\colon \mathcal{C}(F(-), -) \cong \mathcal{C}(G(-), -)$ be a natural equivalence. Thus, for each $f\colon B \to A$ and $g\colon X \to Y$ the diagram

$$
\begin{array}{ccc}
\mathcal{C}(F(A), X) & \xrightarrow{\ \varepsilon_{A,X}\ } & \mathcal{C}(G(A), X) \\
{\scriptstyle \mathcal{C}(F(f), g)}\big\downarrow & & \big\downarrow {\scriptstyle \mathcal{C}(G(f), g)} \\
\mathcal{C}(F(B), Y) & \xrightarrow[\ \varepsilon_{B,Y}\ ]{} & \mathcal{C}(G(B), Y)
\end{array}
$$

is commutative, that is,

$$\varepsilon_{B,Y}(g \cdot \phi \cdot F(f)) = g \cdot \varepsilon_{A,X}(\phi) \cdot G(f).$$

Set

$$\eta_A = \varepsilon_{A, FA}(\mathrm{id}_{FA})\colon G(A) \to F(A).$$

By the procedure from I we see that each η_A is an isomorphism. The system $(\eta_A)_A$ is a natural equivalence: we have

$$
\begin{aligned}
\eta_A \cdot G(f) &= \mathrm{id}_{FA} \cdot \varepsilon_{A, FA}(\mathrm{id}_{FA}) \cdot G(f) \\
&= \varepsilon_{B, FA}(\mathrm{id}_{FA} \cdot \mathrm{id}_{FA} \cdot F(f)) = \varepsilon_{B, FA}(F(f) \cdot \mathrm{id}_{FB} \cdot \mathrm{id}_{FB}) \\
&= F(f) \cdot \varepsilon_{B, FB}(\mathrm{id}_{FB}) \cdot G(\mathrm{id}_{FB}) = F(f) \cdot \eta_B. \qquad \square
\end{aligned}
$$

2.3. Categories of functors. Let K be a small category and $\mathcal{C}$ a general one. Then the collection of functors $F\colon K \to \mathcal{C}$ as objects, and of the natural transformations $\tau\colon F \overset{\cdot}{\to} G$ as morphisms constitute a category which will be denoted by

$$[K, \mathcal{C}]$$

(assuming K small is a reserve preventing clashes with set theory).

2.3.1. Yoneda embedding. For a small category K define a functor

$$\mathsf{Y}\colon K^{\mathrm{op}} \to [K, \mathbf{Set}]$$

by setting $Y(a) = K(a, -)$ and $Y(\alpha) = (Y(\alpha)_c = K(\alpha, 1_c))_c \colon Y(a) \to Y(b)$ for $\alpha \colon b \to a$ (hence $Y(\alpha)_c(\phi) = \phi\alpha$). $Y(\alpha)$ is obviously a natural transformation since $K(\alpha, 1_d)K(1_a, \phi) = K(\alpha, \phi) = K(1_b, \phi)K(\alpha, 1_c)$.

2.3.2. Proposition. Y *is a functor embedding the category* K^{op} *into* $[K, \mathbf{Set}]$ *as a full subcategory.*

Proof. Trivially $Y(a) \neq Y(b)$ if $a \neq b$ and if $\alpha \neq \beta$, $\alpha, \beta \colon b \to a$ then $Y(\alpha)_a(1_a) = \alpha \neq \beta = Y(\beta)_a(1_a)$; hence Y is one-one. Now let $\tau \colon Y(a) \to Y(b)$ be a natural transformation. Set $\alpha = \tau_a(1_a) \in Y(b)(a) = K(b, a)$. Then because of the commuting diagrams

$$
\begin{array}{ccc}
Y(a)(a) & \xrightarrow{\ \tau_a\ } & Y(b)(a) \\
{\scriptstyle K(1_a, \phi)}\big\downarrow & & \big\downarrow {\scriptstyle K(1_b, \phi)} \\
Y(a)(c) & \xrightarrow{\ \tau_c\ } & Y(b)(c)
\end{array}
$$

we have $\tau_c(\phi) = \tau_c(K(1_a, \phi)(1_a)) = K(1_b, \phi)(\tau_a(1_a)) = \phi\alpha = Y(\alpha)_c(\phi)$, and the embedding is full. $\qquad\square$

2.4. Observation and notation. Let $\tau \colon F \to G$ be a natural transformation, $F, G \colon \mathcal{C} \to \mathcal{D}$. Let $H \colon \mathcal{B} \to \mathcal{C}$ and $E \colon \mathcal{D} \to \mathcal{E}$ be functors. Then we have natural transformations

$$\tau H \colon FH \to GH \quad \text{and} \quad E\tau \colon EF \to EG$$

defined by

$$(\tau H)_X = \tau_{H(X)} \quad \text{and} \quad (E\tau)_X = E(\tau_X).$$

3. Some basic constructions

3.1. Products and coproducts (sums). Let A_1, A_2 be objects of $\mathcal{C}$. A *product* of A_1 and A_2 is a pair of morphisms

$$p_i \colon A \to A_i, \quad i = 1, 2$$

such that for every pair $f_i \colon X \to A_i$ in $\mathcal{C}$ there is exactly one $f \colon X \to A$ such that $p_i f = f_i$ for both i:

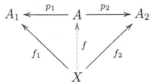

More generally, if A_i, $i \in J$, is a collection of objects of C, their *product* is a system of morphisms

$$p_i \colon A \to A_i, \quad i \in J$$

such that for every system $f_i \colon X \to A_i$, $i \in J$, in C there is exactly one $f \colon X \to A$ such that $p_i f = f_i$ for all $i \in J$.

Products, of course, do not have to exist. If they, or some of them, do exist we say that C has products (finite products, products of pairs). The standard notation for the A in a product is

$$A_1 \times A_2, \quad \prod_{i \in J} A_i, \quad \text{and similarly.}$$

Dually, a *coproduct*, or *sum* of a collection A_i, $i \in J$, is a system of morphisms

$$q_i \colon A_i \to A, \quad i \in J$$

such that for every system $f_i \colon A_i \to X$, $i \in J$, in C there is exactly one $f \colon A \to X$ such that $f q_i = f_i$ for all $i \in J$.

Notes. (1) In particular, the product of a void collection is an object S such that for every $A \in \mathrm{obj}\,C$ there is precisely one morphism $A \to S$; such S is called a *terminal object* or *singleton*. Similarly the coproduct of a void collection is an object I such that for every $A \in \mathrm{obj}\,C$ there is precisely one morphism $I \to A$; such I is called an *initial object*.

(2) In **Set** the product is the cartesian product with p_i the projections. Similarly in typical everyday life concrete categories we have the product carried by the cartesian product of the carriers of the individual A_i, endowed with a suitable structure. The sum in **Set** is the disjoint union, with the embeddings of the summands. Unlike products, the coproducts in everyday concrete categories seldom look like in **Set** and vary in the form.

(3) In a poset $(X, \leq)$ viewed as a category (1.1.1(c)) the products (resp. coproducts) correspond to the suprema (resp. infima).

3.1.1. Exercise. If for $i \in J$, $p_{ik} \colon A_i \to A_{ik}$, $k \in K_i$, are products, and if $q_i \colon A \to A_i$ is a product then

$$p_{ik} q_i \colon A \to A_{ik}, \quad (i,k) \in \bigcup_{j \in J} \{j\} \times K_i$$

is a product. Consequently, if C has products of pairs then it has products of all non-empty finite collections.

3.2. Equalizers and coequalizers. Let $f, g \colon A \to B$ be morphisms. The *equalizer* of the pair, often indicated as $\mathrm{Equ}(f, g)$ is a morphism $e \colon E \to A$ such that

$fe = ge$, and

if $fh = gh$ for an $h\colon X \to A$ then there is exactly one $\overline{h}\colon X \to E$ such that $e\overline{h} = h$:

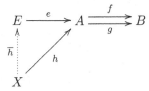

Dually, the *coequalizer* of f, g, indicated by $\mathrm{Coequ}(f, g)$ is a $c\colon B \to C$ such that $cf = cg$, and if $hf = hg$ for an $h\colon B \to X$ then there is exactly one $\overline{h}\colon C \to X$ such that $\overline{h}c = h$.

3.3. Pullbacks and pushouts. Let $f\colon A \to C$, $g\colon B \to C$ be morphisms. A *pullback* of f, g is a commutative diagram

$$
\begin{array}{ccc}
P & \xrightarrow{\ \overline{f}\ } & B \\
{\scriptstyle \overline{g}}\downarrow & & \downarrow{\scriptstyle g} \\
A & \xrightarrow[\ f\ ]{} & C
\end{array}
\qquad\text{(pb)}
$$

such that for every commutative diagram

$$
\begin{array}{ccc}
X & \xrightarrow{\ \alpha\ } & B \\
{\scriptstyle \beta}\downarrow & & \downarrow{\scriptstyle g} \\
A & \xrightarrow[\ f\ ]{} & C
\end{array}
$$

there is precisely one $\gamma\colon X \to P$ making the diagram

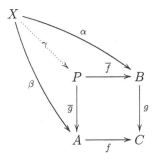

commute, i.e., such that $\overline{f}\gamma = \alpha$ and $\overline{g}\gamma = \beta$.

Similarly, a *pushout* of $f: A \to B$ and $g: A \to C$ is a commutative

$$
\begin{array}{ccc}
A & \xrightarrow{\;f\;} & B \\
\downarrow{\scriptstyle g} & & \downarrow{\scriptstyle \overline{g}} \\
C & \xrightarrow[\;\overline{f}\;]{} & P
\end{array}
\qquad\text{(po)}
$$

such that for every pair $\alpha: C \to X$ and $\beta: B \to X$ satisfying $\alpha g = \beta f$ there is precisely one $\gamma: P \to X$ such that $\gamma \overline{f} = \alpha$ and $\gamma \overline{g} = \beta$.

3.3.1. Proposition. *Let a category have products of pairs of objects, and equalizers. Then it has pullbacks.*

Proof. For $f: A \to C$ and $g: B \to C$ consider, first, a product

$$
A \xleftarrow{\;p_A\;} A \times B \xrightarrow{\;p_B\;} B
$$

and then the equalizer $e: P \to A \times B$ of fp_A and gp_B. It is an easy exercise that we can take $\overline{g} = p_A e$ and $\overline{f} = p_B e$ to obtain a pullback like in (pb) above.   □

3.3.2. Proposition. *Let in the pullback* (pb) g *be a monomorphism. Then* $\overline{g}$ *is a monomorphism.*

   Similarly for pushouts and epimorphisms.

Proof. Let $\overline{g}u = \overline{g}v$. Then $g\overline{f}u = f\overline{g}u = f\overline{g}v = g\overline{f}v$ and since g is a monomorphism, $\overline{f}u = \overline{f}v$. Set $\beta = \overline{g}u = \overline{g}v$ and $\alpha = \overline{f}u = \overline{f}v$. Then $g\alpha = f\beta$ and any of u, v have the property required of the γ in the definition. By the unicity, $u = v$.   □

4.   More special morphisms. Factorization

4.1. Orthogonal morphisms. An ordered pair (f, g) of morphisms of a category is said to be *orthogonal* if for every commutative diagram

$$
\begin{array}{ccc}
A & \xrightarrow{\;f\;} & B \\
\downarrow{\scriptstyle u} & & \downarrow{\scriptstyle v} \\
C & \xrightarrow[\;g\;]{} & D
\end{array}
$$

there is a morphism w such that $gw = v$ and $wf = u$:

We speak of a pair $(\mathcal{E}, \mathcal{M})$ of classes of morphisms as *orthogonal* if for any $f \in \mathcal{E}$, $g \in \mathcal{M}$, (f, g) is orthogonal.

4.2. Extremal, strong and regular. A monomorphism m is *extremal* if in every decomposition $m = f \cdot e$ with e an epimorphism the e is an isomorphism (note that by 1.3.1(b) such an e is always a bimorphism). Dually, one defines an *extremal epimorphism* as an epimorphism e such that in every decomposition $e = m \cdot f$ with m a monomorphism the m is an isomorphism.

A monomorphism m (resp. epimorphism e) is *strong* if for each epimorphism e (resp. monomorphism m) the pair (e, m) is orthogonal.

A monomorphism m (resp. epimorphism e) is *regular* if it is an equalizer (resp. coequalizer) of a pair of morphisms.

4.2.1. Proposition. *We have the implications*

$$\text{regular} \quad \Rightarrow \quad \text{strong} \quad \Rightarrow \quad \text{extremal}.$$

In a category with pushouts,

$$\text{strong} \quad \equiv \quad \text{extremal}.$$

Proof. I. Let $m = \mathrm{Equ}(f, g)$ and let $mu = ve$ for an epimorphism e. Then $fve = fmu = gmu = gve$ so that $fv = gv$ and there is a w such that $mw = v$, by equalization. Now $mwe = ve = mu$ and hence also $we = u$.

II. Let m be strong and let $m = fe$ with e epimorphic. Then we have a commutative diagram

$$
\begin{array}{ccc}
A & \xrightarrow{\ e\ } & C \\
{\scriptstyle \mathrm{id}_A}\big\downarrow & & \big\downarrow{\scriptstyle f} \\
A & \xrightarrow[\ m\ ]{} & B
\end{array}
$$

and the given w satisfies $we = \mathrm{id}_A$, so that also $ewe = e$ and as e is an epimorphism, $ew = \mathrm{id}_C$, and e is an isomorphism.

III. Finally let $\mathcal{C}$ have pushouts and let m be extremal. Consider a commutative

and a pushout

Then $\bar{e}$ is an epimorphism and we have a γ such that $\gamma\bar{e} = m$ so that $\bar{e}$ is an isomorphism and we can set $w = \bar{e}^{-1}\bar{u}$. □

4.2.2. Proposition. *In a category with pushouts, extremal resp. strong monomorphisms are closed under composition.*

Proof. By 4.2.1 it suffices to speak on extremals. Let m_1, m_2 be extremal monomorphisms and let $m_1 m_2 = fe$ with an epimorphism e. Pushing out m_2, e we obtain $\bar{m}_2, \bar{e}$ with $\bar{e}$ epimorphic and $\bar{m}_2 e = \bar{e} m_2$, and a g such that $g\bar{m}_2 = f$ and $g\bar{e} = m_1$; by extremality of m_1, $\bar{e}$ is an isomorphism. Now for $h = \bar{e}^{-1}$ we have $h\bar{m}_2 e = h\bar{e} m_2 = m_2$ and hence, by extremality of m_2, e is an isomorphism. □

4.2.3. Recall 1.3. Extremal resp. strong monomorphisms model embeddings of subobjects inheriting the structure of the bigger object. This is best seen in the definition of extremal monomorphisms claiming that one cannot insert a "strengthening of the structure" (which would be the bimorphism e if it were non-trivial).

Proposition 3.3.2 can be interpreted as obtaining a preimage under f of a (very) weak subobject g. This holds for more genuine subobjects as well. We have

Proposition. *Let*

$$
\begin{array}{ccc}
P & \xrightarrow{\bar{f}} & B \\
{\scriptstyle\bar{m}}\downarrow & & \downarrow{\scriptstyle m} \\
A & \xrightarrow{f} & C
\end{array}
$$

be a pullback in a category with pushouts. Let m be an extremal monomorphism. Then $\bar{m}$ is extremal as well.

Proof. Let $\bar{m} = h\bar{e}$ with an epimorphism $\bar{e}$. Consider a pushout

$$
\begin{array}{ccc}
P & \xrightarrow{\bar{f}} & B \\
{\scriptstyle\bar{e}}\downarrow & & \downarrow{\scriptstyle e} \\
\cdot & \xrightarrow{f'} & \cdot
\end{array}
$$

By the commutativity of the pullback we have started with we have an $\bar{h}$ such that $\bar{h}f' = fh$ and $\bar{h}e = m$. Since e is an epimorphism by 3.3.2, it is by the extremality of m an isomorphism. Now we have $me^{-1}f'\bar{e} = \bar{h}f'\bar{e} = fh\bar{e}$, hence $me^{-1}f' = fh$ and from the pullback property we obtain a ϕ such that in particular $\bar{m}\phi = h$ and hence $\bar{m}\phi\bar{e} = h\bar{e} = \bar{m}$ and finally $\phi\bar{e} = \mathrm{id}_P$ which makes $\bar{e}$ an isomorphism, since it is epimorphic. □

4.3. Factorization systems. A *factorization system* in a category is an orthogonal pair $(\mathcal{E}, \mathcal{M})$ of classes of morphisms such that

(1) each of $\mathcal{E}, \mathcal{M}$ is closed under isomorphisms, and
(2) every morphism $f \colon A \to B$ can be decomposed as

$$A \xrightarrow{\ e\ } C \xrightarrow{\ m\ } B$$

$$\underbrace{\phantom{A \xrightarrow{e} C \xrightarrow{m} B}}_{f}$$

with $e \in \mathcal{E}$ and $m \in \mathcal{M}$.

4.3.1. Fact. Note that in a factorization system each commutative diagram

can be completed, for arbitrary decompositions $f = me$, $f' = m'e'$ as above, into a commutative diagram

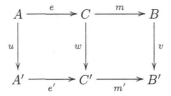

For that, just apply the orthogonality to the commutative diagram

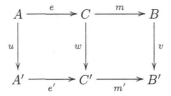

Choosing $u = \mathrm{id}_A$ and $v = \mathrm{id}_B$ we immediately see that

the decomposition $f = me$ from condition (2) is unique up to isomorphism.

5.   Limits and colimits

5.1. Let K be a small category and $\mathcal{C}$ a general one. A K-*diagram* (briefly, *diagram*) in $\mathcal{C}$ is a functor $D\colon K \to \mathcal{C}$.

A *lower* (resp. *upper*) *bound* of D is a collection of morphisms in $\mathcal{C}$

$$(f_a\colon X \to D(a))_{a\in\mathsf{obj}K} \quad \text{resp.} \quad (f_a\colon D(a) \to X)_{a\in\mathsf{obj}K}$$

such that for every $\alpha\colon a \to b$ in K,

$$D(\alpha)f_a = f_b \quad (\text{resp.} \quad f_b D(\alpha) = f_a).$$

A lower bound

$$(p_a\colon L \to D(a))_{a\in\mathsf{obj}K}$$

is a *limit* (an upper bound $(q_a\colon D(a) \to C)_{a\in\mathsf{obj}K}$ is a *colimit*) of D if for every lower (resp. upper) bound as above there is precisely one $f\colon X \to L$ (resp. $f\colon C \to X$) such that

$$\forall a \in \mathsf{obj}K,\ p_a f = f_a \quad (\text{resp.} \quad f q_a = f_a).$$

5.1.1. Proposition. *A (co)limit of a diagram is uniquely determined, up to isomorphism.*

Proof. Let $(p_a\colon L \to D(a))_{a\in\mathsf{obj}K}$, $(p'_a\colon L' \to D(a))_{a\in\mathsf{obj}K}$ be limits. Since they are in particular lower bounds we have $p\colon L \to L'$ and $p'\colon L' \to L$ such that $p'_a p = p_a$ and $p_a p' = p'_a$ for all a. Consequently $p_a p' p = p_a$ and $p'_a p p' = p'_a$ and since these systems of equations are also satisfied by the identities we have, by the unicity, $pp' = \mathrm{id}_{L'}$ and $p'p = \mathrm{id}_L$. $\square$

5.1.2. Notes. (1) Another term for a lower bound is *cone*, similarly *cocone* for an upper bound. Note that in a poset $(X,\leq)$ the standard lower resp. upper bounds correspond to the categorical ones if we think of $(X,\leq)$ in the perspective of 1.1.1(c); the limits then coincide with the infima and colimits with the suprema.

(2) Sometimes one loosely speaks about the L in the $(p_a\colon L \to D(a))_{a\in\mathsf{obj}K}$ above as of the limit. Similarly with colimits.

(3) Recall 2.2. In a slightly more fancy way we can define a lower bound as a natural transformation $(p_a)_a\colon C_L \overset{\cdot}{\to} D$, where C_L is the constant functor sending all the objects to L and all the morphisms to id_L. Note that the isomorphism in 5.1.1 is not just an isomorphism $L \cong L'$; one has more: it is an isomorphism in $[K,\mathcal{C}]$.

5.2. Products, equalizers and pullbacks as limits. Obviously, a product of a collection A_i, $i \in J$, is a limit of a diagram defined on the discrete category with objects $i \in J$ and no non-identical morphisms.

For equalizers consider the category K with two objects a, b and two non-trivial morphisms $\phi, \gamma\colon a \to b$. The $f, g\colon A \to B$ in 3.2 can be viewed as a $D\colon K \to$

$\mathcal{C}$ sending a to A, b to B, ϕ to f and γ to g. An $h\colon X \to A$ has the property that $fh = gh$ iff there exists an $h'\colon X \to B$ such that $h' = D(\phi)h = D(\gamma)h$, that is, iff it can be extended to a lower bound $(h_a = h, h_b)$ of D. Thus, equalizers can be viewed as limits of diagrams $K \to \mathcal{C}$.

For pullbacks consider the category K with three objects a, b, c and non-identical morphisms $\phi\colon a \to c$ and $\gamma\colon b \to c$. Now

$$A \xrightarrow{\ f\ } C \xleftarrow{\ g\ } B \tag{5.2.1}$$

can be viewed as the functor $(a \mapsto A,\ b \mapsto B,\ c \mapsto C,\ \phi \mapsto f,\ \gamma \mapsto g)$ and the $h_a = \beta\colon X \to A$ and $h_b = \alpha\colon X \to B$ can be extended to a lower bound (h_a, h_b, h_c) of D iff $f\beta = g\alpha$. Thus, the pullback can be viewed as the limit of the diagram (5.2.1).

Similarly, coproducts, coequalizers and pushouts are colimits.

5.3. A category $\mathcal{C}$ is said to be *complete* (resp. *cocomplete*) if each diagram in $\mathcal{C}$ has a limit (resp. colimit).

5.3.1. Theorem. [Freyd] *A category $\mathcal{C}$ is complete iff it has all products and pullbacks. Dually, it is cocomplete iff it has all coproducts and pushouts.*

Proof. Let $D\colon K \to \mathcal{C}$ be a diagram. We can assume that

$$\forall\, a \in \mathsf{obj}K \ \ \exists\, \alpha \in M = \mathsf{mph}K, \quad \alpha\colon b \to c \neq a$$

(else there is only one object $a \in \mathsf{obj}K$; then add formally an object b and a single morphism $\alpha\colon a \to b$, and extend the lower bounds by setting $f_b = D(\alpha)f_a$).

Consider the products (the latter one is the power of P, the product of P, M times repeated)

$$\left(p_a\colon P = \prod_{b \in \mathsf{obj}K} D(b) \to D(a) \right)_{a \in \mathsf{obj}K}, \qquad (q_\alpha\colon P^M \to P)_{\alpha \in M}$$

and define $m\colon P \to P^M$ by

$$p_a q_\alpha m = \begin{cases} p_a & \text{if } \alpha\colon b \to c \neq a \\ D(\alpha)p_b & \text{if } \alpha\colon b \to a. \end{cases}$$

Consider the pullback

$$
\begin{array}{ccc}
L & \xrightarrow{\ \overline{m}\ } & P \\
{\scriptstyle \overline{\Delta}}\big\downarrow & & \big\downarrow{\scriptstyle \Delta} \\
P & \xrightarrow{\ m\ } & P^M
\end{array}
$$

where Δ is the diagonal (defined by $q_\alpha\Delta = \mathrm{id}_P$ for all α).

Claim 1. $\overline{\Delta} = \overline{m}$.

Proof of Claim 1. Take an arbitrary a and choose a $\alpha: b \to c \neq a$. We have

$$p_a\overline{\Delta} = p_a q_\alpha m \overline{\Delta} = p_a q_\alpha \Delta \overline{m} = p_a \overline{m}. \qquad \square$$

Now define $\lambda_a: L \to D(a)$ by setting

$$\lambda_a = p_a \overline{\Delta} = p_a \overline{m}.$$

Claim 2. *For every* $\alpha: b \to a$ *in* M *we have* $D(\alpha)\lambda_b = \lambda_a$.

Proof of Claim 2. $D(\alpha)p_b\overline{\Delta} = p_a q_\alpha m \overline{\Delta} = p_a q_\alpha \Delta \overline{m} = p_a \overline{m} = \lambda_a. \qquad \square$

Now we will show that the system $\lambda_a: L \to D(a)$ is a limit of the diagram D.

Indeed, let $f_a: X \to D(a)$ be such that for each $\alpha: b \to a$, $D(\alpha)f_b = f_a$. Define $g: X \to P$ by $p_a g = f_a$. Then we have

$$p_a q_\alpha mg = \begin{cases} p_a g = f_a = p_a q_\alpha \Delta g & \text{(if } \alpha: b \to c \neq a) \\ D(\alpha)p_b g = f_a = p_a q_\alpha \Delta g & \text{(if } \alpha: b \to a) \end{cases}$$

and hence $mg = \Delta g$, and we have an $f: X \to L$ such that $g = \overline{\Delta}f = \overline{m}f$. Now we have $\lambda_a f = p_a \overline{\Delta} f = f_a$. For unicity suppose $\lambda_a h = f_a$ for some $h: X \to L$. Then $p_a \overline{\Delta} h = f_a$ and hence $\overline{\Delta} h = \overline{m} h = g$ because of the unicity in products, and by the pullback unicity we conclude $h = f$. $\qquad \square$

From 3.3.1 we now immediately obtain

5.3.2. Corollary. [Maranda] *A category with arbitrary products and equalizers is complete.* $\qquad \square$

6. Adjunction

Adjunction is a central concept of category theory, and it can be claimed that is one of the most important and pervasive concepts in mathematics.

6.1. Functors $L: \mathcal{C} \to \mathcal{D}$ and $R: \mathcal{D} \to \mathcal{C}$ are said to be *adjoint*, L on the left and R on the right (or, we say that L is a *left adjoint* of R resp. that R is a *right adjoint* of L), symbolically

$$L \dashv R,$$

if there is a natural equivalence

$$\varepsilon_{A,B}: \mathcal{D}(L(A), B) \cong \mathcal{C}(A, R(B)).$$

That is, the $\varepsilon_{A,B}$ are invertible maps and for every $f: A' \to A$ and $g: B \to B'$ the diagram

$$
\begin{array}{ccc}
\mathcal{D}(L(A), B) & \xrightarrow{\varepsilon_{A,B}} & \mathcal{C}(A, R(B)) \\
\downarrow{\scriptstyle \mathcal{D}(L(f),g)} & & \downarrow{\scriptstyle \mathcal{C}(f,R(g))} \\
\mathcal{D}(L(A'), B') & \xrightarrow[\varepsilon_{A',B'}]{} & \mathcal{C}(A', R(B'))
\end{array}
$$

commutes. Explicitly, for each $\phi: L(A) \to B$ we have

$$R(g) \cdot \varepsilon_{A,B}(\phi) \cdot f = \varepsilon_{A',B'}(g \cdot \phi \cdot L(f)). \qquad (\varepsilon)$$

We will write $\bar{\varepsilon}$ for the inverse of ε so that we also have, for $\psi: A \to R(B)$,

$$g \cdot \bar{\varepsilon}_{A,B}(\psi) \cdot L(f) = \bar{\varepsilon}_{A',B'}(R(g) \cdot \psi \cdot f). \qquad (\bar{\varepsilon})$$

Note that by 2.2.2, for a functor there is, up to natural equivalence, at most one right resp. left adjoint.

6.2. Examples. In fact mathematics swarms with adjunction situations in that or other form. Here we will present just a few examples.

(1) Take a fixed set A and consider the functors $L: \textbf{Set} \to \textbf{Set}$ and $R: \textbf{Set} \to \textbf{Set}$ defined by

$$L(X) = X \times A, \; L(f)(x,a) = (f(x), a); \quad R = \textbf{Set}(A, -).$$

Associating with a function $\phi: X \times A \to Y$ the $\tilde{\phi}: X \to \textbf{Set}(A, Y)$ by setting $\tilde{\phi}(x)(a) = \phi(x, a)$ we obtain an adjunction $L \dashv R$. ("Taking one of the variables in a function of two variables for parameter".)

(2) Let $\mathcal{A}$ be a category of algebras in which free algebras exist (any natural category of algebras the reader will think of). That is, we have for every set X an algebra $F(X) \in \text{obj}\mathcal{A}$ and an embedding (of X as a set of generators) $u_X: X \to F(X)$ such that for each algebra $A \in \mathcal{A}$ and each mapping $f: X \to A$ there is precisely one homomorphism $h: F(X) \to A$ such that $h \cdot u_X = f$. In fact, the f is a mapping $X \to UF(X)$ where $U: \mathcal{A} \to \textbf{Set}$ is the forgetful functor (the structure of A does not play any role in it). Thus, we have a one-one correspondence between the homomorphisms $F(X) \to A$ and mappings $X \to U(A)$, making F a left adjoint of U.

(3) Recall Appendix I.5 and 1.1.1(c). Adjoint monotone maps constitute an adjoint situation in the more general categorical sense.

(4) Consider the constant functor $C: \mathcal{C} \to [K, \mathcal{C}]$ defined by $C(A)(\alpha) = \text{id}_A$. Suppose all limits of K-diagrams exist. Define $L(D)$ as the L in the limit $(p_a: L \to$

$D(a))_{a\in\text{obj}K}$. Thus we obtain a functor $L\colon [K,\mathcal{C}] \to \mathcal{C}$ and the definition of limit provides us with an adjunction $C \dashv L$.

Similarly, the definition of colimit yields an adjunction with C on the right.

6.3. Adjunction units. Given an adjunction with a natural equivalence as above, set

$$\lambda_B = \bar{\varepsilon}_{R(B),B}(\text{id}_{R(B)})\colon LR(B) \to B \tag{λ}$$

and

$$\rho_A = \varepsilon_{A,L(A)}(\text{id}_{L(A)})\colon A \to RL(A). \tag{ρ}$$

6.3.1. Proposition. *The systems of morphisms above constitute natural transformations*

$$\lambda = (\lambda_B)_B\colon LR \overset{\cdot}{\to} \text{Id}_{\mathcal{D}} \quad and \quad \rho = (\rho_A)_A\colon \text{Id}_{\mathcal{C}} \overset{\cdot}{\to} RL,$$

and the compositions

$$L \xrightarrow{\ L\rho\ } LRL \xrightarrow{\ \lambda_L\ } L \quad and \quad R \xrightarrow{\ \rho_R\ } RLR \xrightarrow{\ R\lambda\ } R$$

are identical transformations of L resp. R.

Proof. We will often leave out the indices (uniquely determined anyway). For $g\colon B \to B'$ we have, by $(\bar{\varepsilon})$,

$$\lambda_{B'} \cdot LRg = \bar{\varepsilon}(\text{id}) \cdot L(Rg) = \bar{\varepsilon}(R(\text{id}) \cdot \text{id} \cdot Rg)$$
$$= \bar{\varepsilon}(Rg \cdot \text{id} \cdot \text{id}) = g \cdot \bar{\varepsilon}(\text{id}) \cdot L(\text{id}) = g \cdot \lambda,$$

and by (ε), for $f\colon A \to A'$,

$$RLf \cdot \rho_A = R(Lf) \cdot \varepsilon(\text{id}) \cdot \text{id} = \varepsilon(Lf \cdot \text{id} \cdot \text{id}) = \varepsilon(\text{id} \cdot \text{id} \cdot Lf) = R(\text{id}) \cdot \varepsilon(\text{id}) \cdot f = \rho \cdot f.$$

Now for an $A \in \text{obj}\mathcal{C}$ we obtain

$$\lambda_{LA} \cdot L(\rho_A) = \bar{\varepsilon}(\text{id}) \cdot L\rho = \text{id} \cdot \bar{\varepsilon}(\text{id}) \cdot L\rho = \bar{\varepsilon}(R(\text{id}) \cdot \text{id} \cdot \rho) = \bar{\varepsilon}(\varepsilon(\text{id})) = \text{id}.$$

and for a $B \in \text{obj}\mathcal{D}$,

$$R(\lambda_B) \cdot \rho_{RB} = R\lambda \cdot \varepsilon(\text{id}) \cdot \text{id} = \varepsilon(\lambda \cdot \text{id} \cdot L(\text{id})) = \varepsilon(\bar{\varepsilon}(\text{id})) = \text{id}. \qquad \square$$

6.3.2. The natural transformations $\lambda\colon LR \overset{\cdot}{\to} \text{Id}_{\mathcal{D}}$ and $\rho\colon \text{Id}_{\mathcal{C}} \overset{\cdot}{\to} RL$ are called *units of adjunction*. They contain all the information on the adjunction in question. More than that, we have the following

Theorem. *The formulas (λ) and (ρ) from 6.3 constitute a one-one correspondence between adjunctions*

$$\varepsilon\colon \mathcal{D}(L-,-) \cong \mathcal{C}(-,R-)$$

and pairs of natural transformations $\lambda\colon LR \overset{\cdot}{\to} \text{Id}_{\mathcal{D}}$ and $\rho\colon \text{Id}_{\mathcal{C}} \overset{\cdot}{\to} RL$ satisfying

$$(\ L \xrightarrow{\ L\rho\ } LRL \xrightarrow{\ \lambda_L\ } L\ ) = \text{id}_L \quad and \quad (\ R \xrightarrow{\ \rho_R\ } RLR \xrightarrow{\ R\lambda\ } R\ ) = \text{id}_R.$$

Proof. We will provide the inverse formulas.

Suppose a natural transformation satisfying the identities from the statement are given. For $\phi\colon LA \to B$ and $\psi\colon A \to RB$ set

$$\varepsilon(\phi) = R\phi \cdot \rho_A\colon A \to RB \quad \text{and} \quad \bar{\varepsilon}(\psi) = \lambda_B \cdot L\psi\colon LA \to B.$$

We have

$$\varepsilon(\bar{\varepsilon}(\psi)) = R(\bar{\varepsilon}(\psi)) \cdot \rho_A = R\lambda_B \cdot RL\psi \cdot \rho_A = R(\lambda_B) \cdot \rho_{RB} \cdot \psi = \psi$$

and

$$\bar{\varepsilon}(\varepsilon(\phi)) = \lambda_B \cdot L(R\phi \cdot \rho_A) = \lambda_B \cdot LR\phi \cdot L\rho_A = \phi \cdot \lambda_{LA} \cdot L\rho_A = \phi.$$

The mappings ε (and hence also their inverses $\bar{\varepsilon}$) satisfy

$$\varepsilon(g \cdot \phi \cdot Lf) = R(g \cdot \phi \cdot Lf) \cdot \rho_A = Rg \cdot R\phi \cdot RLf \cdot \rho_A$$
$$= Rg \cdot R\phi \cdot \rho \cdot f = Rg \cdot \varepsilon(\phi) \cdot f.$$

Finally, if we start with the natural equivalence ε (and with its inverse), define λ and ρ as in 6.3, and construct a new ε' by the formula above we obtain

$$\varepsilon'(\phi) = R\phi \cdot \rho_A = R\phi \cdot \varepsilon(\mathrm{id}_{LA}) = \varepsilon(\phi \cdot \mathrm{id} \cdot \mathrm{id}) = \varepsilon(\phi).$$

If we start with λ, ρ as assumed, define ε and $\bar{\varepsilon}$ as above, and from that define λ', ρ' as in 6.3 we obtain

$$\lambda'_B = \bar{\varepsilon}(\mathrm{id}_{RB}) = \lambda_B \cdot L(\mathrm{id}) = \lambda_B$$

and

$$\rho'_A = \varepsilon(\mathrm{id}_{LA}) = R(\mathrm{id}) \cdot \rho_A = \rho_A. \qquad \square$$

6.3.3. Note. The units of adjunction are usually much more expedient than the adjunction in the original form: it is easier to work with natural transformations in one variable object than with a natural equivalence in two variables. But the point is not only in the technical advantage. More often than not, the units express important phenomena. Thus for instance in the Example 6.2(1) the

$$\lambda_X\colon X^A \times A \to X$$

is the well-known "evaluation operator" sending (f, a) to the value $f(a)$.

In the Example 6.2(2) the unit $\rho_X\colon X \to UF(X)$ is the embedding of the set of generators into the free algebra; the unit $\lambda_A\colon FU(A) \to A$ expresses the general theorem stating that each algebra (in a nice category of algebras, for instance in any variety of algebras) is a quotient of a free one.

7.   Adjointness and (co)limits

7.1. Proposition. *Right adjoints preserve limits and left adjoints preserve colimits.*

More precisely, if $(f_a\colon A \to D(a))_{a\in\mathrm{obj}K}$ is a limit of a diagram $D\colon K \to \mathcal{D}$ then $(Rf_a\colon RA \to RD(a))_{a\in\mathrm{obj}K}$ is a limit of $RD\colon K \to \mathcal{C}$, and similarly for L and colimits.

Proof. Obviously $(Rf_a)_a$ is a lower bound of RD. Now let $(g_a\colon X \to RD(a))_a$ be a general lower bound of RD. Consider the system

$$(\overline{\varepsilon}(g_a) = \lambda_{D(a)} \cdot L(g_a)\colon L(X) \to D(a))_a.$$

It is a lower bound of D: indeed, we have for $\alpha\colon a \to b$ in K

$$D(\alpha)\cdot\lambda_{D(a)}\cdot L(g_a) = \lambda_{D(b)}\cdot LRD(\alpha)\cdot L(g_a) = \lambda_{D(b)}\cdot L(RD(\alpha)\cdot g_a) = \lambda_{D(b)}\cdot L(g_b).$$

Consequently there is a morphism $h\colon L(X) \to A$ such that

$$\forall a, \quad f_a \cdot h = \overline{\varepsilon}(g_a)$$

and hence

$$g_a = \varepsilon(f_a h) = R(f_a) \cdot R(h) \cdot \rho_X = Rf_a \cdot \varepsilon(h).$$

The unicity of a $\widetilde{h}$ such that $g_a = Rf_a \cdot \widetilde{h}$ is straightforward and can be left to the reader as a simple exercise. $\square$

7.2. We see an analogy with preserving infima and suprema by adjoint monotone maps. We may naturally ask whether the converse fact about partially ordered sets, namely that in case of complete lattices the infima preserving maps are right adjoints and the suprema preserving maps are left ones, could perhaps also extrapolate.

On second thought the idea seems to be overoptimistic. The situation is different: in complete lattices there were limits and colimits (infima and suprema) of the same size as the lattices themselves while here the diagrams are (necessarily) small and the category is typically a proper class. Perhaps surprisingly one does have a desirable converse to 7.1 for a large type of categories, including many of the everyday ones.

We will need some more terminology. A category $\mathcal{C}$ is *wellpowered* (resp. *co-wellpowered*) if for each $A \in \mathcal{C}$ there is up-to isomorphism only a set of monomorphisms $X \to A$ (resp. epimorphisms $A \to X$).

A *generator* (resp. *cogenerator*) of a category $\mathcal{C}$ is an object G such that for any $f \neq g\colon A \to B$ there is a morphism $h\colon G \to A$ such that $fh \neq gh$ (resp. $h\colon B \to G$ such that $hf \neq hg$).

7.2.1. Let $R: \mathcal{D} \to \mathcal{C}$ be a functor, $A \in \mathsf{obj}\mathcal{C}$. A subset $\mathcal{S} \subseteq \mathsf{obj}\mathcal{D}$ is a *solution set for A with respect to R* if for each $B \in \mathsf{obj}\mathcal{D}$ and for each $\beta: A \to R(B)$ in $\mathcal{C}$ there are

$$S \in \mathcal{S}, \ \alpha: S \to B \text{ and } \overline{\beta}: A \to R(S)$$

such that the diagram

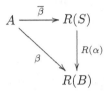

commutes.

A *solution* (resp. *strong solution*) for A with respect to R is an element $S \in \mathsf{obj}\mathcal{D}$ together with a $\rho: A \to R(S)$ such that for each $B \in \mathsf{obj}\mathcal{D}$ and for each $\beta: A \to R(B)$ in $\mathcal{C}$ there is an $\alpha: S \to B$ (resp. precisely one $\alpha: S \to B$) such that

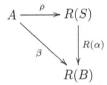

commutes.

7.2.2. Proposition. *Let $R: \mathcal{D} \to \mathcal{C}$ have a strong solution for each $A \in \mathsf{obj}\mathcal{C}$. Then it has a left adjoint.*

Proof. For each A choose a strong solution $\rho_A: A \to R(L(A))$ and define $L(f): L(A) \to L(A')$ for $f: A \to A'$ as the unique α for which

$$
\begin{array}{ccc}
A & \xrightarrow{\rho_A} & R(L(A)) \\
{\scriptstyle f}\downarrow & & \downarrow{\scriptstyle R(\alpha)=R(L(f))} \\
A' & \xrightarrow[\rho_{A'}]{} & R(L(A'))
\end{array}
$$

commutes. This yields a functor $L: \mathcal{C} \to \mathcal{D}$ and $\rho: \mathrm{Id}_{\mathcal{C}} \xrightarrow{\cdot} RL$.

For a $B \in \mathsf{obj}\mathcal{D}$ define $\lambda_B: L(R(B)) \to B$ as the unique morphism for which

$$R(\lambda_B) \cdot \rho_{R(B)} = \mathrm{id}_{R(B)}. \tag{$*$}$$

We easily check that we have obtained a transformation $\lambda: LR \xrightarrow{\cdot} \mathrm{Id}_{\mathcal{D}}$.

By $(*)$ we have $R\lambda \cdot \rho_R = \mathrm{id}$ and

$$R(\lambda_{L(A)} \cdot L\rho_A) \cdot \rho_A = R(\lambda_{L(A)}) \cdot RL(\rho_A) \cdot \rho_A = R(\lambda_{L(A)}) \cdot \rho_{R(L(A))} \cdot \rho_A = \rho_A$$

and hence also $\lambda_L \cdot L\rho = \mathrm{id}$. $\square$

7.2.3. Lemma. *Let $\mathcal{D}$ have products and let R preserve them. Let it admit a solution set for $A \in \text{obj}\mathcal{C}$. Then it admits a solution for A.*

Let, moreover, $\mathcal{D}$ be complete and wellpowered and let R preserve all limits. Then it admits a strong solution for A.

Proof. I. Let $\mathcal{S}$ be the solution set. Consider the product

$$p_{S,\phi} \colon \overline{S} = \prod \{ S^{\mathcal{C}(A,R(S))} \mid S \in \mathcal{S} \} \to S$$

(where, of course, S^M is the product $\prod_{m \in M} S_m$ with $S_m = S$ for all m). Then also $(R(p_{S,\phi}) \colon R(\overline{S}) \to R(S))_{S,\phi}$ is a product and we can define a morphism $\rho \colon A \to R(\overline{S})$ by requiring that $R(p_{S,\phi}) \cdot \rho = \phi$ for each $\phi \in \mathcal{C}(A, R(S))$.

Now, choose for $\beta \colon A \to R(B)$ an $S \in \mathcal{S}$, an $\alpha' \colon S \to B$ and a $\overline{\beta} \colon A \to R(S)$ such that $R(\alpha') \cdot \overline{\beta} = \beta$ and set $\alpha = \alpha' \cdot p_{S,\overline{\beta}}$. Then

$$R(\alpha) \cdot \rho = R(\alpha') \cdot R(p_{S,\overline{\beta}}) \cdot \rho = R(\alpha') \cdot \overline{\beta} = \beta.$$

II. Let $\mathcal{D}$ be complete and wellpowered and let $S, \rho \colon A \to R(S)$ be a solution. Let $M = \{ \mu \colon S_\mu \to S \}$ be a set representing all monomorphisms with codomain S such that $\rho = R(\mu) \cdot \rho_\mu$ for some ρ_μ. Let $\overline{\mu} \colon \overline{S} \to S$ be the limit of M. Then $R(\overline{\mu})$ is the limit of

$$\{ R(\mu) \colon R(S_\mu) \to R(S) \}$$

and since $\{ \rho_\mu \mid \mu \in M \}$ is a lower bound of this diagram there is a $\overline{\rho} \colon A \to R(\overline{S})$ such that $R(\overline{\mu}) \cdot \overline{\rho} = \rho$.

For $\beta \colon A \to R(B)$ we have an α' with $R(\alpha') \cdot \rho = \beta$. Then for $\alpha = \alpha' \cdot \overline{\mu}$ we have $R(\alpha) \cdot \overline{\rho} = \beta$. Finally, if $R(\alpha_i) \cdot \overline{\rho} = \beta$ for $i = 1, 2$, consider the equalizer κ of α_1 and α_2. Then $R(\kappa)$, the equalizer of $R(\alpha_1)$ and $R(\alpha_2)$, is an isomorphism and hence ρ factorizes through $R(\overline{\mu} \cdot \kappa)$; consequently κ is an isomorphism, and $\alpha_1 = \alpha_2$. $\square$

7.2.4. Theorem. *Let $\mathcal{D}$ be complete and wellpowered and let it have a cogenerator. Then a functor $R \colon \mathcal{D} \to \mathcal{C}$ is a right adjoint iff it preserves limits.*

Dually, if $\mathcal{C}$ is cocomplete and co-wellpowered with a generator then a functor $L \colon \mathcal{C} \to \mathcal{D}$ is a left adjoint iff it preserves colimits.

Proof. By 7.2.2 and 7.2.3 it suffices to show that there is a solution set for each $A \in \text{obj}\mathcal{C}$. Let G be a cogenerator of $\mathcal{D}$ and let $\beta \colon A \to R(B)$ in $\mathcal{C}$. Consider the products $G^{\mathcal{C}(A,R(G))} \xrightarrow{p'_\psi} G$ and $G^{\mathcal{D}(B,G)} \xrightarrow{p_\phi} G$ and the pullback

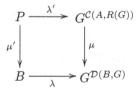

where λ is the unique morphism such that $p_\phi \cdot \lambda = \phi$ for every $\phi \in \mathcal{D}(B, G)$ and μ is the unique morphism such that $p_\phi \cdot \mu = p'_{R(\phi) \cdot \beta}$ for every $\phi \in \mathcal{D}(B, G)$; since G is a cogenerator, λ is a monomorphism, and hence so is λ'. Further, take the diagram

$$
\begin{array}{ccc}
A & \xrightarrow{\gamma} & R(G)^{\mathcal{C}(A, R(G))} \\
\beta \downarrow & & \downarrow R(\mu) \\
R(B) & \xrightarrow[R(\lambda)]{} & R(G)^{\mathcal{D}(B, G)}
\end{array}
$$

where $R(p'_\psi) \cdot \gamma = \psi$ for every $\psi \in \mathcal{C}(A, R(G))$. We have

$$
R(p_\phi) \cdot R(\mu) \cdot \gamma = R(p'_{R(\phi)\beta}) \cdot \gamma = R(\phi) \cdot \beta = R(p_\phi) \cdot R(\lambda) \cdot \beta
$$

so that the diagram commutes and hence there is a $\kappa \colon A \to R(P)$ such that $\gamma = R(\lambda') \cdot \kappa$ and $\beta = R(\mu') \cdot \kappa$. Thus the system of all P such that there is a monomorphism $\lambda' \colon P \to G^{\mathcal{C}(A, R(G))}$ is the required solution set. $\qquad\square$

7.3. As a corollary one obtains another slightly surprising analogon of a fact about ordered sets. A poset that has all suprema has all infima, and vice versa. Again, one hesitates to expect that the same should hold about large categories, that is, that complete categories should be automatically cocomplete and conversely. In the proof of the poset fact one takes, to obtain an infimum, a supremum of a very large subset indeed. But we do have something.

Recall the Example 6.2(4). For any small category K, the constant functor $\mathcal{C} \to [K, \mathcal{C}]$ preserves all limits and colimits. Hence, if we apply Theorem 7.2.4 we obtain right and left adjoints that are in fact limits resp. colimits. Explicitly:

If $\mathcal{D}$ is complete and wellpowered with a cogenerator then it is cocomplete, and if $\mathcal{C}$ is cocomplete and co-wellpowered with a generator then it is complete.

8. Reflective and coreflective subcategories

8.1. A full subcategory $\mathcal{D}$ of a category $\mathcal{C}$ is said to be *reflective* (resp. *coreflective*) if the embedding $J \colon \mathcal{D} \subseteq \mathcal{C}$ is a right (resp. left) adjoint.

For technical reasons one usually assumes that the $\mathcal{D}$ is closed under isomorphisms (which is always the case in everyday examples).

If we write $\overline{A}$ for $L(A)$, where L is the left adjoint, we immediately see that $\mathcal{D}$ being a reflective subcategory amounts to a correspondence

$$
A \quad \mapsto \quad \rho_A \colon A \to \overline{A}
$$

associating with each $A \in \text{obj}\mathcal{C}$ a morphism ρ_A with codomain in $\mathcal{D}$ such that

$$\forall\, f\colon A \to B \in \text{obj}\mathcal{D} \ \exists!\, \overline{f}\colon \overline{A} \to B \ \text{ satisfying } \overline{f} \cdot \rho_A = f:$$

Such a correspondence is usually called a *reflection* (of $\mathcal{C}$ onto $\mathcal{D}$).

Similarly, a *coreflection* of $\mathcal{C}$ (onto $\mathcal{D}$) associates with each A in $\mathcal{C}$ a morphism $\lambda_A\colon \overline{A} \to A$ with $\overline{A} \in \text{obj}\mathcal{D}$ such that

$$\forall\, f\colon B \to A, \ B \in \text{obj}\mathcal{D}, \ \exists!\, \overline{f}\colon B \to \overline{A} \ \text{ satisfying } \lambda_A \cdot \overline{f} = f.$$

8.1.1. Remark. Note that the morphisms ρ_A behave like *epimorphisms with respect to the smaller category* $\mathcal{D}$. That is, if one has $f\rho_A = g\rho_A$ for $f, g\colon \overline{A} \to B$ with $B \in \text{obj}\mathcal{D}$ then $f = g$. They do not have to be epimorphisms in the whole of $\mathcal{C}$, though. If they are, one speaks of an *epireflection* and of an *epireflective subcategory*.

Similarly, a coreflection $(\lambda_A\colon \overline{A} \to A)_A$ is a *monocoreflection* if all the λ_A are monomorphisms in $\mathcal{C}$.

8.2. Examples. Typically, a reflection or a coreflection occurs when qualifying special objects among more general ones in the given type of structure by extra conditions (axioms). Thus for instance, additional equational axioms in varieties of algebras (abelian groups in among groups, or, for that matter, groups among general algebras of type $(2, 1, 0)$) make for a reflective category.

The Stone-Čech compactification is a reflection of completely regular topological spaces onto compact Hausdorff ones.

Completion is a reflection of the category of uniform spaces onto that of complete uniform spaces; in uniform frames, completion is a coreflection.

In the category of binary relations (oriented graphs) the subcategory of symmetric relations is both a reflective and a coreflective one. We have the (identity carried) reflection $(X, R) \to (X, \overline{R})$ with

$$\overline{R} = \{(x, y) \mid (x, y) \in R \text{ or } (y, x) \in R\}$$

and the coreflection $(X, \widetilde{R}) \to (X, R)$ with

$$\widetilde{R} = \{(x, y) \mid (x, y) \in R \text{ and } (y, x) \in R\}.$$

8.3. By 7.1 the embedding of a reflective subcategory preserves all existing limits. But we have much more.

Proposition. *Let $J: \mathcal{D} \subseteq \mathcal{C}$ be an embedding of a reflective subcategory. Let $D: K \to \mathcal{D}$ be a diagram and let JD have a limit $(f_a: A \to D(a))_a$ in $\mathcal{C}$. Then $(\overline{f}_a : \overline{A} \to D(a))_a$ is a limit of D in $\mathcal{D}$. If JD has a colimit $(f_a: D(a) \to A)_a$ in $\mathcal{C}$ then $(\rho_A \cdot f_a: D(a) \to \overline{A})_a$ is a colimit of D in $\mathcal{D}$.*

Consequently, each reflective subcategory of a complete resp. cocomplete category is complete resp. cocomplete. Similarly for coreflective subcategories.

Proof. I. The system $(\overline{f}_a : \overline{A} \to D(a))_a$ is a lower bound since $D(\alpha)\overline{f}_a\rho_A = D(\alpha)f_a = f_b = \overline{f}_b\rho_A$ for all $\alpha: a \to b$ in K, and by the unicity $D(\alpha)\overline{f}_a = \overline{f}_b$. Thus there is a morphism $\phi: \overline{A} \to A$ such that $f_a\phi = \overline{f}_a$ for all a, and hence $f_a\phi\rho_A = \overline{f}_a\rho_A = f_a$ and by the unicity property of limits, $\phi\rho_A = \mathrm{id}_A$. Further, $\rho_A\phi\rho_A = \rho_A$ and since $\rho_A\phi$ targets in $\mathcal{D}$ we conclude that $\rho_A\phi = \mathrm{id}_{\overline{A}}$, and ρ_A is an isomorphism.

II. Let $(g_a: D(a) \to B)_a$ be an upper bound of D in $\mathcal{D}$; then it is an upper bound of JD in $\mathcal{C}$ and hence we have precisely one $g: A \to B$ such that $gf_a = g_a$ for all a. Then the $\overline{g}$ such that $\overline{g}\rho_A = g$ satisfies $\overline{g}\rho_A f_a = g_a$. The unicity is straightforward. $\qquad\square$

> **Note.** This is a useful fact. The existence of limits or colimits in the categories of objects with a given type of structure without axioms is very often obvious while it may be far from obvious that under the construction special properties are preserved.

9. Monads

9.1. Monads in the standard setting.
A *monad* in a category $\mathcal{C}$ is a triple

$$\mathbb{T} = (T, \mu, \eta)$$

where $T: \mathcal{C} \to \mathcal{C}$ is a functor, and $\mu: TT \overset{\cdot}{\to} T$ and $\eta: \mathrm{Id}_{\mathcal{C}} \overset{\cdot}{\to} T$ are natural transformations such that

$$\mu \cdot T\mu = \mu \cdot \mu_T \quad \text{and} \quad \mu \cdot \eta_T = \mu \cdot T\eta = \mathrm{id}_T, \qquad \text{(mon)}$$

i.e., for each object A of $\mathcal{C}$ the diagrams

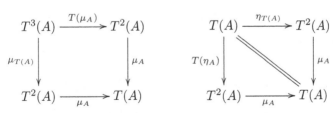

commute.

The motivation and examples will be seen shortly.

9.2. The monad of an adjoint situation. Let

$$L: \mathcal{C} \to \mathcal{D}, \quad R: \mathcal{D} \to \mathcal{C}, \quad L \dashv R$$

be an adjoint situation with adjunction units

$$\lambda: LR \overset{\cdot}{\to} \mathrm{Id}_{\mathcal{D}}, \quad \rho: \mathrm{Id}_{\mathcal{C}} \overset{\cdot}{\to} RL$$

(so that $\lambda_L \cdot L\rho = \mathrm{id}_L$ and $R\lambda \cdot \rho_R = \mathrm{id}_R$).

Proposition. *Setting $T = RL$, $\mu = R\lambda_L$ and $\eta = \rho$ we obtain a monad $\mathbb{T} = (T, \mu, \eta)$ on $\mathcal{C}$.*

Proof. We have

$$(\mu \cdot T\mu)_A = R(\lambda_{L(A)}) \cdot RLR(\lambda_{L(A)}) = R(\lambda_{L(A)} \cdot LR(\lambda_{L(A)})) = R(\lambda_{L(A)} \cdot \lambda_{LRL(A)})$$

(using the λ as a transformation applied for the morphism $\lambda_{L(A)}$), and further

$$\cdots = \mu_A \cdot R\lambda_{LRL_A} = \mu_A \cdot (\mu_T)_A.$$

Finally, $\mu_A \cdot \eta_{TA} = R\lambda_{L(A)}\eta_{RL(A)} = \mathrm{id}_{T(A)}$ and $\mu_A \cdot T\eta_A = R(\lambda_{L(A)}L\eta_A) = R(\mathrm{id}_{L(A)}) = \mathrm{id}_{T(A)}$. $\qquad\square$

9.3. Eilenberg-Moore algebras. On the other hand, each monad results from an adjoint situation (in fact, from various adjoint situations, see below). The one we will describe now is connected with (one of) the original motivation(s) for introducing the concept; the point was to explain the algebra-like behaviour of compact Hausdorff spaces and to provide a natural roof for varieties of algebras and the category of compact spaces.

Let $\mathbb{T} = (T, \mu, \eta)$ be a monad on $\mathcal{C}$. A $(\mathbb{T})$-*Eilenberg-Moore algebra* (briefly, $\mathbb{T}$-algebra) is a pair (A, α) with $\alpha: TA \to A$ satisfying

$$\alpha \cdot T(\alpha) = \alpha \cdot \mu_A \quad \text{and} \quad \alpha \cdot \eta_A = \mathrm{id}_A. \tag{alg}$$

A *homomorphism* $f: (A, \alpha) \to (B, \beta)$ is an $f: A \to B$ in $\mathcal{C}$ such that $\beta \cdot Tf = f \cdot \alpha$. The resulting category will be denoted by

$$\mathcal{C}^{\mathbb{T}}.$$

Define functors

$$F: \mathcal{C} \to \mathcal{C}^{\mathbb{T}}, \quad U: \mathcal{C}^{\mathbb{T}} \to \mathcal{C}$$

by setting

$$F(A) = (TA, \mu_A), \ Ff = Tf \quad \text{and} \quad U(A, \alpha) = A, \ Uf = f$$

(by (mon), $F(A)$ is a $\mathbb{T}$-algebra, and $Tf \cdot \mu_A = \mu_B \cdot T(Tf)$ since μ is a natural transformation).

Proposition. *U is right adjoint of F and the monad resulting from the adjoint situation is the original $\mathbb{T}$.*

Proof. Define $\lambda\colon FU \overset{\cdot}{\to} \mathrm{Id}_{\mathcal{C}^{\mathbb{T}}}$ by setting

$$\lambda_{(A,\alpha)} = \alpha\colon FU(A,\alpha) = (TA, \mu_A) \to (A, \alpha).$$

Thus defined $\lambda_{(A,\alpha)}$ are homomorphisms by the (alg) above, and they constitute a natural transformation by definition of homomorphism. We have $(\lambda_F \cdot F\eta)_A = \lambda_{(TA,\mu_A)} \cdot T\eta_A = \mu_A \cdot T\eta_A = \mathrm{id}_{TA}$ by (mon), and $(U\lambda \cdot \eta_U)_{(A,\alpha)} = U(\lambda_{(A,\alpha)}) \cdot \eta_A = \alpha \cdot \eta_A = \mathrm{id}_A$ by (alg) so that $F \dashv U$.

The second statement is obvious. $\qquad\qquad\qquad\qquad\qquad\qquad\qquad\qquad\square$

9.3.1. Remarks. (1) A variety of algebras can be represented as $\mathbf{Set}^{\mathbb{T}}$ as follows. TX is the set of all formal expressions in symbols of composite operations and the generators $x \in X$. The $\alpha\colon TX \to X$ in a $\mathbb{T}$-algebra is an evaluation of such formulas in a concrete algebra; η_X is the embedding of X as the set of formal generators.

(2) To represent the category of compact Hausdorff spaces use TX the set of all ultrafilters in $\mathfrak{P}(X)$ (the underlying set of the Stone-Čech compactification $\beta D(X)$ of the discrete space carried by X). A compact space A is represented as $(U(A), \alpha)$ where $U(A)$ is the underlying set of A and $\alpha\colon \beta U(A) \to A$ lifts the identical map $U(A) \to U(A)$.

9.4. Kleisli category of a monad. In among the adjunctions yielding a monad $\mathbb{T}$, the one connected with the Eilenberg-Moore algebras is in a sense the largest one. There is also the other extreme, the *Kleisli category* $\mathcal{C}_{\mathbb{T}}$. Here,

$$\mathrm{obj}\mathcal{C}_{\mathbb{T}} = \mathrm{obj}\mathcal{C}, \text{ and morphisms } A \multimap B \text{ are } A \to TB \text{ from } \mathcal{C}.$$

We have a new composition and new identities

$$g \circ f = \mu_C \cdot Tg \cdot f \quad \text{and} \quad 1_A = \eta_A\colon A \multimap A.$$

This is correct: for $f\colon A \multimap B$, $g\colon B \multimap C$ and $h\colon C \multimap D$ we have

$$h \circ (g \circ f) = \mu_D Th \cdot (g \circ f) = \mu_D Th \cdot \mu_C Tg \cdot f = \mu_D \cdot \mu_{TD} \cdot TTh \cdot Tg \cdot f$$

$$= \mu_D \cdot T\mu_D \cdot TTh \cdot Tg \cdot f = \mu_D \cdot T(\mu_D \cdot Th \cdot g) \cdot f = (h \circ g) \circ f$$

and for $f\colon A \multimap B$,

$$\eta_B \circ F = \mu_B T(\eta_B)f = f \quad \text{and} \quad f \circ \eta_A = \mu_B T(f)\eta_A = \mu_B \eta_{TB} f = f.$$

Define functors $L\colon \mathcal{C} \to \mathcal{C}_{\mathbb{T}}$ and $R\colon \mathcal{C}_{\mathbb{T}} \to \mathcal{C}$ by setting

$$L(A) = A, \; L(f) = \eta_B \cdot f \quad \text{and} \quad R(A) = T(A), \; R(f) = \mu_B \cdot Tf.$$

This is correct: we have $L(\mathrm{id}_A) = \eta_A = 1_A$, $R(\eta_A) = \mu_A T\eta_A = \mathrm{id}_{TA}$,

$$Lf \circ Lg = \mu_C T(\eta_C g)\eta_B f = \mu_C T\eta_C Tg \cdot \eta_B gf = \eta_B \cdot g \cdot f = L(g \cdot f) \text{ and}$$

$$R(g \circ f) = R(\mu_C Tg \cdot f) = \mu_C \cdot T\mu_C \cdot TTg \cdot Tf = \mu_C \cdot \mu_{TC} \cdot TTg \cdot Tf$$

$$= \mu_C Tg \cdot \mu_B Tf = Rg \cdot Rf.$$

Now take again the $\eta\colon \mathrm{Id}_{\mathcal{C}} \overset{\cdot}{\to} RL$ and set $\lambda_A = \mathrm{id}_{TA}\colon TA \multimap A$. We have

Proposition. *R is a right adjoint of L and the monad resulting from the adjoint situation is the original $\mathbb{T}$.*

Proof. We have $(\lambda_R \cdot R\eta)_A = \lambda_{RA} \cdot R(\eta_A) = \mathrm{id}_{RA}\mu_A T\eta_A = \mathrm{id}_{RA}$, and

$$(L\lambda \cdot \eta_L)_A = L(\lambda_A) \circ \eta_{LA} = (\eta_{TA} \cdot \mathrm{id}_{TA}) \circ \eta_{LA} = \mu_{TA} \cdot T\eta_{TA} \cdot \eta_{LA} = \eta_{LA} = \mathbf{1}_{LA}.$$

For the second statement it suffices to check the formula for μ: we have $(R\lambda_L)_A = R(\lambda_{LA}) = \mu_A \cdot \mathrm{id}_{TTA} = \mu_A$. $\qquad\square$

9.5. Manes description of a monad.
In a category $\mathcal{C}$ let us have a mapping $T\colon \mathrm{obj}\mathcal{C} \to \mathrm{obj}\mathcal{C}$, a system of morphisms $\eta_A\colon A \to TA$, and a lifting

$$(f\colon A \to TB) \quad \mapsto \quad \left(\widetilde{f}\colon TA \to TB\right)$$

such that

(1) $\widetilde{\eta}_A = \mathrm{id}_{TA}$,

(2) $\widetilde{f}\eta_A = f$, and

(3) $\widetilde{g} \cdot \widetilde{f} = \widetilde{\widetilde{g} \cdot f}$ for $f\colon A \to TB$ and $g\colon B \to TC$.

With such a (Manes) system $(T, \eta, \widetilde{(-)})$ associate

$$\mathbb{T} = (T, \mu, \eta) \quad \text{by setting} \quad T(f) = \widetilde{\eta_B f} \text{ and } \mu_A = \widetilde{\mathrm{id}_{TA}}.$$

Proposition. *$\mathbb{T}$ is a monad. On the other hand, with a monad (T, μ, η) we can associate a Manes system $(T, \eta, \widetilde{(-)})$ by setting $\widetilde{f} = \mu_B \cdot Tf$ thus establishing a one-one correspondence between monads and Manes systems.*

Proof. I. T is a functor: $T(\mathrm{id}_A) = \widetilde{\eta_A} = \mathrm{id}_{TA}$ and $T(g)T(f) = \widetilde{\eta_C g} \cdot \widetilde{\eta_B f} = \widetilde{\eta_C g} \cdot \eta_B f = \widetilde{\eta_C g f} = T(gf)$.

η is a natural transformation $\mathrm{Id}_{\mathcal{C}} \overset{\cdot}{\to} T$: we have $Tf \cdot \eta_A = \widetilde{\eta_B f} \cdot \eta_A = \eta_B f$.

μ is a natural transformation $TT \overset{\cdot}{\to} T$: we have

$$\mu_B TTf = \widetilde{\mathrm{id}_B}(\widetilde{\eta_{TB}\widetilde{\eta_B f}}) = \widetilde{\mathrm{id}\eta_B} \cdot \widetilde{\eta_B f} = \widetilde{\widetilde{\eta_B f}} = \widetilde{\eta_B f} \cdot \mathrm{id} = \widetilde{\eta_B f} \cdot \widetilde{\mathrm{id}} = Tf \cdot \mu_A.$$

Finally, we have

$$\mu_A \cdot T\mu_A = \widetilde{\mathrm{id}_{TA}} \cdot \widetilde{\eta_{TA}\mathrm{id}_{TA}} = \widetilde{\mathrm{id} \cdot \eta} \cdot \widetilde{\mathrm{id}} = \widetilde{\widetilde{\mathrm{id}_{TA}}},$$

$$\mu_A \cdot \mu_{TA} = \widetilde{\mathrm{id}_{TA}} \cdot \widetilde{\mathrm{id}TTA} = \widetilde{\mathrm{id} \cdot \mathrm{id}} = \widetilde{\widetilde{\mathrm{id}_{TA}}},$$

and

$$\mu_A \eta_{TA} = \widetilde{\mathrm{id}} \cdot \eta_A = \mathrm{id}_{TA} = \widetilde{\eta_A} = \widetilde{\mathrm{id} \cdot \eta_A} = \widetilde{\mathrm{id}_{TA}} \cdot \widetilde{\eta_{TA} \eta_A} = \mu_A T \eta_A.$$

II. If (T, μ, η) is a monad and if $\tilde{f} = \mu_B T f$ then $\widetilde{\eta_A} = \mu_A T \eta_A = \mathrm{id}$, $\tilde{f} \cdot \eta_A = \mu_B \cdot T f \cdot \eta_A = \mu_B \eta_T B f = f$, and

$$\widetilde{\tilde{g} \cdot f} = \mu_D \cdot T(\mu_C \cdot Tg \cdot f) = \mu_D \cdot T\mu_D \cdot TTg \cdot TF$$
$$= \mu_D \mu_{TD} \cdot TTg \cdot TF = \mu_D Tg \cdot \mu_C Tf = \tilde{g} \cdot \tilde{f}.$$

III. Start with a monad (T, μ, η), define $(T, \eta, \widetilde{(-)})$ as in the statement, and construct a new monad (T', μ', η) from this system. We have $T'(f) = \widetilde{\eta_B f} = \mu_B \cdot T\eta_B \cdot Tf = Tf$ and $\mu'_A = \widetilde{\mathrm{id}_{TA}} = \mu_A T(\mathrm{id}_A) = \mu_A$.

On the other hand, starting with $(T, \eta, \widetilde{(-)})$, taking the associated monad, and constructing new $(T, \eta, \widetilde{(-)})$ we have for the new lifting $\widehat{f} = \mu_B T f = \widetilde{\mathrm{id}_{TB}} \cdot \widetilde{\eta_B f} = \widetilde{\mathrm{id} \cdot f} = \tilde{f}$. $\qquad \square$

10. Algebras in a category

10.1. For an object A in a category $\mathcal{C}$ with finite products and a natural n write

$$A^n = A \times \cdots \times A \quad (n \text{ times})$$

(A^0 is the void product, that is, the terminal object of $\mathcal{C}$). Generalizing the concepts concerning originally non-structured sets we define an n-ary operation on A as a morphism

$$\alpha \colon A^n \to A$$

(thus, a standard n-ary operation is an n-ary operation in **Set**; for instance in **Top** operations are continuous, like the multiplication in a topological group).

If $\Delta = (n_t)_{t \in T}$ is a *type* (a mapping associating natural numbers with indices from T), an *algebraic structure of a type* Δ is a collection $\alpha = (\alpha_t)_{t \in T}$ of operations on A, the α_t an n_t-ary one. An *algebra of a type* Δ *in the category* $\mathcal{C}$ is a pair (A, α) where α is an algebraic structure of a type Δ on A.

A *homomorphism* $h \colon (A, \alpha) \to (B, \beta)$ is then a morphism $h \colon A \to B$ in $\mathcal{C}$ such that for all $t \in T$ we have

$$\beta \cdot (h \times \cdots \times h) = h \cdot \alpha$$

($h \times \cdots \times h$ is the morphism $A^{n_t} \to B^{n_t}$ defined by $p_i(h \times \cdots \times h) = hp_i$).

10.2. Equations. Similarly like in **Set** one can, further, define composite operations by taking products, composing, and identifying some of the variables. The formula

ω for thus obtained composite operation α is said to be its *pattern*, and the concrete composite operation $\alpha = \overline{\omega}^{\mathcal{C}}$ is called the model of the pattern in the concrete algebra.

An *algebraic theory* $\mathcal{A}$ consists of a type Δ and a set Ω of pairs of patterns, and a *model of* $\mathcal{A}$ in $\mathcal{C}$ is an algebra (A, α) of the type Δ in $\mathcal{C}$ such that for all $(\omega_1, \omega_2) \in \Omega$ one has $\overline{\omega}_1 = \overline{\omega}_2$. The category of such models with the homomorphisms as above will be denoted by

$$\mathcal{C}^{\mathcal{A}}$$

(note that the categories $\mathbf{Set}^{\mathcal{A}}$ are the standard varieties of algebras).

10.3. For proving the following statement we do not have sufficient techniques. What we need in the text is, therefore, proved ad hoc at each instance that occurs. We would like just to inform the reader about the general fact.

Theorem. [Kock] *Each identity that can be proved in the variety* $\mathbf{Set}^{\mathcal{A}}$ *holds in every category* $\mathcal{C}^{\mathcal{A}}$.

Bibliography

[1] J. Adámek, H. Herrlich, and G.E. Strecker. *Abstract and concrete categories: the joy of cats.* Pure and Applied Mathematics (New York). John Wiley & Sons Inc., New York, 1990.

[2] P.S. Alexandroff. Zur Begründung der n-dimensionalen mengentheoretischen Topologie. *Math. Ann.*, 94:296–308, 1925.

[3] C.E. Aull and W.J. Thron. Separation axioms between T_0 and T_1. *Indag. Math.*, 24:26–37, 1963.

[4] D. Baboolal. Connectedness in metric frames. *Appl. Categ. Structures*, 13(2):161–169, 2005.

[5] R.N. Ball and A.W. Hager. On the localic Yosida representation of an Archimedean lattice ordered group with weak order unit. *J. Pure Appl. Algebra*, 70(1-2):17–43, 1991.

[6] R.N. Ball, A.W. Hager and A.T. Molitor. Spaces with filters. In: *Symposium on Categorical Topology* (Rondebosch, 1994), pp. 21–35. Univ. Cape Town, Rondebosch, 1999.

[7] R.N. Ball and J. Walters-Wayland. C- and C^*-quotients in pointfree topology. *Dissertationes Mathematicae (Rozprawy Mat.)*, 412:1–62, 2002.

[8] B. Banaschewski. *Untersuchungen über Filterräume.* Doctoral dissertation, Universität Hamburg, 1953.

[9] B. Banaschewski. Frames and compactifications. In: *Proc. International Symp. on Extension Theory of Topological Structures and its Applications*, pp. 29–33. VEB Deutscher Verlag der Wissenschaften, 1969.

[10] B. Banaschewski. The duality of distributive continuous lattices. *Canad. J. Math.*, 32:385–394, 1980.

[11] B. Banaschewski. Coherent frames. In: *Continuous lattices*, pp. 1–11. Lecture Notes in Mathematics, vol. 871. Springer-Verlag, Berlin, 1981.

[12] B. Banaschewski. Another look at the localic Tychonoff theorem. *Comment. Math. Univ. Carolinae*, 29(4):647–656, 1988.

[13] B. Banaschewski. On pushing out frames. *Comment. Math. Univ. Carolinae*, 31(1):13–21, 1990.

[14] B. Banaschewski. On proving the Tychonoff Product Theorem. *Kyungpook Math. J.*, 30(1):65–73, 1990.

[15] B. Banaschewski. Compactifications of frames. *Math. Nachr.*, 149:105–115, 1990.

[16] B. Banaschewski. *The real numbers in pointfree topology.* Textos de Matemática (Série B), vol. 12. Universidade de Coimbra, Departamento de Matemática, Coimbra, 1997.

[17] B. Banaschewski. A uniform view of localic realcompactness. *J. Pure Appl. Algebra*, 143(1-3):49–68, 1999.

[18] B. Banaschewski. The axiom of countable choice and pointfree topology. *Appl. Categ. Structures*, 9(3):245–258, 2001.

[19] B. Banaschewski. Uniform completion in pointfree topology. In: *Topological and Algebraic Structures in Fuzzy Sets*, pp. 19–56. Trends Log. Stud. Log. Libr., vol. 20, Kluwer Academic Publishers, Boston, Dordrecht, London, 2003.

[20] B. Banaschewski. Ring theory and pointfree topology. *Topology Appl.*, 137(1-3):21–37, 2004.

[21] B. Banaschewski. On the function ring functor in pointfree topology. *Appl. Categ. Structures*, 13(4):305–328, 2005.

[22] B. Banaschewski. On the function rings of pointfree topology. *Kyungpook Math. J.*, 48(2):195–206, 2008.

[23] B. Banaschewski. A new aspect of the cozero lattice in pointfree topology. *Topology Appl.*, 156(12):2028–2038, 2009.

[24] B. Banaschewski, G.C.L. Brümmer and K.A. Hardie. Biframes and bispaces. *Quaest. Math.*, 6:13–25, 1983.

[25] B. Banaschewski, J.L. Frith and C.R.A. Gilmour. On the congruence lattice of a frame. *Pacific J. Math.*, 130(2):209–213, 1987.

[26] B. Banaschewski and C.R.A. Gilmour. Pseudocompactness and the cozero part of a frame. *Comment. Math. Univ. Carolinae*, 37(3):577–587, 1996.

[27] B. Banaschewski and C.R.A. Gilmour. Realcompactness and the cozero part of a frame. *Appl. Categ. Structures*, 9(4):395–417, 2001.

[28] B. Banaschewski, J. Gutiérrez García and J. Picado. Extended real functions in pointfree topology. *J. Pure Appl. Algebra*, to appear.

[29] B. Banaschewski and S.S. Hong. Completeness properties of function rings in pointfree topology. *Comment. Math. Univ. Carolinae*, 44(2):245–259, 2003.

[30] B. Banaschewski and S.S. Hong. Variants of compactness in pointfree topology. *Kyungpook Math. J.*, 45(4):455–470, 2005.

[31] B. Banaschewski, S.S. Hong and A. Pultr. On the completion of nearness frames. *Quaest. Math.*, 21(1-2):19–37, 1998.

[32] B. Banaschewski and C.J. Mulvey. Stone-Čech compactification of locales I. *Houston J. Math.*, 6(3):301–312, 1980.

[33] B. Banaschewski and C.J. Mulvey. Stone-Čech compactification of locales II. *J. Pure Appl. Algebra*, 33(2):107–122, 1984.

[34] B. Banaschewski and C.J. Mulvey. The Stone-Čech compactification of locales III. *J. Pure Appl. Algebra*, 185(1-3):25–33, 2003.

[35] B. Banaschewski and A. Pultr. Cauchy points of metric locales. *Canad. J. Math.*, 41(5):830–854, 1989.

[36] B. Banaschewski and A. Pultr. Samuel compactification and completion of uniform frames. *Math. Proc. Cambridge Phil. Soc.*, 108(1):63–78, 1990.

[37] B. Banaschewski and A. Pultr. A Stone duality for metric spaces. In: *Category Theory* 1991, pp. 33–42. Canadian Math. Soc. Conf. Proc, vol. 13, Amer. Math. Soc., Providence, RI, 1992.

[38] B. Banaschewski and A. Pultr. Distributive algebras in linear categories. *Algebra Universalis*, 30(1):101–118, 1993.

[39] B. Banaschewski and A. Pultr. Paracompactness revisited. *Appl. Categ. Structures*, 1(2):181–190, 1993.

[40] B. Banaschewski and A. Pultr. Variants of openness. *Appl. Categ. Structures*, 2(4):331–350, 1994.

[41] B. Banaschewski and A. Pultr. Booleanization. *Cahiers Topologie Géom. Diff. Catég.*, 37(1):41–60, 1996.

[42] B. Banaschewski and A. Pultr. Cauchy points of uniform and nearness frames. *Quaest. Math.*, 19(1-2):101–127, 1996.

[43] B. Banaschewski and A. Pultr. Universal categories of uniform and metric locales. *Algebra Universalis*, 35:556–569, 1996.

[44] B. Banaschewski and A. Pultr. On Cauchy homomorphisms of nearness frames. *Math. Nachr.*, 183:5–18, 1997.

[45] B. Banaschewski and A. Pultr. A new look at pointfree metrization theorems. *Comment. Math. Univ. Carolinae*, 39(1):167–175, 1998.

[46] B. Banaschewski and A. Pultr. A constructive view of complete regularity. *Kyungpook Math. J.*, 43(2):257–262, 2003.

[47] B. Banaschewski and A. Pultr. A general view of approximation. *Appl. Categ. Structures*, 14(2):165–190, 2006.

[48] B. Banaschewski and A. Pultr. Approximate maps, filter monad, and a representation of localic maps. *Arch. Math. (Brno)*, 46(4):285–298, 2010.

[49] B. Banaschewski and A. Pultr. Pointfree aspects of the T_D axiom of classical topology. *Quaest. Math.*, 33(3):369–385, 2010.

[50] B. Banaschewski and M. Sioen. Ring ideals and the Stone-Čech compactification in pointfree topology. *J. Pure Appl. Algebra*, 214(12):2159–2164, 2010.

[51] B. Banaschewski and J.J.C. Vermeulen. On the completeness of localic groups. *Comment. Math. Univ. Carolinae*, 40(2):293–307, 1999.

[52] B. Banaschewski and J. Walters-Wayland. On Lindelöf uniform frames and the axiom of countable choice. *Quaest. Math.*, 30(2):115–121, 2007.

[53] J. Bénabou. *Treillis locaux et paratopologies*. Séminaire Ehresmann, 1re année, exposé 2. Paris, 1958.

[54] G. Birkhoff and R.S. Pierce. Lattice-ordered rings. *An. Acad. Brasil. Ci.*, 28:41–69 (1956).

[55] F. Borceux. *Handbook of categorical algebra*. Encyclopedia of Mathematics and its Applications, vol. 52. Cambridge University Press, Cambridge, 1994.

[56] N. Bourbaki. *Topologie Générale: Les structures fondamentales de l'analyse, Livre III*. Hermann & Cie, Paris, 1966.

[57] G. Bruns. Darstellungen und Erweiterungen geordneter Mengen II. *J. Reine Angew. Math.*, 210:1–23, 1962.

[58] J.R. Büchi. Representation of complete lattices by sets. *Portugal. Math.*, 11:151–167, 1952.

[59] C. Carathéodory. Über die Begrenzung einfach zusammenhängender Gebiete. *Math. Ann.*, 73:323–370, 1913.

[60] M.C. Charalambous. Dimension theory for σ-frames. *J. London Math. Soc.*, 8:149–160, 1974.

[61] X. Chen. *Closed frame homomorphisms*. Doctoral dissertation, McMaster University, Hamilton, 1991.

[62] X. Chen. On binary coproducts of frames. *Comment. Math. Univ. Carolinae*, 33:699–712, 1992.

[63] M. M. Clementino. *Separation and compactness in categories*. Doctoral dissertation, University of Coimbra, 1991.

[64] M. M. Clementino, E. Giuli and W. Tholen. A functional approach to general topology. In: *Categorical Foundations – Special Topics in Order, Algebra and Sheaf Theory*, pp. 103–163. Encyclopedia of Mathematics and its Applications, vol. 97. Cambridge Univ. Press, Cambridge, 2004.

[65] B.A. Davey and H.A. Priestley. *Introduction to lattices and order*. Second edition. Cambridge University Press, New York, 2002.

[66] J.W.R. Dedekind. *Stetigkeit und irrationale Zahlen* (1872). Translation by W.W. Beman: Continuity and irrational numbers. In: *Essays on the Theory of Numbers by Richard Dedekind*, Dover edition, 1963.

[67] C.H. Dowker and D. Papert. Quotient frames and subspaces. *Proc. London Math. Soc. (3)*, 16:275–296, 1966.

[68] C.H. Dowker and D. Papert. On Urysohn's lemma. In: *General Topology and its Relations to Modern Analysis and Algebra, II*, pp. 111–114. Proc. Second Prague Topological Sympos., 1966. Academia, Prague, 1967.

[69] C.H. Dowker and D.P. Strauss. Separation axioms for frames. In: *Topics in Topology*, pp. 223–240. Proc. Colloq., Keszthely, 1972. Colloq. Math. Soc. Janos Bolyai, vol. 8, North-Holland, Amsterdam, 1974.

[70] C.H. Dowker and D.P. Strauss. Paracompact frames and closed maps. In: *Symposia Mathematica, XVI*, pp. 93–116. Convegno sulla Topologia Insiemistica e Generale, INDAM, Rome, 1973. Academic Press, London, 1975.

[71] C.H. Dowker and D. Strauss. Products and sums in the category of frames. In: *Categorical Topology*, pp. 208–219. Proc. Conf., Mannheim, 1975. Lecture Notes in Mathematics, vol. 540, Springer-Verlag, Berlin, 1976.

[72] C.H. Dowker and D. Strauss. Sums in the category of frames. *Houston J. Math.*, 3(1):7–15, 1977.

[73] C.H. Dowker and D. Strauss. T_1- and T_2-axioms for frames. In: *Aspects of Topology*, pp. 325–335. London Math. Soc. Lecture Note Ser., vol. 93. Cambridge Univ. Press, Cambridge, 1985.

[74] T. Dube. Paracompact and locally fine nearness frames. *Topology Appl.*, 62(3):247–253, 1995.

[75] T. Dube. The Tamano-Dowker type theorems for nearness frames. *J. Pure Appl. Algebra*, 99(1):1–7, 1995.

[76] T. Dube. Strong nearness frames. In: *Symposium on Categorical Topology* (Rondebosch, 1994), pp. 103–112. University of Cape Town, Rondebosch, 1999.

[77] T. Dube. Notes on uniform frames. *Quaest. Math.*, 27(1):9–20, 2004.

[78] T. Dube. On compactness of frames. *Algebra Universalis*, 51(4):411–417, 2004.

[79] T. Dube. An algebraic view of weaker forms of realcompactness. *Algebra Universalis*, 55(2-3):187–202, 2006.

[80] T. Dube. Some notes on *C*- and *C**-quotients of frames. *Order*, 25(4):369–375, 2008.

[81] T. Dube. Notes on pointfree disconnectivity with a ring-theoretic slant. *Appl. Categ. Structures*, 18(1):55–72, 2010.

[82] C. Ehresmann. Gattungen von lokalen Strukturen. *Jber. Deutsch. Math. Verein*, 60:59–77, 1957.

[83] R. Engelking. *General topology*. Sigma Series in Pure Mathematics, vol. 6. Heldermann Verlag, Berlin, 1989.

[84] M. Erné. Lattice representations for categories of closure spaces. In: *Categorical Topology*, pp. 197–222. Proc. Conf. Toledo, Ohio, 1983. Heldermann, Berlin, 1984.

[85] M. Erné. Compact generation in partially ordered sets. *J. Austral. Math. Soc. Ser. A*, 42(1):69–83, 1987.

[86] M. Erné. The ABC of order and topology. In: *Category theory at work*, pp. 57–83. Proc. Workshop Bremen 1990. Res. Exp. Math., vol. 18, Heldermann Verlag, Berlin, 1991.

[87] M. Erné. Distributive laws for concept lattices. *Algebra Universalis*, 30(4):538–580, 1993.

[88] M. Erné. Algebraic ordered sets and their generalizations. In: *Algebras and orders*, pp. 113–192. Montreal, PQ, 1991. NATO Adv. Sci. Inst. Ser. C Math. Phys. Sci., vol. 389, Kluwer Acad. Publ., Dordrecht, 1993.

[89] M. Erné. Adjunctions and Galois connections: origins, history and development. In: *Galois Connections and Applications*, pp. 1–138. Math. Appl., vol. 565, Kluwer Acad. Publ., Dordrecht, 2004.

[90] M. Erné. Closure. In: *Beyond Topology*, pp. 163–238. Contemp. Math., vol. 486, Amer. Math. Soc., Providence, RI, 2009.

[91] M. Erné, M. Gehrke and A. Pultr. Complete congruences on topologies and downset lattices. *Appl. Categ. Structures*, 15(1-2):163–184, 2007.

[92] M. Erné, J. Koslowski, A. Melton and G.E. Strecker. A primer on Galois connections. In: *Papers on general topology and applications*, pp. 103–125. Madison, WI, 1991. Ann. New York Acad. Sci., vol. 704, 1993.

[93] M.J. Ferreira and J. Picado. The Galois approach to uniform structures. *Quaest. Math.*, 28(3):355–373, 2005.

[94] M.J. Ferreira and J. Picado. Functorial quasi-uniformities on frames. *Appl. Categ. Structures*, 13(4):281–303, 2005.

[95] M.J. Ferreira and J. Picado. On point-finiteness in pointfree topology. *Appl. Categ. Structures*, 15(1-2):185–198, 2007.

[96] M.J. Ferreira and J. Picado. The semicontinuous quasi-uniformity of a frame, revisited. *Quaest. Math.*, 34:175–186, 2011.

[97] M.J. Ferreira. J. Gutiérrez García and J. Picado. Completely normal frames and real-valued functions. *Topology Appl.*, 156(18):2932–2941, 2009.

[98] M.J. Ferreira. J. Gutiérrez García and J. Picado. Insertion of continuous real functions on spaces, bispaces, ordered spaces and point-free spaces – a common root. *Appl. Categ. Structures*, 19:469–487, 2011.

[99] P. Fletcher and W. Hunsaker. Entourage uniformities for frames. *Monatsh. Math.*, 112(4):271–279, 1991.

[100] P. Fletcher, W. Hunsaker and W. Lindgren. Characterizations of frame quasi-uniformities. *Quaestiones Math.*, 16(4):371–383, 1993.

[101] P. Fletcher, W. Hunsaker and W. Lindgren. Frame quasi-uniformities. *Monatsh. Math.*, 117(3-4):223–236, 1994.

[102] P. Fletcher, W. Hunsaker and W. Lindgren. A functional approach to uniform and quasi-uniform frames. In: *Topology with applications* (Szekszárd, 1993), pp. 217–222. Bolyai Soc. Math. Stud., vol. 4, János Bolyai Math. Soc., Budapest, 1995.

[103] P. Fletcher and W. Lindgren. *Quasi-uniform spaces.* Marcel Dekker, New York, 1982.

[104] M.P. Fourman and J.M.E. Hyland. Sheaf models for analysis. In: *Applications of Sheaves*, pp. 280–301. Lecture Notes in Mathematics, vol. 753, Springer-Verlag, Berlin, 1979.

[105] H. Freudenthal. Neuaufbau der Endentheorie. *Ann. of Math. (2)*, 43:261–279, 1942.

[106] P.J. Freyd and A. Scedrov. *Categories, allegories.* North-Holland, Amsterdam, 1990.

[107] J.L. Frith. *Structured frames.* Doctoral dissertation. University of Cape Town, 1987.

[108] J.L. Frith. The category of uniform frames. *Cahiers Topologie Géom. Différentielle Catég.*, 31(4):305–313, 1990.

[109] J.L. Frith and Anneliese Schauerte. The Samuel compactification for quasi-uniform biframes. *Topology Appl.*, 156(12):2116–2122, 2009.

[110] H. Gaifman. Infinite Boolean polynomials I. *Fund. Math.*, 54:229–250, 1964; errata ibid. 57:117, 1965.

[111] T.E. Gantner and R.C. Steinlage. Characterizations of quasi-uniformities. *J. London Math. Soc. (2)*, 5:48–52, 1972.

[112] G. Gierz, K.H. Hofmann, K. Keimel, J.D. Lawson, M. Mislove and D.S. Scott. *Continuous lattices and domains.* Encyclopedia of Mathematics and its Applications, vol. 93. Cambridge University Press, Cambridge, 2003.

[113] L. Gillman and M. Jerison. *Rings of continuous functions.* The University Series in Higher Mathematics. D. Van Nostrand Co., Inc., Princeton, 1960.

[114] C.R.A. Gilmour. Realcompact spaces and regular σ-frames. *Math. Proc. Cambridge Philos. Soc.*, 96(1):73–79, 1984.

[115] C.R.A. Gilmour. *Realcompact Alexandroff spaces and regular σ-frames*. Mathematical Monographs of the University of Cape Town, vol. 3. University of Cape Town, Department of Mathematics, 1985.

[116] J. Gutiérrez García and T. Kubiak. General insertion and extension theorems for localic real functions. *J. Pure Appl. Algebra*, 215(6):1198–1204, 2011.

[117] J. Gutiérrez García, T. Kubiak and J. Picado. Monotone insertion and monotone extension of frame homomorphisms. *J. Pure Appl. Algebra*, 212(5):955–968, 2008.

[118] J. Gutiérrez García, T. Kubiak and J. Picado. Lower and upper regularizations of frame semicontinuous real functions. *Algebra Universalis*, 60(2):169–184, 2009.

[119] J. Gutiérrez García, T. Kubiak and J. Picado. Pointfree forms of Dowker's and Michael's insertion theorems. *J. Pure Appl. Algebra*, 213(1):98–108, 2009.

[120] J. Gutiérrez García, T. Kubiak and J. Picado. Localic real-valued functions: a general setting. *J. Pure Appl. Algebra*, 213:1064–1074, 2009.

[121] J. Gutiérrez García, T. Kubiak and J. Picado. An invitation to localic real functions. In: *Applied Topology: Recent progress for Computer Science, Fuzzy Mathematics and Economics*, pp. 119–129. Proc. WiAT'10, Gandia, 2010. Editorial Universitat Politècnica de València, 2010.

[122] J. Gutiérrez García and J. Picado. On the algebraic representation of semicontinuity. *J. Pure Appl. Algebra*, 210(2):299–306, 2007.

[123] J. Gutiérrez García and J. Picado. Rings of real functions in pointfree topology. *Topology Appl.*, 158, 2011 (to appear, doi: 10.1016/j.topol.2011.05.040).

[124] A.W. Hales. On the non-existence of free complete Boolean algebras. *Fund. Math.*, 54:45–66, 1964.

[125] F. Hausdorff. *Grundzüge der Mengenlehre*. Veit & Co., Leipzig, 1914.

[126] H. Herrlich. A concept of nearness. *General Topology and Appl.*, 4:191–212, 1974.

[127] H. Herrlich. On the extendibility of continuous functions. *General Topology and Appl.*, 4:213–215, 1974.

[128] H. Herrlich. *Topologie II: Uniforme Räume*. Heldermann Verlag, Berlin, 1988.

[129] H. Herrlich and M. Hušek. Galois connections categorically. *J. Pure Appl. Algebra*, 68(1-2):165–180, 1990.

[130] H. Herrlich and A. Pultr. Nearness, subfitness and sequential regularity. *Appl. Categ. Structures*, 8(1-2):67–80, 2000.

[131] E. Hewitt. Rings of real-valued continuous functions, I. *Trans. Amer. Math. Soc.*, 64:45–99, 1948.

[132] K.H. Hofmann and J.D. Lawson. The spectral theory of distributive continuous lattices. *Trans. Amer. Math. Soc.*, 246:285–310, 1978.

[133] W. Hunsaker and J. Picado. Frames with transitive structures. *Appl. Categ. Structures*, 10(1):63–79, 2002.

[134] J.M.E. Hyland. Function spaces in the category of locales. In: *Continuous lattices*, pp. 264–281. Lecture Notes in Mathematics, vol. 871. Springer-Verlag, Berlin, 1981.

[135] J.R. Isbell. *Uniform spaces*. Math. Surveys, vol. 12. Amer. Math. Soc., Providence, RI, 1964.

[136] J.R. Isbell. Atomless parts of spaces. *Math. Scand.*, 31:5–32, 1972.

[137] J.R. Isbell. Functions spaces and adjoints. *Math. Scand.*, 36:317–339, 1975.

[138] J.R. Isbell. Product spaces on locales. *Proc. Amer. Math. Soc.*, 81:116–118, 1981.

[139] J.R. Isbell. Direct limits of meet-continuous lattices. *J. Pure Appl. Algebra*, 23(1):33–35, 1982.

[140] J.R. Isbell. Graduation and dimension in locales. In: *Aspects of Topology*, pp. 195–210. London Math. Soc. Lecture Note Ser., vol. 93. Cambridge Univ. Press, Cambridge, 1985.

[141] J.R. Isbell. First steps in descriptive theory of locales. *Trans. Amer. Math. Soc.*, 327(1):353–371, 1991; errata ibid. 341(1):467–468, 1994.

[142] J.R. Isbell, I. Kříž, A. Pultr and J. Rosický. Remarks on localic groups. In: *Categorical Algebra and its Applications*, pp. 154–172. Lecture Notes in Mathematics, vol. 1348. Springer-Verlag, Berlin, 1988.

[143] P.T. Johnstone. Scott is not always sober. In: *Continuous lattices*, pp. 282–283. Lecture Notes in Mathematics, vol. 871. Springer-Verlag, Berlin, 1981.

[144] P.T. Johnstone. Tychonoff's theorem without the axiom of choice. *Fund. Math.*, 113:21–35, 1981.

[145] P.T. Johnstone. *Stone spaces*. Cambridge Studies in Advanced Mathematics, vol. 3. Cambridge University Press, Cambridge 1982.

[146] P.T. Johnstone. The point of pointless topology. *Bull. Amer. Math. Soc. (N.S.)*, 8(1):41–53, 1983.

[147] P.T. Johnstone. Wallman compactification of locales. *Houston J. Math.*, 10(2):201–206, 1984.

[148] P.T. Johnstone. A simple proof that localic groups are closed. *Cahiers Topologie Géom. Différentielle Catég.*, 29:157–161, 1988.

[149] P.T. Johnstone. A constructive "closed subgroup theorem" for localic groups and groupoids. *Cahiers Topologie Géom. Différentielle Catég.*, 30(1):3–23, 1989.

[150] P.T. Johnstone. Fibrewise separation axioms for locales. *Math. Proc. Cambridge Philos. Soc.*, 108:247–256, 1990.

[151] P.T. Johnstone. The art of pointless thinking: a student's guide to the category of locales. In: *Category Theory at Work*, pp. 85–107. Proc. Workshop Bremen 1990. Research and Exposition in Math., vol. 18, Heldermann Verlag, Berlin, 1991.

[152] P.T. Johnstone. Elements of the history of locale theory. In: *Handbook of the history of general topology*, Vol. 3, pp. 835–851. Kluwer Acad. Publ., Dordrecht, 2001.

[153] P.T. Johnstone. *Sketches of an Elephant: a topos theory compendium*, vol. 2. Oxford Logic Guides, vol. 44. The Clarendon Press, Oxford University Press, Oxford, 2002.

[154] P.T. Johnstone and S.-H. Sun. Weak products and Hausdorff locales. In: *Categorical Algebra and its Applications*, pp. 173–193. Lecture Notes in Mathematics, vol. 1348. Springer-Verlag, Berlin, 1988.

[155] P.T. Johnstone and S. Vickers. Preframe presentations present. In: *Proc. 1990 Como Categ. Conference*, pp. 193–212. Lecture Notes in Mathematics, vol. 1348. Springer-Verlag, Berlin, 1990.

[156] A. Joyal. *Nouveaux fondaments de l'analyse*. Lecture notes, Montreal (manuscript), 1973 and 1974.

[157] A. Joyal and M. Tierney. *An extension of the Galois theory of Grothendieck*. Mem. Amer. Math. Soc., vol. 309. Amer. Math. Soc., Providence, RI, 1984.

[158] T. Kaiser. A sufficient condition of full normality. *Comment. Math. Univ. Carolinae*, 37(2):381–389, 1996.

[159] M. Katětov. On real-valued functions in topological spaces. *Fund. Math.*, 38:85-91, 1951; errata ibid. 40:203–205, 1953.

[160] J.L. Kelley. The Tychonoff product theorem implies the axiom of choice. *Fund. Math.*, 37:75–76, 1950.

[161] J.L. Kelley. *General Topology*. The University Series in Higher Mathematics, Van Nostrand, 1955.

[162] J.C. Kelly. Bitopological spaces. *Proc. London Math. Soc. (3)*, 13:71–89, 1963.

[163] I. Kříž. A constructive proof of the Tychonoff's theorem for locales. *Comment. Math. Univ. Carolinae*, 26(3):619–630, 1985.

[164] I. Kříž. A direct description of uniform completion in locales and a characterization of *LT*-groups. *Cahiers Topologie Géom. Différentielle Catég.*, 27(1):19–34, 1986.

[165] I. Kříž and A. Pultr. Products of locally connected locales. *Rend. Circ. Mat. Palermo (2) Suppl.*, 11:61–70, 1985.

[166] I. Kříž and A. Pultr. Systems of covers of frames and resulting subframes. *Rend. Circ. Mat. Palermo (2) Suppl.*, 14:353–364, 1987.

[167] I. Kříž and A. Pultr. Peculiar behaviour of connected locales. *Cahiers Topologie Géom. Différentielle Catég.*, 30(1):25–43, 1989.

[168] I. Kříž and A. Pultr. A spatiality criterion and an example of a quasitopology which is not a topology. *Houston J. Math.*, 15(2):215–234, 1989.

[169] H.-P. Künzi. Nonsymmetric distances and their associated topologies: about the origins of basic ideas in the area of asymmetric topology. In: *Handbook of the history of general topology*, vol. 3, pp. 853–968. Kluwer Acad. Publ., Dordrecht, 2001.

[170] H.-P. Künzi. Quasi-uniform spaces in the year 2001. In: *Recent Progress in General Topology II*, pp. 313–344. Elsevier, Amsterdam, 2002.

[171] R. LaGrange. Amalgamation and epimorphisms in m complete Boolean algebras. *Algebra Universalis*, 4:277–279, 1974.

[172] S. Lal. Pairwise concepts in bitopological spaces. *J. Austral. Math. Soc. (A)*, 26:241–250, 1978.

[173] Y.-M. Li and Z.-H. Li. Constructive insertion theorems and extension theorems on extremally disconnected frames. *Algebra Universalis*, 44(3-4):271–281, 2000.

[174] Y.-M. Li and G.-J. Wang. Localic Katětov-Tong insertion theorem and localic Tietze extension theorem. *Comment. Math. Univ. Carolinae*, 38(4):801–814, 1997.

[175] S. Mac Lane. *Categories for the working mathematician*. Graduate Texts in Mathematics, vol. 5, Second edition. Springer-Verlag, New York, 1998.

[176] S. MacLane and I. Moerdijk. *Sheaves in geometry and logic. A first introduction to topos theory*. Corrected reprint of the 1992 edition. Universitext. Springer-Verlag, New York, 1994.

[177] J. Madden. κ-frames. *J. Pure Appl. Alg.*, 70(1-2):107–127, 1991.

[178] J. Madden and A.T. Molitor. Epimorphisms of frames. *J. Pure Appl. Algebra*, 70(1-2):129–132, 1991.

[179] J. Madden and J. Vermeer. Lindelöf locales and realcompactness. *Math. Proc. Cambridge Philos. Soc.*, 99(3):473–480, 1986.

[180] J. Martínez. The role of frames in the development of lattice-ordered groups: a personal account. In: *Positivity*, pp. 161–195. Trends Math., Birkhäuser, Basel, 2007.

[181] J. Martínez. Archimedean frames, revisited. *Comment. Math. Univ. Carolinae*, 49(1):25–44, 2008.

[182] J.C.C. McKinsey and A. Tarski. The algebra of topology. *Ann. Math.*, 45:141–191, 1944.

[183] K. Menger. Topology without points. *Rice Institute pamphlet*, 27:80–107, 1940.

[184] I. Moerdijk and G.C. Wraith. Connected locally connected toposes are path-connected. *Trans. Amer. Math. Soc.*, 295:849–859, 1986.

[185] C.J. Mulvey and J.W. Pelletier. A quantisation of the calculus of relations. In: *Category Theory* 1991, pp. 345–360. Canadian Math. Soc. Conf. Proc, vol. 13, Amer. Math. Soc., Providence, RI, 1992.

[186] L. Nachbin. Sur les espaces uniformes ordonnés. *C. R. Acad. Sci. Paris*, 226:774–775, 1948.

[187] L. Nachbin. On the continuity of positive linear transformations. In: *Proceedings of the International Congress of Mathematicians* (Cambridge, Mass., 1950), vol. I, pp. 464–465. Providence, RI, 1952.

[188] L. Nachbin. *Topology and order*. D. van Nostrand, Princeton, 1965.

[189] S.B. Niefield and K.I. Rosenthal. Spatial sublocales and essential primes. *Topology Appl.*, 26:263–269, 1987.

[190] S.B. Niefield and K.I. Rosenthal. Constructing locales from quantales. *Math. Proc. Cambridge Phil. Soc.*, 104:215–234, 1988.

[191] D. Papert and S. Papert. *Sur les treillis des ouverts et paratopologies*. Séminaire Ehresmann (1re année, exposé 1). Paris, 1958.

[192] J. Paseka. T_2-separation axioms on frames. *Acta Univ. Carolinae Mat. Phys.*, 28:95–98, 1987.

[193] J. Paseka. Lindelöf locales and N-compactness. *Math. Proc. Cambridge Philos. Soc.*, 109(1):187–191, 1991.

[194] J. Paseka. Paracompact locales and metric spaces. *Math. Proc. Cambridge Philos. Soc.*, 110:251–256, 1991.

[195] J. Paseka. Products in the category of locales: which properties are preserved? *Discrete Math.*, 108(1-3):63–73, 1992.

[196] J. Paseka and B. Šmarda. T_2-frames and almost compact frames. *Czech Math. J.*, 42(117):297–313, 1992

[197] J.W. Pelletier. Locales in functional analysis. *J. Pure Appl. Algebra*, 70(1-1):133–145, 1991.

[198] A.P. Gomes. Introduction to the notion of functional in spaces without points. *Portugaliae Math.*, 5:1–120, 1946; errata ibid. 5:218, 1946.

[199] J. Picado. Weil uniformities for frames. *Comment. Math. Univ. Carolinae*, 36(2): 357–370, 1995.

[200] J. Picado. Frame quasi-uniformities by entourages. In: *Symposium on Categorical Topology* (Rondebosch, 1994), pp. 161–175. Univ. Cape Town, Rondebosch, 1999.

[201] J. Picado. Structured frames by Weil entourages. *Appl. Categ. Structures*, 8(1-2): 351–366, 2000.

[202] J. Picado. The gauge of a uniform frame. *Appl. Categ. Structures*, 10(6):593–602, 2002.

[203] J. Picado. The quantale of Galois connections. *Algebra Universalis*, 52(4):527–540, 2004.

[204] J. Picado. A new look at localic interpolation theorems. *Topology Appl.*, 153(16): 3203–3218, 2006.

[205] J. Picado and A. Pultr. Sublocale sets and sublocale lattices. *Arch. Math. (Brno)*, 42(4):409–418, 2006.

[206] J. Picado and A. Pultr. *Locales mostly treated in a covariant way*. Textos de Matemática (Série B), vol. 41. Universidade de Coimbra, Departamento de Matemática, Coimbra, 2008.

[207] J. Picado and A. Pultr. Cover quasi-uniformities in frames. *Topology Appl.*, 158(7):869–881, 2011.

[208] J. Picado and A. Pultr. Entourages, covers and localic groups. *Appl. Categ. Structures*, to appear (doi: 10.1007/s10485-011-9254-3).

[209] J. Picado, A. Pultr and A. Tozzi. Locales. In: *Categorical Foundations – Special Topics in Order, Algebra and Sheaf Theory*, pp. 49–101. Encyclopedia of Mathematics and its Applications, vol. 97. Cambridge Univ. Press, Cambridge, 2004.

[210] J. Picado, A. Pultr and A. Tozzi. Ideals in Heyting semilattices and open homomorphisms. *Quaest. Math.*, 30:391–405, 2007.

[211] T. Plewe. Localic products of spaces. *Proc. London Math. Soc. (3)*, 73(3):642–678, 1996.

[212] T. Plewe. Quotient maps of locales. *Appl. Categ. Structures*, 8(1-2):17–44, 2000.

[213] T. Plewe. Sublocale lattices. *J. Pure Appl. Algebra*, 168(2-3):309–326, 2002.

[214] T. Plewe, A. Pultr and A. Tozzi. Regular monomorphisms of Hausdorff frames. *Appl. Categ. Structures*, 9(1):15–33, 2001.

[215] A. Pultr. Pointless uniformities I. Complete regularity. *Comment. Math. Univ. Carolinae*, 25(1):91–104, 1984.

[216] A. Pultr. Pointless uniformities II. (Dia)metrization. *Comment. Math. Univ. Carolinae*, 25(1):105–120, 1984.

[217] A. Pultr. Remarks on metrizable locales. *Rend. Circ. Mat. Palermo (2)*, 6:247–258, 1984.

[218] A. Pultr. Diameters in locales: how bad they can be. *Comment. Math. Univ. Carolinae*, 29(4):731–742, 1988.

[219] A. Pultr. Categories of diametric frames. *Math. Proc. Cambridge Phil. Soc.*, 105:285–297, 1989.

[220] A. Pultr. Frames. In: *Handbook of Algebra*, vol. 3, pp. 791–857. North-Holland, Amsterdam, 2003.

[221] A. Pultr and S.E. Rodabaugh. Lattice-valued frames, functor categories, and classes of sober spaces. In: *Topological and algebraic structures in fuzzy sets*, pp. 153–187. Trends Log. Stud. Log. Libr., vol. 20. Kluwer Acad. Publ., Dordrecht, 2003.

[222] A. Pultr and S.E. Rodabaugh. Category theoretic aspects of chain-valued frames I. Categorical and presheaf theoretic foundations. *Fuzzy Sets and Systems*, 159(5):501–528, 2008.

[223] A. Pultr and S.E. Rodabaugh. Category theoretic aspects of chain-valued frames II. Applications to lattice-valued topology. *Fuzzy Sets and Systems*, 159(5):529–558, 2008.

[224] A. Pultr and J. Sichler. Frames in Priestley duality. *Cahiers Topologie Géom. Différentielle Catég.*, 29(3):193–202, 1988.

[225] A. Pultr and A. Tozzi. Notes on Kuratowski-Mrówka theorems in point-free context. *Cahiers Topologie Géom. Différentielle Catég.*, 33(1):3–14, 1992.

[226] A. Pultr and A. Tozzi. Equationally closed subframes and representation of quotient spaces. *Cahiers Topologie Géom. Différentielle Catég.*, 34(3):167–183, 1993.

[227] A. Pultr and A. Tozzi. Separation axioms and frame representation of some topological facts. *Appl. Categ. Structures*, 2(1):107–118, 1994.

[228] A. Pultr and A. Tozzi. Completion and coproducts of nearness frames. In: *Symposium on Categorical Topology* (Rondebosch, 1994), pp. 177–186. Univ. Cape Town, Rondebosch, 1999.

[229] A. Pultr and A. Tozzi. A note on reconstruction of spaces and maps from lattice data. *Quaest. Math.*, 24(1):55–63, 2001.

[230] A. Pultr and J. Úlehla. Notes on characterization of paracompact frames. *Comment. Math. Univ. Carolinae*, 30(2):377–384, 1989.

[231] G.N. Raney. A subdirect-union representation for completely distributive complete lattices. *Proc. Amer. Math. Soc.*, 4:518–522, 1953.

[232] G. Reynolds. On the spectrum of a real representable ring. In: *Applications of sheaves*, pp. 595–611. Lecture Notes in Mathematics, vol. 753. Springer-Verlag, Berlin, 1979.

[233] K.I. Rosenthal. *Quantales and their applications.* Pitman Research Notes in Math., vol. 234. Longman Scientific & Technical, 1990.

[234] J. Rosický and B. Šmarda. T_1-locales. *Math. Proc. Cambridge Philos. Soc.*, 98(1):81–86, 1985.

[235] A. Schauerte. *Biframes.* Doctoral dissertation, McMaster University, Hamilton, 1992.

[236] A. Schauerte. Biframe compactifications. *Comment. Math. Univ. Carolinae*, 34(3):567–574, 1993.

[237] G. Schlitt. N-*compact frames and applications.* Doctoral dissertation, McMaster University, Hamilton, 1990.

[238] G. Schlitt. The Lindelöf-Tychonoff theorem and choice principles. *Math. Proc. Cambridge Phil. Soc.*, 110:57–65, 1991.

[239] G. Schlitt. N-compact frames. *Comm. Math. Univ. Carolinae*, 32:173–187, 1991.

[240] D. Scott. Continuous lattices. In: *Toposes, Geometry and Logic*, pp. 97–136. Lecture Notes in Mathematics, vol. 247. Springer-Verlag, Berlin, 1972.

[241] R.A. Sexton and H. Simmons. An ordinal indexed hierarchy of separation properties. *Topology Proc.*, 30(2):585–625, 2006.

[242] T. Shirota. A class of topological spaces. *Osaka Math. J.*, 4:23–40, 1952.

[243] W. Sierpiński. La notion de derivée comme base d'une théorie des ensembles abstraits. *Math. Ann.*, 97:321–337, 1927.

[244] H. Simmons. A framework for topology. In: *Logic Colloq. '77*, pp. 239–251. Stud. Logic Foundations Math., vol. 96. North-Holland, Amsterdam-New York, 1978.

[245] H. Simmons. The lattice theoretic part of topological separation properties. *Proc. Edinburgh Math. Soc.*, 21:41–48, 1978.

[246] H. Simmons. Spaces with Boolean assemblies. *Colloq. Math.*, 43(1):23–29, 1980.

[247] H. Simmons. Regularity, fitness, and the block structure of frames. *Appl. Categ. Structures*, 14(1):1–34, 2006.

[248] S. Steinberg. *Lattice-ordered rings and modules*. Springer-Verlag, New York, 2010.

[249] M.H. Stone. Boolean algebras and their application in topology. *Proc. Nat. Acad. Sci. USA*, 20:197–202, 1934.

[250] M.H. Stone. The theory of representations for Boolean algebras. *Trans. Amer. Mat. Soc.*, 40:37–111, 1936.

[251] M.H. Stone. Boundedness properties in function-lattices. *Canad. J. Math.*, 1:176–186, 1949.

[252] S.H. Sun. On paracompact locales and metric locales. *Comment. Math. Univ. Carolinae*, 30:101–107, 1989.

[253] W.J. Thron. Lattice-equivalence of topological spaces. *Duke Math. J.*, 29:671–679, 1962.

[254] H. Tong. Some characterizations of normal and perfectly normal spaces. *Duke Math. J.*, 19:289–292, 1952.

[255] C. Townsend. *Preframe techniques in constructive locale theory*. Doctoral dissertation, Imperial College, London, 1996.

[256] C. Townsend. On the parallel between the suplattice and preframe approaches to locale theory. *Ann. Pure Appl. Logic*, 137(1-3):391–412, 2006.

[257] C. Townsend. A categorical proof of the equivalence of local compactness and exponentiability in locale theory. *Cahiers Topol. Géom. Différ. Catég.*, 47(3):233–239, 2006.

[258] C. Townsend. A categorical account of the localic closed subgroup theorem. *Comment. Math. Univ. Carolinae*, 48(3):541–553, 2007.

[259] J.W. Tukey. *Convergence and uniformity in topology*. Annals of Mathematics Studies, vol. 2. Princeton University Press, Princeton, N. J., 1940.

[260] J. Vermeulen. *Constructive techniques in functional analysis*. Doctoral dissertation, University of Sussex, 1987.

[261] J. Vermeulen. Some constructive results related to compactness and the (strong) Hausdorff property for locales. In: *Category Theory*, pp. 401–409. Lecture Notes in Mathematics, vol. 1488. Springer-Verlag, Berlin, 1991.

[262] J. Vermeulen. Weak compactness in constructive spaces. *Math. Proc. Cambridge Philos. Soc.*, 111:63–74, 1992.

[263] J. Vermeulen. Proper maps of locales. *J. Pure Appl. Algebra*, 92(1):79–107, 1994.

[264] J. Vermeulen. A note on stably closed maps of locales. *J. Pure Appl. Algebra*, 157(2-3):335–339, 2001.

[265] S. Vickers. *Topology via Logic*. Cambridge Tracts in Theoretical Computer Science, vol. 5. Cambridge University Press, Cambridge, 1989.

[266] H. Wallman. Lattices and topological spaces. *Ann. Math.*, 39(1):112–126, 1938.

[267] J. Walters-Wayland. *Completeness and nearly fine uniform frames*. Doctoral dissertation, Univ. of Louvain, 1995.

[268] J. Walters-Wayland. A Shirota theorem for frames. *Appl. Categ. Structures*, 7(3):271–277, 1999.

[269] J. Walters-Wayland. Cozfine and measurable uniform frames. *J. Pure Appl. Algebra*, 145(2):199–213, 2000.

[270] D. Wigner. Two notes on frames. *J. Austral. Math. Soc. Ser. A*, 28(3):257–268, 1979.

[271] R.J. Wood. Ordered sets via adjunctions. In: *Categorical Foundations – Special Topics in Order, Algebra and Sheaf Theory*, pp. 5–47. Encyclopedia of Mathematics and its Applications, vol. 97. Cambridge Univ. Press, Cambridge, 2004.

[272] A. Weil. *Sur les espaces à structure uniforme et sur la topologie générale*. Publications de l'Institute Mathématique de l'Université de Strasbourg. Hermann, Paris, 1938.

[273] M. D. Weir. *Hewitt-Nachbin spaces*. North-Holland Mathematics Studies, vol. 17. North-Holland Publishing Co., Amsterdam-Oxford, 1975.

[274] J.T. Wilson. *The assembly tower and some categorical and algebraic aspects of frame theory*. Doctoral dissertation, Carnegie Mellon Univ., Pittsburgh, 1994.

[275] G.C. Wraith. Localic groups. *Cahiers Topologie Géom. Différentielle*, 22(1):61–66, 1981.

[276] G.C. Wraith. Unsurprising results on localic groups. *J. Pure Appl. Algebra*, 67(1):95–100, 1990.

List of Symbols

$(X, \leq)$, 316
$(X, \leq)^{\mathrm{op}}$, 316
$(-, q)$, 270
$(p, -)$, 270
$<$, 316
$\lhd_{\mathcal{A}}$, 149
$\#$, 307
$\overset{.}{\rightarrow}$, 341
$\perp$, 320
$\dashv$, 352
$\downarrow M$, 317
$\downarrow x$, 317
$\rightarrow$, 331
$\langle p, q \rangle$, 270
$\lceil x \rceil$, 323
$\leq$, 315
$\lfloor x \rfloor$, 323
$\lhd_{\varepsilon}$, 231
$\lhd_{\varepsilon}^{i}$, 250
$\lhd_{\mathcal{U}}^{i}$, 242
$\ll$, 134
$\lll$, 142
$\twoheadleftarrow$, 90
$\twoheadleftarrow_{i}$, 241
$\prec$, 88
$\prec_{i}$, 241
$\twoheadrightarrow$, 230
$\overset{\mathcal{U}}{\in}$, 244
$\bigvee M$, 319
$\bigwedge M$, 319
$\top$, 320
$\lhd_{\varepsilon}^{d}$, 204
$\uparrow M$, 317

$\uparrow x$, 317
0, 320
1_{A}, 337
$\bar{1}$, 56
1, 320
$\mathbf{2}$, 10

$A^{\downarrow}$, 161
Ax, 148
$\mathcal{A}^{\downarrow}$, 161
$\widetilde{A}$, 99
a^{*}, 330
$a^{\#}$, 107
$a^{\bullet}$, 240
$a_1 \oplus a_2$, 60
AB, 148
$a \smallsetminus b$, 107
a^{c}, 19, 106
A_{sloc}, 29
$\langle Au\varepsilon \rangle$, 262

$B^{\times}$, 174
B_2, 256
$\mathfrak{b}(a)$, 42
B_L, 40
$\mathfrak{B}L$, 334
$B(x)$, 301

$\widetilde{\mathbf{C}}(L, \mathcal{A})$, 185
$\mathbf{C}(L, \mathcal{A})$, 161
$\mathcal{C}(L)$, 278
$\check{c}$, 82
χ_a, 279
$\mathcal{C}^{\mathcal{A}}$, 366

$\mathfrak{c}(a)$, 33
$\mathcal{C}(A, B)$, 338
$\mathcal{C}^{\mathbb{T}}$, 362
$\mathcal{C}_{\mathbb{T}}$, 363
$\mathcal{C} \times \mathcal{D}$, 338
$\mathfrak{C}(L)$, 31
$\mathfrak{C}^*(L)$, 293
$\mathfrak{c}L$, 287
codom, 337
$\mathrm{Coequ}(f, g)$, 345
$\mathcal{C}^{\mathrm{op}}$, 338
$\mathrm{Coz}\, L$, 284
coz, 284
χ_S, 289
$\mathfrak{c}_S(a)$, 34

Δ, 63
Δ_a, 36
$\delta(f; S)$, 211
$\delta(f; x, y, S)$, 211
d_f, 211
$\mathrm{dc}(\mathbb{B})$, 261
∂S, 110
diam_ρ, 204
d_L, 79
d_L^M, 84
$\mathrm{DMN}(X, \leq)$, 322
dom, 337

$E \circ F$, 228
E^{-1}, 228
E_A, 7
$\varepsilon_{A,B}$, 352
$\mathrm{Equ}(f, g)$, 344
$\mathrm{EssPrEssPr}(a)$, 118
Ex, 31

f^+, 283
f^-, 283
f_{-1}, 29
f^*, 11
$f_1 \oplus f_2$, 63
$\mathcal{F}_J$, 21
ϕ_L, 16

F_p, 14
$\mathcal{F}(X, \mathbb{R})$, 287

$\Gamma_M(L)$, 84

$h[[\mathcal{A}]]$, 153
h_*, 11
$\mathrm{Haus}(L)$, 95

id_A, 337
$\mathrm{Idl}(X, \leq)$, 22
$\inf M$, 319

$J_{\mathcal{F}}$, 21
$\mathfrak{J}(L)$, 97, 131

$\mathfrak{k}(a)$, 161
$[K, \mathcal{C}]$, 342

$L_i(\mathcal{U})$, 242
$L_{\lll}$, 93
$L_{\prec}$, 92
$\mathfrak{L}[p, q]$, 273
$\mathrm{lb}M$, 317
L/E, 31
$\mathcal{L}(L)$, 288
$L \oplus M$, 58
L/R, 44
$\mathfrak{L}(\mathbb{R})$, 269
λ_X, 17

$\mathrm{m}(A)$, 103
$\max_{Ex}$, 44
$\mathrm{MinPr}(a)$, 116
$\mathrm{mph}\mathcal{C}$, 337
μ_R, 44

$\mathbb{N}$, 6
∇, 60
∇_a, 36
$\mathbf{n}$, 59
$\tilde{\mathbf{n}}$, 221
N_L, 300
∇_L, 36

O, 10, 26
$\Omega(X)$, 1
$\mathcal{O}(x;\varepsilon)$, 176
$\mathfrak{o}(a)$, 33
obj$\mathcal{C}$, 337
$\Omega(f)$, 4
$\mathfrak{o}_S(a)$, 34

P, 10
p, 279
$\pi_1(U)$, 64
$\pi_2(U)$, 64
Par(L), 180
p_F, 14
$\phi_{\#}$, 307
Pr(a), 116
PrL, 116
$\prod'$, 56
$\mathfrak{P}(X)$, 19, 316
$\Psi(X, \mathcal{A})$, 190
Ψ_u, 190
$\Psi_{\mathcal{A}}$, 193

Q, 296

R_d, 220
$\widetilde{\mathfrak{R}}(L, \mathcal{A})$, 183
$\mathfrak{R}(L)$, 132
$r_\lambda(S)$, 110

$S_{\mathbb{B}}$, 261
S', 74
Σ_a, 15
$S - A_\alpha$, 112
Set$^{\mathbb{T}}$, 363
Sk(τ), 7
$\mathfrak{S}(L)$, 287
σ_L, 16
$\mathscr{S}\ell(L)$, 28
Sp$'$, 16
$S \smallsetminus T$, 107
st, 242
st$_i$, 242
sup M, 319

$\mathfrak{T}_l$, 287
$\mathfrak{T}_u$, 287
$\mathbb{T}$, 361
$\mathfrak{T}_e$, 288
$\overline{T}^S$, 40

U_ε^d, 203
U^s, 245
$\mathcal{U}(d)$, 203
$\mathcal{U}^\circ$, 147
$U(a)$, 300
ub$M$, 317
$\mathfrak{U}(L)$, 288
$\mathfrak{U}\mathfrak{p}(X, \le)$, 21
U^{s}, 102
$\mathcal{U}(x)$, 3

$V(a)$, 300

$X_{\mathbb{B}}$, 261
$\widetilde{x}$, 13
x/A, 149
x^*, 43
$x*_j$, 56
$x * y$, 308

Y, 342

$\zeta(A, J)$, 262

List of Categories

BiFrm	240	Biframes & biframe homomorphisms
BiTop	240	Bitopological spaces & bicontinuous maps
ContFrm	139	Continuous frames & frame homomorphisms
CRegFrm	93	Completely regular frames & frame homomorphisms
CRegLoc	93	Completely regular locales & localic maps
CSNearFrm	165	Complete strong nearness frames & uniform homomorphisms
CUniSp	196	Complete uniform spaces & uniformly continuous maps
CUniFrm	165	Complete uniform frames & uniform homomorphisms
DFrm	208	Metric frames & contractive homomorphisms
DLoc	208	Metric locales & contractive localic maps
$\mathbf{D_M Frm}$	208	Metric frames (with metric diameters) & contractive homomorphisms
$\mathbf{D_M Loc}$	208	Metric locales (with metric diameters) & contractive localic maps
$\mathbf{D_S Frm}$	208	Metric frames (with star-additive diameters) & contractive homomorphisms
$\mathbf{D_S Loc}$	208	Metric locales (with star-additive diameters) & contractive localic maps
EQUniFrm	250	Entourage quasi-uniform frames & uniform homomorphisms
EUniFrm	232	Entourage uniform frames & uniform homomorphisms
FitLoc	94	Fit locales & localic maps
Frm	10	Frames & frame homomorphisms
HausLoc	94	I-Hausdorff locales & localic maps
Loc	11	Locales & localic maps
LKSob	139	Sober locally compact spaces & continuous maps
Metr	207	Metric spaces & contractive homomorphisms
NearFrm	153	Nearness frames & uniform homomorphisms
ParFrm	181	Paracompact frames & frame homomorphisms

Pos   317   Posets & monotone maps

P$_S$Frm   220   Star-additive prediametric frames & contractive
                          homomorphisms

QUnif   237   Quasi-uniform spaces & uniformly continuous maps

QUniFrm   244   Quasi-uniform frames & uniform homomorphisms

RegFrm   93   Regular frames & frame homomorphisms

RegKFrm   133   Regular compact frames & frame homomorphisms

RegLoc   93   Regular locales & localic maps

Set   337   Sets & functions

SLat$_1$ 51   (Meet-) Semilattices & $(\wedge, 1)$-homomorphisms

SNearFrm   153   Strong nearness frames & uniform homomorphisms

Sob   9   Sober spaces & continuous maps

Top   9   Topological spaces & continuous maps

UniFrm   153   Uniform frames & uniform homomorphisms

UniFrm$_{EC}$ 198   Uniform frames with enough Cauchy points & uniform
                          homomorphisms

UniSp   146   Uniform spaces & uniformly continuous maps

Index

3-chain, 50

$\mathcal{A}$-Cauchy filter, 189
$\mathcal{A}$-regular filter, 189
$\mathcal{A}$-regular ideal, 183
additive inverse, 279
adjoint functors, 352
adjunction, 352
adjunction units, 354
admissible
  diameter, 203
  set of paircovers, 243
  strong inclusion, 177
  system
    of covers, 150
    of entourages, 231
Alexandroff space, 5
algebra
  Boolean $\sim$, 333
  complete Boolean $\sim$ with
    property (P), 263
  Eilenberg-Moore $\sim$, 362
  Heyting $\sim$, 331
  $\mathbb{T}$-$\sim$, 362
algebraic
  frame, 142
  model of an $\sim$ theory, 366
  structure of a type Δ, 365
  theory, 366
almost inverse function, 323
α-compact
  frame, 125
  locale, 125
α-epimorphism, 72

α-residua, 110
assembly space, 117
atom, 19
atomic Boolean algebra, 19
axiom
  Hausdorff $\sim$, 79
  of choice, 318
  of Dowker and Strauss, 83
  separation $\sim$s, 73
  T_D $\sim$, 5

basis
  of nearness, 151
  of quasi-uniformity, 243
  of uniformity, 151
  regular $\sim$, 215
  σ-admissible $\sim$, 214
  σ-stratified $\sim$, 215
bicontinuous map, 240
biframe, 240
  completely regular $\sim$, 241
  homomorphism, 240
  induced $\sim$, 246
  regular $\sim$, 241
bimorphism, 65
Birkhoff's theorem, 330
bispace, 239
bitopological space, 239
  induced $\sim$, 236
Boolean
  frame, 11
  sublocale, 43
  ultrafilter theorem, 334
Boolean algebra, 333

atomic $\sim$, 19
complete $\sim$ with property (P), 263
κ-complete $\sim$, 70
Booleanization, 334
bottom, 320
bound
  lower $\sim$, 317
  upper $\sim$, 317
bounded
  f-ring, 283
  join-semilattice, 320
  lattice, 320
  meet-semilattice, 320
  real function, 293

Cantor-Bernstein Theorem, 321
category, 337
  $\mathcal{C}^{\mathcal{A}}$, 366
  $\mathcal{C}^{\mathbb{T}}$, 362
  $\mathcal{C}_{\mathbb{T}}$, 363
  co-wellpowered $\sim$, 356
  cocomplete $\sim$, 351
  complete $\sim$, 351
  dual $\sim$, 338
  κ-filtered $\sim$, 70
  Kleisli $\sim$, 363
  locally small $\sim$, 339
  of $\mathbb{T}$-algebras, 362
  of functors, 342
  of relations, 338
  opposite $\sim$, 338
  small $\sim$, 339
  sub$\sim$, 339
  wellpowered $\sim$, 356
Cauchy
  complete nearness frame, 194
  complete uniform frame, 194
  completion, 200
  filter, 189
  fully $\sim$ map, 167
  generalized $\sim$ point, 201
  map, 160
  point, 159, 189
    center of a $\sim$, 159

regular $\sim$ map, 160
spectrum, 193
Čech complete space, 313
chain, 315
  3-$\sim$, 50
  xy-$\sim$, 211
characteristic function, 279, 289
choice
  axiom of $\sim$, 318
closed
  (closed extremal epi, dense)
    factorization, 302
  interval frame, 273
  localic map, 38
  set
    join-irreducible $\sim$, 2
  subgroup theorem, 306
  sublocale, 33
closure, 39
  in a sublocale, 40
co-atom, 19
co-dense
  frame homomorphism, 77
  localic map, 77
co-frame, 10
  of sublocales, 28
co-group, 298
co-linear element, 106
co-wellpowered category, 356
cocomplete category, 351
cocone, 350
codiagonal homomorphism, 60
codomain, 337
coequalizers, 345
  in **Frm**, 55
  in **Loc**, 54
  in **NearFrm**, 156
cogenerator, 356
colimits, 350
  creation of $\sim$, 70
  directed $\sim$ in **Frm**, 67
compact
  frame, 125
  locale, 125

compactification
    Samuel ∼, 183
    Stone-Čech ∼, 133
complement, 106, 331
complete
    Boolean algebra
        with property (P), 263
    category, 351
    lattice, 321
    strong nearness frame, 160
    sublocale, 86
    uniform frame, 160
    uniform space, 159
completely
    below relation, 90
    proper ideal, 295
    regular
        biframe, 241
        frame, 91
        locale, 91
completion
    Cauchy ∼, 200
    Dedekind–MacNeille ∼, 322
    of a uniform frame, 160
    of the rationals, 277
components
    connected ∼, 255
composition law, 337
cone, 350
congruence, 30
    frame, 31
    tower, 67
conjugate pair of covers, 236
connected
    components, 255
    element, 255
    frame, 254
    locale, 254
    set, 211
constant function, 279
continuous
    extension, 292
    lattice, 135
    metric diameter, 206

real function, 278
contractive
    frame homomorphism, 208
    map, 207
contravariant functor, 340
coproducts, 344
    in **Frm**, 58
    in **Loc**, 54
    in **NearFrm**, 156
    in **SLat$_1$**, 56
coreflection, 360
coreflective subcategory, 359
cover, 125
    admissible system of ∼s, 150
    left system of ∼s, 300
    metric ∼s of $\mathfrak{L}(\mathbb{R})$, 276
    normal ∼, 152
    of a set, 145
    open ∼, 125
    real-uniform ∼, 294
    right system of ∼s, 300
cozero
    element, 284
    map, 284
creation of colimits, 70
cut
    Dedekind ∼, 274
    open Dedekind ∼, 274

De Morgan formulas, 333
De Morgan law, 34, 331, 332
Dedekind cut, 274
    open ∼, 274
Dedekind–MacNeille completion, 322
dense
    frame homomorphism, 40
    join, 331
    localic map, 39
    sublocale, 39
density theorem, 40
development, 113
    theorem, 115
diagonal localic map, 61
diagram, 350

diameter, 203
    admissible ∼, 203
    continuous metric ∼, 206
    metric ∼, 205
    pre-∼, 211
    star-additive ∼, 204
difference, 109
directed colimits
    in **Frm**, 67
directed set, 317
disconnected
    element, 255
    frame, 254
    locale, 254
discrete subset, 173
disjoint sublocales, 254
distributive lattice, 326
domain, 337
doubly dense sublocale, 111
down-set, 317
    functor, 51
DS-Hausdorff
    frame, 83
    locale, 83
dual category, 338
dual order, 316

Eilenberg-Moore algebra, 362
element
    co-linear ∼, 106
    connected ∼, 255
    cozero ∼, 284
    disconnected ∼, 255
    greatest ∼, 320
    least ∼, 320
    linear ∼, 106
    maximal ∼, 317
    meet-irreducible ∼, 13
    minimal ∼, 317
    prime ∼, 13
    saturated ∼, 44
    semiprime ∼, 87
    supercompact ∼, 141

embedding
    uniform ∼ surjection, 154
    Yoneda ∼, 342
entourage, 145
    admissible system of ∼s, 231
    of a frame, 228
    quasi-uniform frame, 250
    quasi-uniformity, 250
    symmetric ∼, 228
    uniform frame, 231
    uniformity, 231
epimorphism, 339
    extremal ∼, 347
    extremal ∼ in **Loc**, 50
    in **Loc**, 50
    regular ∼, 347
    regular ∼ in **Loc**, 50
    strong ∼, 347
    strong ∼ in **Loc**, 50
epireflection, 360
epireflective subcategory, 360
equalizers, 344
    in **Frm**, 54
    in **Loc**, 55
essential prime, 118
evaluation operator, 355
extended
    real functions, 294
    reals, 294
extension
    continuous ∼, 292
extremal
    epimorphism, 24, 347
        in **Frm**, 25
        in **Loc**, 50
    monomorphism, 24, 347
        in **Frm**, 50
        in **Loc**, 25
extremally disconnected
    frame, 257
    locale, 257

f-ring, 283
    bounded ∼, 283

strong ∼, 283
factorization
    (closed extremal epi, dense) ∼, 302
    system, 349
filter, 329
    𝒜-Cauchy ∼, 189
    𝒜-regular ∼, 189
    maximal ∼, 329
    prime ∼, 2, 329
    proper ∼, 329
    slicing ∼, 5
fine uniformity, 152
fit
    frame, 74
    locale, 74
frame, 10
    α-compact ∼, 125
    (PC)-∼, 173
    algebraic ∼, 142
    Boolean ∼, 11
    closed interval ∼, 273
    co-dense ∼ homomorphism, 77
    compact ∼, 125
    complete uniform ∼, 160
    completely regular ∼, 91
    congruence, 30
    congruence ∼, 31
    connected ∼, 254
    cover, 125
    disconnected ∼, 254
    DS-Hausdorff ∼, 83
    entourage of a ∼, 228
    extremally disconnected ∼, 257
    fit ∼, 74
    fully normal ∼, 170
    H-realcompact ∼, 295
    homomorphism, 10
        contractive ∼, 208
        dense ∼, 40
        quasi-uniform ∼, 244
        uniform ∼, 152, 232
    I-Hausdorff ∼, 79
    Lindelöf ∼, 125
    locally connected ∼, 258

metric ∼, 203
metrizable ∼, 172, 204
nearness ∼, 150
normal ∼, 91
of extended reals, 294
of ideals, 131
of reals, 269
of sublocales, 287
paircover of a ∼, 241
paracompact ∼, 173
point of a ∼, 13, 14
product-connected ∼, 260
quasi-uniform ∼, 243
quotient ∼, 31, 44
real-complete ∼, 294
realcompact ∼, 295
regular ∼, 89
spatial ∼, 10
spectrum of a ∼, 15
strongly Hausdorff ∼, 79
subfit ∼, 73
superalgebraic ∼, 142
supercompact ∼, 141
supercontinuous ∼, 142
uniform ∼, 150
uniformity on a ∼, 150
weakly Hausdorff ∼, 83
zero-dimensional ∼, 105
frames described by generators
    and relations, 53
free
    construction, 53
    frame, 53
    functor, 53
front topology, 7
full subcategory, 339
fully Cauchy map, 167
fully normal frame, 170
function
    almost inverse ∼, 323
    characteristic ∼, 279, 289
    constant ∼, 279
    real ∼, 288
functor, 340

Lc, 12
$\mathfrak{C}$, 66
Ω, 10
Par, 181
Ψ, 196
$\mathfrak{R}$, 133
$\widetilde{\mathfrak{R}}$, 183
Sp, 15
category of $\sim$s, 342
contravariant $\sim$, 340
down-set $\sim$, 51
hom-$\sim$, 340
spectrum $\sim$, 15

Galois
    adjoint, 323
    connection, 323
gauge, 147
G_δ-sublocale, 141
generalized
    Cauchy point, 201
    point, 201
generator, 356
greatest element, 320
group
    in a category, 297
    localic $\sim$, 298
    LT-$\sim$, 313
    topological $\sim$, 297

H-realcompact frame, 295
Heyting
    algebra, 331
    homomorphism, 332
    semilattice, 331
hom-functor, 340
hom-set, 338
homomorphism
    biframe $\sim$, 240
    codiagonal $\sim$, 60
    frame $\sim$, 10
    Heyting $\sim$, 332
    localic group $\sim$, 302
    sublocale $\sim$, 30

uniform $\sim$, 244
hypercomplete uniform space, 195

i-completely below relation, 241
I-Hausdorff
    frame, 79
    locale, 79
i-rather below relation, 241
ideal, 131, 329
    $\mathcal{A}$-regular $\sim$, 183
    completely proper $\sim$, 295
    maximal $\sim$, 329
    prime $\sim$, 329
    proper $\sim$, 329
    regular $\sim$, 132
    σ-proper $\sim$, 295
identity morphism, 337
image, 29
induced
    biframe, 246
    bitopological space, 236
    sublocale, 102
infimum, 319
initial object, 344
insertion theorem
    Katětov-Tong $\sim$, 292
interior of a paircover, 244
interpolative relation, 90
isolated point, 7
isomorphism, 339

join, 319
    -irreducible, 2
    -semilattice, 320
    dense $\sim$, 331

κ-complete Boolean algebra, 70
κ-directed poset, 70
κ-filtered category, 70
κ-frame, 70
κ-subset, 70
Katětov-Tong
    Insertion Theorem, 292
Kleisli category, 363

Knaster-Tarski Fixed
    Point Theorem, 321
Kuratowski-Mrówka Theorem, 128

lattice, 320
    bounded ∼, 320
    complete ∼, 321
    continuous ∼, 135
    distributive ∼, 326
    modular ∼, 326
lattice-ordered ring, 283
law
    composition ∼, 337
    first De Morgan ∼, 331, 332
    second De Morgan ∼, 34
least element, 320
left system of covers, 300
left uniformity, 310
lemma
    Urysohn's separation ∼, 292
    Zorn's ∼, 318
limit, 350
Lindelöf
    frame, 125
    locale, 125
linear
    element, 106
    order, 315
linearly ordered set, 315
locale, 11
    α-compact ∼, 125
    compact ∼, 125
    completely regular ∼, 91
    connected ∼, 254
    disconnected ∼, 254
    DS-Hausdorff ∼, 83
    extremally disconnected ∼, 257
    fit ∼, 74
    I-Hausdorff ∼, 79
    Lindelöf ∼, 125
    locally connected ∼, 258
    metric ∼, 203
    nearness ∼, 150
    normal ∼, 91

one-point ∼, 10
point of a ∼, 13, 14
product-connected ∼, 260
regular ∼, 89
spectrum of a ∼, 15
strongly Hausdorff ∼, 79
subfit ∼, 73
uniform ∼, 150
void ∼, 10
weakly Hausdorff ∼, 83
localic
    group, 298
        homomorphism, 302
    map, 12
        closed ∼, 38
        co-dense ∼, 77
        dense ∼, 39
        diagonal ∼, 61
        open ∼, 37
        proper ∼, 39
locally
    compact
        space, 135
    connected
        frame, 258
        locale, 258
    finite
        refinement, 173
        subset, 173
    small category, 339
lower bound, 317
lower semicontinuous
    real function, 288
LT-group, 313

Manes system, 364
map
    bicontinuous ∼, 240
    contractive ∼, 207
    cozero ∼, 284
    fully Cauchy ∼, 167
    localic ∼, 12
    monotone ∼, 317
    regular Cauchy ∼, 160

maximal
   element, 317
   filter, 329
   ideal, 329
meet, 319
   -irreducible element, 2, 13
   -semilattice, 51, 320
metric
   covers of $\mathfrak{L}(\mathbb{R})$, 276
   diameter, 205
   frame, 203
   locale, 203
   quasi-uniformity, 296
   sublocale, 218
   uniformity, 204
      of $\mathfrak{L}(\mathbb{R})$, 276
metrizable frame, 172, 204
minimal
   element, 317
   prime, 116
model of an algebraic theory, 366
modified Birkhoff's theorem, 137
modular lattice, 326
monad, 361
monocoreflection, 360
monocoreflective subcategory, 360
monomorphism, 23, 339
   extremal $\sim$, 347
   extremal $\sim$ in **Frm**, 50
   in **Frm**, 50
   regular $\sim$, 347
   regular $\sim$ in **Frm**, 50
   strong $\sim$, 347
   strong $\sim$ in **Frm**, 50
monotone map, 317
morphism, 337
   epi$\sim$, 339
   identity $\sim$, 337
   iso$\sim$, 339
   orthogonal pair of $\sim$s, 346

natural
   equivalence, 341
   transformation, 341

nearness, 150
   basis of $\sim$, 151
   Cauchy complete $\sim$, 194
   frame, 150
      with enough
         Cauchy points, 191
   locale, 150
   regular $\sim$, 194
   strong $\sim$, 151
   subbasis of $\sim$, 151
   weak $\sim$, 158
Noetherian poset, 22
normal
   cover, 152
   frame, 91
   locale, 91
nucleus, 31
   generated by a prenucleus, 47

object, 337
   initial $\sim$, 344
   terminal $\sim$, 344
one-point
   locale, 10
   sublocale, 42
open
   cover, 125
   Dedekind cut, 274
   localic map, 37
   paircover, 238
   set
      meet-irreducible $\sim$, 2
   sublocale, 33
opposite
   category, 338
   order, 316
order
   dual $\sim$, 316
   linear $\sim$, 315
   partial $\sim$, 315
   total $\sim$, 315
orthogonal pair of morphisms, 346

paircover
  interior of a ∼, 244
  of a frame, 241
  of a set, 236
  open ∼, 238
  strong ∼, 236, 241
paracompact
  (co)reflection, 180
  frame, 173
partial order, 315
partially ordered set, 315
(PC)-frame, 173
point, 13, 14
  Cauchy ∼, 159, 189
  generalized ∼, 201
  generalized Cauchy ∼, 201
  isolated ∼, 7
  weakly isolated ∼, 7
poset, 315
  κ-directed ∼, 70
  Noetherian ∼, 22
  separated ∼, 143
  sober ∼, 21
prediameter, 211
  strong ∼, 211
prenucleus, 47
preorder, 315
preordered set, 315
prime
  element, 13
  essential ∼, 118
  filter, 2, 329
  ideal, 329
  minimal ∼, 116
product, 343
  -connected
    frame, 260
    locale, 260
products
  in **Frm**, 54
  in **Loc**, 58
proper
  filter, 329
  ideal, 329

localic map, 39
property (P)
  complete Boolean algebra
    with ∼, 263
pseudocomplement, 330
pseudodifference, 107, 108
pullback, 345
pushout, 346

quasi
  -discrete space, 5
  -uniform
    frame, 243
    space, 236
  -uniform frame
    entourage ∼, 250
  -uniformity
    basis, 243
    entourage ∼, 250
    metric ∼, 296
    on a frame, 243
    semicontinuous ∼, 296
  -uniformly below relation, 242
quasitopology, 86
quotient frame, 31

rather below relation, 88
real
  uniform cover, 294
  uniformity, 294
real function, 288
  bounded ∼, 293
  continuous ∼, 278
  extended ∼, 294
real-complete frame, 294
realcompact
  frame, 295
  space, 295
reals
  construction, 275
  frame of ∼, 269
refinement, 148
  locally finite ∼, 173
  star-∼, 150

reflection, 360
reflective subcategory, 359
regular
  basis, 215
  biframe, 241
  Cauchy map, 160
  epimorphism, 347
  epimorphism in **Loc**, 50
  filter, 189
  frame, 89
  ideal, 132
  locale, 89
  monomorphism, 347
  monomorphism in **Frm**, 50
  nearness, 194
relation, 230
  completely below ∼, 90
  i-completely below ∼, 241
  i-rather below ∼, 241
  interpolative ∼, 90
  quasi-uniformly below ∼, 242
  rather below ∼, 88
  super-way-below ∼, 142
  uniformly below ∼, 149, 231
  well below ∼, 134
relative complement, 108
right system of covers, 300
right uniformity, 310
ring
  lattice-ordered ∼, 283
  of continuous
    real functions, 278, 283

Samuel compactification, 183
saturated element, 44
saturation conditions (R1)–(R2), 161
scale, 278
scattered space, 120
Scott
  open in a frame, 137
  topology, 21
    in a frame, 137
semicontinuous
  quasi-uniformity, 296

real function
  lower ∼, 288
  upper ∼, 288
semigroup operation, 308
semilattice
  bounded join-∼, 320
  bounded meet-∼, 320
  Heyting ∼, 331
  join-∼, 320
  meet-∼, 51, 320
semiprime element, 87
separated poset, 143
set
  directed ∼, 317
  down-∼, 317
  hom-∼, 338
  linearly ordered ∼, 315
  meet-irreducible open ∼, 2
  partially ordered ∼, 315
  preordered ∼, 315
  solution ∼, 357
  totally ordered ∼, 315
  up-∼, 317
σ-admissible basis, 214
σ-discrete subset, 173
σ-frame, 284
σ-locally finite subset, 173
σ-proper ideal, 295
σ-stratified basis, 215
Skula
  closure operator, 123
  interior operator, 123
  topology, 7
slicing filter, 5
small category, 339
smallest dense sublocale, 110
sober
  poset, 21
  space, 2
sobrification, 20
solution set, 357
space
  assembly ∼, 117
  bitopological ∼, 239

hypercomplete uniform ∼, 195
   quasi-uniform ∼, 236
spatial
   frame, 10
   sublocale, 102
spectrum, 15
   Cauchy ∼, 193
   functor, 15
star-additive diameter, 204
star-refinement, 150
Stone-Čech compactification, 133
strong
   epimorphism, 347
   epimorphism in **Loc**, 50
   f-ring, 283
   inclusion, 177
   monomorphism, 347
   monomorphism in **Frm**, 50
   nearness, 151
   paircover, 236, 241
   prediameter, 211
   solution, 357
   ue-surjection, 166
strongly Hausdorff
   frame, 79
   locale, 79
subbasis
   of nearness, 151
subcategory, 339
   coreflective ∼, 359
   epireflective ∼, 360
   full ∼, 339
   monocoreflective ∼, 360
   reflective ∼, 359
subcover, 125
subfit
   frame, 73
   locale, 73
subframe, 24
sublocale, 26
   Boolean ∼, 43
   closed ∼, 33
   closure, 39
   complete ∼, 86

dense ∼, 39
development of a ∼, 113
disjoint ∼, 254
doubly dense ∼, 111
G_δ-∼, 141
homomorphism, 30
induced ∼, 102
metric ∼, 218
one-point ∼, 42
open ∼, 33
smallest dense ∼, 110
spatial ∼, 102
subset
   σ-discrete ∼, 173
   σ-locally finite ∼, 173
super-way-below relation, 142
superalgebraic frame, 142
supercompact
   element, 141
   frame, 141
supercontinuous frame, 142
supremum, 319
surjection
   strong uniform embedding ∼, 166
   uniform embedding ∼, 154
symmetric entourage, 228

T_D-space, 5
terminal object, 344
theorem
   Birkhoff's ∼, 330
   Boolean Ultrafilter ∼, 334
   Cantor-Bernstein ∼, 321
   Closed Subgroup ∼, 306
   Isbell's Density ∼, 40
   Isbell's Development ∼, 115
   Katětov-Tong Insertion ∼, 292
   Knaster-Tarski Fixed Point ∼, 321
   Kuratowski-Mrówka ∼, 128
   Modified Birkhoff's ∼, 137
   Tietze's Extension ∼, 292
top, 320
topological group, 297

topological space
    Alexandroff ∼, 5
    assembly ∼, 117
    Čech complete ∼, 313
    locally continuous ∼, 135
    quasi-discrete ∼, 5
    realcompact ∼, 295
    scattered ∼, 120
    sober ∼, 2
    T_D-∼, 5
    weakly scattered ∼, 120
topology
    front ∼, 7
    Scott ∼, 21
    Skula ∼, 7
    uniform ∼, 145, 146
total order, 315
totally bounded uniformity, 188
tower construction, 67
𝕋-algebra, 362
two-sided uniformity, 311
type, 365

ue-surjection, 154
ultrafilter, 334
uniform
    continuity, 146
    embedding surjection, 154
    frame, 150
        completion of a ∼, 160
        entourage ∼, 231
    homomorphism, 152, 232
    locale, 150
    space
        complete ∼, 159
    topology, 145, 146
uniformity
    basis of ∼, 151
    Cauchy complete ∼, 194
    entourage ∼, 231
    fine ∼, 152
    left ∼, 310
    metric ∼, 204
    metric ∼ of $\mathfrak{L}(\mathbb{R})$, 276

on a frame, 150
on a set, 145, 146
real ∼, 294
right ∼, 310
totally bounded ∼, 188
two-sided ∼, 311
uniformly
    below relation, 149, 231
    continuous map, 146
units
    of adjunction, 354
up-set, 317
upper bound, 317
upper semicontinuous
    real function, 288
Urysohn's Separation Lemma, 292

void locale, 10

weak nearness, 158
weakly
    Hausdorff
        frame, 83
        locale, 83
    isolated point, 7
    scattered space, 120
well below relation, 134
wellpowered category, 356

xy-chain, 211

Yoneda embedding, 342

zero-dimensional frame, 105
Zorn's lemma, 318